Lecture Notes in Artificial Intelligence 13029

Subseries of Lecture Notes in Computer Science

More information about this subseries at http://www.springer.com/series/1244

Lu Wang · Yansong Feng ·
Yu Hong · Ruifang He (Eds.)

Natural Language Processing and Chinese Computing

10th CCF International Conference, NLPCC 2021
Qingdao, China, October 13–17, 2021
Proceedings, Part II

Editors
Lu Wang
University of Michigan
Ann Arbor, MI, USA

Yansong Feng
Peking University
Beijing, China

Yu Hong
Soochow University
Suzhou, China

Ruifang He
Tianjin University
Tianjin, China

ISSN 0302-9743 ISSN 1611-3349 (electronic)
Lecture Notes in Artificial Intelligence
ISBN 978-3-030-88482-6 ISBN 978-3-030-88483-3 (eBook)
https://doi.org/10.1007/978-3-030-88483-3

LNCS Sublibrary: SL7 – Artificial Intelligence

This Springer imprint is published by the registered company Springer Nature Switzerland AG
The registered company address is: Gewerbestrasse 11, 6330 Cham, Switzerland

Preface

Welcome to NLPCC 2021, the tenth CCF International Conference on Natural Language Processing and Chinese Computing. Following the success of previous conferences held in Beijing (2012), Chongqing (2013), Shenzhen (2014), Nanchang (2015), Kunming (2016), Dalian (2017), Hohhot (2018), Dunhuang (2019), and Zhengzhou (2020), this year's NLPCC was held in Qingdao, a beautiful coastal city in East China. As a premier international conference on natural language processing and Chinese computing, organized by the CCF-NLP (Technical Committee of Natural Language Processing, China Computer Federation, formerly known as Technical Committee of Chinese Information, China Computer Federation), NLPCC 2021 serves as an important forum for researchers and practitioners from academia, industry, and government to share their ideas, research results, and experiences, and to promote their research and technical innovations in the fields.

The fields of natural language processing (NLP) and Chinese computing (CC) have boomed in recent years. Following NLPCC's tradition, we welcomed submissions in ten areas for the main conference: Fundamentals of NLP; Machine Translation and Multilinguality; Machine Learning for NLP; Informtion Extraction and Knowledge Graph; Summarization and Generation; Question Answering; Dialogue Systems; Social Media and Sentiment Analysis; NLP Applications and Text Mining; Multimodality and Explainability. On the submission deadline, we were thrilled to have received a record number of 446 valid submissions to the main conference.

After a rigid review process, out of 446 submissions (some of which were withdrawn or rejected without review due to format issues or policy violations), 104 papers were finally accepted to appear in the main conference, where 89 were written in English and 15 in Chinese, resulting in an acceptance rate of 23.3%. Among them, 72 submissions were accepted as oral papers and 32 as poster papers. Specifically, ten papers were nominated by our area chairs for the best paper award. An independent best paper award committee was formed to select the best papers from the shortlist. This proceedings includes only the accepted English papers; the Chinese papers will appear in the ACTA Scientiarum Naturalium Universitatis Pekinensis. In addition to the main proceedings, three papers were accepted to the Student workshop, 22 papers were accepted to the Evaluation workshop, and two papers were accepted to the Explainable AI (XAI) workshop.

We were honored to have four internationally renowned keynote speakers—Rada Mihalcea (University of Michigan), Yanchao Bi (Beijing Normal University), Sebastian Riedel (University College London and Facebook AI Research), and Graham Neubig (Carnegie Mellon University)—share their findings on recent research progress and achievements in natural language processing.

We would like to thank all the people who have contributed to NLPCC 2021. First of all, we would like to thank our 20 area chairs for their hard work recruiting reviewers, monitoring the review and discussion processes, and carefully rating and recommending submissions. We would like to thank all 432 reviewers for their time and efforts to review the submissions. We are very grateful to Tim Baldwin, Chin-Yew Lin, Kang Liu, Deyi Xiong, and Yue Zhang for their participation in the best paper committee. We are also grateful for the help and support from the general chairs, Tim Baldwin and Jie Tang, and from the organization committee chairs, Zhumin Chen, Pengjie Ren, and Xiaojun Wan. Special thanks go to Yu Hong and Ruifang He, the publication chairs, for their great help. We greatly appreciate all your help!

Finally, we would like to thank all the authors who submitted their work to NLPCC 2021, and thank our sponsors for their contributions to the conference. Without your support, we could not have such a strong conference program.

We were happy to see you at NLPCC 2021 in Qingdao and hope you enjoyed the conference!

October 2021

Lu Wang
Yansong Feng

Organization

NLPCC 2021 was organized by the China Computer Federation (CCF), and hosted by Shandong University. The proceedings were published in Lecture Notes on Artificial Intelligence (LNAI), Springer-Verlag, and ACTA Scientiarum Naturalium Universitatis Pekinensis.

Organization Committee

General Chairs

Tim Baldwin	University of Melbourne, Australia
Jie Tang	Tsinghua University, China

Program Committee Chairs

Lu Wang	University of Michigan, USA
Yansong Feng	Peking University, China

Student Workshop Chairs

Fang Kong	Soochow University, China
Meishan Zhang	Tianjin University, China

Evaluation Workshop Chairs

Jiajun Zhang	Institute of Automation, Chinese Academy of Sciences, China
Yunbo Cao	Tencent, China

Tutorial Chairs

Minlie Huang	Tsinghua University, China
Zhongyu Wei	Fudan University, China

Publication Chairs

Yu Hong	Soochow University, China
Ruifang He	Tianjin University, China

Journal Coordinator

Yunfang Wu Peking University, China

Conference Handbook Chair

Xiaomeng Song Shandong University, China

Sponsorship Chairs

Dongyan Zhao Peking University, China
Zhaochun Ren Shandong University, China

Publicity Chairs

Wei Jia Beijing Academy of Artificial Intelligence, China
Jing Li Hong Kong Polytechnic University, China
Ruifeng Xu Harbin Institute of Technology, China

Organization Committee Chairs

Zhumin Chen Shandong University, China
Pengjie Ren Shandong University, China
Xiaojun Wan Peking University, China

Area Chairs

Fundamentals of NLP

Liang Huang Oregon State University, USA
Kewei Tu ShanghaiTech University, China

Machine Translation and Multilinguality

Derek Wong University of Macau, Macau
Boxing Chen Alibaba Group, China

Machine Learning for NLP

Yangfeng Ji University of Virginia, USA
Lingpeng Kong The University of Hong Kong, Hong Kong

Information Extraction and Knowledge Graph

Lifu Huang Virginia Tech, USA
Shizhu He Chinese Academy of Sciences, China

Summarization and Generation

Rui Zhang	Pennsylvania State University, USA
Jin-ge Yao	Microsoft Research, China

Question Answering

Huan Sun	Ohio State University, USA
Yiming Cui	Harbin Institute of Technology, China

Dialogue Systems

Jessy Li	University of Texas at Austin, USA
Wei Wu	Meituan, China

Social Media and Sentiment Analysis

Fei Liu	University of Central Florida, USA
Duyu Tang	Tencent AI Lab, China

NLP Applications and Text Mining

Wenpeng Yin	Salesforce Research, USA
Zhunchen Luo	PLA Academy of Military Science, China

Multimodality and Explainability

Xin Wang	University of California, Santa Cruz, USA
Zhongyu Wei	Fudan University, China

Treasurer

Yajing Zhang	Soochow University, China
Xueying Zhang	Peking University, China

Webmaster

Hui Liu	Peking University, China

Program Committee

Ayana	Inner Mongolia University of Finance and Economics, China
Bo An	Institute of Software, Chinese Academy of Sciences, China
Xiang Ao	Institute of Computing Technology, Chinese Academy of Sciences, China
Jiaqi Bai	Beihang University, China
Guangsheng Bao	Westlake University, China
Junwei Bao	JD AI Research, China
Qiming Bao	University of Auckland, New Zealand

Paul Buitelaar	Data Science Institute, National University of Ireland Galway, Ireland
Xuanting Cai	Facebook, USA
Xiangrui Cai	Nankai University, China
Yi Cai	South China University of Technology, China
Yixin Cao	Nanyang Technological University, Singapore
Yuan Cao	Google Brain, USA
Shuyang Cao	University of Michigan, USA
Pengfei Cao	Institute of Automation, Chinese Academy of Sciences, China
Ziqiang Cao	Soochow University, China
Hailong Cao	Harbin Institute of Technology, China
Shuaichen Chang	Ohio State University, USA
Zhangming Chan	Peking University, China
Zhumin Chen	Shandong University, China
Sihao Chen	University of Pennsylvania, USA
Xiuying Chen	Peking University, China
Jiajun Chen	Nanjing University, China
Junkun Chen	Oregon State University, USA
Sanxing Chen	University of Virginia, USA
Hongshen Chen	JD.com, China
Zhipeng Chen	iFLYTEK Research, China
Jie Chen	Ahu University, China
Qingcai Chen	Harbin Institute of Technology, China
Lei Chen	Fudan University, China
Wenliang Chen	Soochow University, China
Shuang Chen	Harbin Institute of Technology, China
Xinyu Chen	Soochow University, China
Ruijun Chen	Yunnan University, China
Bo Chen	Institute of Software, Chinese Academy of Sciences, China
Yulong Chen	Zhejiang University/Westlake University, China
Huadong Chen	ByteDance, China
Pei Chen	Texas A&M University, USA
Lu Chen	Shanghai Jiao Tong University, China
Guanyi Chen	Utrecht University, The Netherlands
Yidong Chen	Xiamen University, China
Liwei Chen	Peking University, China
Xinchi Chen	Amazon Web Services, USA
Hanjie Chen	University of Virginia, USA
Wei Chen	Fudan University, China
Hannah Chen	University of Virginia, USA
Yubo Chen	Institute of Automation, Chinese Academy of Sciences, China
Kehai Chen	National Institute of Information and Communications Technology, Japan

Mingda Chen	Toyota Technological Institute at Chicago, USA
Jiangjie Chen	Fudan University, China
Jifan Chen	University of Texas at Austin, USA
Jiaao Chen	Georgia Institute of Technology, USA
Jiayang Cheng	Fudan University, China
Xin Cheng	Peking University, China
Qikai Cheng	Wuhan University, China
Zewen Chi	Beijing Institute of Technology, China
Zewei Chu	Google, USA
Chenhui Chu	Kyoto University, Japan
James Cross	Facebook, USA
Peng Cui	Harbin Institute of Technology, China
Yiming Cui	Harbin Institute of Technology, China
Mao Cunli	Kunming University of Science and Technology, China
Junqi Dai	Fudan University, China
Xinyu Dai	Nanjing University, China
Xiang Deng	Ohio State University, USA
Xiao Ding	Harbin Institute of Technology, China
Chenchen Ding	NICT, Japan
Mengxing Dong	Soochow University, China
Zhicheng Dou	Renmin University of China, China
Dejing Dou	University of Oregon, USA
Ziyi Dou	UCLA, USA
Rotem Dror	University of Pennsylvania, Israel
Wanyu Du	University of Virginia, USA
Jinhua Du	Investments AI, AIG, UK
Xinya Du	Cornell University, USA
Junwen Duan	Central South University, China
Chaoqun Duan	Harbin Institute of Technology, China
Xiangyu Duan	Soochow University, China
Philipp Dufter	Ludwig-Maximilians-Universität München, Germany
Kevin Duh	Johns Hopkins University, USA
Ismail El Maarouf	Fortia Financial Solutions, France
Alexander Fabbri	Yale University, USA
Chuang Fan	Harbin Institute of Technology, China
Zhihao Fan	Fudan University, China
Zhiyuan Fang	Arizona State University, USA
Xiachong Feng	Harbin Institute of Technology, China
Zhangyin Feng	Harbin Institute of Technology, China
Xiaocheng Feng	Harbin Institute of Technology, China
Shi Feng	Northeastern University, China
Jiazhan Feng	Peking University, China
Xingyu Fu	University of Pennsylvania, USA
Yao Fu	University of Edinburgh, UK

Guohong Fu	Soochow University, China
Zihao Fu	The Chinese University of Hong Kong, Hong Kong Special Administrative Region of China
Yi Fung	University of Illinois at Urbana Champaign, USA
Jun Gao	Harbin Institute of Technology, China
Shen Gao	Peking University, China
Yang Gao	Royal Holloway, University of London, UK
Yifan Gao	The Chinese University of Hong Kong, Hong Kong Special Administrative Region of China
Michaela Geierhos	Universität der Bundeswehr München, Germany
Ruiying Geng	Alibaba Group, China
Erfan Ghadery	KU Leuven, Belgium
Heng Gong	Harbin Institute of Technology, China
Yeyun Gong	Microsoft Research Asia, China
Jiachen Gu	University of Science and Technology of China, China
Yu Gu	Ohio State University, USA
Jian Guan	Tsinghua University, China
Yuhang Guo	Beijing Institute of Technology, China
Daya Guo	Sun Yat-Sen University, China
Qipeng Guo	Fudan University, China
Junliang Guo	University of Science and Technology of China, China
Jiale Han	Beijing University of Posts and Telecommunications, China
Fangqiu Han	Facebook, Inc., USA
Xudong Han	University of Melbourne, Australia
Lifeng Han	Dublin City University, Ireland
Wenjuan Han	Beijing Institute for General Artificial Intelligence, China
Tianyong Hao	South China Normal University, China
Shizhu He	Institute of Automation, Chinese Academy of Sciences, China
Zhongjun He	Baidu, Inc., China
Ziwei He	Shanghai Jiao Tong University, China
Ben He	University of Chinese Academy of Sciences, China
Ruifang He	Tianjin University, China
Yanqing He	Institute of Scientific and Technical Information of China, China
Yu Hong	Soochow University, China
Xudong Hong	Saarland University / MPI Informatics, Germany
Lei Hou	Tsinghua University, China
Yacheng Hsu	The University of Hong Kong, Taiwan, China
Huang Hu	Microsoft STCA NLP Group, China
Zhe Hu	Baidu, China
Zikun Hu	National University of Singapore, China
Minghao Hu	PLA Academy of Military Science, China
Baotian Hu	Harbin Institute of Technology, China

Wenpeng Hu	Peking University, China
Zechuan Hu	ShanghaiTech University, China
Xinyu Hua	Northeastern University, USA
Zhen Huang	National University of Defense Technology, China
Jiangping Huang	Chongqing University of Posts and Telecommunications, China
Shujian Huang	Nanjing University, China
Qingbao Huang	Guangxi University/South China University of Technology, China
Zhongqiang Huang	Alibaba Group, USA
Yongjie Huang	Sun Yat-sen University, China
Junjie Huang	Beihang University, China
Yi-Ting Huang	Academia Sinica, Taiwan, China
Julia Ive	Imperial College London, UK
Lei Ji	Microsoft Research Asia, China
Menglin Jia	Cornell University, USA
Zixia Jia	ShanghaiTech University, China
Ping Jian	Beijing Institute of Technology, China
Xuhui Jiang	Institute of Computing Technology, Chinese Academy of Sciences, China
Yong Jiang	Alibaba DAMO Academy, China
Xin Jiang	Huawei Noah's Ark Lab, Hong Kong Special Administrative Region of China
Ming Jiang	University of Illinois at Urbana-Champaign, USA
Yichen Jiang	University of North Carolina at Chapel Hill, USA
Wenbin Jiang	Baidu Inc., China
Peng Jin	Leshan Normal University, China
Xisen Jin	University of Southern California, USA
Zhijing Jin	Max Planck Institute for Intelligent Systems, Germany
Zhiling Jin	Soochow University, China
Lin Jun	Alibaba Group, China
Zixuan Ke	University of Illinois at Chicago, USA
Dongjin Kim	KAIST, South Korea
Fang Kong	Soochow University, China
Bernhard Kratzwald	ETH Zurich, Switzerland
Tuan Lai	University of Illinois at Urbana-Champaign, USA
Viet Lai	University of Oregon, USA
Yuxuan Lai	Peking University, China
Man Lan	East China Normal University, China
Chongshou Li	Southwest Jiaotong University, China
Maoxi Li	Jiangxi Normal University, China
Bin Li	Nanjing Normal University, China
Mingzhe Li	Peking University, China
Hongzheng Li	Beijing Institute of Technology, China
Chenliang Li	Wuhan University, China
Zhongyang Li	Harbin Institute of Technology, China

Manling Li	University of Illinois Urbana-Champaign, USA
Dongfang Li	Harbin Institute of Technology, China
Shasha Li	National University of Defense Technology, China
Xinyi Li	National University of Defense Technology, China
Zejun Li	Fudan University, China
Peifeng Li	Soochow University, China
Irene Li	Yale University, USA
Jian Li	Huawei Noah's Ark Lab, China
Piji Li	Tencent AI Lab, China
Xin Li	Alibaba Group, China
Lishuang Li	Dalian University of Technology, China
Fei Li	Wuhan University, China
Weikang Li	Peking University, China
Yachao Li	Soochow University, China
Liangyou Li	Huawei Noah's Ark Lab, Hong Kong Special Administrative Region of China
Qintong Li	Shandong University, China
Xinhang Li	Tsinghua University, China
Linjie Li	Microsoft, USA
Yanran Li	The Hong Kong Polytechnic University, Hong Kong Special Administrative Region of China
Jing Li	The Hong Kong Polytechnic University, Hong Kong Special Administrative Region of China
Junhui Li	Soochow University, China
Zuchao Li	Shanghai Jiao Tong University, China
Lin Li	Qinghai Normal University, China
Zhenghua Li	Soochow University, China
Ruizhe Li	University of Sheffield, UK
Xilai Li	Amazon, USA
Peng Li	Institute of Information Engineering, CAS, China
Liunian Harold Li	UCLA, USA
Binyang Li	University of International Relations, China
Xiaolong Li	SAS Institute Inc., USA
Zhixu Li	Soochow University, China
Zichao Li	McGill University, Canada
Xiao Li	University of Reading, UK
Zujie Liang	Sun Yat-sen University, China
Paul Pu Liang	Carnegie Mellon University, USA
Ying Lin	Apple, USA
Junyang Lin	Alibaba Group, China
Zhaojiang Lin	The Hong Kong University of Science and Technology, Hong Kong Special Administrative Region of China
Zi Lin	Google, USA
Zhouhan Lin	Shanghai Jiao Tong University, China
Ye Liu	National University of Singapore, Singapore

Wei Wang	Tsinghua University, China
Ruize Wang	Fudan University, China
Xing Wang	Tencent, China
Tao Wang	King's College London, UK
Jingjing Wang	Soochow University, China
Le Wang	Chinese Academy of Military Science, China
Hongwei Wang	University of Illinois Urbana-Champaign, USA
Hongling Wang	Soochow University, China
Shaonan Wang	Institute of Automation, Chinese Academy of Sciences, China
Sijia Wang	Virginia Tech, USA
Xun Wang	University of Massachusetts Amherst, USA
Zhen Wang	Ohio State University, USA
Rui Wang	Shanghai Jiao Tong University, China
Siyuan Wang	Fudan University, China
Yaqiang Wang	Chengdu University of Information Technology, China
Dingmin Wang	University of Oxford, UK
Zijian Wang	Amazon Web Services AI, USA
Qingyun Wang	University of Illinois at Urbana-Champaign, USA
Ke Wang	Peking University, China
Kexiang Wang	Peking University, China
Pancheng Wang	National University of Defence Technology, China
Ge Wang	ShanghaiTech University, China
Hong Wang	University of California, Santa Barbara, USA
Jianyou Wang	University of California, San Diego, USA
Zhuoyu Wei	Yuanfudao, China
Zhongyu Wei	Fudan University, China
Haoyang Wen	Carnegie Mellon University, USA
Shuangzhi Wu	Tencent, China
Sixing Wu	Peking University, China
Junshuang Wu	Beihang University, China
Yu Wu	Microsoft Research Asia, China
Lianwei Wu	Xi'an Jiaotong University, China
Chiensheng Wu	Salesforce, USA
Changxing Wu	East China Jiaotong University, China
Congying Xia	University of Illinois at Chicago, USA
Rui Xia	Nanjing University of Science and Technology, China
Wen Xiao	University of British Columbia, Canada
Tong Xiao	Northeastern University, China
Yuqiang Xie	Institute of Information Engineering, Chinese Academy of Sciences, China
Deyi Xiong	Tianjin University, China
Hao Xiong	Alibaba Group, China
Can Xu	Microsoft STCA NLP Group, China
Ruijian Xu	Peking University, China
Silei Xu	Stanford University, USA
Jingjing Xu	ByteDance AI Lab, China
Tong Xu	University of Science and Technology of China, China

Yiheng Xu	Microsoft Research Asia, China
Yan Xu	Hong Kong University of Science and Technology, Hong Kong Special Administrative Region of China
Jinan Xu	Beijing Jiaotong University, China
Yadollah Yaghoobzadeh	University of Tehran, Iran
Yuanmeng Yan	Beijing University of Posts and Telecommunications, China
Lingyong Yan	Baidu Inc., China
Songlin Yang	ShanghaiTech University, China
Liang Yang	Dalian University of Technology, China
Muyun Yang	Harbin Institute of Technology, China
Yaqin Yang	Paypal, USA
Jingxuan Yang	Beijing University of Posts and Telecommunications, China
Qiang Yang	King Abdullah University of Science and Technology, Saudi Arabia
Ziqing Yang	iFLYTEK Research, China
Haoran Yang	The Chinese University of Hong Kong, Hong Kong Special Administrative Region of China
Baosong Yang	Alibaba DAMO Academy, China
Cheng Yang	Beijing University of Posts and Telecommunications, China
Zixiaofan Yang	Columbia University, USA
Zhiwei Yang	Jilin University, China
Kai Yang	Peking University, China
Zhewei Yao	University of California, Berkeley, USA
Jianmin Yao	Soochow University, China
Yiqun Yao	University of Michigan, USA
Wenlin Yao	Tencent AI Lab, USA
Zhe Ye	Microsoft, China
Xiaoyuan Yi	Tsinghua University, China
Pengcheng Yin	Carnegie Mellon University, USA
Da Yin	University of California, Los Angeles, USA
Zhiwei Yu	Microsoft Research Asia, China
Adams Yu	Google Brain, USA
Tiezheng Yu	The Hong Kong University of Science and Technology, Hong Kong Special Administrative Region of China
Heng Yu	Alibaba Group, China
Pengfei Yu	Department of Computer Science, University of Illinois at Urbana-Champaign, USA
Dian Yu	Tencent AI Lab, USA
Xiaodong Yu	University of Pennsylvania, USA
Tao Yu	Yale University, USA
Chunyuan Yuan	Institute of Information Engineering, Chinese Academy of Sciences, China

Zheng Yuan	University of Cambridge, UK
Xiang Yue	Ohio State University, USA
Zhiyuan Zeng	Beijing University of Posts and Telecommunications, China
Qi Zeng	University of Illinois at Urbana-Champaign, USA
Shuang (Sophie) Zhai	University of Oklahoma, USA
Danqing Zhang	Amazon, USA
Jiajun Zhang	Institute of Automation, Chinese Academy of Sciences, China
Zixuan Zhang	University of Illinois Urbana-Champaign, USA
Han Zhang	Vispek, China
Peng Zhang	Tianjin University, China
Meishan Zhang	Tianjin University, China
Zhifei Zhang	Tongji University, China
Wenxuan Zhang	The Chinese University of Hong Kong, Hong Kong Special Administrative Region of China
Hongming Zhang	Hong Kong University of Science and Technology, Hong Kong Special Administrative Region of China
Biao Zhang	University of Edinburgh, UK
Qi Zhang	Fudan University, China
Liwen Zhang	ShanghaiTech University, China
Chengzhi Zhang	Nanjing University of Science and Technology, China
Iris Zhang	Cornell University, USA
Zhirui Zhang	Alibaba DAMO Academy, China
Shuaicheng Zhang	Virginia Tech, USA
Tongtao Zhang	Siemens Corporate Technology, USA
Dongxu Zhang	University of Massachusetts, Amherst, USA
Xuanyu Zhang	Du Xiaoman Financial, China
Fan Zhang	Tianjin University, China
Hao Zhang	Agency for Science, Technology and Research, Singapore
Zhuosheng Zhang	Shanghai Jiao Tong University, China
Mengjie Zhao	Ludwig-Maximilians-Universität München, Germany
Yufan Zhao	Microsoft, China
Yang Zhao	Institute of Automation, Chinese Academy of Sciences, China
Xiang Zhao	National University of Defense Technology, China
Zhenjie Zhao	Nanjing University of Information Science and Technology, China
Sanqiang Zhao	University of Pittsburgh, USA
Lulu Zhao	Beijing University of Posts and Telecommunications, China
Kai Zhao	Google, USA
Jie Zhao	Amazon, USA
Renjie Zheng	Baidu Research, USA
Chen Zheng	Michigan State University, USA
Ming Zhong	University of Illinois at Urbana-Champaign, USA

Wanjun Zhong	Sun Yat-sen University, China
Junsheng Zhou	Nanjing Normal University, China
Wenxuan Zhou	University of Southern California, USA
Guangyou Zhou	Central China Normal University, China
Qingyu Zhou	Tencent, China
Ben Zhou	University of Pennsylvania, USA
Xian Zhou	Information Research Center of Military Science, China
Jie Zhu	Soochow University, China
Muhua Zhu	Tencent, China
Junnan Zhu	Institute of Automation, Chinese Academy of Sciences, China
Yazhou Zhang	Zhengzhou University of Light Industry, China
Shi Zong	Nanjing University, China
Wangchunshu Zhou	Beihang University, China

Organizers

Organized by

China Computer Federation, China

Hosted by

Shandong University

In cooperation with

Lecture Notes in Computer Science

Springer

ACTA Scientiarum Naturalium Universitatis Pekinensis

Sponsoring Institutions

Diamond Sponsors

China Mobile

AISpeech

AISPEECH 思必驰
专 注 人 性 化 的 智 能 语 音

Alibaba

ARC

Platinum Sponsors

Microsoft

Baidu

ByteDance

GTCOM

GTCOM

Huawei

Data Grand

LaiYe

Hisense

Tencent AI Lab

Golden Sponsors

Gridsum

Sougou

Xiaomi

Silver Sponsors

NiuTrans

Leyan Technologies

Contents – Part II

Summarization and Generation

Question Answering

Dialogue Systems

Social Media and Sentiment Analysis

NLP Applications and Text Mining

Multimodality and Explainability

Explainable AI Workshop

Student Workshop

Evaluation Workshop

Contents – Part I

Oral - Fundamentals of NLP

Machine Translation and Multilinguality

Machine Learning for NLP

Information Extraction and Knowledge Graph

Question Answering

Dialogue Systems

Social Media and Sentiment Analysis

NLP Applications and Text Mining

Multimodality and Explainability

Posters - Fundamentals of NLP

Syntax and Coherence - The Effect on Automatic Argument Quality Assessment

Xichen Sun[1], Wenhan Chao[1], and Zhunchen Luo[2](✉)

[1] School of Computer Science and Engineering, Beihang University, Beijing, China
{sunxichen,chaowenhan}@buaa.edu.cn
[2] Information Research Center of Military, PLA Academy of Military Science, Beijing, China

Abstract. In this paper, we focus on the task of automatic argument quality assessment. Prior empirical methods largely ignore the syntax structure in one argument or depend on handcrafted features that have shallow representation ability. In contrast, we proposed a method that directly models syntax and topic coherence. Our method can acquire both topic coherence and syntactic information from an argument that explicitly utilizes various types of relationships among words, thus can help with argument quality assessment. Experimental results suggest that our method significantly outperforms the previous state-of-the-art method and strongly indicates syntax and coherence correlate with argument quality.

Keywords: Argument mining · Argument quality assessment · Syntax · Coherence

1 Introduction

Arguments are used to resolve a conflict of opinions concerning controversial topics, usually involving two or more parties, which are the foundations of reasoning [23]. With the rapid growth of the research field of computational argumentation, the task of automatic argument quality assessment is attracting more and more special attention. There are many envisioned applications built on top of automatic quality assessment such as automatic essay scoring systems [16], argument search engines [22] and legal disputes resolution [1].

However, due to the highly subjective inheritance of this task, it is hard to assign an objective score to a single argument according to its quality. A pairwise approach is proposed to tackle this problem, in which two arguments under the same topic are compared and the higher quality one should be selected [7]. Based on this insight, the quality of an individual argument is a derivative of its relative quality. More recent works [2,6,17,18] all adopt this assessment framework to settle this subjective bias issue. An example of the pair preference can be seen in Fig. 1.

L. Wang et al. (Eds.): NLPCC 2021, LNAI 13029, pp. 3–12, 2021.
https://doi.org/10.1007/978-3-030-88483-3_1

Topic: we should support information privacy laws

Arg A: letting people choose what others know about them is important to mental well-being and personal autonomy.

Arg B: **personal information** should not be widely available and is subject to misuse and hence **privacy laws** are needed.

Label: Arg B

Fig. 1. An example of argument pairs. The colored spans are two 2-grams that maintain good coherence with the topic because their semantic information is closely related to the topic *information privacy law*. The goal of the task is to select the higher quality one from a pair of arguments.

Generally, we find that two factors are crucial to argument quality. We argue that syntactic information and the coherence between an argument and its topic helps the author to express his ideas more logically and effectively and can affect the judgments of the preference. Syntax refers to the arrangement of the words of a sentence that can help us understand the meaning of a sentence by knowing which words modify which other word to get the correct interpretation. It can be involved in many aspects of arguments, such as style of writing, grammar, and word usage, etc. [9]. Those aspects can naturally reflect arguments cogency, reasonableness and clarity. On the other hand, topic coherence measures how relevant the argument's standpoint to the given topic. Many previous studies have studied coherence to improve the performance of topic models [13–15]. Both syntax and coherence can naturally reflect arguments cogency, reasonableness and clarity.

In this paper, we propose a pairwise model that directly use syntax and coherence information to compare the qualities of an argument pair. We develop a Graph Attention Network (GAT) based neural model which can operate on the dependency tree of an argument that can capture long term syntactic dependency and preserve syntax information effectively. Besides, a well coherent sentence is characterized by the tendency of semantically similar words to be grouped together [4,10], which inspire us to employ an N-gram based model which is capable of grouping nearby words information together to represent topic coherence information [24]. Experimental results show our model achieves great performances and strongly indicates those two factors can help with automatic argument quality decision.

2 Related Work

The recently related works can be categorized into theory studies and empirical methods.

2.1 Theory Studies

Wachsmuth et al. [21] delivered the first comprehensive survey of research on argumentation quality assessment and suggested a taxonomy of major quality dimensions of argumentation. As follow-ups of their work, some corpus-based studies [12,20] have been conducted, suggesting that the theory-based and practical argument quality assessment approaches coincide to a large extent. Some quality dimensions in their taxonomy reflect the importance of syntax and topic coherence to some extent, therefore, we choose to model syntax and coherence directly in order to be more specific and strengthen the interpretability.

2.2 Empirical Methods

In the first attempt, two feature-based methods [7] were proposed that rely on a 64k dimensions feature vector. However, there is plenty of redundant information in the feature set as shown in Chalaguine and Schulz [2], where they achieved comparable performances using only a small fraction of the feature set. Besides pairwise approaches, a pointwise approach based on Gaussian Process Preference Learning [17] has been proposed, which can obtain a quality score per argument but still needs a huge set of linguistic features. Unlike these methods, our method does not require time-consuming and laborious feature engineering steps and is end-to-end learning. And unlike other end-to-end neural network approaches [6,18] that only use a generic deep model like BERT and does not consider what factors may affect argument quality, our method directly model syntax and topic coherence that can reveal actual properties that might have an effect on argument quality.

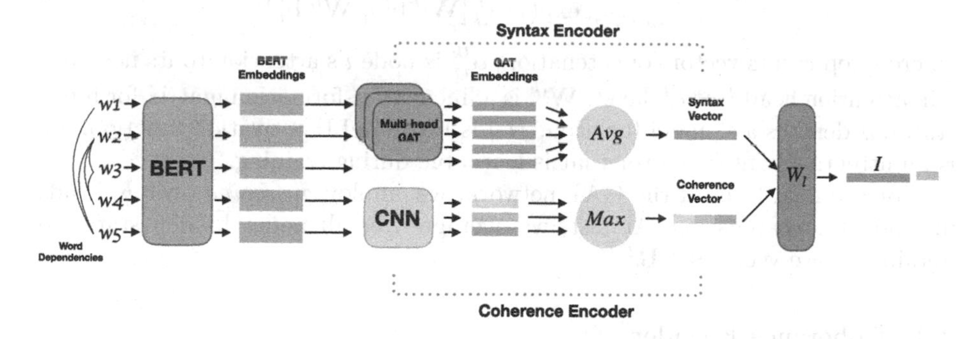

Fig. 2. Overview of our proposed model. I is the final argument representation.

3 Methodology

The architecture of our model is presented in Fig. 2. The model mainly consists of two components, *Syntax Encoder* and *Coherence Encoder*.

3.1 Input

The initial word embeddings used in our model is generated through BERT's Base Uncased English pre-trained model and we average the subword embeddings to get word embeddings.

3.2 Syntax Encoder

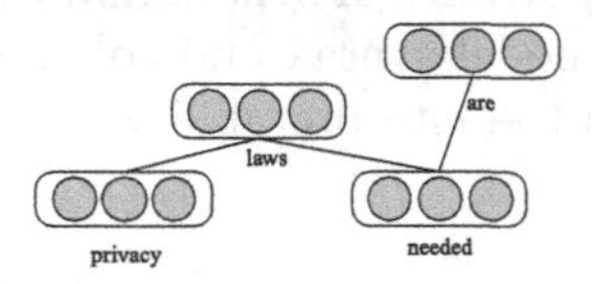

Fig. 3. The dependency graph for "privacy laws are needed". Semantic information can be propagated along with edges.

The encoder takes the dependency tree of an argument (as shown in Fig. 3) as input and the dependency information flows along the undirected dependency paths between words using vanilla graph attention network (GAT) [19]. The dependency tree can be interpreted as a graph with N nodes, where nodes corresponding to words in an argument. Each node is initially associated with BERT embeddings $\mathbf{x} \in \mathbf{R}^D$ as described above. One GAT layer computes each node representation via aggregating the neighborhood's hidden states using a multi-head attention mechanism. With an L-layer GAT network, dependency information within L hops can be propagated together. The detailed hidden states update formulation of K attention heads is as follows:

$$\mathbf{h}_i^{l+1} = \|_{k=1}^K \sigma(\sum_{j \in \hat{\mathcal{N}}(i)} \alpha_{ij}^{lk} \mathbf{W}^{lk} \mathbf{h}_j^l) \tag{1}$$

$$\alpha_{ij}^{lk} = \frac{\exp(f(\mathbf{a}_{lk}^T [\mathbf{W}^{lk} \mathbf{h}_i^l \parallel \mathbf{W}^{lk} \mathbf{h}_j^l]))}{\sum_{u \in \hat{\mathcal{N}}(i)} \exp(f(\mathbf{a}_{lk}^T [\mathbf{W}^{lk} \mathbf{h}_i^l \parallel \mathbf{W}^{lk} \mathbf{h}_u^l]))} \tag{2}$$

where $\|$ represents vector concatenation, α_{ij}^{lk} is node i's attention to its neighbor j in attention head k at l^{th} layer. $\mathbf{W}^{lk}$ is a linear transformation matrix for input states. σ denotes a sigmoid function. $f(\cdot)$ is LeakyReLU activation function. $\mathbf{a}_{lk}$ is an attention context vector that is learnable during training.

For the final layer of the GAT network, we employ *averaging* over K heads instead of concatenation. At last, we average over all nodes hidden states to obtain a dense vector $\mathbf{s} \in \mathbf{R}^D$.

3.3 Coherence Encoder

We use N-gram Convolutional Neural Network [11] to model arguments coherence. An argument consisted of N words is represented as a stack of its word embeddings $\mathbf{x} \in \mathbf{R}^D$. A convolution operation then is applied to the argument embedding matrix via a $F \times D$ filter, which group a window of F words together to produce a new feature h_i. This filter is applied to each window of words in the argument and produce a feature map $\mathbf{h}$.

$$h_i = \text{filter}(\mathbf{x}_{i:i+F-1}; \mathbf{W}) \tag{3}$$

$$\mathbf{h} = [h_1, h_2, ..., h_{N-F+1}] \in \mathbf{R}^{N-F+1} \quad (4)$$

We then apply a max-pooling operation over the feature map to get the maximum value $\hat{h} = \max(\mathbf{h})$ as the final feature value of this specific filter. In a word, one feature is extracted using one filter. The coherent encoder uses multiple filters with varying window size to get multiple features as well as the final coherent feature vector $\mathbf{c}$.

$$\mathbf{c} = [\hat{h_1}, \hat{h_2}, ..., \hat{h_m}] \in \mathbf{R}^m \quad (5)$$

where m is the total number of filters used in the coherent encoder.

3.4 Classification

Then we multiply a learnable matrix $\mathbf{W}_l$ to get a single vector $\mathbf{I}$ which is linear combinations of syntax vector $\mathbf{s}$ and coherence vector $\mathbf{c}$:

$$\mathbf{I} = \mathbf{W}_l[\mathbf{s} \parallel \mathbf{c}] \quad (6)$$

All weights of the model are shared within the pair. We can follow the procedure to get the final argument pair representation vectors, for example, $\mathbf{I}_a$ and $\mathbf{I}_b$. At last, we apply a fully-connected layer and the softmax operation to get the preference probability and the training loss is cross-entropy loss with L_2 regularization.

$$\mathbf{p} = \text{softmax}(fc([\mathbf{I}_a \parallel \mathbf{I}_b \parallel \mathbf{I}_a - \mathbf{I}_b \parallel \mathbf{I}_a \odot \mathbf{I}_b])) \quad (7)$$

$$\text{loss} = -\sum_{i \in C} y_i \cdot \log(p(y_i)) + (1 - y_i) \cdot \log(1 - p(y_i)) + \lambda \|\mathbf{W}\|^2 \quad (8)$$

where $\odot$ means the element-wise product, C is the collection of argument pair and y_i is the ground truth of argument pair i which takes the value either 1 that indicating argument A is more of high quality or 0 otherwise. $p(y_i)$ is the model's prediction probability of label y_i and $\mathbf{W}$ is the set of all the parameters in our model.

4 Experiments

4.1 Dataset

We evaluate our model on the dataset released by Toledo et al. [18]. The dataset (*IBMPairs*) has $9.1k$ argument pairs as well as cleaner texts that are also more homogeneous in terms of length compared to previous datasets [6,7], then we can expect its quality label to be less biased relative to argument length, holding more potential to reveal other characteristics that contribute to argument quality as mentioned in their dataset description. One sample of the dataset has been revealed in Fig. 1.

4.2 Settings

We use the dependency parser from AllenNLP to get the dependency tree [5]. All the embeddings dimension D is set to 768. For syntax encoder, we use a 3-Layers GAT with 3 attention heads for each layer. For the coherence encoder, we use three sizes of window and each size has 256 filters, thus, 768 filters in total. Specifically, we set F equals 3, 4 and 5 respectively. As for training, we use stochastic gradient descent as our optimizer and the learning rate is 0.001. Moreover, we set weight decay as 0.01. The model is trained for 10 epochs.

4.3 Baselines

We employ three previous successful methods in this task which includes: **Feature-based SVM** [2] utilizes 35 sets of linguistic features such as word mean length, named entities, part of speech and so on; **Siamese BiLSTM** [6] employs siamese neural network architecture and each leg is a Bidirectional LSTM network which is a function of an input argument A and has two outputs that contain a quality score. Then a softmax layer is applied over two arguments' quality scores. The original model use GloVe as word embeddings, but we also use BERT for fairness comparison; **Fine-tuned BERT** [18] achieves state-of-the-art results in this task which concatenate two arguments together and encode them using pre-trained model BERT [3], then apply a task-specific weighted matrix to do a binary classification.

4.4 Results and Discussions

Table 1. Evaluation of several models on IBMPairs using metrics accuracy and ROC area under curve(AUC).

Model	Accuracy	AUC
Feature-based SVM	0.76	0.76
Siamese BiLSTM + GloVe	0.79	0.80
Siamese BiLSTM + BERT	0.83	0.84
Fine-tuned BERT	0.80	0.86
Syntax encoder only	0.86	0.86
Coherence encoder only	0.89	0.89
Our full model	**0.92**	**0.92**

Table 2. Top n-grams scored by one filter sampled from filter size 5. In bold are words that deliver coherent topic semantics and their corresponding slot scores.

Rank	Top 5 5-grams										
	ngram					Score	Slot scores				
1	**privacy**	**laws**	improve	our	society	1.874	**0.601**	**0.302**	0.475	0.220	0.276
2	information	**privacy**	and	the	security	1.800	0.459	**0.539**	0.255	0.165	0.382
3	to	**privacy**	and	the	protection	1.770	0.608	**0.569**	0.233	0.079	0.281
4	**privacy**	**laws**	will	increase	people	1.733	**0.6**	**0.292**	0.218	0.177	0.446
5	**privacy**	**laws**	protect	our	clients	1.691	**0.683**	**0.153**	0.424	0.217	0.214

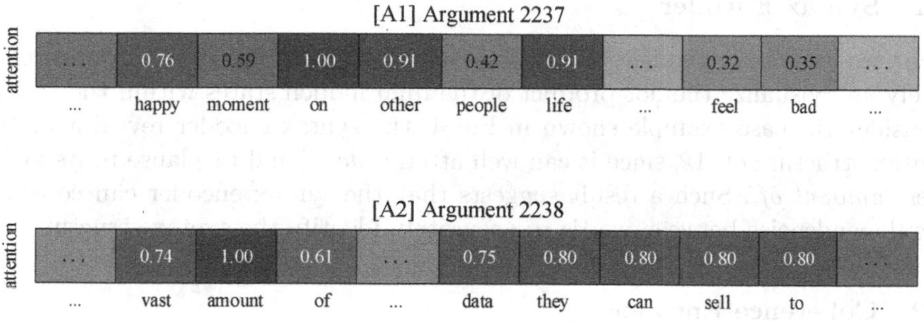

Fig. 4. Word dot product attention scores concerning the final embedding of the syntax encoder.

Main Results. As we can see in Table 1, feature-based SVM still can achieve competitive performance, indicating the importance of feature engineering and linguistic knowledge. The comparison between GloVe based and BERT based Siamese BiLSTM shows the strong power of large-scale pre-trained contextualized word embeddings. BERT based Siamese BiLSTM achieves higher accuracy over Fine-tuned BERT model. We believe the reason lies in the increase of learnable parameters so that the learning ability of the model is improved. Finally, our result on the full model reveals that incorporating both syntactic and coherence information can boost the classification performance compared to the models without considering them.

Ablation Study. To investigate the separate effects of the two components of our model in the task, we conduct an ablation study. The experimental results indicate that the coherence encoder contributes more to our full model than the other component. Furthermore, the results also show that strong performance can be achieved even if only one of the syntax or coherence encoder is used which has similar total numbers of learnable parameters as the Siamese LSTM model does, thus stressing the value of both information in argument quality assessment.

5 Case Study

In this section, we study the behavior of the syntax encoder and the coherence encoder on the case examples. To do this, we have proposed different methods for the two encoders, respectively. Regarding the syntax encoder, we present visualizations showing the attention that GAT places on the word embeddings. For the coherence encoder, we follow the work of Jacovi et al. [8] and present the top n-grams scored by the encoder.

5.1 Syntax Encoder

Here, we feed the dependency trees of an argument pair into the encoder respectively and visualize the dot product of the final hidden states within the nodes. Consider the case example shown in Fig. 4, the syntax encoder reveal a better syntax structure of *A2*, since it can well attend *'data'* and its clause to its modifier *'amount of'*. Such a result suggests that the syntax encoder can consider the dependencies between words to accurately identify the syntax structure.

5.2 Coherence Encoder

Suppose that we have a ℓ-gram $u = [w_1, \ldots, w_\ell]$ and a ℓ-size filter $f \in \mathbf{R}^{\ell \times D}$, we calculate the *slot score* as:

$$slot_i = w_i \odot f(i), i \in \{1, 2, \ldots, \ell\} \tag{9}$$

where $f(i)$ is the *slot* i (i.e. the i-th row) of the filter weights. The slot score $slot_i$ can be considered as how much the ith word in the n-gram contributes to the topic coherence. Then the total score of this ℓ-gram is the sum of these ℓ slot scores. For each filter size F, we sample some filters from a total of 256 filters. For each filter, we observed the top n-grams scored by this filter, and we found some highly ranked n-grams present similar semantics, as shown in Table 2. As we can see, the example filter may specialize in detecting n-grams that discuss topics of privacy law. This observation indicates that filters can often be specialized to different topic semantics, and can well capture topic coherent semantic information.

6 Conclusions and Future Work

In this paper, we study the task of automatic argument quality assessment. We represent a method using syntax and coherence information to help with the argument quality preference decision. We demonstrate the feasibility and effectiveness of the proposed method. Moreover, our experimental results significantly outperformed prior methods suggesting such semantic information is worthy of attention. Future work could consider using dependency relation types since this work ignores them. We would also like to explore other graph-based models to avoid potential noise caused by dependency parsing errors.

Acknowledgments. This work was supported by National Natural Science Foundation of China (No. 61976221), and State Key Laboratory of Software Development Environment (SKLSDE-2013ZX-15).

References

1. Bush, R.A.B.: Defining quality in dispute resolution: taxonomies and anti-taxonomies of quality arguments. Denv. UL Rev. **66**, 335 (1988)
2. Chalaguine, L.A., Schulz, C.: Assessing convincingness of arguments in online debates with limited number of features. In: Proceedings of the Student Research Workshop at the 15th Conference of the European Chapter of the Association for Computational Linguistics, Valencia, Spain, pp. 75–83. Association for Computational Linguistics, April 2017. https://www.aclweb.org/anthology/E17-4008
3. Devlin, J., Chang, M.W., Lee, K., Toutanova, K.: BERT: pre-training of deep bidirectional transformers for language understanding. arXiv preprint arXiv:1810.04805 (2018)
4. Eneva, E., Hoberman, R., Lita, L.V.: Learning within-sentence semantic coherence. In: Proceedings of the 2001 Conference on Empirical Methods in Natural Language Processing (2001)
5. Gardner, M., et al.: AllenNLP: a deep semantic natural language processing platform (2017)
6. Gleize, M., et al.: Are you convinced? Choosing the more convincing evidence with a Siamese network. In: Proceedings of the 57th Annual Meeting of the Association for Computational Linguistics, Florence, Italy, pp. 967–976. Association for Computational Linguistics, July 2019. https://doi.org/10.18653/v1/P19-1093. https://www.aclweb.org/anthology/P19-1093
7. Habernal, I., Gurevych, I.: Which argument is more convincing? Analyzing and predicting convincingness of web arguments using bidirectional LSTM. In: Proceedings of the 54th Annual Meeting of the Association for Computational Linguistics (Volume 1: Long Papers), Berlin, Germany, pp. 1589–1599. Association for Computational Linguistics, August 2016. https://doi.org/10.18653/v1/P16-1150. https://www.aclweb.org/anthology/P16-1150
8. Jacovi, A., Shalom, O.S., Goldberg, Y.: Understanding convolutional neural networks for text classification. arXiv preprint arXiv:1809.08037 (2018)
9. Janda, H.K., Pawar, A., Du, S., Mago, V.: Syntactic, semantic and sentiment analysis: the joint effect on automated essay evaluation. IEEE Access **7**, 108486–108503 (2019)
10. Jurafsky, D., Martin, J.H.: Speech and Language Processing: An Introduction to Natural Language Processing, Computational Linguistics, and Speech Recognition. Pearson Prentice Hall, Upper Saddle River (2009)
11. Kim, Y.: Convolutional neural networks for sentence classification. arXiv preprint arXiv:1408.5882 (2014)
12. Lauscher, A., Ng, L., Napoles, C., Tetreault, J.: Rhetoric, logic, and dialectic: advancing theory-based argument quality assessment in natural language processing. arXiv preprint arXiv:2006.00843 (2020)
13. Mimno, D., Wallach, H., Talley, E., Leenders, M., McCallum, A.: Optimizing semantic coherence in topic models. In: Proceedings of the 2011 Conference on Empirical Methods in Natural Language Processing, pp. 262–272 (2011)

14. Newman, D., Lau, J.H., Grieser, K., Baldwin, T.: Automatic evaluation of topic coherence. In: Human Language Technologies: The 2010 Annual Conference of the North American Chapter of the Association for Computational Linguistics, pp. 100–108 (2010)
15. Nokel, M., Loukachevitch, N.: Accounting ngrams and multi-word terms can improve topic models. In: Proceedings of the 12th Workshop on Multiword Expressions, Berlin, Germany, pp. 44–49. Association for Computational Linguistics, August 2016. https://doi.org/10.18653/v1/W16-1806. https://www.aclweb.org/anthology/W16-1806
16. Persing, I., Ng, V.: Modeling argument strength in student essays. In: Proceedings of the 53rd Annual Meeting of the Association for Computational Linguistics and the 7th International Joint Conference on Natural Language Processing (Volume 1: Long Papers), Beijing, China, pp. 543–552. Association for Computational Linguistics, July 2015. https://doi.org/10.3115/v1/P15-1053. https://www.aclweb.org/anthology/P15-1053
17. Simpson, E., Gurevych, I.: Finding convincing arguments using scalable Bayesian preference learning. Trans. Assoc. Comput. Linguist. **6**, 357–371 (2018)
18. Toledo, A., et al.: Automatic argument quality assessment - new datasets and methods. In: Proceedings of the 2019 Conference on Empirical Methods in Natural Language Processing and the 9th International Joint Conference on Natural Language Processing (EMNLP-IJCNLP), Hong Kong, China, pp. 5625–5635. Association for Computational Linguistics, November 2019. https://doi.org/10.18653/v1/D19-1564. https://www.aclweb.org/anthology/D19-1564
19. Veličković, P., Cucurull, G., Casanova, A., Romero, A., Lio, P., Bengio, Y.: Graph attention networks. arXiv preprint arXiv:1710.10903 (2017)
20. Wachsmuth, H., et al.: Argumentation quality assessment: theory vs. practice. In: Proceedings of the 55th Annual Meeting of the Association for Computational Linguistics (Volume 2: Short Papers), Vancouver, Canada, pp. 250–255. Association for Computational Linguistics, July 2017. https://doi.org/10.18653/v1/P17-2039. https://www.aclweb.org/anthology/P17-2039
21. Wachsmuth, H., et al.: Computational argumentation quality assessment in natural language. In: Proceedings of the 15th Conference of the European Chapter of the Association for Computational Linguistics: Volume 1, Long Papers, Valencia, Spain, pp. 176–187. Association for Computational Linguistics, April 2017. https://www.aclweb.org/anthology/E17-1017
22. Wachsmuth, H., Stein, B., Ajjour, Y.: "PageRank" for argument relevance. In: Proceedings of the 15th Conference of the European Chapter of the Association for Computational Linguistics: Volume 1, Long Papers, Valencia, Spain, pp. 1117–1127. Association for Computational Linguistics, April 2017. https://www.aclweb.org/anthology/E17-1105
23. Walton, D.N.: What is reasoning? What is an argument? J. Philos. **87**(8), 399–419 (1990)
24. Wang, S.I., Manning, C.D.: Baselines and bigrams: simple, good sentiment and topic classification. In: Proceedings of the 50th Annual Meeting of the Association for Computational Linguistics (Volume 2: Short Papers), pp. 90–94 (2012)

ExperienceGen 1.0: A Text Generation Challenge Which Requires Deduction and Induction Ability

Hu Zhang[1], Pengyuan Liu[1,2(✉)], Dong Yu[1], and Sanle Zhang[1]

[1] Beijing Language and Culture University, Beijing, China
{201921198263,201921198128}@stu.blcu.edu.cn, liupengyuan@blcu.edu.cn
[2] Chinese National Language Monitoring and Research Center (Print Media), Beijing, China

Abstract. This paper introduces a novel commonsense generation task ExperienceGen 1.0, which is used to test whether the current models have deduction and induction capabilities. It includes two subtasks, both are used to generate commonsense knowledge expressed in natural language. The difference is that the first task is to generate commonsense using causal sentences that contain causal relationships, the second is to generate commonsense with the sentence which is the major premise of the syllogism reconstructed from the original causal sentence. ExperienceGen 1.0 is challenging because it essentially requires the model to have 1) abundant commonsense knowledge, 2) the ability of deduction and induction, and 3) relational reasoning with commonsense. We selected webtext 2019 (https://github.com/brightmart/nlp_chinese_corpus) as the data source, filtered causal sentences and got major premise of the syllogism with manual annotations. ExperienceGen 1.0 contains 2000 items which include causal sentence, major premise of the syllogism and their corresponding commonsense. It is worth noting that the ExperienceGen 1.0 is the product of deduction and induction based on commonsense knowledge from people, which is different from existed commonsense knowledge base. Experiments have shown that even the current best-performing generative models still performs poorly. We are currently releasing an initial version which is publicly available at https://github.com/NLUSoCo/ExperienceGen to inspire work in the field along with feedback gathered from the research community.

1 Introduction

Commonsense knowledge, is typically unstated, as it is considered obvious to most humans [1], and consists of universally accepted beliefs about the world [10]. Although commonsense is rarely mentioned directly in language, commonsense is everywhere in people's lives. When people are communicating or thinking, commonsense knowledge plays a role in secret, allowing us to reach agreement and summarize when we see unknown things. Commonsense knowledge also consists of everyday assumptions about the world [24], and is generally learned through

L. Wang et al. (Eds.): NLPCC 2021, LNAI 13029, pp. 13–24, 2021.
https://doi.org/10.1007/978-3-030-88483-3_2

one's own experience of the world, but can also be inferred by generalizing from common knowledge [17]. While common knowledge can vary by region, culture, and other factors, we expect that commonsense knowledge should be roughly typical to all humans. Most recent commonsense reasoning challenges, such as CommonsenseQA [18], SocialIQA [16], WinoGrande [14] and HellaSwag [27], have been framed as discriminative tasks. AI systems are required to choose the correct option from a set of choices based on a given context. Significant progress has been made on these discriminative tasks. Lin [6] present a constrained task, COMMONGEN associated with a benchmark dataset, to explicitly test machines for the ability of generative commonsense reasoning. They ask the machine to generate a coherent sentence describing an everyday scenario using several concepts according to a given set of common concepts. They require relational reasoning with background commonsense knowledge and compositional generalization ability to work on unseen concept combinations. But we argue that commonsense generation from natural language sentences poses a unique and more comprehensive challenge. In the process of generating commonsense and its reliance on deduction and induction, we want to know whether these models that already contain a large amount of information can directly obtain commonsense from a sentence containing commonsense information.

Through knowledge learning and life experience accumulation, humans can deductively reason and induct the phenomena in life, and obtain abstract and applicable knowledge or commonsense, which could be a perfect reference when we encounter similar situations. The ability of deductive reasoning and induction will directly affect people's way of thinking, thereby affecting all aspects of people's study, work and other life [19]. For example, when we are told that "所有技艺精湛的师傅捏出的泥人都是惟妙惟肖，非常逼真的" (the clay figurines made by all the skilled masters are always vivid and lifelike), then we should think of those experienced people who can perform very well in their own industry as well. Although we have not been told these facts directly, and these incidents cannot be fully enumerated. At this time, we are relying on deduction and induction to get some universal truth. This kind of ability requires a high understanding of world knowledge instead of overfitting shallow lexical features, which has been pointed out as a major drawback in natural language inference [13,28]. Can the machine obtain such deduction and induction capabilities? So in order to initiate the investigation, we proposed ExperienceGen 1.0 - a novel commonsense knowledge generation task. This task requires machine to automatically generate the implied commonsense knowledge according to a sentence which describes a type of event or a type of phenomenon in real life. Given two sentences "所有技艺精湛的师傅捏出的泥人都是惟妙惟肖，非常逼真的" (the clay figurines made by all the skilled masters are always vivid and lifelike) and "由于他技艺精湛,所以他捏的泥人总是惟妙惟肖，非常逼真" (because of his superb skills, the clay figurines he squeezed are always vivid and lifelike), the task is to generate a sentence, which is the commonsense or social experience implied in the given sentence "技术好的人创造的作品是优秀的" (the works created by people with good skills are excellent). Task form is shown in Fig. 1. In

Fig. 1. a. an example showing the structure of the first figure of syllogism; b. given a sentence "所有技艺精湛的师傅捏出的泥人都是惟妙惟肖，非常逼真的" (the clay figurines made by all the skilled masters are always vivid and lifelike), machine are required to generate a sentence such as "技术好的人创造的作品是优秀的" (the works created by people with good skills are excellent)

order to successfully solve the task, the model needs to have 1) abundant commonsense knowledge, 2) the ability of deduction and induction, and 3) relational reasoning with commonsense. Grammatically sound sentences may not always be realistic as they might violate our commonsense, such as "玩具喜欢玩小孩子" (toys like to play children). The model must be able to understand the concepts involved in the sentence. Due to experience is summarized and abstracted normally, therefore, in order to get a piece of commonsense knowledge or experience from a given sentence, the model needs to have a high degree of abstraction and generalization, which is more profound than the language itself. Models additionally need compositional generalization ability to infer unseen concept compounds. This encourages the model to abstract a variety of similar concepts into higher-level concepts to deal with a potentially unlimited number of unfamiliar concepts.

In this paper, we proposed a dataset consisting of 2000 sentences and the their commonsense to support the ExperienceGen 1.0 task. We explicitly designed our data collection process. We gain the commonsense according to the syllogism conducted based on causal sentence. We provide comprehensive benchmark performance for the most advanced language generation models through extensive automatic evaluation and manual comparison. There is still a big gap between the results of the machine and the commonsense summarized by humans. Our major contributions in this paper are twofold:

- We present a commonsense generation task and constructed a high-quality commonsense dataset.
- Our extensive experiments systematically examine recent pretrained language generation models (BERT, WoBERT etc.) on the tasks, and find that their performance is poor.

2 Related Work

Commonsense Benchmark Dataset. Commonsense knowledge, is considered obvious to most humans, and not likely to be explicitly stated [1]. There has been a long effort in capturing and encoding commonsense knowledge, and various knowledge bases have been developed for this from different angles, such as commonsense extraction [4,21], next situation prediction (SWAG [26], CODAH [2], HellaSWAG [27]), cultural and social understanding [5,15], visual scene comprehension [25], and general commonsense question answering [3,18,20]. However, the success of fine-tuning pretrained language models for these tasks does not necessarily mean machines can produce novel assumptions in a more open, realistic, generative setting. We believe that ExperienceGen 1.0 is a novel commonsense generation benchmark task, which not only requires the model to have a large amount of commonsense knowledge, but also requires the model to have deep abstraction and generalization ability. This task can be used to detect whether the machine has abstraction and generalization ability.

Natural Language Generation. NLG means that the computer generates natural language text based on the information provided. The NLG field has developed rapidly from the simplest information merging. Recently, some people have incorporated commonsense knowledge into the task of language generation. Such as short text generation [22], image captioning [8], video storytelling [23], and conversational systems [28]. These works suggest that generative commonsense reasoning has a great potential to benefit downstream applications. As far as we know, the ExperienceGen 1.0 task we proposed is the first novel task that automatically generates commonsense based on natural sentences. It can be used to detect whether the model has a lot of commonsense and whether it has abstract and generalization ability.

3 Task Formulation

We formulate the task with mathematical notations and discuss its inherent challenges with concrete examples. The input is a natural sentence with complete semantics we call x. X is always a universal proposition(singular propositions are also classified as universal propositions), describing a certain type of event or a certain point of view. There is always some commonsense or experience implied in x. Our expectation is to output a grammatically correct simple sentence y, which describes a kind of commonsense or life experience, not necessarily using the words that appears in x, but x and y are semantically related, and the specific expression is that y is summarized and abstracted from x. The ExperienceGen 1.0 task is to learn a function f: X → Y, which maps the given sentence X to sentence Y.

4 Dataset Construction

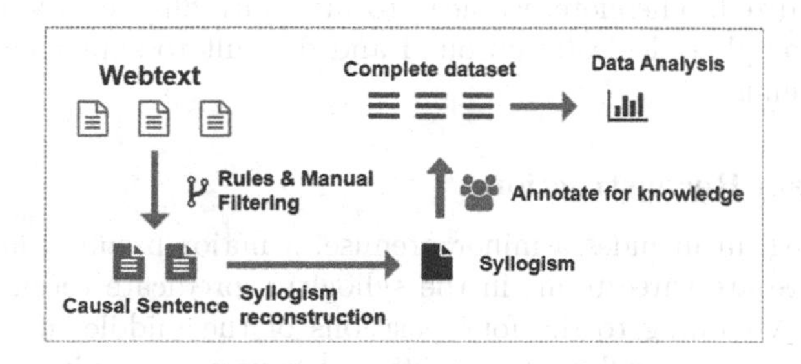

Fig. 2. Dataset construction workflow overview.

Figure 2 illustrates the overall workflow of data construction. We found that some commonsense is implicit in the major premise of syllogism, and causal sentences are a type of syllogism reasoning, so we chose webtext 2019, which has more causal sentences compared with news and other corpus, as the data source. After causal candidate sentence selection, syllogism reconstruction and commonsense sentence extraction, a commonsense generation data set is finally formed.

4.1 Causal Sentences and Syllogism

Causal sentences are a type of deductive reasoning, which is a syllogism that omits the major premises [9], The omitted part is often self-evident and can be regarded as commonsense. When we see the example, "由于他技艺精湛，所以他捏的泥人总是惟妙惟肖，非常逼真" (because of his superb skills, the clay figurines he squeezed are always vivid and lifelike). We will not doubt the correctness of this sentence, because there is an omitted premise, "所有技艺精湛的师傅捏出的泥人都是惟妙惟肖，非常逼真" (the clay figurines made by all the skilled masters are always vivid and lifelike), which is the consensus among most people. The supplemented sentence and the corresponding causal sentence can form a syllogism. We get commonsense from the major premise. We have obtained a set of methods for obtaining reliable commonsense from causal sentences.

4.2 Get Candidate Sentences

Most of the causal sentences in natural language texts match its related words [31], Zhang [30] proposed to determine the type of semantic relationship between sentences according to the related word indicator rules, which includes causality, and their experiment has achieved good results. so we prepared a causal related word set to match causal sentences. Our word set contains 13 typical related words that are frequently used to match in the community encyclopedia corpus, and got 126504 sentences using causal related words. Since the community encyclopedia corpus does not follow strict grammar, the punctuation is not standardized. Therefore, we need to manually filter candidate sentences to remove incomplete, logically unsound and difficult to supplement sentences with major premises.

4.3 Syllogism Reconstruction

A typical syllogism includes a minor premise, a major premise, and a conclusion [12]. There are three terms in the syllogism, predicate term, subject and middle term. According to the four positions of the middle term in the two premises, the syllogism will have four different figures, each with different characteristics and functions. We mainly reconstruct the first syllogism, which is the most widely used in daily life. The middle term of the first figure is the major term and predicate term in the major premise and minor premise respectively, as shown in the Fig. 3.

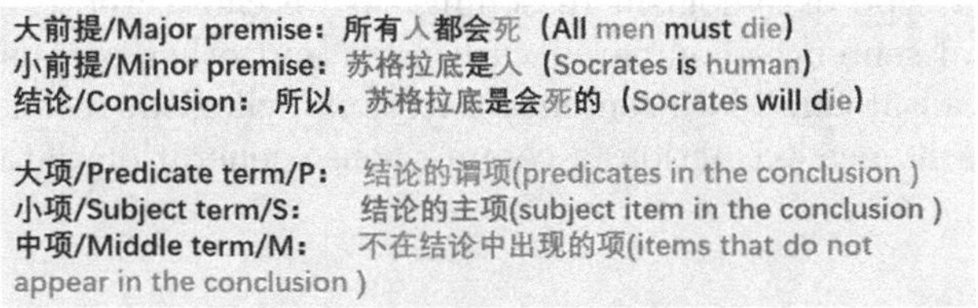

大前提/Major premise：所有人都会死（All men must die）
小前提/Minor premise：苏格拉底是人（Socrates is human）
结论/Conclusion：所以，苏格拉底是会死的（Socrates will die）

大项/Predicate term/P：结论的谓项(predicates in the conclusion)
小项/Subject term/S：结论的主项(subject item in the conclusion)
中项/Middle term/M：不在结论中出现的项(items that do not appear in the conclusion)

Fig. 3. Dataset construction workflow overview.

4.4 Commonsense Knowledge Extraction

We found that it is not entirely correct to directly treat the major premise as commonsense, because the major premise obtained by our reconstruction is often for a special object, and commonsense knowledge should be a sentence that is general and belongs to a higher level in semantics. Therefore, we analogize the three closely related parts in the major premise sentence. These three parts are called the qualified component, the qualified object, and the supplementary component. After the expansion has obtained multiple related components, we will perform the induction. The sentence obtained by the induction is the commonsense of the major premise. For more details, please see Fig. 1b.

4.5 Quality Inspection of Dataset

The whole annotation work was completed by 6 annotators with professional knowledge of linguistics. All the annotators have been trained in advance, and they can not formally annotate until they pass the training. In the final commonsense generation, each major premise was separately completed by two annotators. The three parts are analogized and summarized to obtain commonsense sentences. Finally, the third person judged whether the summarized commonsense sentences is semantically and logically correct. Under strict control, we finally got a pass rate of **99.35%** for commonsense, which shows that the quality of our commonsense generated dataset is extremely high.

4.6 Dataset Statistics

Table 1. Statistics of the ExperienceGen 1.0 dataset.

Statistics	Causal sentence	Major premise	Commonsense
Sentence	2000	2000	2000
Words	23949	16558	13467
Ave sentence length	36.46	17.59	14.05
Type of words	8497	4497	3193
Ave num of words per sentence	11.97	8.28	6.73

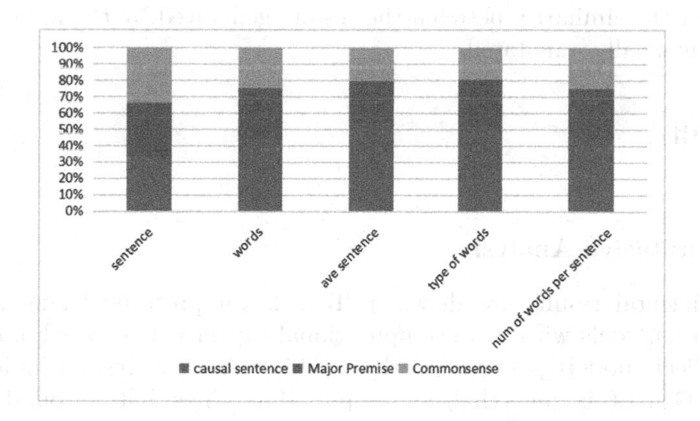

Fig. 4. The distribution of various indicators.

Here we use statistical methods to analyze our dataset to prove that this ExperienceGen 1.0 task is quite difficult. The statistics are shown in Table 1. The distribution of various indicators of the two sentences is shown in Fig. 4.

As shown in the table above, our dataset contains 2000 sentence pairs (each pair includes a causal sentence, a major premise and a commonsense sentence). We calculated the number the words, the average length of sentence and the types of words in each kind of sentence. The more type of word the sentence uses, the richer of words. We find that the average length of causal sentences is the longest, and that of the commonsense is the shortest. This is in line with our expectations, because in daily life people are more accustomed to using simple sentences to express, and the words used tend to be common words. This shows that most commonsense knowledge in life can be summarized by using short sentences and common words.

5 Experiments and Analysis

We divided the experimental data into 1000 and 2000 to observe the influence of the training data in the ExperienceGen 1.0 on the experimental results. We used several pretrained language models, BERT-base[1], Wobert[2], UniLM[3]. Specifically, we constructed the input sequence for each original sentence and commonsense sentence with a special token [SEP]. We then concatenated the input sequences for output sentence with [SEP] and fed it into a BERT-like model. Finally, we used the hidden representation of the first token [CLS] to output the final prediction with a linear layer. We fine-tuned the models on the training set and reported the results on the development and test sets. We also used Bi-directional Long Short-Term Memory(BiLSTM)[4], which does not contain any information. We use several widely-used automatic metrics to automatically assess the performance, namely BLEU [11] and ROUGE [7], which mainly focus on measuring surface similarities. These two indicators can effectively detect the similarity between the results generated by the model and the results we manually annotated.

6 Result

6.1 Quantitative Analysis

The experimental results are shown in Table 2, for pretrained language models, large-size models with more computational capacity. It is worth noting that these excellent models perform poorly on this task. The results indicate that ExperienceGen 1.0 is quite challenging, providing a new battleground for next-generation pretrained language models. For BiLSTM, the experimental performance is even worse. In addition, we also analyzed the influence of the abstract

[1] https://github.com/CLUEbenchmark/CLGE.
[2] https://github.com/ZhuiyiTechnology/WoBERT.
[3] https://github.com/percent4/UniLM_Chinese_DEMO.
[4] https://github.com/mryuan0428/Title_Generator_CN/tree/master/TG_BiLSTM.

Table 2. Experimental results on ExperienceGen 1.0

Model/Metrics	ROUGE-1		ROUGE-2		ROUGE-L		BLEU	
	Task1	Task2	Task1	Task2	Task1	Task2	Task1	Task2
BiLSTM-1000	0.68	1.67	0.00	0.00	0.66	1.67	0.30	0.32
BiLSTM-2000	1.08	1.72	0.00	0.00	1.08	1.70	0.15	0.11
BERT-1000	34.30	53.75	17.27	37.63	33.03	54.16	10.10	26.36
BERT-2000	34.58	54.55	18.12	38.47	33.94	54.73	10.92	28.30
UniLM-1000	10.68	35.43	4.21	15.50	10.30	35.25	4.19	20.50
UniLM-2000	26.16	56.49	8.83	32.15	25.84	56.08	9.65	**39.37**
WoBERT-1000	**40.22**	**57.32**	**25.65**	**42.24**	**40.78**	**55.04**	**15.05**	**29.59**
WoBERT-2000	**40.98**	**56.52**	24.14	**40.71**	**40.34**	**56.18**	**15.87**	30.41

Table 3. The effect of abstraction on WoBert about Task-2.

Percentage/metrics	ROUGE-1	ROUGE-2	ROUGE-L	BLEU
0–75%	50.97	34.02	52.28	21.93
75%–90%	**58.60**	40.88	**57.70**	29.80
90%–100%	57.33	**40.94**	56.87	**30.11**

degree of commonsense sentences on the experimental results, the results can be found in Table 3. The commonsense sentences are obtained from the major premise sentences. Generally speaking, the degree of abstraction of each concept is different, and there are high and low levels. In fact, the logic in terms of the degree of abstraction, the so-called superior concepts and subordinate concepts in the academic study are always higher in abstraction than the latter [1982?]. So we use the lower concept (major premise) to abstract the upper concept(commonsense), and the sentence obtained after induction should be more concise and clear. So we hypothesis the proportion of the length of the commonsense sentence to the sentence length of the major premise sentences can be roughly regarded as the degree of abstraction. We selected 500 sentences that accounted for 0–75%, 75%–90%, and 90%–100% respectively. Experiment separately, and count the average number of co-occurring words and the number of word categories of these three types of sentences. As Fig. 5 shows, we found that as the proportion of sentence length increases, the performance of the model first increases then decreases, while other indicators also increase slowly. This shows that when the degree of sentence abstraction is high (0–75%), the difficulty of model generation commonsense will increase.

6.2 Qualitative Analysis

For example "所有地震高发国的防震体系都是立体的" (All earthquake-prone countries have three-dimensional anti-seismic systems). We hope it generates commonsense "自然灾害频繁发生的国家的防灾体系都是立体的" (The disaster prevention systems of countries where natural disasters frequently

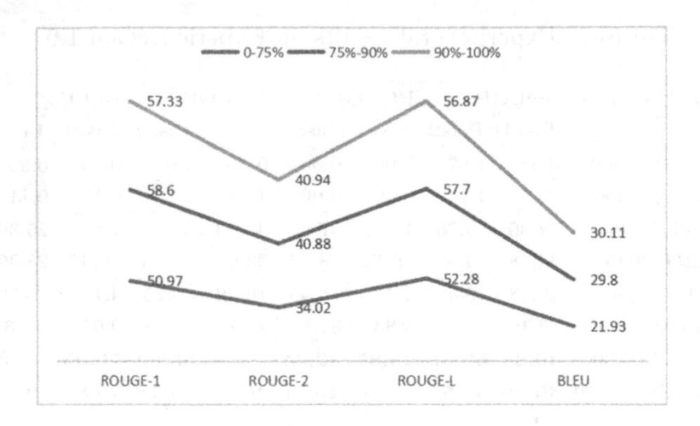

Fig. 5. The effect of abstraction on the result.

occur are all three-dimensional). But the sentence generate by BERT is "自然灾害的防御体系是立体的" (The defense system against natural disasters is three-dimensional). We find the subject, predicate, and object of the sentence in this generative sentence, but their combination is illogical. There is no disaster prevention system for natural disasters. And the model cannot associate all natural disasters based on earthquakes. This shows that the model still lacks commonsense and reasoning ability.

7 Conclusion

In this paper, we propose a novel commonsense generation task consisting of two subtasks to detect whether the existing generative model has the ability of deduction and induction. We experimented with BERT, WoBERT, UniLM and BiLSTM, and the results show that the model performed poorly on these two task.

There are several deficiencies in our work in this article that need to be improved. The first is that the amount of data is not sufficient, and this may cause the models to not be fully trained and learned. Then, when we manually obtained commonsense, we only selected the one that the third annotator thought was the most correct as the result, but multiple commonsense may be more reasonable. Finally, we are not very satisfied with the evaluation indicators of the machine-generated results. This can only calculate the similarity of the vocabulary at a shallow level, and cannot evaluate whether the semantics are reasonable. These are also some points that we need to improve in the future.

Acknowledgments. The research is supported by Beijing Natural Science Foundation (4192057) and Science Foundation of Beijing Language and Culture University (the Fundamental Research Funds for the Central Universities: 21YJ040005). We thank anonymous reviewers for their helpful feedback and suggestions.

References

1. Cambria, E., Song, Y., Wang, H., Hussain, A.: Isanette: a common and common sense knowledge base for opinion mining. In: 2011 IEEE 11th International Conference on Data Mining Workshops, pp. 315–322. IEEE (2011)
2. Chen, M., D'Arcy, M., Liu, A., Fernandez, J., Downey, D.: CODAH: an adversarially authored question-answer dataset for common sense. arXiv preprint arXiv:1904.04365 (2019)
3. Huang, L., Bras, R.L., Bhagavatula, C., Choi, Y.: Cosmos QA: machine reading comprehension with contextual commonsense reasoning. arXiv preprint arXiv:1909.00277 (2019)
4. Li, X., Taheri, A., Tu, L., Gimpel, K.: Commonsense knowledge base completion. In: Proceedings of the 54th Annual Meeting of the Association for Computational Linguistics (Volume 1: Long Papers), pp. 1445–1455 (2016)
5. Lin, B.Y., Xu, F.F., Zhu, K., Hwang, S.w.: Mining cross-cultural differences and similarities in social media. In: Proceedings of the 56th Annual Meeting of the Association for Computational Linguistics (Volume 1: Long Papers), pp. 709–719 (2018)
6. Lin, B.Y., et al.: CommonGen: a constrained text generation challenge for generative commonsense reasoning. arXiv preprint arXiv:1911.03705 (2019)
7. Lin, C.Y.: ROUGE: a package for automatic evaluation of summaries. In: Text Summarization Branches Out, pp. 74–81 (2004)
8. Lu, J., Yang, J., Batra, D., Parikh, D.: Neural baby talk. In: Proceedings of the IEEE Conference on Computer Vision and Pattern Recognition, pp. 7219–7228 (2018)
9. Ming-Chen, L.V., Ding, X.F., University, J.: A contrastive study of the causal and causation relationship from the perspective of logic reasoning and semantic category. J. Northeast Normal Univ. (Philos. Soc. Sci.) (2019)
10. Nunberg, G.: Position paper on common-sense and formal semantics. In: Theoretical Issues in Natural Language Processing 3 (1987)
11. Papineni, K., Roukos, S., Ward, T., Zhu, W.J.: BLEU: a method for automatic evaluation of machine translation. In: Proceedings of the 40th Annual Meeting of the Association for Computational Linguistics, pp. 311–318 (2002)
12. Peng, S., Liu, L., Liu, C., Yu, D.: Exploring reasoning schemes: a dataset for syllogism figure identification. In: Liu, M., Kit, C., Su, Q. (eds.) CLSW 2020. LNCS (LNAI), vol. 12278, pp. 445–451. Springer, Cham (2021). https://doi.org/10.1007/978-3-030-81197-6_37
13. Poliak, A., Naradowsky, J., Haldar, A., Rudinger, R., Van Durme, B.: Hypothesis only baselines in natural language inference. arXiv preprint arXiv:1805.01042 (2018)
14. Sakaguchi, K., Le Bras, R., Bhagavatula, C., Choi, Y.: WinoGrande: an adversarial Winograd schema challenge at scale. In: Proceedings of the AAAI Conference on Artificial Intelligence, vol. 34, pp. 8732–8740 (2020)
15. Sap, M., et al.: ATOMIC: an Atlas of machine commonsense for if-then reasoning. In: Proceedings of the AAAI Conference on Artificial Intelligence, vol. 33, pp. 3027–3035 (2019)
16. Sap, M., Rashkin, H., Chen, D., LeBras, R., Choi, Y.: SocialIQA: commonsense reasoning about social interactions. arXiv preprint arXiv:1904.09728 (2019)
17. Speer, R., Havasi, C., Lieberman, H.: AnalogySpace: reducing the dimensionality of common sense knowledge. In: AAAI, vol. 8, pp. 548–553 (2008)

18. Talmor, A., Herzig, J., Lourie, N., Berant, J.: CommonsenseQA: a question answering challenge targeting commonsense knowledge. arXiv preprint arXiv:1811.00937 (2018)
19. Tincoff, R., Jusczyk, P.W.: Some beginnings of word comprehension in 6-month-olds. Psychol. Sci. **10**(2), 172–175 (1999)
20. Wang, C., Liang, S., Zhang, Y., Li, X., Gao, T.: Does it make sense? And why? A pilot study for sense making and explanation. arXiv preprint arXiv:1906.00363 (2019)
21. Xu, F.F., Lin, B.Y., Zhu, K.Q.: Automatic extraction of commonsense located near knowledge. arXiv preprint arXiv:1711.04204 (2017)
22. Yang, P., Li, L., Luo, F., Liu, T., Sun, X.: Enhancing topic-to-essay generation with external commonsense knowledge. In: Proceedings of the 57th Annual Meeting of the Association for Computational Linguistics, pp. 2002–2012 (2019)
23. Yang, P., et al.: Knowledgeable storyteller: a commonsense-driven generative model for visual storytelling. In: IJCAI, pp. 5356–5362 (2019)
24. Zang, L.J., Cao, C., Cao, Y.N., Wu, Y.M., Cun-Gen, C.: A survey of commonsense knowledge acquisition. J. Comput. Sci. Technol. **28**(4), 689–719 (2013)
25. Zellers, R., Bisk, Y., Farhadi, A., Choi, Y.: From recognition to cognition: visual commonsense reasoning. In: Proceedings of the IEEE/CVF Conference on Computer Vision and Pattern Recognition, pp. 6720–6731 (2019)
26. Zellers, R., Bisk, Y., Schwartz, R., Choi, Y.: SWAG: a large-scale adversarial dataset for grounded commonsense inference. arXiv preprint arXiv:1808.05326 (2018)
27. Zellers, R., Holtzman, A., Bisk, Y., Farhadi, A., Choi, Y.: HellaSwag: can a machine really finish your sentence? arXiv preprint arXiv:1905.07830 (2019)
28. Zhang, H., Liu, Z., Xiong, C., Liu, Z.: Grounded conversation generation as guided traverses in commonsense knowledge graphs. arXiv preprint arXiv:1911.02707 (2019)
29. 周昌忠: 讲究抽象的程度. 逻辑与语言学习. **2**(10) (1982)
30. 张牧宇, 宋原, 秦兵, 刘挺: 中文篇章级句间语义关系识别. 中文信息学报. **27**(6), 51–58 (2013)
31. 李为政: 近代汉语因果句研究. Ph.D. thesis 北京大学 (2013)

Machine Translation and Multilinguality

SynXLM-R: Syntax-Enhanced XLM-R in Translation Quality Estimation

Bin Ni[1], Xiaolei Lu[2(✉)], and Yiqi Tong[3,4(✉)]

[1] Xiamen Data Intelligence Academy of ICT, CAS, Xiamen, China
nibin@casxm.cn
[2] College of Foreign Languages and Cultures, Xiamen University, Xiamen, China
luxiaolei@xmu.edu.cn
[3] Department of Artificial Intelligence, School of Informatics, Xiamen University, Xiamen, China
[4] Institute of Artificial Intelligence, Beihang University, Beijing, China
yqtong@stu.xmu.edu.cn

Abstract. Quality estimation (QE) is an NLP downstream task that aims to determine the quality of translations without given references. Existing QE models mainly draw on neural-based or pre-trained models, overlooking syntactic information. In this paper, we propose a syntax-enhanced model based on GAT and XLM-R. Our model (SynXLM-R) introduces GAT to extract graph-structured features derived from dependency parsing. As for semantics, we utilize XLM-R to obtain the representations of the input sentences. Our model achieves higher correlations (Pearson's r) with human judgments than previous models both on mid-resource (Ro-En) and high-resource (En-De, En-Zh) language pairs in WMT20 QE sentence-level tasks. Subsequent analysis shows that different parsers result in different performances of our model. Furthermore, our model achieves the best result when the number of attention head is three.

Keywords: Quality estimation · Graph Attention Network · XLM-R · Dependency parsing

1 Introduction

With the development of Machine Translation, Quality Estimation (QE, see Blatz [1], Specia [2]), a task that seeks to evaluate the quality of MT systems without reference translations, has attracted much attention. Existing QE methods can be roughly grouped into three categories:

This work was supported by the Fundamental Research Funds for the Central Universities (20720191053).

L. Wang et al. (Eds.): NLPCC 2021, LNAI 13029, pp. 27–40, 2021.
https://doi.org/10.1007/978-3-030-88483-3_3

Feature-Based. Feature-based QE models (e.g., QuEst [3] and QuEst++ [4]) combined pre-defined linguistic features with traditional machine learning algorithms (such as support vector regression or randomized decision tree). Though adequately interpretable, feature-based QE systems rely heavily on feature engineering and thus have difficulty in dealing with new linguistic phenomenon.

Neural-Based. With the advent of neural machine translation, feature-based models are being replaced by neural-based systems. Since WMT 2017, most of the best-performing QE systems are neural-based (e.g., Predictor-Estimator [5]), neural bilingual expert model [6]). Neural-based QE models are often based on a two-stage architecture. At the first stage, the high level joint latent representation of the source and the target in a parallel pair would capture the alignment or semantic information. Then a bidirectional LSTM is built to estimate translation quality.

Pre-training-Based. Pre-trained language models, especially multi-language models (e.g., mBERT [7] and XLM-R [8]), have become state-of-the-art in QE tasks more recently. Self-supervised learning in pre-trained models greatly reduced the complexity of the model. Some scholars utilized pre-trained models for QE tasks and achieved fairly good results. For instance, Lee [9] proposed an XLM-R-based neural network architecture with a few additional parameters for word and sentence-level QE tasks and reported positive results.

While recent study [10] showed that some pre-trained models, such as BERT, does learn the structure information from sentence, most of the existing QE models treated a sentence as a word sequence, ignoring the syntactic structure of sentence. The relations between words could still benefit QE tasks, as syntactic (grammatical) features are crucial indicators of sentence coherence(Specia [4], Comelles [11]).

Our proposed model, entitled SynXLM-R, is a syntax-enhanced model based on GAT (Graph Attention Network) and XLM-R. It contains two parallel modules, i.e., syntax-aware extractor and contextualized embedding module. Specifically, in the syntax-aware extractor module, a syntax extractor was designed using dependency parser, which can capture dependency relations of an input sentence. Several different parsers were applied to generate dependency graph (Fig. 2). To exploit rich syntactic information in the graphs, we transformed parsed sentences into lists of triplets and then encoded them using GAT. After that, the syntactic features were fed into a CNN to detect complex features, obtain fixed size vectors, and clear noisy information in the graphs [12]. In contextualized embedding module, we utilized XLM-R to obtain the representations of the input sentences and later fed them into a fully connected layer to calculate the score of translation quality. Since the parser plays a critical role in our model, an experiment was carried out to compare the effect of different dependency parsers. We also conducted experiments on the influence of attention heads in GAT.

The contributions of our work are summarized as follows::

- We proposed a syntax-enhanced QE framework based on dependency graph and GAT. To the best of our knowledge, the present study is the first of its kind that attempted to apply GAT to translation quality estimation.
- Our proposed model outperformed the state-of-the-art model (MTransQuest-base) by 1.96% (En-Zh), 3.17% (En-De) and 0.06% (Ro-En) in Pearson's r, respectively.
- Different parsing models were compared to evaluate their impact on our model's performance.
- An additional experiment was conducted to show how the number of attention heads of GAT affects result.

2 Related Works

Pre-trained Multi-language Model in QE: Prior works have proved the effectiveness of multilingual masked language models (e.g., multilingual BERT [7] and XLM [8]) in QE tasks. TransQuest,the winning solution in all language pairs in the WMT20 QE tasks, relyed on fine-tuned XLM-R (XLM-RoBERTa) model to generate crosslingual embeddings, and provided links between languages [13,14]. Hu [15] fine-tuned the mBERT and XLM-R in both domain-adaptive and task-adaptive manners on the parallel corpora. Fomicheva [16] built a black-box multi-task learning QE model (MTL) with a shared BERT or XLM-R encoder. Wu [17] proposed XLM-based Predictor-Estimator architecture, which introduced the cross-lingual language model to QE task via transfer learning technique. Overall, pre-trained multi-language models enhanced the robustness of QE models, and contextualized word representations showed great potentials in QE tasks.

GAT with Dependency Graph: GAT [18] and GCN (Graph Conventional Networks, Kipf [19]), two typical variants of GNN (Graph neural networks, Scarselli [20]), are neural network architectures that operate on graph-structured data. Over the years, Graph-based models have been employed in a wide array of NLP downstream tasks, such as machine translation [21], semantic role labeling [22], text classification [23], and achieved relatively good results. Specifically, Zhang [24] and Sun [25] proposed to use GCN to learn node representations from dependency graph and used them together with other features for sentiment classification. For a similar purpose, Huang [12] used GAT to establish the dependency relationships between words for aspect-level sentiment analysis. However, little research has utilized graph-based models for QE tasks. In the current study, we attempt to use GAT models to extract the syntactic dependency relations among words, and provide a syntactic enhanced way of assessing machine translation quality.

3 Methodology

This section is devoted to the basic architecture of our SynXLM-R. As is shown in Fig. 1, our model is composed of two modules: XLM-R and Syntax-aware Extractor.

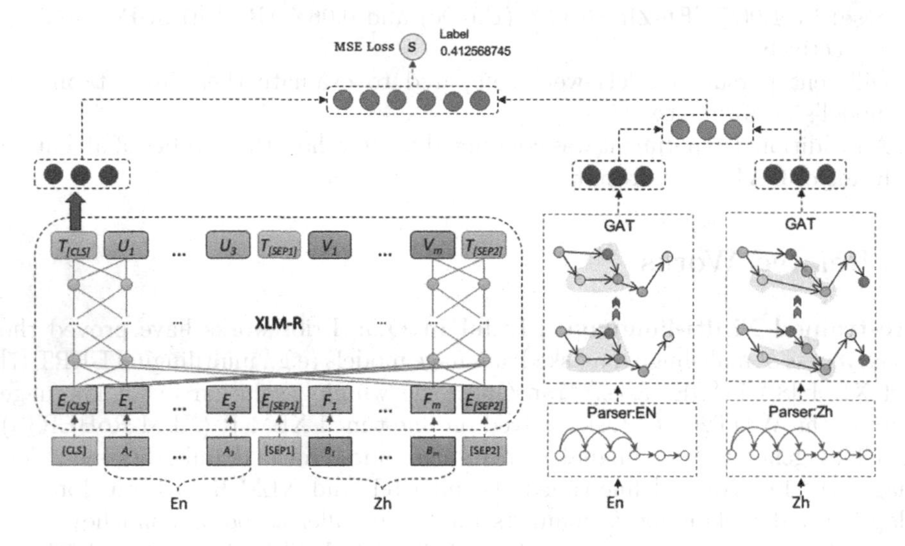

Fig. 1. the architecture of SynXLM-R

3.1 XLM-R

We used XLM-R-base (˜300M parameters) with its original parameters, which is a module that extracts contextual information from sentences. In our model, the tokenized source language and target language sentences were concated with three special tokens: [CLS] (an indicator of the start of a sequence), [SEP1] (an indicator for the separator of two inputs), [SEP2] (an indicator for the end of inputs). The outputs of XLM-R can be [CLS] embedding (representing the whole sequence) or embeddings of each word. Since the early experiments carried out by Ranasinghe [14] demonstrated that the [CLS] embedding achieved better results, the outputs of [CLS] embedding were fed into a simple upper fully connected layer and to calculate QE score with syntactic vector (see 3.2). We used the open-source library transformers [26] to build our XLM-R module.

3.2 Syntax-Aware Extractor

This section is about how to encode syntactic information from sentences. We did not build the whole dataset into one single graph (as GCN does); instead, we build a specific graph for each sentence.

Building Text Graphs: There are many ways to transform text to graph, such as external knowledge graph [23] and dependency parsing [12]. In our study, we used dependency parser to generate a dependency graph to represent the syntactic structure. We used lightweight and off-the-shelf NLP library spaCy to build text graphs. The syntactic relationships between words in dependency trees can be denoted with directed edges and labels, which are then represented as triplets. For example, in Fig. 2a, an input sentence "*Dudley was very fat and hated exercise*" would be transformed to a relation path with several dependency heads and dependency relation labels. These heads and edges are then stored in triplets (*was, nsubj, Dudley*), (*fat, advmod, very*) etc. These triplets can be treated as graphs (Fig. 2b) and then be processed by graph neural networks.

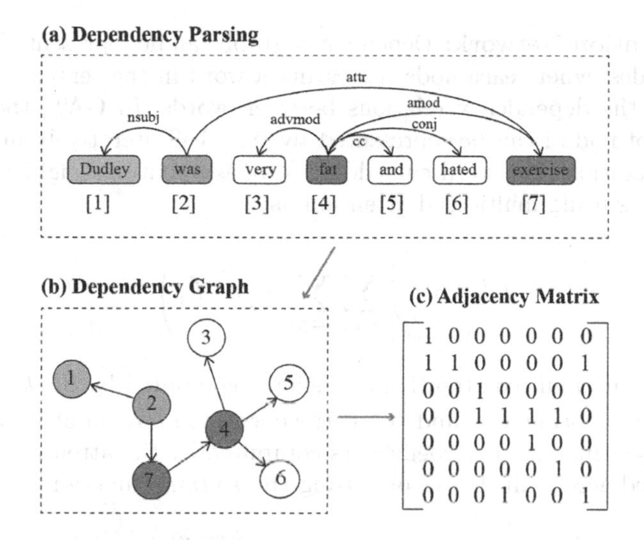

Fig. 2. (a): the dependency parsing for sentence "Dudley was very fat and hated exercise". nsubj: nominal subject; attr: attributive; advmod: adverbial modifier; amod: adjectival modifier; cc: coordination; conj: conjunct. (b): dependency graph derived from dependency parser. (c): adjacency matrix. Note that element in main diagonal of the matrix is 1.

Adjacency Matrix: We then used adjacent matrix to transform triplets into matrix. Adjacency matrix is a two-dimensional array of size $V \times V$, where V is the number of nodes in a graph G. Let the Adjacency matrix be $adj[][]$, as the following equation set:

$$adj[i][j] = \begin{cases} 1, & (v_i, v_j) \in G \\ 1, & i = j \\ 0, & (v_i, v_j) \notin G \end{cases} \tag{1}$$

Where (v_i, v_j) indicates the edge from node i to node j. If the edge (v_i, v_j) in graph G and $i = j$ (which means node i and node j is the same one), we set a slot $adj[i][j] = 1$.

Node Embedding: Given a sentence s = $[w_1, w_2, \ldots, w_n]$ with length n, we first map each word into a low-dimensional word embedding vector. For each word w_i, we get one vector $X_i \in R^d$ where d is the dimension of the word embedding space. We try two embedding methods including 300-dimensional *GloVe* embeddings and we use static vector from pre-trained XLM-R-base as the same in XLM-R module mentioned above. Once the word representations are initialized, they are frozen during training.

Graph Attention Network: Dependency graph can be represented by a graph G with n nodes, where each node represents a word in the sentence. The edges of G denote the dependency relations between words. In GAT, the neighborhood nodes of node i can be represented by N_i. GAT iteratively updates each node representation (i.e., word embeddings) by aggregating neighborhood node representations using multi-head attention as:

$$\vec{h_i}' = \sigma\left(\frac{1}{k}\sum_{k=1}^{N}\sum_{j\in Ni} a_{ij}^{k} W^{k} \vec{h}_j\right) \tag{2}$$

Where α_{ij}^k are normalized attention coefficients computed by the k-th attention mechanism (See Formula 3), and W^k is the corresponding input linear transformation's weight matrix. The coefficients computed by the attention mechanism are normalized across all choices of j using the *softmax* function:

$$\alpha_{ij} = softmax_j(e_{ij}) = \frac{exp(e_{ij})}{\sum_{k\in N_i} exp(e_{ik})} \tag{3}$$

The attention mechanism is shared with all edges in the graph. In this way, when using GAT, we don't need to go through the entire graph, but only to visit the adjacent ones that matter. The features generated by GAT are effective in that: (1) they can handle directed graphs; (2) they can be used directly for inductive learning.

Finally, the semantic and syntactic features extracted by XLM-R and GAT were fed into a fully connected layer to calculate the final QE score.

4 Experiments

4.1 Data Preparation

We used WMT20 dataset for Sentence-Level Direct Assessment Task (7000 sentence pairs for training, 1000 sentence pairs for validation and 1000 sentence pairs for testing). Each translation was rated with a score between 0 and 100 according to the perceived translation quality by at least three raters [16]. The DA (Direct Assesment) scores were standardized as z-scores (also called standard scores). The quality estimation systems have to predict the mean DA z-scores of the test sentence pairs [27]. Because of the lack of dependency parsers in low-resource language pairs, the dataset we use in this study is composed of the high-resource English-German (En-De) pairs, English-Chinese (En-Zh) pairs and the medium-resource Romanian-English (Ro-En) pairs.

4.2 Baselines

We chose neural-based PredEst and pre-training-based Bert-BiRNN as our baseline.

PredEst: As one of the baselines of WMT20, the PredEst consisted of a predictor and an estimator and is implemented by OpenKiwi [28]. A bidirectional LSTM was used in predictor to encode the source sentence, and two unidirectional LSTMs were used to process the target in both left-to-right and right-to-left directions. The above sequence features were then fed into the estimator to obtain the score of translation quality.

Bert-BiRNN: This model used a large scale pre-trained BERT model to obtain token-level representations, which were then fed into two independent bidirectional RNNs to encode both the source sentence and its translation independently. Through an attention mechanism, these two sentence representations were then concatenated as a weighted sum of their word vectors. The final sentence-level representation was finally fed to a sigmoid layer to produce the sentence-level quality estimates. During training, BERT was fine-tuned by unfreezing the weights of the last four layers along with the embedding layer [29].

MTransQuest-Base: Ranasinghe [14] presented TransQuest (i.e. MTransQuest and STransQuest), which was the winning solution of WMT20 QE tasks. XLMR-base and XLMR-large pre-trained language model was also implemented and the default setting of TransQuest was XLMR-base.

4.3 Training Details

The SynXLM-R was implemented by Python 3.6 and Pytorch 1.7.1.

We used the same set of configurations for all language-pairs evaluated in this paper to ensure consistency among language pairs. We set the batch-size as 4 and utilized AdamW optimizer with a learning rate 8e−6. Gradient clip by norm was applied using 1.0 and the max sentence length was 128. The word embedding dimension was 300. As for GAT, we set the number of attention heads as 3.

Since GAT is not capable of batch learning, we concatenated the N (the number of batch size) adjacency matrixes to a bigger batch matrix in order to boost the progress of computation.

There are two versions of the pre-trained XLM-R models, i.e., XLM-R-base and XLM-R-large. We chose XLM-R-base, which consists of 7 hidden layers, 768 hidden states of models, 3072 dimensional feed-forward layers and 270M parameters in total. We used K-fold cross-validation method to gauge the performance of our model. The average value of outputs and losses over the K folds would be the final results.

4.4 Correlation with DA Scores

Table 1 shows Pearson's correlation of SynXLM-R with DA scores. Part I depicts the results of baseline models: PredEst, BERT-BiRNN and MTransQuest-base; Part II shows the results of our models: XLM-R and SynXLM-R (with 3 attention heads) and Syn-*GloVe*, Syn-XLM-R*static*; Part III is the results of SynXLM-R with other numbers of attention heads for comparison (different number of asterisks indicate the number of attention heads).

Overall, our model SynXLM-R outperformed MTransQuest-base in all language pairs. To be specific, we got 0.005 (0.6%) improvement in mid-resource language pair (Ro-En); 0.014 (3.2%) in high-resource language pair (En-De); 0.009 (2.0%) in En-Zh, which confirmed our hypothesis that adding syntactic information could improve the performance of QE models. We observed that the improvement in different language pairs rank as En-De>En-Zh>Ro-En, which preliminarily corresponded to the performances of our model in varied levels of language resources.

The highest correlation was achieved for the mid-resource languages, whereas for high-resource language pairs, the performance of the model become worse. The main reason for this difference may be a lower variability in translation quality for high-resource language pairs, i.e., Ro-En has a substantial number of high-quality sentences, but the distribution of En-De is highly skewed (which is also discussed in the study of Fomicheva [16]). The improvement across all language pairs showed that our model did not significantly outperform in mid-resource language pair where XLM-R already performs competitively. The reason for the higher improvement of En-De than En-Zh may be that English and Germany share more similar grammatical and syntactic structures than English and Chinese.

We also conducted a further ablation experiment on the performance of the syntactic extractor and GAT without fine-tuned XLM-R. Table 1, Part II shows the result of Syn-*GloVe* and Syn-XLM-R*static*. The word embedding in GAT is in *GloVe*, and XLM-R static embeddings is in Syn-XLM-R*static*. We found that introducing syntactic features without XLM-R component achieved no better result than the SOTA QE model MTransQuest-base, which may imply that semantic information plays a more important role in the task of machine translation quality estimation.

5 Discussion

An initial objective of the study was to utilize syntax and semantics to estimate the quality of translation in a more comprehensive way. According to the results, our proposed model makes great improvement in Pearson's correlation with human DA scores.

Table 1. Pearson's correlation of baselines (PredEst, BERT-BiRN and MTransQuest-base in Part I)/our proposed model (SynXLM-R, Syn-GloVe and Syn-XLM-Rstatic in Part II) with human DA scores. Part III shows the results of SynXLM-R with different numbers of attention heads (* indicates different numbers of attention heads. The attention heads in Part II is 3). Best results for each language pair are marked in bold.

	Methods	Ro-En	En-De	En-Zh
Part I	PredEst [28]	0.685	0.146	0.19
	BERT-BiRNN [16]	0.763	0.273	0.371
	MTransQuest-base [14]	0.872	0.442	0.459
Part II	SynXLM-R	**0.877**	**0.456**	**0.468**
	Syn-GloVe	0.856	0.418	0.438
	Syn-XLM-Rstatic	0.843	0.407	0.429
Part III	*	0.563	0.294	0.325
	**	0.812	0.435	0.443
	****	0.825	0.426	0.377
	*****	0.635	0.387	0.353
	******	0.601	0.369	0.338

In the previous section, we analyzed the performance of our syntax-enhanced model. In this section we would discuss the effect of different dependency parsers and various numbers of attention heads in GAT.

5.1 Effect of Different Parsers

Dependency parser plays a critical role in our method. To evaluate the impact of parsing result, we conducted an experiment based on different parsing models. Table 2 shows the performance of the five parsing methods in English-Chinese language pair in WMT20. Penn Treebank is a well-known open-source dataset in the fields of dependency parsing. These methods achieved pretty good result in both *UAS* and *LAS*. From the table, we can find that Label Attention Layer + HPSG + XLNet achieved the best UAS and LAS both in English and Chinese, and our SynXLM-R got the best Pearson score. As the parsing performance decreases, the performance of our model reduces. This indicates that the performance of dependency parsing has an immense impact on our model. Moreover, it further implies that our model has the potential to be improved along with the advances of parsing techniques.

Table 2. Pearson's correlation using different methods in our model. *UAS*: Unlabeled attachment score (without dependency relation label); *LAS*: labeled attachment score (with correct head and the correct dependency relation label).

Methods	English PTB-SD 3.3.0		Chinese PTB 5.1		Pearson
	UAS	*LAS*	*UAS*	*LAS*	
Label Attention Layer + HPSG + XLNet [30]	**97.42**	**96.26**	**94.56**	**89.28**	**0.468**
Deep Biaffine [31]	95.74	94.08	89.30	88.23	0.460
BiLSTMs [29]	93.9	91.9	87.6	86.1	0.457
BiAtt-DP [32]	94.10	91.49	88.1	85.7	0453
Arc-hybrid [33]	93.56	91.42	87.65	86.21	0.450

5.2 Attention Heads in GAT

Multi-head attention allows the model to jointly attend to information from different representation subspaces at varied positions [34]. It is adequately interpretable because attention weights indicate the weights of a particular word when computing the next representation for the current word. The number of multi-heads attention (MHAs) may negatively/positively affect the final results to a large extent, which can serve as an alternative tuning space for the model. Some researchers have explored the optimal number of heads in transformer (see Clark [35]). However, little research has been conducted to examine the impact of the number of attention heads in GAT. To study the influence of MHAs, we carried out a set of experiments with different numbers of MHAs ranged from 1 to 6. Figure 3 shows that SynXLM-R performed best for all the language pairs when the number of MHAs was closer to 3. To be more specific, when the number of MHAs ranges from 1 to 3, we could observe an upward trend in all language pairs. However, in En-De and Ro-En, model's performance dropped dramatically when the number was more than 4. Based on these results, we could draw a preliminary conclusion that more MHAs don't necessarily mean better performance, which echos Michel's [36] observation that a large percentage of attention heads can be eliminated at test time without significantly impacting performance.

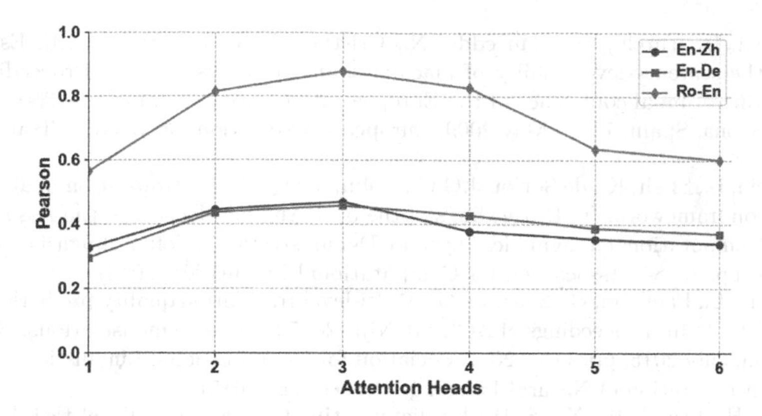

Fig. 3. Pearson's correlation between SynXLM-R and human DA judgment with different attention heads in three language pairs.

5.3 Limitations and Suggestions

The SynXLM-R is based on graph-structured relationship derived from dependency parsing. The performance of the model is therefore related to the performance of the parser. Although dependency parsing techniques were developing rapidly in recent years, there are still some complicated sentences beyond their capabilities. Besides, dependency parsing is not available in some mid- or low-resource languages (such as Romanian, Nepalese, etc.), which may limit the usage scope of our model. Potential improvement could be made through the amelioration of parsing techniques. Besides, all the relations in the graph are treated as the same in this study, which ignores the different types of relations. The various types of relations could be assigned with different weights in the adjency matrix in the future works.

6 Conclusion

In this paper, we proposed a novel syntax-enhanced model for translation quality estimation, which levaraged the syntactic dependency graph and the contextual information of sentences. Compared with previous works, our model employed GAT to extract the graph-based relation features in a sentence, which achieved 0.6%-3.2% improvement compared with the state-of-the-art MTransQuest-base in WMT20 QE tasks. Furthermore, we compared different dependency parsing methods to evaluate the influence of parsers on syntax-based models. We also conducted an experiment to evaluate the impact of the number of the attention heads in GAT on SynXLM-R.

References

1. Blatz, J., et al.: Confidence estimation for machine translation. In: COLING 2004: Proceedings of the 20th International Conference on Computational Linguistics, Geneva, Switzerland, 23–27 August 2004, pp. 315–321. COLING (2004)

2. Specia, L., Turchi, M., Cancedda, N., Cristianini, N., Dymetman, M.: Estimating the sentence-level quality of machine translation systems. In: Proceedings of the 13th Annual conference of the European Association for Machine Translation, Barcelona, Spain, 14–15 May 2009. European Association for Machine Translation (2009)
3. Specia, L., Shah, K., de Souza, J.G.C., Cohn, T.: QuEst - a translation quality estimation framework. In: Proceedings of the 51st Annual Meeting of the Association for Computational Linguistics: System Demonstrations, Sofia, Bulgaria, August 2013, pp. 79–84. Association for Computational Linguistics (2013)
4. Specia, L., Paetzold, G., Scarton, C.: Multi-level translation quality prediction with QuEst++. In: Proceedings of ACL-IJCNLP 2015 System Demonstrations, Beijing, China, July 2015, pp. 115–120. Association for Computational Linguistics and The Asian Federation of Natural Language Processing (2015)
5. Kim, H., Lee, J.-H., Na, S.-H.: Predictor-estimator using multilevel task learning with stack propagation for neural quality estimation. In: Proceedings of the Second Conference on Machine Translation, Copenhagen, Denmark, September 2017, pp. 562–568. Association for Computational Linguistics (2017)
6. Fan, K., Wang, J., Li, B., Zhou, F., Chen, B., Si, L.: "Bilingual expert" can find translation errors. In: Proceedings of the AAAI Conference on Artificial Intelligence, vol. 33, pp. 6367–6374 (2019)
7. Devlin, J., Chang, M.-W., Lee, K., Toutanova, K.: BERT: pre-training of deep bidirectional transformers for language understanding. In: Proceedings of the 2019 Conference of the North American Chapter of the Association for Computational Linguistics: Human Language Technologies, Volume 1 (Long and Short Papers), Minneapolis, Minnesota, June 2019, pp. 4171–4186. Association for Computational Linguistics (2019)
8. Conneau, A., et al.: Unsupervised cross-lingual representation learning at scale. In: Proceedings of the 58th Annual Meeting of the Association for Computational Linguistics, Online, July 2020, pp. 8440–8451. Association for Computational Linguistics (2020)
9. Lee, D.: Two-phase cross-lingual language model fine-tuning for machine translation quality estimation. In: Proceedings of the Fifth Conference on Machine Translation, Online, November 2020, pp. 1024–1028. Association for Computational Linguistics (2020)
10. Jawahar, G., Sagot, B., Seddah, D.: What does BERT learn about the structure of language? In: Proceedings of the 57th Annual Meeting of the Association for Computational Linguistics, Florence, Italy, July 2019, pp. 3651–3657. Association for Computational Linguistics (2019)
11. Comelles, E., Atserias, J.: VERTa: a linguistic approach to automatic machine translation evaluation. Lang. Resour. Eval. **53**(1), 57–86 (2019)
12. Huang, B., Carley, K.: Syntax-aware aspect level sentiment classification with graph attention networks. In: Proceedings of the 2019 Conference on Empirical Methods in Natural Language Processing and the 9th International Joint Conference on Natural Language Processing (EMNLP-IJCNLP), Hong Kong, China, November 2019, pp. 5469–5477. Association for Computational Linguistics (2019)
13. Ruder, S., Vulić, I., Søgaard, A.: A survey of cross-lingual word embedding models. J. Artif. Intell. Res. **65**, 569–631 (2019)
14. Ranasinghe, T., Orasan, C., Mitkov, R.: TransQuest: translation quality estimation with cross-lingual transformers. In: Proceedings of the 28th International Conference on Computational Linguistics, Barcelona, Spain (Online), December 2020, pp. 5070–5081. International Committee on Computational Linguistics (2020)

15. Hu, C., et al.: The NiuTrans system for the WMT20 quality estimation shared task. In: Proceedings of the Fifth Conference on Machine Translation, Online, November 2020, pp. 1018–1023. Association for Computational Linguistics (2020)
16. Fomicheva, M., et al.: Unsupervised quality estimation for neural machine translation. Trans. Assoc. Comput. Linguist. **8**, 539–555 (2020)
17. Wu, H., et al.: Tencent submission for WMT20 quality estimation shared task. In: Proceedings of the Fifth Conference on Machine Translation, Online, November 2020, pp. 1062–1067. Association for Computational Linguistics (2020)
18. Veličković, P., Cucurull, G., Casanova, A., Romero, A., Lio, P., Bengio, Y.: Graph attention networks. arXiv preprint arXiv:1710.10903 (2017)
19. Kipf, T.N., Welling, M.: Semi-supervised classification with graph convolutional networks. arXiv preprint arXiv:1609.02907 (2016)
20. Scarselli, F., Gori, M., Tsoi, A.C., Hagenbuchner, M., Monfardini, G.: The graph neural network model. IEEE Trans. Neural Netw. **20**(1), 61–80 (2008)
21. Bastings, J., Titov, I., Aziz, W., Marcheggiani, D., Sima'an, K.: Graph convolutional encoders for syntax-aware neural machine translation. In: Proceedings of the 2017 Conference on Empirical Methods in Natural Language Processing, Copenhagen, Denmark, September 2017, pp. 1957–1967. Association for Computational Linguistics (2017)
22. Marcheggiani, D., Titov, I.: Encoding sentences with graph convolutional networks for semantic role labeling. In: Proceedings of the 2017 Conference on Empirical Methods in Natural Language Processing, Copenhagen, Denmark, September 2017, pp. 1506–1515. Association for Computational Linguistics (2017)
23. Yao, L., et al.: Incorporating knowledge graph embeddings into topic modeling. In: Proceedings of the AAAI Conference on Artificial Intelligence, vol. 31 (2017)
24. Zhang, C., Li, Q., Song, D.: Aspect-based sentiment classification with aspect-specific graph convolutional networks. In: Proceedings of the 2019 Conference on Empirical Methods in Natural Language Processing and the 9th International Joint Conference on Natural Language Processing (EMNLP-IJCNLP), Hong Kong, China, November 2019, pp. 4568–4578. Association for Computational Linguistics (2019)
25. Sun, K., Zhang, R., Mensah, S., Mao, Y., Liu, X.: Aspect-level sentiment analysis via convolution over dependency tree. In: Proceedings of the 2019 Conference on Empirical Methods in Natural Language Processing and the 9th International Joint Conference on Natural Language Processing (EMNLP-IJCNLP), Hong Kong, China, November 2019, pp. 5679–5688. Association for Computational Linguistics (2019)
26. Wolf, T., et al.: Transformers: state-of-the-art natural language processing. In: Proceedings of the 2020 Conference on Empirical Methods in Natural Language Processing: System Demonstrations, Online, October 2020, pp. 38–45. Association for Computational Linguistics (2020)
27. Specia, L., et al.: Findings of the WMT 2020 shared task on quality estimation. In: Proceedings of the Fifth Conference on Machine Translation, Online, November 2020, pp. 743–764. Association for Computational Linguistics (2020)
28. Kepler, F., Trénous, J., Treviso, M., Vera, M., Martins, A.F.T.: OpenKiwi: an open source framework for quality estimation. In: Proceedings of the 57th Annual Meeting of the Association for Computational Linguistics: System Demonstrations, Florence, Italy, July 2019, pp. 117–122. Association for Computational Linguistics (2019)

29. Kiperwasser, E., Goldberg, Y.: Simple and accurate dependency parsing using bidirectional LSTM feature representations. Trans. Assoc. Comput. Linguist. **4**, 313–327 (2016)
30. Mrini, K., Dernoncourt, F., Tran, Q.H., Bui, T., Chang, W., Nakashole, N.: Rethinking self-attention: towards interpretability in neural parsing. In: Findings of the Association for Computational Linguistics: EMNLP 2020, Online, November 2020, pp. 731–742. Association for Computational Linguistics (2020)
31. Dozat, T., Manning, C.D.: Deep biaffine attention for neural dependency parsing. arXiv preprint arXiv:1611.01734 (2016)
32. Cheng, H., Fang, H., He, X., Gao, J., Deng, L.: Bi-directional attention with agreement for dependency parsing. arXiv preprint arXiv:1608.02076 (2016)
33. Ballesteros, M., Goldberg, Y., Dyer, C., Smith, N.A.: Training with exploration improves a greedy stack-LSTM parser. arXiv preprint arXiv:1603.03793 (2016)
34. Vaswani, A., et al.: Attention is all you need. arXiv preprint arXiv:1706.03762 (2017)
35. Clark, K., Khandelwal, U., Levy, O., Manning, C.D.: What does BERT look at? An analysis of BERT's attention. arXiv preprint arXiv:1906.04341 (2019)
36. Michel, P., Levy, O., Neubig, G.: Are sixteen heads really better than one? arXiv preprint arXiv:1905.10650 (2019)

Machine Learning for NLP

Memetic Federated Learning for Biomedical Natural Language Processing

Xinya Zhou[1], Conghui Tan[2], Di Jiang[2], Bosen Zhang[1], Si Li[1], Yajing Xu[1], Qian Xu[2], and Sheng Gao[1,3](✉)

[1] Beijing University of Post and Telecommunications, Beijing, China
{zhouxinya,zhangbosen,lisi,xyj,gaosheng}@bupt.edu.cn
[2] WeBank Co., Ltd., Shenzhen, China
{martintan,dijiang,qianxu}@webank.com
[3] GuiZhou University, Guizhou Provincial Key Laboratory of Public Big Data, Guiyang, Guizhou, China

Abstract. Privacy protection is an essential issue in biomedical natural language processing (BioNLP). Recently, some researchers apply federated learning (FL) in BioNLP to protect the privacy of biomedical data. However, their methods are only applicable for small NLP models, whose effectiveness is heavily limited in processing biomedical data. In this paper, we propose a novel memetic federated learning framework named Mem-Fed, which is tailored for federated learning of large-scale NLP models in the biomedical scenario. Experiments with large-scale BioNLP model on the public dataset show that the proposed framework significantly outperforms the state-of-the-art counterparts.

Keywords: Federated learning · Biomedical natural language processing · Memetic algorithm

1 Introduction

With the breakthroughs of the language representation models made in the field of natural language processing, more and more researchers attempt to apply these large models into biomedical natural language processing (BioNLP). Such works like BioBERT [10], BLUE [18] and BioMegatron [21] achieved state-of-the-art results in several BioNLP tasks by pre-training the models on large biomedical corpus [8,20,24]. These large models are more powerful in capturing the information in the complex unstructured biomedical text than the lightweight models, and achieve better results on most of BioNLP tasks.

However, with the rising needs in privacy protection and the introduction of the General Data Protection Regulation (GDPR), it is more and more difficult to gather biomedical data. Biomedical data are usually scattered among different medical institutions, who are unwilling to share it with the risk of exposing privacy.

L. Wang et al. (Eds.): NLPCC 2021, LNAI 13029, pp. 43–55, 2021.
https://doi.org/10.1007/978-3-030-88483-3_4

As a solution, federated learning (FL) is emerging in recent years [14]. In FL frameworks, there is no need for the data owners to share their private data. Instead, each data owner locally trains a model, and the parameters of local models are uploaded to the central server for model aggregation. After that, the aggregated model is sent back to the data owners, who can further fine-tune this model with the local data again. By repeating this process, the models will finally converge after several rounds of training and aggregation.

Many researchers have applied FL into BioNLP to protect data privacy [3, 4, 9, 11]. However, most of these attempts to combine BioNLP and federated learning just use simple models, such as SVM, logistic regression, or some simple neural network layers. These models are too primitive compared to the SOTA ones, i.e., pre-trained language representation models in BioNLP. Though there are some works which apply FL to train large BioNLP model [12], they simply apply existing FL technique without any improvement in training efficiency.

To tackle the aforementioned issues, we propose memetic federated learning (Mem-Fed), a novel federated learning framework inspired by memetic evolutionary algorithm [15]. This framework works in a privacy-preserving way, which is suitable for model training over sensitive biomedical data. To improve the efficiency for training large models, we develop local searching and model aggregation approaches, where the former one can reduce both the communication and computation cost at the central server, while the latter one exploits the locally trained models in a better way and generate superior model than existing FL approaches. We conduct experiments to verify the effectiveness of our proposed framework, which suggests that Mem-Fed can improve the trained model performance and reduce the training cost at the same time.

The contributions of this paper are summarized as follows:

- To the best of our knowledge, this is the first method especially designed for optimizing large BioNLP models under federated learning framework.
- We propose a local searching strategy, which can alleviate the communication congestion of central server during model uploading stage, and reduce the workload of later aggregation step.
- The aggregation method based on memetic evolutionary algorithm is developed. Instead of simply averaging all the locally trained models, it combines the models in an adaptive way so that better aggregated global models can be discovered.

2 Related Work

BioNLP. Early researchers of BioNLP focused on some tasks such as diagnosis and prediction [4, 16], document classification [2, 17], biomedical named entity recognition [23] and relation extraction [1]. SVM [23], logistic regression, CNN [7], and LSTM [17] are commonly used models for these tasks. However, due to the complexity of biomedical texts, recently researchers apply large pre-trained language models such as BERT [6] and ELMo [19] to BioNLP and claimed that these models have better performance [8, 10, 18, 20, 24].

Federated Learning in BioNLP. The first FL framework for neural networks is FedAvg that is proposed by McMahan et al. [14]. Most of the FL attempts in BioNLP are directly applying existing FL method to simple prediction or phenotyping tasks, using machine learning models such as SVM, logistic regression, random forest, XGBoost or some simple neural network layers [4,9,13]. But we still note that Liu and Miller [12] applied FL to fine-tuning BERT in FedAvg.

Incorporate Evolutionary Algorithms into FL. Recently, some researchers incorporate evolutionary algorithms into model aggregation in federated learning. Zhu and Jin [25] use evolutionary algorithms during the optimization of CNNs with FL, which decreases the communication cost by removing part of the connections of the models. SOMA [22] is also a work in this way, which used genetic algorithms into acoustic model optimization under federated learning setting.

Memetic Algorithms. Memetic algorithm [15] is one of evolutionary algorithms. Memetic algorithm realizes information exchange between each individual through the alternation of local searching and global searching [5]. Different from most evolutionary algorithms which focus on the transmission of genes between generations, memetic algorithms let the offspring imitate their parents to obtain more positive traits. Memetic algorithm conducts local searching for outstanding individuals, regards them as the parents of evolution, and performs operations such as crossover and reproduction without applying mutation. Without poorly mutated offspring, memetic algorithm could converge to high-quality solutions more efficiently than conventional evolutionary counterparts.

3 Mem-Fed Framework

3.1 Overview

We consider the scenario where private data are distributed among N parties, and there is a central server which communicates with all the parties for model exchange. Like typical horizontal federated learning settings, these N data owners are responsible for the training of local models. The server is responsible for the collection, aggregation and distribution of models to be trained. Figure 1 illustrates the overall process of the Mem-Fed framework. The working mechanism of Mem-Fed is essentially an iterative process, and each round of training can be summarized as three steps: local training, local searching and model aggregation.

- In the first step, the central server broadcast the unified model to all the participants. Each party utilizes its local private data to fine-tune the received model.
- In the local searching step, a subset of models are selected as the agents, and only these agents rather than all the locally trained models are uploaded to the central server.

- Finally, the agents are aggregated by memetic algorithms into the a single global model, which will be broadcast back to all the participants in next iteration.

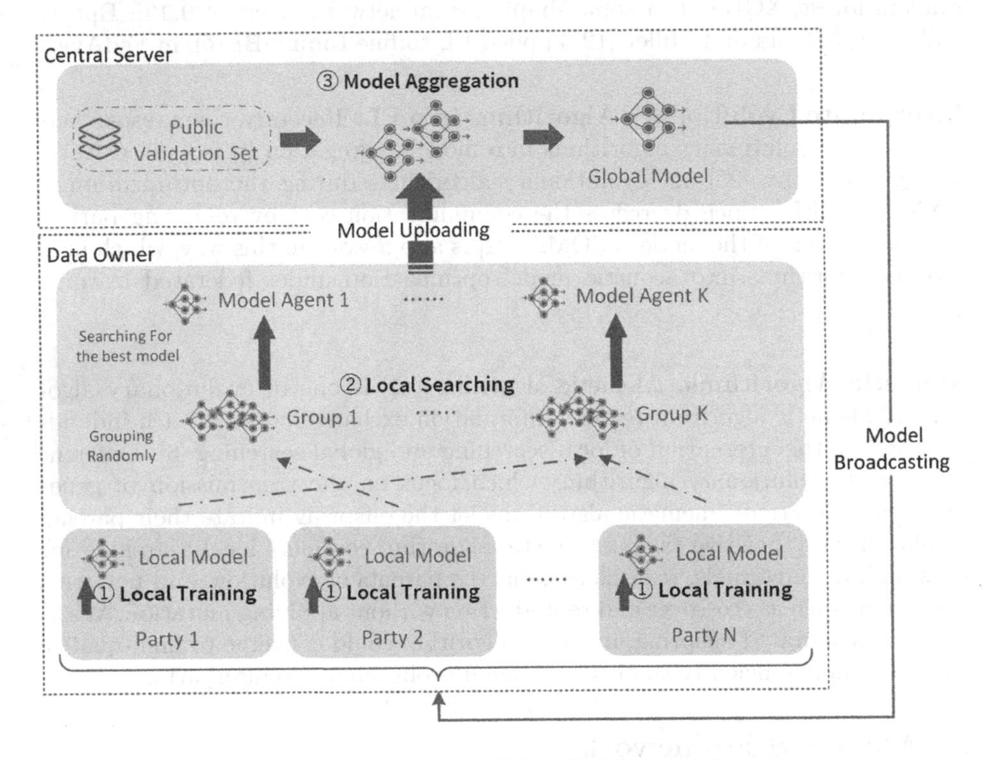

Fig. 1. Illustration of Mem-Fed framework.

Different from the existing FL method, we creatively propose a local searching step after local training to reduce communication and computation costs. Meanwhile, we replaced the model aggregation strategy of the common parameter average method with a generative method based on the memetic algorithm. These special designs are the key to make our approach more efficient and effective than existing FL methods.

In the next, we introduce the details of these steps.

3.2 Local Training

In the first step, each party would receive a unified model from the central server, i.e., the initial model for the first round, or the latest aggregated global model for later rounds. Different from the normal federeted learning process, the private data of each party are randomly divided into two parts: training set and validation set, according to a preset ratio. Then each party utilizes its local training set to fine-tune the received model and uses the validation set to evaluated the model performance. The performances of the local models are reported to the central server to facilitate the later step.

3.3 Local Searching

Inspired by the theory of memetic evolution, we propose a step called local searching, in which a subset of local models are selected as the agents. In the memetic theory proposed by Moscato et al. [15], the local searching step is responsible for finding the local optimal solution in each area. In this way, a large population can quickly reach a better initial state. Outstanding individuals are selected as evolutionary parents. Then, the evolutionary method can be used to generate the global optimal solution more efficiently. In Mem-Fed, local searching also undertakes this kind of work.

With local searching, we only need to upload the agents instead of all the locally trained models to the central server for model aggregation, which can greatly reduce the complexity of the aggregation step. This is crucial for optimizing large-scale BioNLP models. These models might be too large, which makes it difficult for the server to load many locally trained models into memory at the same time, or, too computationally expensive to aggregate them together.

In our framework, a simple and light-weight strategy is adopted for local searching: in each round, we randomly divide all the local models into K groups, and select the best-performing model in each group as the agent according to the model performances evaluated in the local training step.

In this way, we can reduce the number of inputs in aggregation stage from N to K, where K is a hyper-parameter. Besides, the number of models needed to upload is also reduced, which can alleviate the network congestion in the central server. Actually, we also proposed some alternative ways for local searching. To save space, we defer the related discussion to Sect. 4.5.

3.4 Model Aggregation

In model aggregation step, the agents are aggregated into a single global model and then this global model is broadcast to all the participants.

In Mem-Fed, we develop an aggregation method, which generates better global model and speedups the convergence of the whole FL process.

We apply memetic algorithm into model aggregation, regarding each input model as an individual, and letting the population of models evolve by generating offspring and conducting natural selection repeatedly.

In memetic algorithms, the parameters of each layer in the model are considered to be the genes of this individual. Reproduction, crossover and linear interpolation operators are applied on these genes, to generate offspring.

Reproduction: Directly copies the existing parameters into next generation.

$$\theta_{new} = \theta \tag{1}$$

Crossover: Randomly draws an integer l ($1 \leq l \leq L$, where L is the number of the layers of the whole model), and swap the parameters of the first l layers of two parent models as the next generation.

$$\begin{aligned}\{\theta_{new,1}^{1}, \ldots, \theta_{new,1}^{L}\} &= \{\theta_2^1, \ldots, \theta_2^l, \theta_1^{l+1}, \ldots, \theta_1^L\} \\ \{\theta_{new,2}^{1}, \ldots, \theta_{new,2}^{L}\} &= \{\theta_1^1, \ldots, \theta_1^l, \theta_2^{l+1}, \ldots, \theta_2^L\}\end{aligned} \tag{2}$$

Linear Interpolation: Linearly combines all the parameters of two parent models in a weighted way. The parameter of the generated model satisfies:

$$\theta_{new} = \lambda\theta_1 + (1-\lambda)\theta_2 \quad (0 < \lambda < 1) \tag{3}$$

where θ_k^l is the parameters of layer l of the parent models, and θ_{new}^l is the corresponding parameter of the generated offspring.

Inspired by memetic theory, we did not adopt the mutation operator, since it greatly increases the complexity of aggregation once the input models have large number of parameters.

As for natural selection, it is to evaluate the performance of the generated model, and give some of them the right to generate offspring. We select the best offspring according to their performance on the public validation set. If the global optimal solution is not found, the evolutionary process will continue, and the models with excellent performance are selected as the parents of the next generation. Then the operations mentioned above will be repeat until convergence.

However, it is well-known that evolutionary algorithm is slow, which is intolerable for training large-scale models. Hence, we follow Tan et al. [22] to transfer the running process of the genetic algorithm into an optimization problem. In detail, if we assume θ_k^l is the parameter of the l-th layer of input model k ($1 \leq k \leq K$), we force the parameter on the corresponding layer of the aggregated model, say θ^l, to satisfy the pattern:

$$\theta^l = \sum_{k=1}^{K} \alpha_k^l \cdot \theta_k^l \tag{4}$$

for some free variables α_k^l. Then we determine the values of α_k^l and the corresponding model parameters by solving the following optimization problem:

$$\begin{aligned}\min_{\theta^l, \alpha_k^l} \quad & \text{loss}(\theta^1, \ldots, \theta^L) \\ \text{s.t.} \quad & \theta^l = \sum_k \alpha_k^l \cdot \theta_k^l \\ & \alpha_k^l \geq 0, \sum_k \alpha_k^l = 1,\end{aligned} \tag{5}$$

where $\text{loss}(\theta^1, \ldots, \theta^L)$ refers to the loss of the model with parameters $\theta^1, \ldots, \theta^L$.

Algorithm 1. Model Aggregation Algorithm

Input: θ_k^l: parameter of the l-th layer of the k-th model agent
1: Initialize $\alpha_k^l = 1/K$ for all k and l , where K is the number of local model agents
2: **while** not converged **do**
3: **for** each layer l **do**
4: $\theta^l = \sum_k \alpha_k^l \cdot \theta_k^l$
5: **end for**
6: draw a batch of public validation data and back-propagate to compute $\frac{\partial \text{loss}}{\partial \theta^l}$ for all layers
7: **for** each layer l **do**
8: **for** each model k **do**
9: $\alpha_k^l = \alpha_k^l - \eta \cdot \frac{\partial \text{loss}}{\partial \theta^l} \cdot \theta_k^l$
10: **end for**
11: $\alpha^l = \text{softmax}(\alpha^l)$
12: **end for**
13: **end while**
Output: aggregated global model with parameters $\theta^1, \ldots, \theta^L$

Algorithm 1 illustrates that how this problem is solved and how the global model is generated therefore. The rough idea of this algorithm is: we first use the latest values of α_k^l to update the model parameters θ^l, according to pattern (4). Then we draw of a mini-batch of data to compute the gradients of the current model. After that, the gradient of α_k^l can be derived by chain rule:

$$\frac{\partial \text{loss}}{\partial \alpha_k^l} = \frac{\partial \text{loss}}{\partial \theta^l} \cdot \frac{\partial \theta^l}{\partial \alpha_k^l} = \frac{\partial \text{loss}}{\partial \theta^l} \cdot \theta_k^l, \tag{6}$$

and we can do gradient descent for α_k^l. Finally, the softmax operation is conducted to make sure the constraints in (5) still hold.

In this way, the aggregation algorithm turns from simply averaging the parameters into the optimization of α_k^l, and a better global model is efficiently generated.

4 Experiments

4.1 Experimental Setup

We choose multilabel classification in biomedical field as the testbed and the Hallmarks of Cancers (HoC) corpus as the dataset. HoC [2] consists of 1,580 texts annotated with ten currently known hallmarks of cancer. We use 10% of the texts as the public validation data, 20% as the public test data, and the remaining 70% as private data. The private data are scattered to 10 parties, and are randomly divided into 90% private training set and 10% private validation set in each round. We choose to optimize a BERT-like model [18] which is pre-trained on large biomedical corpus. We report the F1-score on the public test data as the evaluation metric.

4.2 Quantitative Comparison

We quantitatively compare the following three methods in terms of effectiveness and communication efficiency:

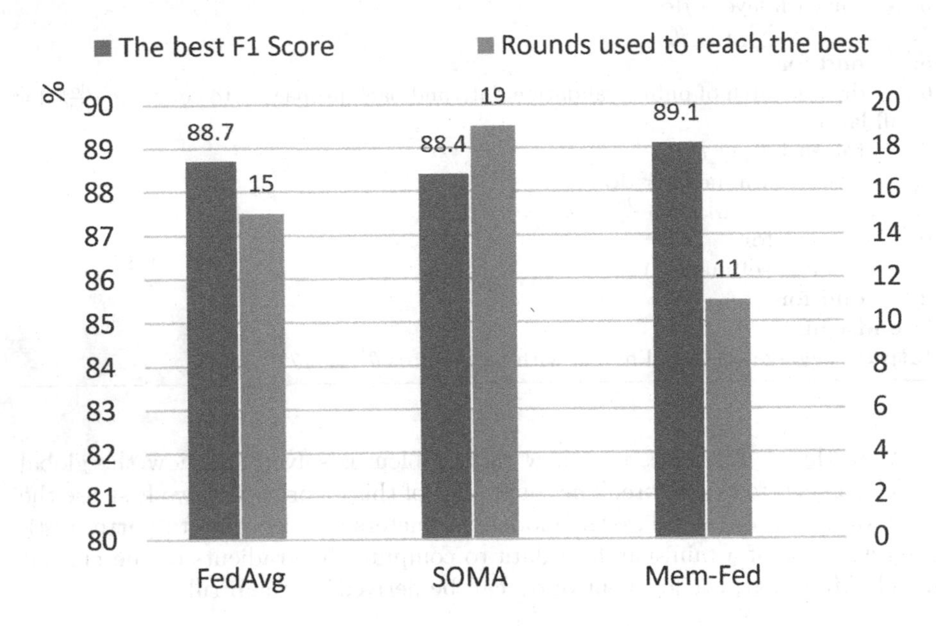

Fig. 2. Comparison of effectiveness and communication efficiency.

FedAvg: The FL framework with FedAvg aggregation [14], which is the most widely used FL method.

SOMA: A recent method using genetic algorithm in model aggregation in the FL setting [22]. We make two changes to adapt this method for our experiment setting: first, the original SOMA only applies one round of model aggregation, but we use it in iteratively to facilitate the comparison with our method. Second, SOMA is not suitable for processing large models. To make the experiments on large models possible, we also incorporate the local searching step we proposed into it. So, the SOMA compared here is actually an improved version of it.

Mem-Fed: The proposed framework. The number of groups of local searching K is set to be 3 by default in this experiment.

The comparison results are presented in Fig. 2. We observe that Mem-Fed reached 89.1 after only 11 rounds of training and aggregation, i.e., it used the least time to achieve the best results among all methods. This result proves that our proposed framework and aggregation algorithm are superior to these baselines in terms of both efficiency and effectiveness. Besides, we find that SOMA takes more rounds than others. This is because its mutation operation greatly increases the complexity of aggregating large models.

4.3 Group Number in Mem-Fed

We further investigate how the number of groups affects the performance of Mem-Fed. We vary the number of groups from $K = 2$ to $K = 4$. The results are shown in Fig. 3. It can be seen that larger number of groups can result in models with higher F1 values, meanwhile, the number of rounds required to achieve the best performance becomes larger. Therefore, there is a trade-off in selecting the number of groups K: smaller K can make the algorithm faster, in terms of both per-round cost and number of rounds needed for converging, while larger K is effective for training a better model.

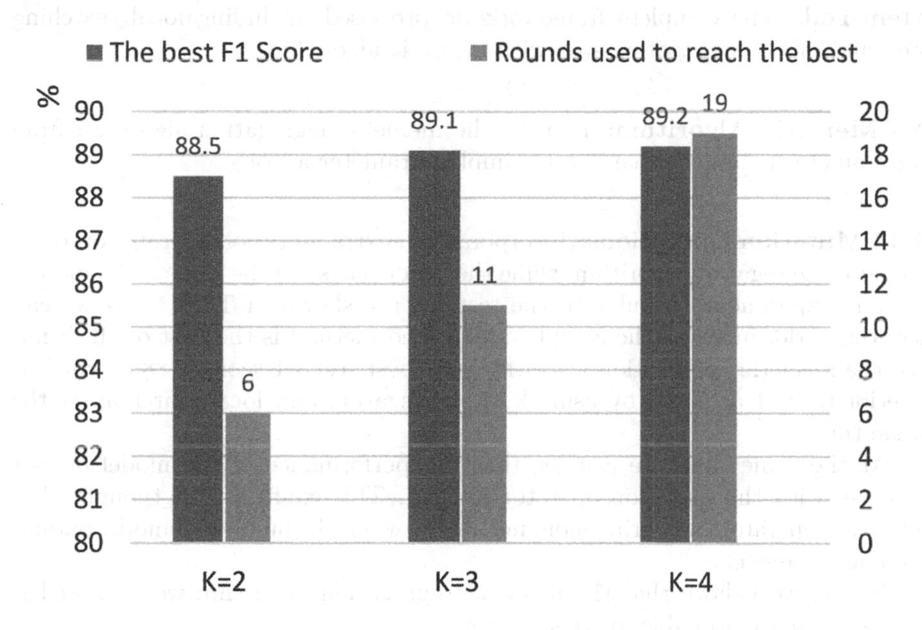

Fig. 3. Results of different number of groups.

Table 1. Results of ablation study.

Model	F1
Complete Mem-Fed model	**89.1**
No memetic algorithm	88.2
Add mutation operations	88.4

Table 2. Results of different local searching strategies.

Strategies	F1
Select K best performing models among all models	88.6
Randomly divide all models into K groups, and choose the best-performing model in each group	**89.1**
Randomly divide all models into K groups, and generate a model by averaging the model parameters in each group	88.8

4.4 Ablation Study on Memetic Aggregation

To evaluate the rationality of memetic aggregation algorithm, we conduct an ablation study, evaluating three different approaches:

Mem-Fed: The complete framework we proposed, including local searching step and model aggregation step with memetic algorithm.

No Memetic Algorithm: Remove the memetic aggregation algorithm from the framework, and replace it with simple parameter average.

Add Mutation Operations: Incorporate an extra mutation operator onto the memetic aggregation algorithm, while the other parts are the same as Mem-Fed.

The experimental results on the test set are shown in Table 1. As we can see, the performance of the complete Mem-Fed method is the best of all. If not use the memetic aggregation algorithm, the best test F1 score is 88.2, which is inferior to 89.1 achieved by using both aggregation and local searching at the same time.

At the same time, we can see that the performance of the model do not increase with the mutation operator applied. This confirms our thought that mutation operators will bring more instability when the number of model parameters are massive.

Hence, we believe that the memetic aggregation algorithm we proposed is necessary and reasonable for this system.

4.5 Local Searching Strategies of Mem-Fed

We proceed to discuss the effect of local searching strategies. When using gradient-based methods to aggregate multiple large models, it is impossible aggregate too many models in the same time due to the limitation of memory or computation resources. The introduction of local searching is for tackling this challenge, by reducing the number of models to aggregate from N to K. Therefore, the local searching step is an indispensable component of our framework when dealing with large models. Because of this, we do not provide ablation study of local searching step in our experiments. Instead, we evaluate several different local searching strategies.

We present different strategies with the corresponding experimental results in Table 2. We can see that the second strategy (i.e., random grouping and choosing the best one in each group) is the best strategy. Comparing to the first strategy, the strategy of grouping and searching creates some randomness and avoids that the same parties are always selected and the models from the others are always wasted. While the last strategy treats each model equally, which can be improved with evaluation and selection.

5 Conclusion

In this paper, we propose a novel memetic federated learning framework named Mem-Fed. Compared with existing methods, the proposed Mem-Fed framework supports effective training of large-scale BioNLP models in federated scenario, which effectively protects the privacy of biomedical data. We wish that Mem-Fed would pave the way for the next-generation computing paradigm for biomedicine.

Acknowledge. This work was supported by National Natural Science Foundation of China (61872338) and the Foundation of Guizhou Provincial Key Laboratory of Public Big Data (No. 2019BDKFJJ002).

References

1. Alicante, A., Corazza, A., Isgrò, F., Silvestri, S.: Unsupervised entity and relation extraction from clinical records in Italian. Comput. Biol. Med. **72**, 263–275 (2016)
2. Baker, S., et al.: Automatic semantic classification of scientific literature according to the hallmarks of cancer. Bioinform. **32**(3), 432–440 (2016)
3. Boughorbel, S., Jarray, F., Venugopal, N., Moosa, S., Elhadi, H., Makhlouf, M.: Federated uncertainty-aware learning for distributed hospital EHR data. CoRR abs/1910.12191 (2019)
4. Brisimi, T.S., Chen, R., Mela, T., Olshevsky, A., Paschalidis, I.C., Shi, W.: Federated learning of predictive models from federated electronic health records. Int. J. Med. Inform. **112**, 59–67 (2018)
5. Cotta, C., Mathieson, L., Moscato, P.: Memetic algorithms. In: Martí, R., Pardalos, P.M., Resende, M.G.C. (eds.) Handbook of Heuristics, pp. 607–638 (2018)
6. Devlin, J., Chang, M.W., Lee, K., Toutanova, K.: BERT: pre-training of deep bidirectional transformers for language understanding. In: Proceedings of the 2019 Conference of the North American Chapter of the Association for Computational Linguistics: Human Language Technologies, Volume 1 (Long and Short Papers), pp. 4171–4186, June 2019
7. Gehrmann, S., et al.: Comparing rule-based and deep learning models for patient phenotyping. arXiv preprint arXiv:1703.08705 (2017)
8. Hafiane, W., Legrand, J., Toussaint, Y., Coulet, A.: Experiments on transfer learning architectures for biomedical relation extraction. CoRR abs/2011.12380 (2020)

9. Ju, C., et al.: Privacy-preserving technology to help millions of people: federated prediction model for stroke prevention. CoRR abs/2006.10517 (2020)
10. Lee, J., et al.: BioBERT: a pre-trained biomedical language representation model for biomedical text mining. Bioinformatics **36**(4), 1234–1240 (2020)
11. Liu, D., Dligach, D., Miller, T.: Two-stage federated phenotyping and patient representation learning. In: Proceedings of the 18th BioNLP Workshop and Shared Task, pp. 283–291, August 2019
12. Liu, D., Miller, T.A.: Federated pretraining and fine tuning of BERT using clinical notes from multiple silos. CoRR abs/2002.08562 (2020)
13. Liu, D., Miller, T.A., Sayeed, R., Mandl, K.D.: FADL: federated-autonomous deep learning for distributed electronic health record. CoRR abs/1811.11400 (2018)
14. McMahan, B., Moore, E., Ramage, D., Hampson, S., y Arcas, B.A.: Communication-efficient learning of deep networks from decentralized data. In: Singh, A., Zhu, X.J. (eds.) Proceedings of the 20th International Conference on Artificial Intelligence and Statistics, AISTATS 2017, Fort Lauderdale, FL, USA, 20–22 April 2017. Proceedings of Machine Learning Research, vol. 54, pp. 1273–1282 (2017)
15. Moscato, P., et al.: On evolution, search, optimization, genetic algorithms and martial arts: towards memetic algorithms. Caltech concurrent computation program, C3P Report 826, 1989 (1989)
16. Mullenbach, J., Wiegreffe, S., Duke, J., Sun, J., Eisenstein, J.: Explainable prediction of medical codes from clinical text. In: Proceedings of the 2018 Conference of the North American Chapter of the Association for Computational Linguistics: Human Language Technologies, Volume 1 (Long Papers), June 2018
17. Ormerod, M., Martínez-del Rincón, J., Robertson, N., McGuinness, B., Devereux, B.: Analysing representations of memory impairment in a clinical notes classification model. In: Proceedings of the 18th BioNLP Workshop and Shared Task, pp. 48–57 (2019)
18. Peng, Y., Yan, S., Lu, Z.: Transfer learning in biomedical natural language processing: an evaluation of BERT and ELMo on Ten benchmarking datasets. In: Proceedings of the 18th BioNLP Workshop and Shared Task, pp. 58–65, August 2019
19. Peters, M., et al.: Deep contextualized word representations. In: Proceedings of the 2018 Conference of the North American Chapter of the Association for Computational Linguistics: Human Language Technologies, Volume 1 (Long Papers), pp. 2227–2237, June 2018
20. Schumacher, E., Mulyar, A., Dredze, M.: Clinical concept linking with contextualized neural representations. In: Proceedings of the 58th Annual Meeting of the Association for Computational Linguistics, pp. 8585–8592, July 2020
21. Shin, H.C., et al.: BioMegatron: larger biomedical domain language model. In: Proceedings of the 2020 Conference on Empirical Methods in Natural Language Processing (EMNLP), pp. 4700–4706, November 2020
22. Tan, C., Jiang, D., Peng, J., Wu, X., Xu, Q., Yang, Q.: A de novo divide-and-merge paradigm for acoustic model optimization in automatic speech recognition. In: Bessiere, C. (ed.) Proceedings of the Twenty-Ninth International Joint Conference on Artificial Intelligence, IJCAI 2020, pp. 3709–3715 (2020)
23. Tang, B., Cao, H., Wu, Y., Jiang, M., Xu, H.: Recognizing clinical entities in hospital discharge summaries using structural support vector machines with word representation features. BMC Med. Inform. Decis. Mak. **13**, 1–10 (2013). BioMed Central

24. Vaidhya, T., Kaushal, A.: IITKGP at W-NUT 2020 shared task-1: domain specific BERT representation for named entity recognition of lab protocol. In: Xu, W., Ritter, A., Baldwin, T., Rahimi, A. (eds.) Proceedings of the Sixth Workshop on Noisy User-Generated Text, W-NUT@EMNLP 2020 Online, 19 November 2020, pp. 268–272 (2020)
25. Zhu, H., Jin, Y.: Multi-objective evolutionary federated learning. IEEE Trans. Neural Netw. Learn. Syst. **31**(4), 1310–1322 (2019)

Information Extraction and Knowledge Graph

Event Argument Extraction via a Distance-Sensitive Graph Convolutional Network

Lu Dai[1], Bang Wang[1(✉)], Wei Xiang[1], and Yijun Mo[2]

[1] School of Electronic Information and Communications, Huazhong University of Science and Technology (HUST), Wuhan, China
{dailu18,wangbang,xiangwei}@hust.edu.cn

[2] School of Computer Science and Technology, Huazhong University of Science and Technology (HUST), Wuhan, China
moyj@hust.edu.cn

Abstract. Event argument extraction (EAE) is a challenging task due to that some event arguments are far away from event triggers. Recently there have been many works using dependency trees to capture long-range dependencies between event triggers and event arguments. However, not all information in the dependency trees are useful for the EAE task. In this paper, we propose a Distance-Sensitive Graph Convolutional Network (DSGCN) for the EAE task. Our model can not only capture long-range dependencies via the graph convolution over the dependency trees, but also keep the information relevant to EAE via our designed distance-sensitive attention. Furthermore, dependency relation type information is utilized to enrich the representation of each word to further distinguish the roles of event arguments playing in an event. Experiment results on the ACE 2005 English dataset show that the proposed model achieves superior performance than the peer state-of-the-art methods.

Keywords: Event argument extraction · Graph Convolutional Network · Dependency tree · Distance-sensitive attention

1 Introduction

As a subtask of event extraction, event argument extraction (EAE) refers to identifying event arguments and determining the roles they play in an event [9,22]. For example, given the sentence "*The judge ordered Ranjha to pay a **fine** of 50,000 rupees.*", event detection (ED) as another subtask of event extraction aims to identify that the word "fine" triggers a "Fine" event. Then EAE aims to identify "The judge", "Ranjha" and "50,000 rupees" as the "Adjudicator",

This work is supported in part by National Natural Science Foundation of China (Grant No: 62172167).

L. Wang et al. (Eds.): NLPCC 2021, LNAI 13029, pp. 59–72, 2021.
https://doi.org/10.1007/978-3-030-88483-3_5

"Entity" and "Money" of the "Fine" event, respectively. Recently, neural networks have already been applied in ED [3,14,18,23], which have achieved promising results. EAE becomes the key to improving the overall performance of EE and has attracted growing interests [11,19–21].

The EAE task can be divided into two sub-tasks: (I) Argument Identification (AI) to identify the text spans of event arguments;(ii) Role Classification (RC) to classify event arguments into predefined role types. In the Automatic Context Extraction (ACE) evaluation, event argument is defined as an entity mention[1] that participates in an event [4]. As the ACE dataset provides the golden annotation for entity mentions, EAE can be modeled as a multi-class classification problem [12,13]: Given an event trigger with the event type and an entity mention, the EAE task is to predict the role label that the entity mention plays in the event. The label can be one of the predefined role types in the dataset or "NONE" to indicate that the entity mention is not an event argument.

Existing work on event argument extraction can be divided into feature-based and neural network-based ones. Feature-based methods [5–8] rely on handcrafted features, such as syntactic features and entity features, which, however, suffer from the problem of extensive human labor and poor flexibility. Early neural network-based methods utilize convolutional neural networks [13] or recurrent neural networks [12] to encode sentence semantics as low-dimensional vectors to perform EAE, which are hard to capture long-range dependencies between event triggers and event arguments. As the distance between an event trigger and its event arguments can be shortened via syntactic shortcut arcs, some dependency-based models have been proposed to leverage syntactic information to capture the long-range syntactic relations that are not directly indicated in a sentence sequence structure [11,19–21]. For example, Liu et al. [10] model a sentence as a graph based on a dependency tree and use Graph Convolutional Networks (GCNs) to learn the sentence-level representation. Veyseh et al. [19] utilize Graph Transformer Networks (GTNs) to combine semantic and syntactic structures to learn effective sentence representations for EAE.

However, there are two problems in the above methods: (i) They do not consider to distinguish the importance of information contained in different words. It is often argued that the information contained in the words on the shortest dependency path between the event trigger and the entity mention is more important than the other words. (ii) They ignore the dependency relation type information, which often conveys rich and useful linguistic knowledge. For example, when an entity mention and an event trigger "injured" is connected by a syntactic arc of type "nsubj", the entity mention is likely to play the role of "Victim" in the event triggered by the word "injured".

In this paper, we propose a *Distance-Sensitive Graph Convolutional Network* (DSGCN) to tackle the above problems for EAE. Compared with the traditional Graph Convolutional Network, we design a novel distance-sensitive attention to learn the importance of each node in the graph to EAE, where the weight of each

[1] For convenience, entities, time and value expressions are presented as entity mentions in the paper.

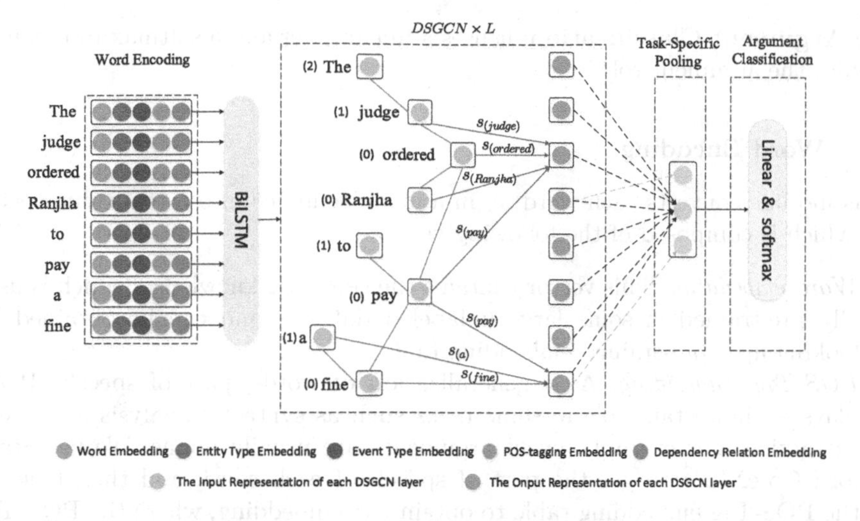

Fig. 1. The overall architecture of our model shown with an example predicting the argument role that the entity mention "Ranjha" plays in the event triggered by the word "fine". In the DSGCN module, the yellow solid line represents the syntactic arc between two words. The number (x) next to a word represents that the distance between the word and the shortest dependency path constructed by the event trigger "fine" and the entity mention "Ranjha" is x. The blue solid line represents the convolution process, where only the convolution process of word "ordered" and "fine" is drawn. $s_{(w)}$ represents the weight of the word w in the convolution process. (Color figure online)

node is sensitive to the distance between the node and the shortest dependency path constructed by the given event trigger and the entity mention. We argue that the information along the shortest dependency path is very important to EAE. The closer the node to the shortest dependency path, the more important the information contained in the node to EAE. After convolution, we design a pooling specific to the given event trigger and the entity mention to aggregate the representation of all nodes into a single representation for argument role prediction. Additionally, we propose to utilize dependency relation type information by integrating the *dependency relation type embedding* into the representation of each word. We evaluate the proposed DSGCN on the ACE 2005 English dataset and experiments show that our model can achieve the state-of-the-art performance.

2 The DSGCN Model

The overall architecture of the DSGCN model is presented in Fig. 1. It consists of four modules: (i) **Word Encoding**: it is to encode a word as a vector. (ii) **Distance-Sensitive Graph Convolutional Network**: it is to capture the syntactic contextual information through performing the convolution over dependency trees. (iii) **Task-Specific Pooling**: it is to aggregate the representation of all nodes into a single representation for argument role classification.

(iv) **Argument Classification**: it uses a linear layer and a softmax function to predict the argument role.

2.1 Word Encoding

This module converts each word w_i in a given sentence into a real-value vector $\mathbf{x}_i$, which is composed of the following vectors:

- *Word embedding*: This vector contains the semantic knowledge, which is usually pre-trained in some large unlabeled datasets and can be obtained by looking up a pre-trained embedding table.
- *POS-Tag Embedding*: As a generalization of words, part of speech (POS) plays an important role in some tasks such as syntactic analysis and event extraction. For example, event arguments are usually nouns. We use Stanford CoreNLP[2] to get the part of speech of each word, and then look up the POS-Tag embedding table to obtain the embedding, where the POS-Tag embedding table is randomly initialized and will be optimized along with network training.
- *Dependency Relation Type Embedding*: We use Stanford CoreNLP to obtain the dependency relation between each word and its parent word. According to the size of the dependency relation type embedding, a unique vector is randomly generated and assigned to each dependency relation.
- *Entity Type Embedding*: The entity type of the given entity mention is important to EAE. For example, the entity type of the "Victim" argument of an "Injure" event can only be "PER". We obtain the entity type embedding by looking up the entity type embedding table, in which the entity type of the word that does not belong to the given entity mention is "None".
- *Event type Embedding*: The event type of the given event trigger is also important. For example, the "Victim" argument only appears in the "Injure" events and the "Die" events. Similar to the entity type embedding, we obtain the event type embedding for each word.

2.2 Distance-Sensitive Graph Convolutional Network

Graph Convolutional Network. Graph Convolutional Network is an improvement of Convolution Neural Network to encode graph structure. Graph Convolutional Network learns the representation of syntactic context for each node by aggregating the information of its first-order neighbors. Let $\mathbf{H}^{l-1} = \{\mathbf{h}_1^{l-1}, \mathbf{h}_2^{l-1}, ..., \mathbf{h}_n^{l-1}\}$ represent the input representation of each node in the l-th layer GCN, where n is the number of nodes. The update process of the i-th node in the l-th layer GCN is as follows:

$$\mathbf{h}_i^l = \sigma(\sum_{j \in \mathcal{N}(i)} \mathbf{W}^l \mathbf{h}_j^{l-1} + \mathbf{b}^l), \tag{1}$$

[2] https://stanfordnlp.github.io/CoreNLP/.

where σ is a nonlinear activation function, such as Relu, $\mathcal{N}(i)$ is a set of nodes directly adjacent to the i-th node, and $\mathbf{W}^l$ and $\mathbf{b}^l$ are learnable parameters.

In order to apply Graph Convolutional Network to the EAE task, we construct an undirected graph based on the dependency tree of a given sentence, where each node represents a word in the sentence and an edge exists if there is a syntactic arc between two words. In addition, we add an self-loop edge for each node to ensure that each node can participate in its own convolution.

In the GCN, each word only interacts with the words directly adjacent to it in the syntactic structure. In order to capture the contextual information of each word, the input representation of a given sentence obtained by the **Word Encoding** module is input to a bidirectional Long Short Term Memory neural network (BiLSTM) to generate the initial representation of the GCN.

Distance-Sensitive Attention. For the event argument extraction task, words in the input sentence are not equally important. For the example illustrated in Fig. 1, the word "Ranjha" and the word "fine" contain more important information than the word "a". When aggregating information on the "fine" node, we should pay more attention on the "fine" node and the "Ranjha" node than the "a" node. In addition, we observe that the information contained in the node closer to the shortest dependency path constructed by the given event trigger and the entity mention is more important for event argument extraction in most cases. We thus propose distance-sensitive attention to maximally keep the information relevant to the event argument extraction while ignoring irrelevant information. For the distance-sensitive attention, the *distance embedding* $\mathbf{d}_j$ of the j-th node is first obtained by looking for the vector corresponding to the distance between the j-th node and the shortest dependency path in the distance embedding table. Then the importance weight of the j-th node is computed according to its representation $\mathbf{h}_j^{l-1}$ and distance embedding $\mathbf{d}_j$:

$$\mathbf{s}_j^l = \sigma(\mathbf{W}_h^l \mathbf{h}_j^{l-1} + \mathbf{W}_d^l \mathbf{d}_j + \mathbf{b}_s^l), \quad (2)$$

where $\mathbf{W}_h^l$, $\mathbf{W}_d^l$ and $\mathbf{b}_s^l$ are learnable parameters. For a dependency-tree based graph constructed by a sentence containing n words, the time complexity of finding out the shortest path between any two words in the graph is $o(n)$. Therefore, this shortest path identification step does not consume too much time.

There is another problem exiting in the information aggregation shown in Eq. 1: the higher the degree of a node, the larger the value of its representation tending to be. When generating the sentence-level representation, the model will pay more attention to the node with a lot of first-order neighbors regardless of the information carried in the node. In order to solve this problem, we need to perform the normalization according to the degree of the node in the convolution process. Then the update process of i-th node in the l-th layer GCN is as follows:

$$\mathbf{h}_i^l = \sigma(\frac{1}{N} \sum_{j \in \mathcal{N}(i)} \mathbf{s}_j^l \mathbf{W}^l \mathbf{h}_j^{l-1} + \mathbf{b}^l), \quad (3)$$

where N is the number of nodes in the set $\mathcal{N}(i)$.

2.3 Task-Specific Pooling

The purpose of pooling is to aggregate the representation of all nodes into a single sentence-level representation. For the traditional pooling mechanism [13], the sentence-level representation $\mathbf{H}_S$ is obtained by taking the element-wise max operation over the representation of all nodes:

$$\mathbf{H}_S = pooling(\mathbf{H}^L), \tag{4}$$

where L is the number of DSGCN layers.

Given the importance of the information contained in the event trigger and entity mention for EAE, we propose to utilize the representation of the event trigger and entity mention to obtain the sentence-level representation. Specifically, the element-wise max operation is performed not only on the representation of all nodes, but also on the representation of the nodes corresponding to the event trigger (entity mention). Let E denote the set of the entity mention's composing word nodes. The representation of the entity mention $\mathbf{H}_E$ is computed by

$$\mathbf{H}_E = pooling(\mathbf{h}_i^L | \forall i \in E). \tag{5}$$

Similarly, we can get the representation of the event trigger $\mathbf{H}_T$. Then the final sentence-level representation $\mathbf{H}$ is obtained by concatenating the sentence representation, the entity mention representation and the event trigger representation:

$$\mathbf{H} = [\mathbf{H}_S; \mathbf{H}_E; \mathbf{H}_T]. \tag{6}$$

2.4 Argument Classification

The representation $\mathbf{H}$ output from the pooling layer is put into a linear layer to obtain the confidence score for each argument role type. Then a softmax function is adopted to compute the probability distribution $p(t|\mathbf{H})$ over all argument role types:

$$p(t|\mathbf{H}) = softmax(\mathbf{W}_t\mathbf{H} + \mathbf{b}_t), \tag{7}$$

where $\mathbf{W}_t$ and $\mathbf{b}_t$ are learnable parameters. After softmax, we choose the argument role label with the highest probability as the classification result.

For training, the loss function is defined as the negative log-likelihood of the correct argument role and all argument role types:

$$Loss = -\sum_{i=1}^{M} \log(p(t_i|\mathbf{H}_i)), \tag{8}$$

where M is the number of training instances.

3 Experiments

Experimental Setup: The performance of our DSGCN model is evaluated on the ACE 2005 English dataset ([2,4]), which has been widely used in the event extraction field. The dataset consists of 599 documents, 33 event subtypes and 35 argument roles. For a fair comparison for experiments, we used the same training/development/test set as those in the previous work [19–21], which respectively contains 529/30/40 documents. Since our work does not focus on event detection, we also follow previous work [19–21] to use the event triggers predicted by the event detection model in [1] to conduct event argument extraction.

We use GloVe [15] vectors with 200 dimensions to initialize each word embedding, while POS-tag, dependency relation, entity type, event type and distance embeddings are 30-dimensional, respectively. The dimension of the BiLSTM and DSGCN hidden layers is set to 200. The number of the DSGCN layer T is set to 2. Stanford CoreNLP toolkit is used to parse a sentence as a dependency tree. To alleviate overfitting, we apply the Dropout mechanism [17] on both input representation of the model with a rate of 0.5 and output representation of each hidden layer with a rate of 0.55. In addition, the network is trained via stochastic gradient descent (SGD) with a batch size of 50, a learning rate of 1 and a gradient clipping of 5.0.

Baselines: We compare our DSGCN model with the following state-of-the-art baselines on the ACE 2005 English dataset.

DMCNN [1] leverages CNNs on massive hand-craft features to perform event extraction and proposes a dynamic multi-pooling mechanism to aggregate features.

JRNN [12] uses a bidirectional Gated Recurrent Unit (GRU) model to jointly predict triggers and extract event arguments.

dbRNN [16] designs a dependency bridge recurrent neural network to utilize sequence information and syntactic information for event extraction.

JMEE [10] models a dependency parsing tree as a graph and adopts an attention-based Graph Convolutional Network to exploit syntactic information.

HMEAE [21] adopts the neural module networks (NMNs) to leverage the knowledge about the hierarchical concept correlation among argument roles.

NGS [20] focuses on the correlation among event arguments and proposes a Neural Gibbs Sampling (NGS) model to jointly extract event arguments.

SemSynGTN [19] employs the Graph Transformer Networks (GTNs) to exploit both semantic and syntactic structures of input sentences for event argument extraction.

Evaluation: In the previous work, an argument is correctly classified if its related event subtype, offsets and argument role exactly match the golden argument. The golden argument is a reference to an entity which may have one or more entity mentions in the given sentence. If any entity mention from the

Table 1. Experiment results of our proposed model and comparison models. "*" denotes our re-implementation. "+ new evaluation" denotes that the model is evaluated by the new rule, where an argument is correctly classified if its related event subtype and argument role match the golden argument and its offsets match any of the entity mentions co-referenced with the golden argument. For these comparison models, as only the HMEAE model provides the running source code, we reproduce the HMEAE model and perform the "new evaluation" on HMEAE.

Model	Argument identification			Role classification		
	P	R	F	P	R	F
DMCNN	68.8	51.9	59.1	62.2	46.9	53.5
JRNN	61.4	64.2	62.8	54.2	56.7	55.4
dbRNN	71.3	64.5	67.7	66.2	52.8	58.7
JMEE	71.4	65.6	68.4	66.8	54.9	60.3
HMEAE	61.2*	60.4*	60.8*	56.8*	56.1*	56.4*
+ new evaluation	63.5*	61.4*	62.4*	58.8*	56.8*	57.8*
NGS	–	–	–	61.3	51.3	55.9
SemSynGTN	–	–	–	68.4	55.4	61.2
DSGCN (ours)	62.0	75.4	68.1	56.9	69.3	62.5
+ new evaluation	61.8	80.2	69.8	56.6	73.5	64.0

co-reference chain is classified as the role of the golden argument, whether the event argument is considered to be correctly extracted. For example, consider the following sentences:

S1: *Holding a framed picture of her son, serving with the Army's 3rd Infantry Division in Iraq, she said she didn't know whether he was* ***dead*** *or alive.*

S2: Hamas *vowed to continue its* ***attacks****, while the Palestinian Authority accused Israel of trying to disrupt top-level Palestinian meetings to discuss reforms of Arafat's administration.*

For the death event triggered by the word "dead" in S1, entity mention "her son" rather than "he" is labeled as the "Victim" argument. However, in S2, entity mention "its" rather than "Hamas" is labeled as the "Attacker" argument in the attack event triggered by the word "attacks". According to the entity annotations in the ACE 2005 English dataset, "her son" and "he" refer to the same entity. "Hamas" and "its" also refer to the same entity. Therefore, when evaluating the performance of our model, if the model recognizes "he" rather than "her son" as the "Victim" argument of the death event triggered by "dead", we still consider the model to have correctly extracted the "Victim" argument. That is to say, when evaluating the performance of a model, an argument is correctly classified if its related event subtype and argument role match the golden argument and its offsets match any of the entity mentions co-referenced with the golden argument.

Similar to previous work, we use standard micro-averaged Precision (P), Recall (R), and F1-score (F1) to evaluate our proposed model.

Overall Results: The overall results of all models on the ACE 2005 English dataset are presented in Table 1. From this table, we have the following observations and analyses.

(1) Even using the same evaluation rule as the previous work, our model DSGCN achieves the best performance on the ACE 2005 English dataset, especially outperforms the state-of-the-art SemSynGTN that utilizes semantic and syntactic information by 1.3% increase of F1 score in term of role classification, which validates the superiority of our proposed DSGCN model. It is notable that DSGCN achieves a remarkable high recall, indicating that DSGCN does identify a number of event arguments which previous methods can not.
(2) Dependency-based models making use of syntactic information dbRNN, JMEE, SemSynGTN and DSGCN are obviously superior to those without syntactic information DMCNN, JRNN, HMEAE and NGS. In addition, compared with the existing dependency-based models which only use the syntactic structure information, our model DSGCN achieves the better performance in term of role classification by leveraging the dependency relation type information to encode each word of the input sentence and screening out the dependency relation type information relevant to the event argument extraction via the designed attention mechanism. These results show that syntactic information including syntactic structure and dependency relation type is important for event argument extraction.
(3) Comparing with the evaluation rule adopted in previous work, the new rule described in the Sect. 3 achieves 1% to 2% F1-score improvements in term of argument identification and role classification. For our model DSGCN, the new evaluation rule further gains 4.8% and 4.2% recall improvements on argument identification task and role classification task respectively. These results indicate that the trained argument extraction models can indeed identify the entity mention co-referenced with the golden argument rather than the golden argument as the event argument. With the new evaluation rule, more event arguments are considered to be correctly identified and classified, which is also more reasonable to evaluate the trained argument extraction models. Therefore, we use the new evaluation rule to compute the performance metrics of a model in the following experiments.

Ablation Study: Compared with the existing dependency-based models, The main improvements of our model DSGCN are as follows: (i) Dependency relation type information is utilized to enrich the information contained in each word representation; (2) The distance-sensitive attention is designed to distinguish the importance of different words for event argument extraction. To verify the effectiveness of each improvement, an ablation study is performed on the ACE 2005 English development dataset. From the results displayed in Table 2, it can be seen that:

Table 2. An ablation study of our proposed model, where DRT is short for dependency relation type and DSA is short for distance-sensitive attention.

Model	Argument identification			Role classification		
	P	R	F	P	R	F
DSGCN (ours)	63.9	71.1	67.3	58.4	64.9	61.4
-DRT	62.5	65.3	63.8	58.9	61.6	60.2
-DSA	62.9	66.5	64.7	57.8	61.2	59.4
-DRT & DSA	65.8	62.0	63.8	61.4	57.9	59.6

(1) The DSGCN with dependency relation type information outperforms itself without dependency relation type information by 3.5% F1 score in term of argument identification and 1.2% F1 score in term of role classification. However, when removing the distance-sensitive attention, the performance of the model with dependency relation type information DSGCN-DSA is decreased by 0.2% F1 score for role classification compared with that of the model without dependency relation type information DSGCN-DRT & DSA. These results suggest that dependency relation type can provide key information for event argument extraction, but not all dependency relation type information is useful. While using dependency relation type information, we need to find some ways to filter out useful information, such as leveraging the attention mechanism; Otherwise, the performance of event argument extraction will be compromised. This suggests that dependency relation type information needs to be exploited well in event argument extraction.
(2) Compared with the model DSGCN, the performance of the model without distance-sensitive attention DSGCN-DSA is decreased by 2.6% F1 score in term of argument identification and 2.0% F1 score in term of role classification. For the model DSGCN-DRT, removing its distance-sensitive attention module degrades the result by 0.6% F1 score for role classification. These results illustrate that the distance-sensitive attention module is effective for event argument extraction.

Table 3. Performance of different models on the subset $D \leq 5$ and the subset $D > 5$ in term of role classification.

Model	$D \leq 5$			$D > 5$		
	P	R	F	P	R	F
HMEAE	65.3	65.4	65.3	35.2	30.1	32.5
DSGCN (ours)	63.5	80.3	71.0	37.3	52.2	43.5
-DRT	62.4	78.8	69.6	35.3	47.3	40.5
-DSA	63.8	79.1	70.7	39.6	44.1	41.7

Performance Against the Distance Between the Event Trigger and Argument: To evaluate the ability of a model to capture long-range dependencies between event triggers and event arguments, we divide the test dataset into two subsets according to the distance D between the event trigger and argument, and evaluate the performance of the model on each subset. If an event trigger and an entity mention is separated by no more than 5 words, the instance is classified into the subset $D \leq 5$; Otherwise, into the subset $D > 5$. Table 3 presents the performance of different models on the subset $D \leq 5$ and the subset $D > 5$ in terms of role classification. We can observe that the F1 score on the subset $D \leq 5$ is higher than the F1 score the subset $D > 5$ by about 30% for each model, which indicates that the existing methods are still not good at identifying event arguments that are far away from the event trigger. Compared with the baseline HMEAE, our model DSGCN improve the performance of role classification in terms of 5.7% and 11.0% F1 score on he subset $D \leq 5$ and $D > 5$ respectively. These results prove that the ability of our model to capture long-range dependencies is better than that of the basic model. Comparing DSGCN with DSGCN-DRT and DSGCN-DSA, we can see that dependency relation type information and our designed distance-sensitive attention can also improve the ability of capturing long-range dependencies.

Table 4. Predictions of two event arguments for different models.

Event Trigger: **killed**, Event Type: **Die**				
Sentence: *Australian commandos*, who have been operating deep in Iraq, destroyed a command and control post and **killed** *a number of soldiers*, according to the country's defense chief, Gen. Peter Cosgrove				
Classification results:				
Event argument	HMEAE	DSGCN (ours)	DSGCN - DRT	DSGCN - DSA
Australian commandos	None (×)	Agent (✓)	Victim (×)	Victim (×)
A number of soldiers	Victim (✓)	Victim (✓)	Victim (✓)	Victim (✓)

Case Study: We sample a sentence from the test dataset to illustrate how DSGCN improves EAE performance by comparing the predictions of two event arguments for different models. From the Table 4, it can be seen that the "Victim" argument "a number of soldiers" is correctly classified by all models, while only our model DSGCN can correctly classify the entity mention "Australian commandos" into the "Agent" role.

The dependency syntactic tree of the sample sentence is illustrated in Fig. 2(a), where the distance between the entity mention "Australian commandos" and the event trigger "killed" is shortened from 16 to 1 via the dependency tree structure. With the help of syntactic structure information, DSGCN, DSGCN-DRT and DSGCN-DSA can recognize the entity mention "Australian commandos" as an event argument. From the Fig. 2(a), it can be clearly seen that dependency relation type information is useful for determining the role that the event argument plays. For the "Die" event triggered by "killed", we can judge whether the role of the event argument is the "Agent" or the "Victim" based on whether the dependency relations on the syntactic arcs between the event trigger and the event argument contains the "nsubj" or the "obj". Without dependency relation type information, the model DSGCN-DRT misclassifies event argument "Australian commandos" into "Victim".

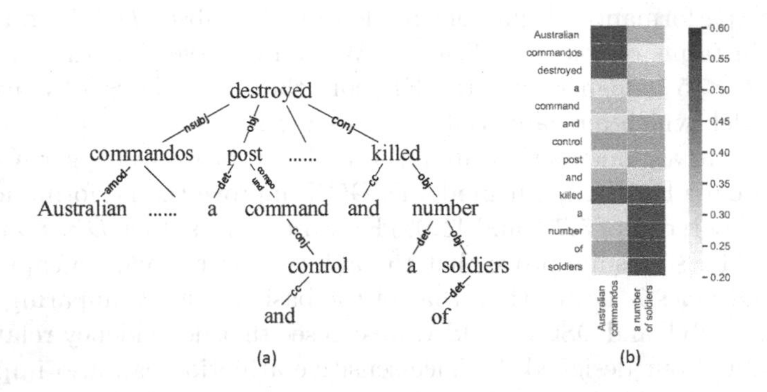

Fig. 2. (a)The dependency syntactic tree of the sample sentence in Table 4, where the orange word represent the event trigger, green words represent event arguments and blue words on the edge represent dependency relations.(b) The visualization of attention weights of some words in the distance-sensitive attention for the classification of "Australian commandos" and "a number of soldiers". (Color figure online)

Figure 2(b) shows the attention weights of some words in the distance-sensitive attention when determining the role of the entity mention "Australian commandos" and the entity mention "a number of soldiers" respectively. Since the attention weight obtained by the distance-sensitive attention is a high-dimensional vector, the weight shown in Fig. 2(b) is the average value of the weights of all dimensions. We can see that the words (e.g., "killed", "soldiers") that are useful to the classification of a given entity mention "a number of soldiers" usually obtain lager attention weights than the other non-effective words (e.g., "command", "and"). It can be also observed that the same word has different importances to the classification of different entity mentions, which is useful to distinguish the roles that different entity mentions play in the same event. For example, without the distance-sensitive attention, it is difficult to distinguish that "Australian commandos" and "a number of soldiers" play different roles, causing that the model DSGCN-DSA misclassifies event argument "Australian commandos" into "Victim". The above analysis of the case in the Table 4 reveals that syntactic structure information, dependency relation type information and our designed distance-sensitive attention are useful for EAE.

4 Conclusion and Future Work

In this paper, we propose a Distance-Sensitive Graph Convolutional Network (DSGCN) for event argument extraction. In our model, syntactic structure information is used to shorten the distance between event trigger and event argument to better capture long-range dependencies. In addition, a novel distance-sensitive attention is designed to quantify the importance of each word so as to learn an effective representation for event argument extraction during convolution. Experiment results have shown that our model can achieve the best results among a variety of the state-of-the-art models. In our future work, we will explore the potentials of joint learning of named entity recognition and event argument extraction.

References

1. Chen, Y., Xu, L., Liu, K., Zeng, D., Zhao, J.: Event extraction via dynamic multi-pooling convolutional neural networks. In: ACL-IJCNLP, pp. 167–176 (2015)
2. Consortium, L.D.: ACE (automatic content extraction) English annotation guidelines for events (2005)
3. Cui, S., Yu, B., Liu, T., Zhang, Z., Wang, X., Shi, J.: Edge-enhanced graph convolution networks for event detection with syntactic relation. In: EMNLP, pp. 2329–2339 (2020)
4. Doddington, G., Mitchell, A., Przbocki, M., Ramshaw, L., Strassel, S., Weischedel, R.: The automatic content extraction (ACE) program-tasks, data, and evaluation. In: LREC (2004)
5. Hong, Y., Zhang, J., Ma, B., Yao, J., Zhou, G., Zhu, Q.: Using cross-entity inference to improve event extraction. In: ACL, pp. 1127–1136 (2011)
6. Ji, H., Grishman, R.: Refining event extraction through cross-document inference. In: ACL-HLT, pp. 254–262 (2008)
7. Li, Q., Ji, H., Huang, L.: Joint event extraction via structured prediction with global features. In: ACL, pp. 73–82 (2013)
8. Liao, S., Grishman, R.: Using document level cross-event inference to improve event extraction. In: ACL, pp. 789–797 (2010)
9. Liu, K., Chen, Y., Liu, J., Zuo, X., Zhao, J.: Extracting event and their relations from texts: a survey on recent research progress and challenges. AI Open **1**, 22–39 (2020)
10. Liu, X., Luo, Z., Huang, H.: Jointly multiple events extraction via attention-based graph information aggregation. In: EMNLP, pp. 1247–1256 (2018)
11. Ma, J., Wang, S.: Resource-enhanced neural model for event argument extraction. In: EMNLP, pp. 3554–3559 (2020)
12. Nguyen, T.H., Cho, K., Grishman, R.: Joint event extraction via recurrent neural networks. In: NAACL, pp. 300–309 (2016)
13. Nguyen, T.H., Grishman, R.: Event detection and domain adaptation with convolutional neural networks. In: ACL-IJCNLP, pp. 365–371 (2015)
14. Nguyen, T.H., Grishman, R.: Graph convolutional networks with argument-aware pooling for event detection. In: AAAI, pp. 5900–5907 (2018)
15. Pennington, J., Socher, R., Manning, C.D.: GloVe: global vectors for word representation. In: EMNLP, pp. 1532–1543 (2014)

16. Sha, L., Qian, F., Chang, B., Sui, Z.: Jointly extracting event triggers and arguments by dependency-bridge RNN and tensor-based argument interaction. In: AAAI, pp. 5916–5923 (2018)
17. Srivastava, N., Hinton, G., Krizhevsky, A., Sutskever, I., Salakhutdinov, R.: Dropout: a simple way to prevent neural networks from overfitting. J. Mach. Learn. Res. **15**(1), 1929–1958 (2014)
18. Tong, M., et al.: Improving event detection via open-domain trigger knowledge. In: ACL, pp. 5887–5897 (2020)
19. Veyseh, A.P.B., Nguyen, T.N., Nguyen, T.H.: Graph transformer networks with syntactic and semantic structures for event argument extraction. In: EMNLP, pp. 3651–3661 (2020)
20. Wang, X., et al.: Neural Gibbs sampling for joint event argument extraction. In: AACL, pp. 169–180 (2020)
21. Wang, X., et al.: HMEAE: hierarchical modular event argument extraction. In: EMNLP-IJCNLP, pp. 5777–5783 (2019)
22. Xiang, W., Wang, B.: A survey of event extraction from text. IEEE Access **7**, 173111–173137 (2019)
23. Yan, H., Jin, X., Meng, X., Guo, J., Cheng, X.: Event detection with multi-order graph convolution and aggregated attention. In: EMNLP-IJCNLP, pp. 5766–5770 (2019)

Exploit Vague Relation: An Augmented Temporal Relation Corpus and Evaluation

Liang Wang, Sheng Xu, Peifeng Li(✉), and Qiaoming Zhu

School of Computer Science and Technology, Soochow University, Jiangsu, China
docy@vip.qq.com, sxu@stu.suda.edu.cn, {pfli,qmzhu}@suda.edu.cn

Abstract. Temporal relation classification is a challenging task in Natural Language Processing (NLP) that faces many difficulties, such as imbalanced distribution of instances, and ambiguity of the vague instances. To address the above issues, this paper proposes a novel data augmentation method on the TimeBank-Dense (TBD) corpus to distinguish those vague instances, which can provide more evidence to identify temporal relation clearly. Specifically, we additionally annotate those VAGUE instances in the form of multiple labels, to further distinguish varied semantic phenomena in VAGUE circumstances. Experimental results show that the models trained on our augmented corpus VATBD significantly outperform those trained on the original TBD corpus, and these results verify the effectiveness of our augmented corpus on temporal relation classification.

Keywords: Temporal relation classification · Data augmentation · VAGUE relation · VATBD

1 Introduction

Temporal relation classification focuses on the occurrence order (e.g., *Before*, *After* and *Include*) of two given events. As a crucial component of relation extraction in the field of Natural Language Processing (NLP), temporal relation classification can help many downstream NLP tasks, such as question answering [1], timeline construction [2] and summarization [3].

Currently, almost all temporal relation classification corpora only annotated one label for each event mention pair. However, there is a large portion of ambiguous mention pairs in the documents because most event mentions do not have explicit temporal cues, or the annotators cannot capture the evidence in the whole document or use common sense to give a unique relation label. In this case, previous work simply labeled those ambiguous mention pairs as vague or no-link relation. Such annotation scheme retains both well-defined and ill-defined instances in an annotated corpus, which brings two issues, i.e., the one-pair-one-label limitation and the ambiguous training dataset.

The first issue in those existing corpora is that they only assigned one label to each event mention pair. However, many pairs may contain more than one relation. For simplification, most corpora annotated those ambiguous mention pairs as vague or no-link

L. Wang et al. (Eds.): NLPCC 2021, LNAI 13029, pp. 73–85, 2021.
https://doi.org/10.1007/978-3-030-88483-3_6

relation, which is useless and uninformative for the task of temporal relation classification. For example, the percentage of the *VAGUE* instances in TimeBank-Dense (TBD) corpus [4] reaches 46.5%, and 72% of them are caused by two annotators' different assigning labels to one instance according to the annotation rules.

Besides, those instances with the vague relation are difficult to be distinguished from those instances with a specific relation label (e.g., *Before*, *After* and *Include*). Hence, the ambiguous training dataset will harm the performance of temporal relation classification. For example, almost 20% of the *VAGUE* instances in TBD maybe belong to *Before*, *After* or *Simultaneous*. These instances will mislead the model to identify those instances whose relations are *Before*, *After* or *Simultaneous* as *VAGUE*. Actually, identifying an event mention pair as *VAGUE* is useless for the temporal relation classification task.

Lots of methods are proposed to address the above issues, such as reducing the number of classification labels [5], filtering insignificant temporal relations in the dataset [6], and adding more manual linguistic features or external knowledge [7], etc. However, the problem of vague labeling still exists, and judging a well-defined pair is quite a tough issue.

To address the above issues, we propose a novel multi-label annotation scheme to handle those *VAGUE* instances. We believe that those *VAGUE* instances have huge semantic differences and label their different relations as many as possible. For this motivation, we annotated an extended TBD dataset, called **V**ague-**A**ugmented-**T**ime**B**ank-**D**ense (VATBD). Moreover, experimental results show that the models trained on our VATBD significantly outperform those trained on TBD, and these results verify the effectiveness of our VATBD on temporal relation classification. In summary, the contributions of our paper are as follows.

- We first annotated a multi-label temporal relation corpus, which can provide more definite temporal relation annotation.
- The temporal relation classification models trained on our multi-label VATBD are better than those trained on single-label TBD, which shows the success of our multi-label scheme.

2 Related Work

There are a few corpora on the task of temporal relation classification. TimeBank [8] is an early work for collecting temporal relations of events. TimeBank-Dense, the extended version of TimeBank, densely annotated all event pairs within adjacent sentences. RED [9] combined temporal, coreference, and causal relation in one corpus. MATRES [6] simplified the labeling scheme (the label number is reduced to 4) and added more data sources to improve the corpus quality. TDDiscourse [10] is a large-scale corpus, which is based on TBD by removing those vague instances. In summary, the temporal relation definitions in the above corpora are shown in Table 1.

Currently, the temporal relation classification models are mainly based on neural networks. Cheng and Miyao [11] used BiLSTM (Bi-directional LSTM) to encode the Shortest Dependency Path (SDP) of the event mentions to efficiently obtain the representation of the contextual information. Zhou et al. [12] added the time expression to

Table 1. Temporal relations on current corpora.

TimeBank	TBD	RED	MATRES	TDDiscourse
Before	Before	Before	Before	Before
IsIncluded	After	Before/causes	After	After
After	Include	Before/precondition	Equal	Include
Simultaneous	IsIncluded	Overlap	Vague	IsIncluded
Identical	Simultaneous	Overlap/causes		Simultaneous
Included	Vague	Overlap/precondition		
Holds		Contains		
Ends		Contains-subevent		
Begins		Begins-on		
I-After		Ends-on		
I-Before		Simultaneous		

SPD. Cheng [13] constructed a central event chain, and used the pre-trained transformer model BERT [14] to encode the event mentions, then dynamically updates the representation of the central event through GRU. Inspired by Liu et al. [15], they also performed a single-task batch multi-task learning strategy to train the model which achieved the state-of-the-art performance.

3 Data Annotation

In this section, we first introduce the dataset and then propose our annotation process. Finally, we report the statistics of the annotated corpus and evaluate its quality.

3.1 Data

We consider that when facing semantic ambiguity, different temporal relations can be compatible, i.e., additionally assigning multi-label to each pair of the 'vague' instances. For example in Table 2, *e1* and *e2* in E1 are two event mentions with short duration. Hence, the *Include* and *IsIncluded* relation can be excluded, while the *Before*, *After*, and *Simultaneous* are all possible.

In TBD, the instance numbers of the *Before*, *After*, *Include*, *IsIncluded* and *Simultaneous* relations are 1348, 1120, 276, 347 and 93, respectively. The majority class in TBD is *VAGUE*, which has 2904 instances, accounting for 47.7%. VATBD dataset is an extended version of TBD, which annotated the 2904 temporal instances.

Our annotation scheme is to re-annotate those *VAGUE* instances with multi-label in the set of *Before*, *After*, *Include*, *IsIncluded* and *Simultaneous*. That is, the annotated relations of those *VAGUE* instances are a five-tuple vector. For example, the label (1, 1, 0, 0, 1) in Table 2 indicates that there are three relations (i.e., *Before*, *After* and *Simultaneous*) between the event mentions *e1* and *e2* in E1.

Table 2. The samples of the vague instances and our annotations in VATBD.

Example	Label
E1: Soviet officials also said (*e1*) Soviet women, children and invalids would be allowed to leave Iraq. President Bush today denounced (*e2*) Saddam's "ruinous policies of war," and said the United States is "striking a blow for the principle that might does not make right."	(1,1,0,0,1)
E2: One of the scenarios widely advanced (*e3*) before the visit is that through the Pope, Cuba, still led (*e4*) by Castro, can begin a more productive relation with the world	(1,0,0,1,0)
E3: He declined (*e5*) to say (*e6*) if the weapons included missiles, but the Iraqis have them in their arsenal	(0,0,0,0,0)

3.2 Annotation Process

To improve the annotation quality, we propose the follow annotation process and rules.

Input: an event mention pair *e1* and *e2* whose relation is *VAGUE* in TBD
Output: a five-tuple vector such as (1, 1, 0, 0, 1) in which each column refers to whether there is the relation *Before*, *After*, *Include*, *IsIncluded* and *Simultaneous*, respectively (1-**Yes**; 0-**No**)

1. Determining whether *e1* and *e2* are actual events.
- If at least one of *e1* and *e2* is non-actual event (e.g., intention, hypothesis, opinion, negation, generic events[1], etc.), it should be labeled as (0, 0, 0, 0, 0) (this type often has a strong vague characteristic).
- Otherwise, *e1* and *e2* are both factual events (including future events), go to step 2.
2. According to contextual semantics, determining the duration span of the two event mentions, denoted as $lsat\_e_i (1 \leq i \leq 2)$, divided into the minute, hour-day, week-and-longer levels (0–2), and it is allowed to be non-exclusive, such as **running** at the minute or hour level ($lsat\_e_{running} = \{0,1\}$).
- If $last\_e_1 = last\_e_2 = \{0\}$, that result means *e1* and *e2* are both transient events, then we exclude the *Include* and *IsIncluded* columns, namely (?,?,0,0,?).
- If $max(last\_e_1) \leq min(last\_e_2)$, meaning the duration of *e1* is significantly shorter than *e2*, then we exclude the *Include* and *Simultaneous* column, namely (?, ?, 0, ?, 0).
- If $max(last\_e_2) \geq min(last\_e_1)$, meaning the duration of *e2* is significantly shorter than *e1*, then we exclude the *Include* and *Simultaneous* column, namely (?, ?, ?, 0, 0).
- Otherwise, go to step 3.
3. Judging whether *e1* and *e2* have ended (-1), is happening (0) or not happened yet (1) by DCT (Document Creation Time), denoted as $anchor\_e_i (1 \leq i \leq 2)$, which still

[1] E.g., a plane **taking off** at 9 o'clock every day.

can be non-exclusive. For example, *a series of* ***homicides*** began in the past, but it is uncertain whether it is still last, thus, $anchor\_e_{homicides} = \{0, -1\}$.

- If $max(anchor\_e_1) < min(anchor\_e_2)$, then we exclude the *After* and *Simultaneous* column, namely (?, 0, ?, ?, 0).
- else if $min(anchor\_e_1) > max(anchor\_e_2)$, then we exclude the *Before* and *Simultaneous* column, namely (0, ?, ?, ?, 0).
- Otherwise, go to step 4.

4. The remaining relation is further excluded according to $last\_e_i$ together with $anchor\_e_i$, taking all columns that could not be excluded into the final annotation, e.g., if $max(last\_e_1) <= min(last\_e_2)$ and $anchor\_e_1 = anchor\_e_2$, we label it as (1, 0, 0, 1, 0), as E2 shown in Table 2.

3.3 Data Statistics

Finally, we annotated 2904 event-event *VAGUE* instances in the TBD corpus by two annotators. The count of each specific annotation is shown in Table 3. From Table 3, we can find out that 727 (25.03%) instances still cannot be assigned multiple labels because most of them are non-actual events and we cannot identify their occurrence time. Fortunately, the remaining 2177 (74.97%) instances can be assigned multiple labels, in

Table 3. Annotation statistics. The bottom row shows the accumulating sum of each column, where b, a, i, ii and s refer to before, after, include, isincluded, simultaneous, respectively.

b	a	i	ii	s	Count
0	0	0	0	0	727
1	1	0	0	1	561
1	0	1	0	0	373
0	1	0	1	0	319
1	1	0	0	0	217
1	1	1	0	0	134
1	0	0	1	0	133
0	1	1	0	0	111
1	1	0	1	0	94
0	0	1	1	0	43
1	0	0	0	1	31
...					
1675	1564	780	708	676	2904

which (1, 1, 0, 0, 1), (1, 0, 1, 0, 0) and (0, 1, 0, 1, 0) are the majority classes. In this case, we also can find out that the numbers of different classes are in a relatively balanced situation.

3.4 Annotation Quality

Table 4 shows the inter-annotator agreements (IAA) of each category between two annotators. We reach a comparable accuracy on the relations *Include*, *IsIncluded* and *Simultaneous*, while a little lower on *Before* and *After* because the start time of an event is relatively hard to anchor.

The overall Cohen's Kappa is 0.51 which is not appropriate for the quality measurement, as our annotation is binary for each column in the vector and the 0–1 distribution of each column is quite different, like *Simultaneous (Kappa = .54, Acc = .85)* has much less **1** than *Before* and *After* because it needs an event pair of equal duration. For those inconsistent labels given by two annotators, they make a further discussion to decide the final annotation.

Table 4. IAA between two annotators, where $accuracy = \frac{count(matched\ label)}{count(total\ examples)}$.

	b	a	i	ii	s	Overall
Distribution	.31	.29	.14	.13	.13	1
IAA: accuracy	.66	.64	.75	.84	.85	.71

4 Temporal Relation Classification and Corpus Usage Strategy

In this section, we first describe the temporal relation classification model and then propose two corpus usage strategies.

4.1 Temporal Relation Classification Model

We use the pre-trained Transformer model [16] as the temporal relation classification model to obtain the embedding of the sentence and the representation of the corresponding two events mentions. Specially, we use *BERT-base* for the domain transfer. Finally, we obtain the distribution of label probability through a fully connected layer.

4.2 Corpus Usage Strategy

The additional multi-label annotation of the *VAGUE* instances is quite different from other non-*VAGUE* labels, as TBD is a single label multi-classification task. Hence, how to make the usage of the multi-label information reasonably is quite a critical issue. Therefore, we propose two strategies for corpus usage as follows.

- **Multi-label to Single-label (M2S).** We transfer the single-label task on TBD to the multi-label classification on VATBD, that is, both of the gold and predicted labels are multiple labels, such as (1, 1, 0, 0, 1). Hence, we use the binary cross entropy loss to train our model, with sigmoid as the activation function of the output layer. To compare with the existing studies on TBD, we transfer the results on VATBD to those on TBD for fair comparison. Take the threshold $c = 0.5$, finally we map the 5 columns output to a single label $\hat{y}$ to match the original classification task on TBD, i.e., for the $output = (l_1, l_2, l_3, l_4, l_5)$, if $count(l_i >= c) = 1$, then $\hat{y} = Argmax(output)$; otherwise, $\hat{y} = vague$.
- **N-To-6.** Based on the statistics of our additional annotations, we choose top $\boldsymbol{k}$ ($\boldsymbol{k}$ $>= 1$) unique vague annotations as independent labels, then the original single-label 6 classification task is transformed into a single-label N ($N = k + 5$) classification task. For example, if $\boldsymbol{k} = 4$, then the top 4 cases in Table 3 would be tagged as the label ID 6, 7, 8, 9, respectively. The remaining cases with a lower proportion are unified into (0, 0, 0, 0, 0) which tagged as the class ID 6. For inference, any prediction $\hat{y} >= 6$ will be mapped to *VAGUE*[2].

Each labeling situation that accounts for the majority is regarded as a category of unique semantic phenomenon. For example, the instances labeled as (0, 0, 0, 0, 0) are mostly negative, hypotheses, ideas, and so on, which own significant vague characteristics (E3 in Table 2). For the annotated vector (1, 1, 0, 0, 1), this type of the instances often contains two actual events with a short duration and an indeterminable [start, end] point (E1 in Table 2). In the case of (1, 0, 0, 1, 0), the target event has an earlier start time point and a larger duration than the source event (E2 in Table 2).

5 Experimentation

In this section, we first introduce the experimental settings and then provide the experimental results. Finally, we discuss the corpus generalization and errors.

5.1 Experimental Settings

We implement several experiments on both the TBD and VATBD datasets. The data split of the training set, development set, and test set follows previous work[3]. We adopt the pre-trained BERT-base to get the representation of two event mentions, and their concatenation is fed into the final classifier. Besides, we set $batch\_size = 32$, AdamW optimizer ($lr = 3\text{e-}5$) and $epoch = 15$. The other hyper parameters are determined based on the validation scores.

M2S. To be consistent with the vague relation, we treat the non-vague labels as multi-label, such as the *Before* label is equivalently converted to (1, 0, 0, 0, 0). For those multi-label instances to be equal to the other categories, they are normalized by unit length, e.g., (1, 0, 1, 0, 0) should be pre-processed as (0.5, 0, 0.5, 0, 0). We adopt BCE

[2] 1–5 corresponds to five non-vague relations, respectively.

[3] http://www.usna.edu/Users/cs/nchamber/c.

loss, and the activation function is sigmoid. Due to the huge difference of the training strategy between VATBD and TBD task, we also implement M2S on the TBD corpus to directly compare the performance.

N-To-6. In the implementation of N-To-6 strategy, we use CrossEntropy loss during training step. We select the top k largest proportion types (i.e., five-tuple vector) as additional unique labels, where the value of k is 2–10 (when $k = 1$, it means all multi-label instances belong to one class which equals to the original TBD task), and the rest of the sparse types are all regarded as (0, 0, 0, 0, 0). Finally, we choose $k = 5$ tuned on the development set, as shown in Fig. 1.

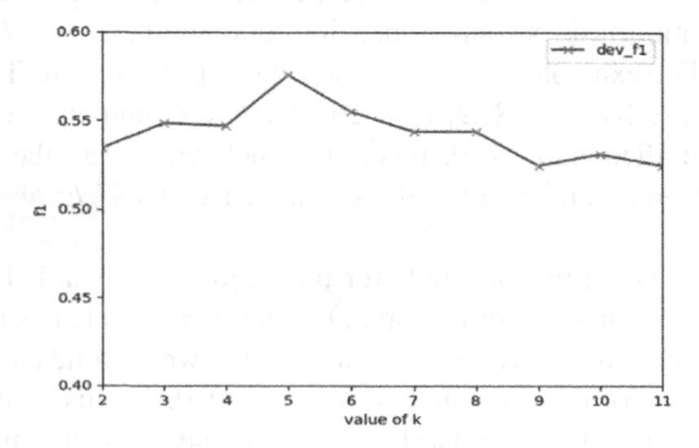

Fig. 1. Performances of different k on development set.

5.2 Experimental Results

Table 5 shows the results of the M2S and 10-To-6 models on both TBD and VATBD. We can find out that both of the training strategies on VATBD outperform those on TBD, with the gains in F1 score by 3.7% and 5.2%, respectively. From Table 5, we can find out that:

Table 5. P/R/F1 scores of two strategies on TBD and VATBD.

Strategy	Corpus	b	a	i	ii	s	v	F1
M2S	TBD	.68/.57/.62	.71/.41/.52	.50/.02/.03	-	–	.52/.78/.62	.578
	VATBD	.66/.68/.67	.73/.54/.62	1./.03/.06	.62/.14/.23	–	.56/.71/.63	.615
N-To-6	TBD[4]	.67/.54/.60	.79/.52/.63	1./.03/.06	.20/.02/.03	–	.54/.81/.65	.598
	VATBD	.65/.68/.67	.76/.66/.71	.29/.11/.16	.48/.19/.27	–	.63/.73/.67	.650

[4] TBD is a 6 classes classification task, here, N = 6.

- The performance of M2S is slightly lower than 10-To-6 both on TBD and VATBD, which may be attributed to the training difference between our proposal and the original task. This result also indicates that one-vs-others classifier is worse than multi-class classifier in this task.
- Both two strategies on VATBD can improve the F1 scores of the five non-vague relations significantly, in comparison with those on TBD. Our annotated data and multi-label scheme can provide more clear evidence to identify those non-vague instances, because the multi-label scheme can increase the numbers of positive instances in those non-vague classes and then improve their recall, with the large gains from 11% to 14%, except the *Include* relation for its low increment.
- In comparison with those non-vague relations, the improvements on the *VAGUE* relation are relatively slight. This reason is that most of the *VAGUE* instances are labeled as other relations in VATBD. Hence, the recall scores of M2S and 10-To-6 on VATBD are much lower than those on TBD. However, their precision scores are increased significantly by 4% and 9%.

5.3 Generalization

To illustrate the generalization of our proposed VATBD corpus, we selected several classic temporal relation classification models for comparison as follows. 1) AC-GCN [12]: it incorporated the temporal cues to AC-GCN; 2) Multi-task [13]: it proposed a multi-task learning method on TBD and achieves the state-of-the-art performance; 3) SEC [13]: it dynamically updates the representation of the central event within a Source Event Centric (SEC) chain via a global RNN. Recently, since Span-BERT [17] has achieved good results on multiple NLP tasks (e.g., QA and coreference resolution), we also adopt this model for evaluation.

The results of the above model both on TBD and VATBD are shown in Table 6. We can find out that all models trained on VATBD achieve higher F1 scores than those on TBD, which indicates that VATBD is effective to train a temporal relation classifier.

Table 6. F1-scores of variant existing models on TBD and VATBD. For fair comparison, we re-implement all of the models (some values are lower than those reported in their papers).

Model	TBD	VATBD
Span-BERT	0.594	0.666 (+7.2%)
AC-GCN	0.620	0.637 (+1.7%)
Multi-task	0.609	0.667 (+5.8%)
SEC(TBD)	0.603	0.645 (+4.2%)

5.4 Error Analysis

Table 7 shows the confusion matrix of error numbers created by 10-To-6 on VATBD. We can find that the main errors are due to the vague and non-vague Conflict issue where

4.3% (17 of 395) of the *Before* instances and 8.5% (26 of 303) of the *After* instances are incorrectly predicted as an extra added vague class in Table 3, while the number of the *VAGUE* instances is 21.7% (145 of 669). Besides, 10 *Before* and 22 *After* instances are predicted as v2 (1, 1, 0, 0, 1) which means both before and after are possible. It further reveals that those existing neural models still could not completely capture that in-direct contextual information.

Table 7. Confusion matrix of errors (10-To-6 on VATBD) where v1-v5 corresponds to 5 extra vague classes in Table 3.

	b	a	i	ii	s	v1	v2	v3	v4	v5
b	274	4	3	0	0	98	10	4	2	1
a	16	186	1	5	0	68	22	0	4	0
i	11	2	9	0	0	28	1	6	0	0
ii	4	3	0	12	0	30	1	0	5	0
s	3	4	0	0	0	12	2	0	2	0
v	104	44	10	14	0	352	66	31	28	20

5.5 Downsampling Training Set

To better understand the rationality and effectiveness of our 10-To-6 method, we deleted part of the training set instances as follows: 1) we randomly delete a fixed percentage of the vague instances; 2) we randomly delete a fixed percentage of all instances. Figure 2 shows the relations between the F1 scores and vague percentages, and Fig. 3 shows the relations between the F1 scores and the percentages of all instances.

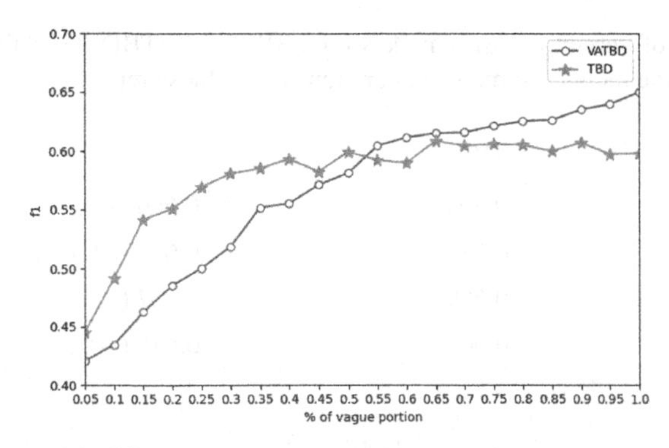

Fig. 2. F1 scores with different percentages of the vague instances on TBD and VATBD.

As shown in Fig. 2 and Fig. 3, the TBD performance reaches a bottleneck when the model uses about 50% training data, while VATBD's learning curve is still climbing. It

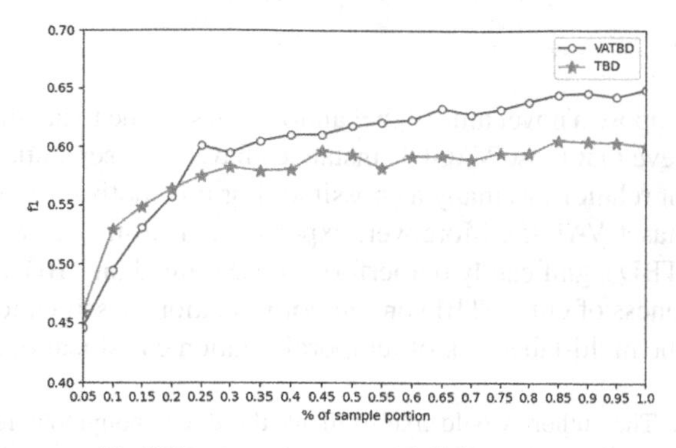

Fig. 3. F1 scores with different percentages of all instances on TBD and VATBD.

is obvious that VATBD can make better use of the entire data. Remarkably, taking only 25% of the entire training data in Fig. 3, VATBD outperforms the best F1-score of TBD, which further indicates the effectiveness of our annotation scheme.

We also evaluate our corpus on different label sets as shown in Table 2, in which each row represents a specific set. Table 8 shows the F1 scores on the training set using different label sets.

In Table 8, it shows that the set (0, 0, 0, 0, 0) is the most important for the model, which is not difficult to understand. This type of vague instance owns the most significant vague features which contain the semantics of negation, hypothesis, opinion, and so on. Similarly, the category *others* is another group with significant vague characteristics, even human is difficult to exclude some relations during the annotation step.

Different from the first two sets, the remaining annotation sets do not contain significant vague features. However, they could help to predict those non-vague instances. As Table 8 shows, after adding these categories, the relation *Before* (b) obtains a 5% gain while *After* (a) obtains a 6% gain.

Table 8. F1 scores on VATBD for subsets cumulatively added, where **No-vague** denotes that all vague instances are deleted, and **Others** refers to the rest of the sparse types.

Added subset	b	a	i	ii	s	v	F1
No-vague	.59	.64	.31	.35	–	–	.435
+(0,0,0,0,0)	.63	.68	.31	.31	–	.50	.563(+10.8%)
+Others	.63	.65	.16	.27	–	.65	.615 (+5.2%)
+(1,1,0,0,0)	.65	.64	.13	.23	–	.66	.623 (+0.8%)
+(1,1,0,0,1)	.66	.66	.11	.21	–	.66	.632 (+0.9%)
+(0,1,0,1,0)	.70	.67	.14	.29	–	.66	.643 (+1.1%)
+(1,0,1,0,0) (Total VATBD)	.68	.71	.16	.27	–	.67	.650 (+0.6%)

6 Conclusion

In this paper, we propose a novel multi-label annotation scheme to handle those VAGUE instances. We believe that those VAGUE instances have huge semantic differences and label their different relations as many as possible. For this motivation, we annotated an extended TBD dataset VATBD. Moreover, experimental results show that the models trained on our VATBD significantly outperform those trained on TBD, and these results verify the effectiveness of our VATBD on temporal relation classification. In the future, we will focus on the multi-label task of temporal relation classification.

Acknowledgments. The authors would like to thank the three anonymous reviewers for their comments on this paper. This research was supported by the National Natural Science Foundation of China (No. 61772354, 61836007 and 61773276.), and the Priority Academic Program Development of Jiangsu Higher Education Institutions (PAPD).

References

1. Mo, Y., Wenpeng, Y., Kazi, S., et al.: Improved neural relation detection for knowledge base question answering. In: Proceedings of ACL 2017, pp. 571–581 (2017)
2. Sakaguchi, T., Kurohashi, S.: Timeline generation based on a two-stage event-time anchoring model. In: Gelbukh, A. (ed.) CICLing 2017. LNCS, vol. 10762, pp. 535–545. Springer, Cham (2018). https://doi.org/10.1007/978-3-319-77116-8_40
3. Zhou, G., Qian, L., Fan, J.: Tree kernel-based semantic relation extraction with rich syntactic and semantic information. Inf. Sci. **180**(8), 1313–1325 (2010)
4. Nathanael, C., Taylor, C., Bill, M., Steven, B.: Dense event ordering with a multi-pass architecture. TACL **2**, 273–284 (2014)
5. Marc, V., Robert, G., Frank, S., et at.: SemEval-2007 Task 15: TempEval temporal relation identification. In: Proceedings of SemEval 2007, pp. 75–80 (2007)
6. Qiang, N., Hao, W., Dan, R.: A multi-axis annotation scheme for event temporal relations. In: Proceedings of ACL 2018, pp. 1318–1328 (2018)
7. Ben, Z., Qiang, N., Daniel, K., Dan, R.: Temporal common sense acquisition with minimal supervision. In: Proceedings of ACL 2020, pp. 7579–7589 (2020)
8. James, P., Patrick, H., Roser, S., et al.: The TIMEBANK corpus. Corpus Linguistics, pp. 647–656 (2003)
9. Tim, O., Kristin, W., Martha, P.: Richer event description: integrating event conference with temporal, causal and bridging annotation. In: Proceedings of the 2nd Workshop on Computing News Storylines (CNS 2016), pp. 47–56 (2016)
10. Aakanksha, N., Luke, B., Carolyn, R.: Tddiscourse: a dataset for discourse-level temporal ordering of events. In: Proceedings of the 20th Annual SIGdial Meeting on Discourse and Dialogue, pp. 239–249 (2019)
11. Cheng, F., Miyao, Y.: Classifying temporal relations by bidirectional LSTM over dependency paths. In: Proceedings of ACL 2017, pp. 1–6 (2017)
12. Zhou, X., Li, P., Zhu, Q., Kong, F.: Incorporating temporal cues and AC-GCN to improve temporal relation classification. In: Proceedings of NLPCC 2019, pp. 580–592 (2019).
13. Cheng, F., Asahara, M., Kobayashi, I., Kurohashi, S.: Dynamically updating event representations for temporal relation classification with multi-category learning. In: Findings of the Association for Computational Linguistics: EMNLP 2020, pp. 1352–1357 (2020)

14. Jacob, D., Ming, W., Kenton, L., Kristina, T.: BERT: pre-training of deep bidirectional transformers for language understanding. In: Proceedings of NAACL 2019, pp. 4171–4186 (2019)
15. Liu, X., He, P., Chen, W., Gao, J.: Multi-task deep neural networks for natural language understanding. In: Proceedings of ACL 2019, pp. 4487–4496 (2019)
16. Ashish, V., Noam, S., Niki, P., et al.: Attention is all you need. In: Proceedings of NeurIPS 2017, pp. 5998–6008 (2017)
17. Joshi, M., Chen, D., Liu, Y., et al.: SpanBERT: improving pre-training by representing and predicting spans. TACL **8**, 64–77 (2019)

Searching Effective Transformer for Seq2Seq Keyphrase Generation

Yige Xu, Yichao Luo, Yicheng Zhou, Zhengyan Li, Qi Zhang, Xipeng Qiu(✉), and Xuanjing Huang

School of Computer Science, Fudan University, Shanghai, China
{ygxu18,ycluo18,yczou18,lzy19,qz,xpqiu,xjhuang}@fudan.edu.cn

Abstract. Keyphrase Generation (KG) aims to generate a set of keyphrases to represent the topic information of a given document, which is a worthy task of Natural Language Processing (NLP). Recently, the Transformer structure with fully-connected self-attention blocks has been widely used in many NLP tasks due to its advantage of parallelism and global context modeling. However, in KG tasks, Transformer-based models can hardly beat the recurrent-based models. Our observations also confirm this phenomenon. Based on our observations, we state the *Information Sparsity Hypothesis* to explain why Transformer-based models perform poorly in KG tasks. In this paper, we conducted exhaustive experiments to confirm our hypothesis, and search for an effective Transformer model for keyphrase generation. Comprehensive experiments on multiple KG benchmarks showed that: (1) In KG tasks, uninformative content abounds in documents while salient information is diluted globally. (2) The vanilla Transformer equipped with a fully-connected self-attention mechanism may overlook the local context, leading to performance degradation. (3) We add constraints to the self-attention mechanism and introduce direction information to improve the vanilla Transformer model, which achieves state-of-the-art performance on KG benchmarks.

Keywords: Keyphrase Generation · Transformer · Seq2Seq

1 Introduction

Keyphrase Generation (KG) is a classic and challenging task in Natural Language Processing (NLP) that aims at predicting a set of keyphrases for the given document. As keyphrases contain the core idea of the document, it is useful in various downstream tasks such as information retrieval [13], document clustering [12], opinion mining [2,28], and text summarization [27]. In most cases, keyphrases can be found in the given document, which means it is a substring of the source text (aka *present keyphrases*). In other challenging cases, some

Y. Xu and Y. Luo—Contribute equally.

L. Wang et al. (Eds.): NLPCC 2021, LNAI 13029, pp. 86–97, 2021.
https://doi.org/10.1007/978-3-030-88483-3_7

Table 1. Sample document with keyphrase labels. In this example, there are some *present keyphrases* (in red color) and some *absent keyphrases* (in blue color).

Document: Rental software valuation in it investment decisions. The growth of application service providers (asps) is very rapid, leading to a number of options to organizations interested in developing new information technology services. ... Some of the more common capital budgeting models may not be appropriate in this volatile marketplace. However, option models allow for many of the quirks to be considered. ...
Keyphrase labels: (present keyphrases) application service providers; options; capital budgeting; (absent keyphrases) information technology investment; stochastic processes; risk analysis

keyphrases may not appear in the given document (aka *absent keyphrases*). An examplar document along with the reference keyphrases can be found in Table 1.

In recent years, Transformer [25] has become prevailing in various NLP tasks, such as machine translation [25], summarization [19], language modeling [6], and pre-trained models [7]. Compared to recurrent-based models, such as GRU [5] and LSTM [10], the Transformer module introduces fully-connected self-attention blocks to incorporate global contextual information and capture the long-range semantic dependency [18]. Meanwhile, Transformer has a better parallelism ability than RNN due to the module structure.

However, the vanilla Transformer has a poor performance in the KG task, and can hardly beat the approaches based on RNN [4,21]. We carefully tuned Transformer models on the KG task, and our observations have confirmed this conclusion. Since Transformer-based models have been proved their success in many other NLP tasks in recent years [18], it is worth exploring why Transformer performs poorly in the KG task.

Considering the example shown in Table 1, the keyphrase "*application service providers*" is mentioned in the sentence "*The growth of application service providers...*", which may not have a close relationship with the other sentences in the example document. The RNN captures features by modeling the text sequence [10], and the Transformer focus more on the global context modeling [25]. Based on the module architecture and previous observations, we state the **Information Sparsity Hypothesis**: *In KG tasks, uninformative content abounds in documents while salient information is diluted in the global context.*

For a vanilla Transformer encoder, every two tokens from different segments can be attended by each other, while it may probably bring uninformative content to each other. Based on the property of Transformer and our hypothesis, we introduce two adaptations towards the objective Transformer encoder.

The first adaptation is to reduce the context that can be attended by the self-attention mechanism. To analyze the influence of the attention mechanism, we firstly manually separate the input sequence into several segments while tokens from different segments cannot be attended by each other. Experiments show that this modification leads to a slight improvement. Thus, we further add some constraints on the attention mechanism by sparsing the attention mask matrix.

The second adaptation is introducing direction information. Typically, previous models usually use BiRNN as encoder and decoder, distinguishing which side the context information comes from. But the vanilla Transformer only uses a position embedding to distinguish tokens from different positions, which does not significantly contain direction information. Thus, Transformer-based model is usually unaware of the distance between different tokens. However, it is more important to model the relationship between tokens and their neighbors in the KG task, which mainly decides they should be or should not be part of the keyphrase. Following [24] and [6], we employ relative multi-head attention to the KG task. Our experiment shows that relative multi-head attention can bring improvement than multi-head attention.

Our main observations are summarized in: (1) The most informative content usually comes from its neighborhood tokens, which provide empirical evidence to our proposed *Information Sparsity Hypothesis.* (2) Due to this phenomenon, we confirm that Transformer performs poorly in KG tasks because the fully-connected self-attention mechanism on vanilla may overlook the local context. (3) We adapt Transformer module by reducing attention to uninformative content and adding direction information. Our proposed model achieves state-of-the-art results on present keyphrase generation.

2 Methodology

2.1 Reduce Attention to Uninformative Content

As mentioned in Sect. 1, we try to use Information Sparsity Hypothesis to explain why Transformer does not work on KG tasks. In RNN models, the attention mechanism mainly focuses on target keyphrase tokens. In Transformer models, the attention distribution is more sparse and tend to receive more information from different tokens. Hence, reducing attention to uninformative content is a considerable solution. In this section, we will introduce two methods to constraint the attention mechanism: mandatory prohibition, and soft prohibition.

Segmentation. Considering a an input sequence $\mathbf{x} = [x_1, x_2, \dots, x_{l_\mathbf{x}}]$, token x_1 and $x_{l_\mathbf{x}}$ can be attended by each other due to the self-attention mechanism. For convenience, firstly we manually separate the input sequence $\mathbf{x}$ into several segments, for example, $\mathbf{x} = [\hat{x}_1, \dots, \hat{x}_K]$:

$$\hat{x}_i = [x_{N\cdot(i-1)+1}, x_{N\cdot(i-1)+2}, \dots, x_{N\cdot i}], \tag{1}$$

$$\hat{\mathbf{h}}_i = \text{Transformer}(\hat{x}_i), \tag{2}$$

$$\mathbf{H} = \text{concat}[\hat{\mathbf{h}}_1; \hat{\mathbf{h}}_2; \dots; \hat{\mathbf{h}}_K], \tag{3}$$

where K denotes the number of segments, and N denotes the sequence length of each segment. The attention between tokens from different segments are prohibit.

Sparsing the Matrix of Attention Mask. In addition to the approaches mentioned above, another efficient way to reduce attention to uninformative content is by adding constraints to the attention mask. If the input sequence $\mathbf{x}$ has a sequence length of n, the attention mask matrix $\mathbf{M} \in \{0, 1\}^{n \times n}$ is:

$$\mathbf{M}_{i,j} = \begin{cases} 0, \text{ if } x_i \text{ does not attends } x_j \\ 1, \text{ if } x_i \text{ attends } x_j \end{cases}, \quad (4)$$

In general, the attention mechanism is fully-connected, which means any two tokens x_i and x_j can be attended by each other. Thus, the attention mask in the vanilla Transformer encoder is a matrix filled with one: $\mathbf{M}^{full} = 1^{n \times n}$.

However, based on our Information Sparsity Hypothesis, a given token x_i does not need to attend the uninformative tokens. Typically, the informative tokens usually come from three parts, including lead tokens that contain the core topic of the whole document, neighborhood tokens that contain the contextual information, and tokens with high attention scores that is highly revalent to x_i.

For attending lead tokens, the lead g tokens are selected to the interaction, so that the attention mask matrix can be formula as:

$$\mathbf{M}_{i,j}^{lead} = \begin{cases} 1, \text{ if } i \leq g \text{ or } j \leq g \\ 0, \text{ if } i > g \text{ and } j > g \end{cases}, \quad (5)$$

For neighborhood attention, each token x_i can mostly attend w tokens, which includes its previous $\lfloor \frac{w}{2} \rfloor$ tokens and its next $\lfloor \frac{w}{2} \rfloor$ tokens:

$$\mathbf{M}_{i,j}^{neigh} = \begin{cases} 1, \text{ if } |i-j| \leq \lfloor \frac{w}{2} \rfloor \\ 0, \text{ if } |i-j| > \lfloor \frac{w}{2} \rfloor \end{cases}, \quad (6)$$

For highly relationship attention, we firstly compute the attention score matrix and then select top k tokens to attend:

$$\mathbf{M}_{i}^{topk} \in [0,1]^n, \text{ where } \sum_{j=1}^{n} \mathbf{M}_{i,j}^{topk} = k \quad (7)$$

Similar to BigBird [32], the attention mask matrices can be mixed up by:

$$\bar{\mathbf{M}}_{i,j} = (\alpha \mathbf{M}_{i,j}^{lead}) \circ (\beta \mathbf{M}_{i,j}^{neigh}) \circ (\gamma \mathbf{M}_{i,j}^{topk}). \quad (8)$$

where α, β, γ are hyperparameters in $\{0, 1\}$ to control which attention mask will be used and which will not, and $\circ$ means element-wised "or" operation.

2.2 Relative Multi-head Attention

Inspired by the success of reducing attention to uninformative content, we further consider what else properties the Transformer lacks compared to RNN-based models. One empirical observation is that Transformer is not sensitive to the direction information because Transformer is not easy to distinguish whether the context comes from the left side or the right side. Hence, following the successful experience in Named Entity Recognition (NER) task [30], we introduce relative multi-head attention to improve our model. The relative multi-head attention

Table 2. Summary statistics of four scientific article benchmark.

Dataset	#Train	#Validation	#Test	#Avg. present	#Avg. absent
Inspec	–	1,500	500	7.64	2.10
Krapivin	–	1,903	400	3.27	2.57
SemEval	–	144	100	6.28	8.12
KP20k	509,818	20,000	20,000	3.32	1.93

uses a relative positional encoder $\mathbf{R}_{i-j}$ to replace the position encoding $\mathbf{P}_j$ in the vanilla self-attention mechanism [6,30], which can be formula as:

$$\mathbf{A}_{i,j}^{abs} = \mathbf{Q}_i\mathbf{K}_j^T = \mathbf{H}_i\mathbf{W}_q(\mathbf{H}_j\mathbf{W}_k)^T + \mathbf{H}_i\mathbf{W}_q(\mathbf{R}_{i-j}\mathbf{W}_k)^T + \mathbf{u}(\mathbf{H}_j\mathbf{W}_k)^T + \mathbf{v}(\mathbf{R}_{i-j}\mathbf{W}_k)^T. \quad (9)$$

where $\mathbf{R}$ is a not learnable sinusoid encoding matrix [25], $\mathbf{u}$ and $\mathbf{v}$ are two learnable parameters used to substitute the position-based query terms.

3 Experiment Settings

3.1 Notations and Problem Definition

In this paper, we use bold lowercase and uppercase letters to denote vectors and matrices, respectively. We use calligraphy letters to indicate the sets and $\mathbf{W}$ to represent a parameter matrix. Given an input document $\mathbf{x}$, the KG task aims to generate a set of ground-truth keyphrases $\mathcal{Y} = \{\mathbf{y}^1, \mathbf{y}^2, \ldots, \mathbf{y}^{|\mathcal{Y}|}\}$, where $|\mathcal{Y}|$ indicates the keyphrase number of $\mathbf{x}$. Both the source document $\mathbf{x} = [x_1, \ldots, x_{l_\mathbf{x}}]$ and each target keyphrase $\mathbf{y}^i = [y_1^i, \ldots, y_{l_{\mathbf{y}_i}}^i]$ are word sequences, where $l_\mathbf{x}$ and $l_{\mathbf{y}_i}$ indicates the word numbers of $\mathbf{x}$ and $\mathbf{y}^i$ respectively.

3.2 Datasets

To verify and analyze the information sparsity hypothesis, we conduct experiments on four public datasets: **Inspec** [11], **Krapivin** [16], **SemEval-2010** [14], **KP20k** [21]. Documents from these four benchmarks are all scientific articles. In Table 2, we describe the detailed statistics of each dataset. Following [4,31], an article with title and abstract are included as the source data and a set of Keyphrases are included as the target data.

3.3 Evaluation Metrics

We use $F_1@5$ and $F_1@M$ as evaluation metric, which is the same as [4]. As for $F_1@5$, we cut off the top 5 keyphrases for calculating. When the number of predicted keyphrases is less than 5, we randomly append incorrect keyphrases until it obtains 5 keyphrases. Unlike $F_1@5$ that using a fixed number for predictions, $F_1@M$ [31] compares all predictions with target keyphrases. Therefore, the effect of the evaluation metric caused by the different number of predictions should be considered. In this paper, we use the macro $F_1@M$ and $F_1@5$ scores to report results. Before calculating the scores, we should stem all keyphrases and remove all duplicated keyphrases.

3.4 Implementation Details

Following previous work [4,21,31], for data preprocessing, we lowercase the characters, tokenize the sequences and replace the digits into "⟨digit⟩" token. For the training stage, we use the source code from [4]. When training, we sort the present keyphrase targets in the order of their first occurrences. The absent keyphrase targets are then appended at the end of ordered present targets. We use a preprocessed vocabulary with 50,000 tokens, which is shared between the encoder and decoder. For RNN models, the dimension of encoder and decoder are kept as 300, and we use a 2-layer encoder and 1-layer decoder. For Transformer models, we carefully tuned the hyperparameters. All learnable parameters are randomly initialized by a uniform distribution in $[-0.1, 0.1]$ before the training stage. Due to the average length of about 180 tokens, it requires high GPU memory in Transformer models. Thus we set the batch size as 16 or 24. Following [4], we set the max gradient norm as 1.0 and the initial learning rate as $1e-3$ for RNN models and $1e-4$ for Transformer models, and remove the dropout. During the training stage, we use Adam as the optimizer. We save checkpoints every 5,000 steps for evaluation.

4 Results and Discussions

4.1 Applying Transformer to Keyphrase Generation

Table 3. Present keyphrase prediction results of different standard encoders.

Model	Inspec		Krapivin		SemEval		KP20k	
	$F_1@M$	$F_1@5$	$F_1@M$	$F_1@5$	$F_1@M$	$F_1@5$	$F_1@M$	$F_1@5$
ExHiRD-h [4]	0.291	0.253	0.347	0.286	0.335	0.284	0.374	0.311
ExHiRD-h (Our implementation)								
w/BiGRU encoder	0.288	0.248	0.344	0.281	0.326	0.274	0.374	0.311
w/BiLSTM encoder	0.285	0.245	0.346	0.278	0.328	0.278	0.374	0.311
w/CNN encoder	0.284	0.247	0.347	0.277	0.324	0.269	0.371	0.306
w/Transformer encoder	0.267	0.240	0.325	0.271	0.304	0.258	0.355	0.299

Typically, previous model applied BiGRU on both encoder and decoders. Thus, we firstly replace the BiGRU encoder with other widely-used components: BiLSTM, CNN, and Transformer. The results are shown in Table 3. Our observation confirms that Transformer encoder can hardly beat the recurrent-based models, while CNN encoder has comparable performance.

In this experiment, we note that the accurate of absent keyphrase generation is very poor. On KP20k, only about 1,000 absent keyphrases are correctly predicted, and the ground truth contains 38,531 keyphrases. The score on absent keyphrase prediction is not only inadequate but also in a high variance. We hypothesis that there are two main reasons: (1) the average number of keyphrases in each document is limited (mainly about 3 keyphrases), thus it is not easy to predict the correct results; (2) the evaluation metric is strict that only the accurate predictions are counted. Due to this reason, we mainly focus on present keyphrase generation in the following experiments.

4.2 Tuning Transformer Model

As shown in Sect. 4.1 and [4], Transformer performs poorly in the KG tasks. However, tuning Transformer with different hyperparameters affects the performance. Therefore, in this section, we will carefully fine-tune the Transformer model with different hyperparameters.

Table 4. Present keyphrase prediction results of different encoder-decoder pairs on KP20k. "#$C@M$" and "#$C@5$" indicates the number of correctly predictions in $F_1@M$ and $F_1@5$, respectively. "#Avg. Len" is the average number of predictions to present keyphrases. "TF" is standard Transformer module.

Encoder	Decoder	$F_1@M$	$F_1@5$	#$C@M$	#$C@5$	#Avg. Len
BiGRU	BiGRU	0.374	0.311	25,275	24,435	3.88
BiGRU	TF	0.372	0.298	24,252	23,603	3.84
TF	BiGRU	0.363	0.290	23,337	22,933	3.81
TF	TF	0.359	0.290	22,912	23,314	3.69

Effective of Different Encoder-Decoder Pairs. As shown in Table 4, we firstly explore the effectiveness of different encoder-decoder pairs in this section. According to our experiment, the $F_1@M$ score drops significantly if we use a standard Transformer encoder to replace the BiGRU encoder. Meanwhile, our experiment shows that the $F_1@M$ score will also obtain a slight change when we use a standard Transformer decoder to replace the BiGRU decoder.

In our experiments, we notice that the average length of target present keyphrases is 3.32, but models usually predict more than 3.5 keyphrases on average. We will randomly append keyphrases until there are five predicted keyphrases when we compute the $F_1@5$ score. Due to this reason, predicting more keyphrases tend to improve the $F_1@5$ score because a maybe-incorrect prediction is better than an absolutely-incorrect prediction. In contrast, when computing the $F_1@M$ score, a maybe-incorrect keyphrase is worse than predicting nothing because the overall accuracy is at a low level from about 0.3 to about

0.4. The average number of predicted keyphrases is also an important indirect factor for evaluating the ability of keyphrase generation. Hence, we will have a joint consideration between the $F_1@5$ score and the $F_1@M$ score for comparison in the following experiments.

Table 5. Present keyphrase results of different hyperparameters on KP20k.

$<\eta, N, A, H>$	$F_1@M$	$F_1@5$	#Avg. Len
$<1e-3, 2, 6, 300>$	0.348	0.278	3.57
$<1e-4, 2, 6, 300>$	0.359	0.290	3.69
$<1e-4, 3, 6, 300>$	0.359	0.284	3.86
$<1e-4, 3, 8, 512>$	0.362	0.299	3.89
$<1e-4, 4, 6, 300>$	0.363	0.296	3.83
$<1e-4, 4, 8, 512>$	0.364	0.300	3.96
$<1e-4, 6, 12, 768>$	0.364	0.304	4.15
$<1e-4, 12, 12, 768>$	0.361	0.293	3.77

Tuning Transformer Model with Different Hyperparameters. As mentioned in Sect. 4.2, Transformer encoder performs poorly in KP20k with general hyperparameters. Thus, we will carefully tune the Transformer models with different hyperparameters. The results are shown in Table 5. We mainly tuned four hyperparameters: learning rate η, number of layers N, number of attention heads A, and hidden size H. Though the hidden size of 300 is usually set in RNN-based models, it is not suitable in Transformer-based models. Meanwhile, more attention heads and layers are required. However, when we stack more layers (e.g., $N = 12$) into our model, we find that: (1) The training stage has become more difficult that obtains a slower convergence; (2) The validation score as well as test evaluation score are lower than a 6-layer or a 4-layer Transformer model. Thus, we set $N = 4$ or $N = 6$ in the following experiments.

4.3 Adapting Transformer to Keyphrase Generation

As shown in Sect. 4.2, we confirm that Transformer performs poorly in KG tasks. Therefore, we apply the methods shown in Sect. 2.

Reduce Attention to Uninformative Content. As shown in Table 6, chunking brings slight improvement on the KG task. It is worth noting that $N = 250$ and $N = 500$ have a similar result to not applying the chunking because the average length of the training set is about 180 tokens.

Table 6. Effects of segment length N on present keyphrase prediction on KP20k. $N = 0$ means do not apply segmentation.

N	0	25	50	100	250	500
$F_1@M$	0.364	0.366	0.368	0.362	0.363	0.362
$F_1@5$	0.300	0.303	0.304	0.297	0.298	0.297
#Avg. Len	3.96	4.05	4.00	3.96	3.92	3.92

Table 7. Statistics of present keyphrases predictions with different attention mask matrices. α, β, and γ indicates hyperparameters defined in Eq. (8). "baseline" indicates standard Transformer with attention mask $\mathbf{M}^{full}$.

# Layer	α	β	γ	$F_1@M$	$F_1@5$	$C@M$	$C@5$	#Avg. Len
4	Baseline			0.364	0.300	24,154	23,827	3.96
4	1	1	1	0.372	0.304	24,812	24,042	3.85
			0	0.367	0.302	24,905	23,929	3.95
		0		0.363	0.298	25,051	23,662	4.05
	0			0.370	0.298	24,315	23,632	3.73
6	1	1	1	0.372	0.304	24,562	23,979	3.82
			0	0.364	0.306	25,112	24,208	4.13
		0		0.366	0.296	24,177	23,347	3.90
	0			0.368	0.302	24,468	23,770	3.87

Typically, the mandatory prohibition promotes the encoder to focus more on modeling tokens within the same segment while forcibly prohibiting attending tokens from other segments containing uninformative noises. In contrast to this, a CNN model mainly captures features from the local context due to spatial convolutional mechanism, while a RNN model models the input sequentially. Moreover, previous work [8,15] has shown that in RNN models each token can only perceive approximately 50 to 100 tokens before it. In summary, this observation also provides empirical evidence to the information sparsity hypothesis.

As shown in Table 7, sparsing the attention mask matrix can boost the performance compared to a standard Transformer with a $\mathbf{M}^{full}$ attention mask. Compared to $\mathbf{M}^{full}$, attention with mask matrix $\bar{\mathbf{M}}$ can not only predict more correct keyphrases but also predict fewer keyphrases on average.

According to our ablation experiments, we found that the evaluation score drops the least when $\alpha = 0$, leading tokens containing the least important information among three types of tokens. Meanwhile, prohibiting the neighborhood tokens will lead to a significant drop, which shows that the most informative content usually comes from neighborhood tokens. This observation provides empirical evidence that enhancing the ability to capture local information can make Transformer performs better on KG tasks.

Table 8. Comparison of present keyphrase prediction results. "SM" indicates sparsing the mask matrix of self-attention, and "RMHA" indicates using relative multi-head attention. We **bold** the best result.

Model	Inspec		Krapivin		SemEval		KP20k	
	$F_1@M$	$F_1@5$	$F_1@M$	$F_1@5$	$F_1@M$	$F_1@5$	$F_1@M$	$F_1@5$
catSeq [31]	0.276	0.233	0.344	0.269	0.313	0.262	0.368	0.295
ExHiRD [4]	0.291	0.253	0.347	**0.286**	0.335	0.284	0.374	0.311
ExHiRD with RNN (our implementation)	0.288	0.248	0.344	0.281	0.326	0.274	0.374	0.311
(Ours) ExHiRD with TF	0.278	0.232	0.329	0.272	0.310	0.258	0.364	0.300
+ SM only	0.280	0.235	0.334	0.275	0.319	0.266	0.372	0.304
+ RMHA only	0.289	0.244	0.336	0.277	0.325	0.278	0.372	0.313
+ SM + RMHA	**0.293**	**0.254**	**0.351**	**0.286**	**0.337**	**0.289**	**0.375**	**0.316**

Applying Relative Multi-head Attention. As shown in Table 8, applying relative multi-head attention can improve the performance of Transformer models. Meanwhile, it also shows that our model with a sparse self-attention mask matrix can be further improved by applying relative multi-head attention.

4.4 Observations and Findings

In summary of our experiment results, we have four main findings: (1) Our experiments have confirmed that the vanilla Transformer encoder-decoder model can hardly beat the recurrent-based models. (2) In KG tasks, the most informative content usually neither comes from leading tokens nor comes from tokens with the highest attention score, but comes from neighborhood tokens. (3) In KG tasks, direction information is also important, which is not significantly in the position embedding. We boost our model with relative multi-head attention. (4) Our experiments provide empirical evidence to our *information sparsity hypothesis*. Our hypothesis also encourages the adaptation of Transformer models to achieve SOTA results.

5 Related Work

Keyphrases are short phrases that contain the core information summarized from the given document. Traditional extractive approaches [22,29] aim to select salience phrase presents in a document. The existing methods generally adopt two steps. First, the identify keyphrase candidates are extracted by heuristic methods [17]. Afterward, the candidates are ranked by either unsupervised methods [26] or supervised learning techniques [11,23]. In order to generate both present and absent keyphrase for a document, [21] introduced CopyRNN, which consists of an attentional encoder-decoder model [1] and a copy mechanism [9]. CopyRNN and its variants use the beam search to generate a fixed-size keyphrase. [31] introduced a new Seq2Seq model that predicts multiple keyphrases for the given document, which enable a generative model to generate variable numbers of keyphrases. Lately, [3,20] proposed reinforcement learning based fine-tuning method for generating both sufficient and accurate keyphrases. Furthermore, [4] proposes an exclusive hierarchical decoding framework to avoid repeated keyphrases and enhance the diversity of the generated keyphrases.

6 Conclusion

In this paper, we confirm the poorly performance Transformer models have in KG tasks and seek to explore the reason. We state the Information Sparsity Hypothesis and conduct experiments to confirm our hypothesis. Based on our hypothesis, we adapt the Transformer model by sparsing the attention mask matrix and introducing relative multi-head attention, which achieves SOTA results on KG benchmarks. Ablation study also proves that in KG task the most informative content usually comes from neighborhood tokens. Our detailed observations also provide more hints for the follow-up researchers to understand the Transformer architecture and design more powerful models.

Acknowledgments. This work was supported by the National Key Research and Development Program of China (No. 2020AAA0106700) and National Natural Science Foundation of China (No. 62022027).

References

1. Bahdanau, D., Cho, K., Bengio, Y.: Neural machine translation by jointly learning to align and translate. In: ICLR (2015)
2. Berend, G.: Opinion expression mining by exploiting keyphrase extraction. In: IJCNLP, pp. 1162–1170, November 2011
3. Chan, H.P., Chen, W., Wang, L., King, I.: Neural keyphrase generation via reinforcement learning with adaptive rewards. In: ACL, pp. 2163–2174, July 2019
4. Chen, W., Chan, H.P., Li, P., King, I.: Exclusive hierarchical decoding for deep keyphrase generation. In: ACL, pp. 1095–1105, July 2020
5. Cho, K., et al.: Learning phrase representations using RNN encoder-decoder for statistical machine translation. In: EMNLP, pp. 1724–1734, October 2014
6. Dai, Z., Yang, Z., Yang, Y., Carbonell, J., Le, Q., Salakhutdinov, R.: Transformer-XL: attentive language models beyond a fixed-length context. In: ACL (2019)
7. Devlin, J., Chang, M.W., Lee, K., Toutanova, K.: BERT: pre-training of deep bidirectional transformers for language understanding. In: NAACL, June 2019
8. Domhan, T.: How much attention do you need? A granular analysis of neural machine translation architectures. In: ACL, pp. 1799–1808, July 2018
9. Gu, J., Lu, Z., Li, H., Li, V.O.: Incorporating copying mechanism in sequence-to-sequence learning. In: ACL, pp. 1631–1640, August 2016
10. Hochreiter, S., Schmidhuber, J.: Long short-term memory. Neural Comput. **9**, 1735–1780 (1997)
11. Hulth, A.: Improved automatic keyword extraction given more linguistic knowledge. In: EMNLP, pp. 216–223 (2003)
12. Hulth, A., Megyesi, B.B.: A study on automatically extracted keywords in text categorization. In: ACL, pp. 537–544. ACL-44 (2006)

13. Jones, S., Staveley, M.S.: Phrasier: a system for interactive document retrieval using keyphrases. In: SIGIR, pp. 160–167 (1999)
14. Kim, S.N., Medelyan, O., Kan, M.Y., Baldwin, T.: SemEval-2010 task 5: automatic keyphrase extraction from scientific articles. In: SemEval 2010, pp. 21–26 (2010)
15. Koehn, P., Knowles, R.: Six challenges for neural machine translation. In: Proceedings of the First Workshop on NMT, pp. 28–39, August 2017
16. Krapivin, M., Autaeu, A., Marchese, M.: Large dataset for keyphrases extraction. Technical report, University of Trento (2009)
17. Le, T.T.N., Nguyen, M.L., Shimazu, A.: Unsupervised keyphrase extraction: introducing new kinds of words to keyphrases. In: Kang, B.H., Bai, Q. (eds.) AI 2016: Advances in Artificial Intelligence, pp. 665–671 (2016)
18. Lin, T., Wang, Y., Liu, X., Qiu, X.: A survey of transformers. arXiv preprint arXiv:2106.04554 (2021)
19. Liu, Y., Lapata, M.: Text summarization with pretrained encoders. arXiv preprint arXiv:1908.08345 (2019)
20. Luo, Y., Xu, Y., Ye, J., Qiu, X., Zhang, Q.: Keyphrase generation with fine-grained evaluation-guided reinforcement learning. arXiv preprint arXiv:2104.08799 (2021)
21. Meng, R., Zhao, S., Han, S., He, D., Brusilovsky, P., Chi, Y.: Deep keyphrase generation. In: ACL, pp. 582–592, July 2017
22. Mihalcea, R., Tarau, P.: TextRank: bringing order into text. In: EMNLP (2004)
23. Nguyen, T.D., Kan, M.Y.: Keyphrase extraction in scientific publications. In: International Conference on Asian Digital Libraries, pp. 317–326 (2007)
24. Shaw, P., Uszkoreit, J., Vaswani, A.: Self-attention with relative position representations. In: Walker, M.A., Ji, H., Stent, A. (eds.) NAACL-HLT (2018)
25. Vaswani, A., et al.: Attention is all you need. In: NeurIPS (2017)
26. Wan, X., Xiao, J.: Single document keyphrase extraction using neighborhood knowledge. In: AAAI, pp. 855–860 (2008)
27. Wang, L., Cardie, C.: Domain-independent abstract generation for focused meeting summarization. In: ACL, pp. 1395–1405, August 2013
28. Wilson, T., Wiebe, J., Hoffmann, P.: Recognizing contextual polarity in phrase-level sentiment analysis. In: EMNLP, pp. 347–354, October 2005
29. Witten, I.H., Paynter, G.W., Frank, E., Gutwin, C., Nevill-Manning, C.G.: KEA: practical automatic keyphrase extraction. In: Proceedings of the Fourth ACM Conference on Digital Libraries, DL 1999, New York, NY, USA, pp. 254–255 (1999)
30. Yan, H., Deng, B., Li, X., Qiu, X.: TENER: adapting transformer encoder for name entity recognition. arXiv preprint arXiv:1911.04474 (2019)
31. Yuan, X., et al.: One size does not fit all: generating and evaluating variable number of keyphrases. In: ACL, pp. 7961–7975, July 2020
32. Zaheer, M., et al.: Big bird: transformers for longer sequences. arXiv preprint arXiv:2007.14062 (2020)

Prerequisite Learning with Pre-trained Language and Graph Embedding Models

Bangqi Li, Boci Peng, Yifeng Shao, and Zhichun Wang(✉)

School of Artificial Intelligence, Beijing Normal University, Beijing 100875, People's Republic of China
zcwang@bnu.edu.cn

Abstract. Prerequisite learning is to automatically identify prerequisite relations between concepts. This paper proposes a new prerequisite learning approach based on pre-trained language model and graph embedding model. In our approach, pre-trained language model BERT is fine-tuned to encode latent features from concept descriptions; graph embedding model Node2Vec is first pre-trained on citation graph of concepts, and then is fine-tuned for generating discriminative features for prerequisite learning. Two models are jointly optimized to get latent features containing both textual and structural information of concepts. Experiments on manually annotated datasets show that our proposed approach achieves better results than the state-of-the-art prerequisite learning approaches.

Keywords: Prerequisite learning · Pre-trained language model · Graph embedding

1 Introduction

Prerequisite chains identify learning orders of concepts or skills, which form the basis of AI-based educational applications, including intelligent tutoring systems, curriculum planning, learning materials generation and recommendation. Given two concepts A and B, if A is a prerequisite of B, then A should be learned before B. For example, the mathematical concept *Limit* is a prerequisite concept of *Derivative*. Without the knowledge of *Limit*, students find it difficult to understand the concept *Derivative*. Prerequisite chains can be created by domain experts, resources like TutorialBank [3], LectureBank [8] and MOOCCube [15] contain manually annotated prerequisite relations. However, manual annotation is expensive and inefficient, and the number of prerequisites in the above resources are limited.

In recent years, much research work has been done on automatic prerequisite learning. Clues in different kinds of learning materials have been explored to discover prerequisite relations between concepts, including Wikipedia, MOOCs, textbooks and scientific corpora. Typically, features of concept pairs are extracted from different resources, which are then fed into machine learning models to determine whether the concept pairs hold a prerequisite relation. For

L. Wang et al. (Eds.): NLPCC 2021, LNAI 13029, pp. 98–108, 2021.
https://doi.org/10.1007/978-3-030-88483-3_8

example, Pan et al. [12] defined 7 features of concept pairs from MOOCs, and investigated SVM, Naive Bayes, Logistic Regression, and Random Forest classifiers to predict the prerequisite relations. The quality of predicted prerequisite relations is highly dependent on the used learning resources and handcrafted features. Most recently, representation learning models and neural networks have been applied in the task of prerequisite learning. For example, Roy et al. [13] proposed an approach PREREQ based on Pairwise Latent Dirichlet Allocation model and Siamese network; it learns latent representations of concepts from MOOC playlists, which are then used for prerequisite prediction. Li et al. [7] proposed a relational-variational graph autoencoder R-VGAE for unsupervised prerequisite learning.

Although much progress has been achieved for the task of prerequisite learning, we found there are still several challenging problems:

- The quality of predictions needs further improvement for the usage in real-life applications. According to recent published work, the precisions of prerequisite predictions are below 70%. It shows that prerequisite learning is a difficult task, and we need more accurate models for solving it.
- Resources like transcripts in MOOCs, contents of textbooks, and dependencies between university courses are difficult to collect. The coverage of concepts in these resources is also limited.

To solve the above challenges, this paper proposes a new prerequisite learning approach which uses pre-trained language model and graph model to predict the dependencies between concepts. In detail, contributions of this work include:

- We propose a pre-trained language model and a graph embedding model in the task of prerequisite learning. Specifically, BERT [2] and Node2Vec [5] are used for handling textual and link information of concepts, respectively. Both models are pre-trained based on data from Wikipedia, which has high coverage of concepts and is easy to acquire.
- We propose a fine-tuning method which use the pre-trained BERT and Node2Vec models for the task of predicting prerequisite relations.
- Experiments on two datasets show that our proposed approach achieves better results than the state-of-the-art approaches.

The rest of this paper is organized as follows, Sect. 2 reviews some related work, Sect. 3 describes our proposed approach, Sect. 4 presents the evaluation results, Sect. 5 is the conclusion.

2 Related Work

The problem of mining prerequisite relations of concepts has been studied for years, and some approaches have been proposed for the task. Most approaches treat the prerequisite mining problem as a classification task, which is to classify a given concept pair into prerequisite and non-prerequisite.

Early work on prerequisite learning mainly explores Wikipedia to discover the dependencies between concepts. Partha and William [14] investigated reliable features to predict prerequisite relations from Wikipedia. They found that features based on hyperlinks, edits, and page content in Wikipedia can work well with standard machine learning models for the task. Liang et al. [9] propose a link-based metric called reference distance (RefD), which can effectively measure the prerequisite relations between concepts. To compute the RefD, page links in Wikipedia are used in their work. RefD models prerequisite relation by measuring how differently two concepts refer to each other. Liang et al. propose to use active learning for discovering concept prerequisite [10]. They compare many features for the task, including graph-based features and text-based ones.

With the development of MOOCs, some research tried to extract prerequisite relations from MOOCs. Liangming et al. proposed to mine prerequisites based on features extracted from MOOCs [12], including video references, sentence references and average position distance. They also use features computed from Wikipedia, including reference distance. ALSaad et al. [1] proposed BEM (Bridge Ensemble Measure) and GDM (Global Direction Measure), which are two measures based on lecture transcripts of MOOCs for prerequisite prediction. BEM captures concept dependencies based on lecture bridging concepts, sliding windows, and the first lecture indicator; GDM incorporates time directly by analyzing the concept time ordering both globally and within lectures. Roy et al. [13] proposed PREREQ, which uses playlists in MOOCs to infer prerequisite relations. PREREQ learns latent representations based on Pairwise Latent Dirichlet Allocation model and Siamese network.

There are also some work which explore textbooks, scientific corpus, and university course dependencies resources to discover prerequisite relations. Gordon et al. [4] proposes methods for discovering concepts and prerequisite relations among them from scientific text corpus. Concepts are first identified by using a LDA model, and prerequisite relations are discovered based on cross entropy or information flow. CPR-Recover [11] recovers concept prerequisite relations from course dependencies by using an optimization based framework. Information and dependencies of courses are collected from different universities, which are consumed by CPR-Recover to get the concept-level prerequisite relations. EMRCM [6] builds a concept map containing multiple relations between concepts including the prerequisite relations. Student question logs, textbooks, and Wikipedia are used by EMRCM to infer prerequisite relations. A number of features are defined and computed by EMRCM, which are finally fed to binary classifiers to discover concept relations.

3 Proposed Approach

This paper proposes a new prerequisite learning approach based on pre-trained language model and graph embedding model. There are two component modules in our approach, one is a text-based module based on BERT, and the other is a graph-based module based on Node2Vec. The text-based module takes concept descriptions as inputs, and generates textual latent features for prerequisite learning. The graph-based module takes a citation graph of concepts, and generates structural latent features for prerequisite learning. These two modules are jointly optimized in the model fine-tuning step. In the following, we introduce the proposed approach in detail.

3.1 Text-Based Module

We fine-tune BERT in the text-based module to extract textual latent features for concept pairs, which are then used for predicting prerequisite relations. The text-based module is shown in Fig. 1. Pairs of concepts are represented as connections of their definitions to form the inputs of fine-tuned BERT model. Concatenate two definitions s and e as a single input token sequence (s, e). A classification token [CLS] should be added as the first input token, and a separation token [SEP] is added after each of the two definitions. The two definitions are each split into tokens $\langle T_1, \ldots, T_m \rangle$ when used as input. Here m is the number of words in the definition.

The input word vector of BERT is the sum of three vectors, token embedding, segment embedding and position embedding. For given tokens, segment embedding indicates which definition the word belongs to, and position embedding are learned vectors. Segment embedding on the same side of the [CLS] token is the same. For the input sequence (s, e), its segment embedding can be expressed as $\langle E_{[CLS]}, E_s, \ldots, E_s, E^s_{[SEP]}, E_e, \ldots, E_e, E^e_{[SEP]} \rangle$. Position embedding is determined by the position of the word, words at different positions have different position embedding. It can be expressed as $\langle E_1, \ldots, E_n \rangle$, where n denotes the length of the whole input.

The vector generated $\langle I_{[CLS]}, I^s_1, \ldots, I^s_{m_1}, I^s_{[SEP]}, I^e_1, \ldots, I^e_{m2}, I^e_{[SEP]} \rangle$ is used as the input of the BERT model.

Let the output corresponding to [CLS] token in the BERT model be $t \in \mathbb{R}^H$, where H denotes the size of the hidden state in BERT model. The output corresponding to each token are $C_1, \ldots, C_n \in \mathbb{R}^H$. These output vectors form the result of the text feature extraction in text-based module, and are used in the Joint Learning module to train the BERT model here.

3.2 Graph-Based Module

The graph-based module is shown in Fig. 2. We construct a graph $G = \{V, E\}$, in which V is the set of all nodes, and E is the set of all links between them. Nodes consist of all concepts and inbound/outbound links of them. Inbound links can be acquired from *What links here* in Wikipedia tools, and outbound links are acquired from the page of concepts in Wikipedia. Node v_i links to node v_j means that v_i is an inbound link of v_j, and v_j is an outbound link of v_i. At the same time, we sample a sub-graph $G' = \{V', E'\}$, in which $V' \subset V, E' \subset E$ and V' is the smallest set including all concepts.

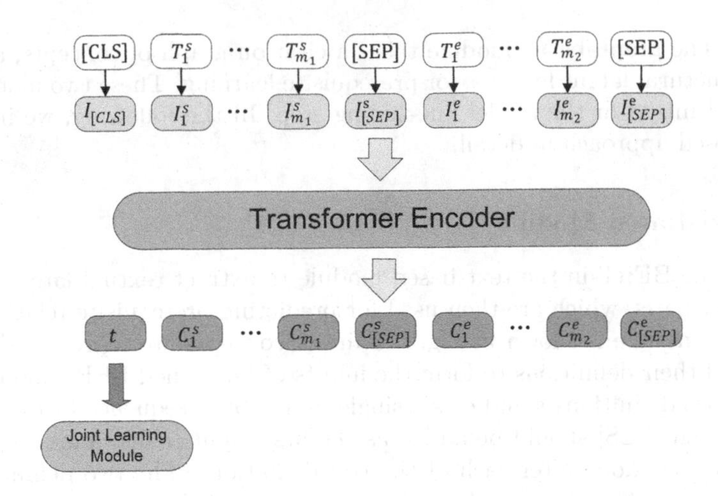

Fig. 1. Text-based module

We use Node2Vec for the graph-based module to generate continuous vector representations for the nodes based on the structure of the graph. First, a biased random walk is adopted to convert the network graph G' mentioned above into a node sequence $S = \langle v_1, v_2, \ldots, v_N \rangle$, where N is the number of nodes. Then S is fed into the Skip-Gram language model, in which the co-occurrence probability of the target node and the neighbor node in the sequence window is maximized. The random gradient is degraded for the update of the node representation. Each node v_i in graph G' can be represented by a 64-dimension vector $u^{(i)} = (u_1^{(i)}, u_2^{(i)}, \ldots, u_{64}^{(i)})^T$. And finally the 128-dimension vectors can be made by joining two 64-dimension vectors together.

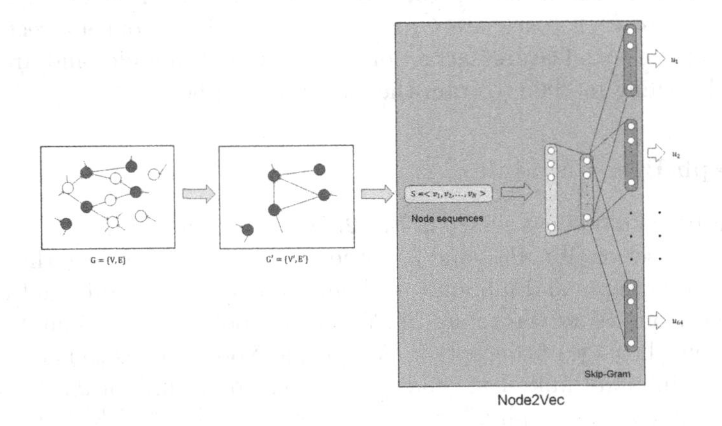

Fig. 2. Graph-based module

3.3 Joint Learning of Two Modules

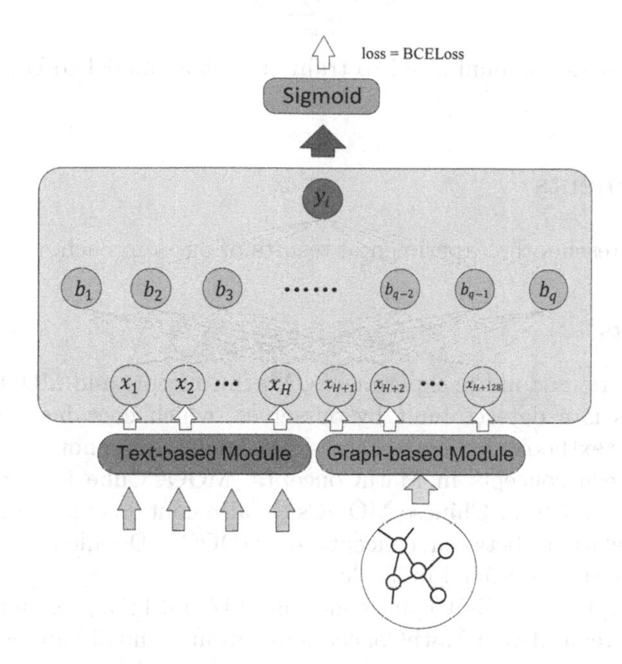

Fig. 3. Joint learning of text-based and graph-based modules

The text-based module and graph-based module are jointly optimized in the training process (as shown in Fig. 3). We denote the output of the text-based module as $t \in \mathbb{R}^H$, and the output of the graph-based Module is denoted as $u \in \mathbb{R}^{128}$. Then we concatenate t and u, which can be denoted as $X \in \mathbb{R}^{(H+128)}$. Then we feed X into a fully connected network, whose parameter is $W \in \mathbb{R}^{(H+128)}$ formed by weights. Let the activation function be sigmoid, the scoring function for X can be ι_X $=\text{sigmoid}(XW^T)$. ι_X is a real number which denotes the probability of X being 1, and let it be p. With a batch size of N, we can get a target vector $y = (y_1, y_2, \ldots, y_N)$ from the labeled dataset, where y_i denotes the label of the $i - th$ data. If there is a prerequisite relationship, y_i should be 1, otherwise it should be 0. We build an output vector $x = (p_1, p_2, \ldots, p_i)$, where p_i denotes the output probability of the $i - th$ data being 1. We compute the Binary Cross Entropy Loss ℓ as follows:

$$\ell(x, y) = (l_1, l_2, \ldots, l_N)^T \tag{1}$$

$$l_k = -(y_k \cdot \log p_k + (1 - y_k) \cdot \log(1 - p_k)) \tag{2}$$

The average of all the loss function values is used as the overall loss function value L. With n denotes the number of data, the overall loss can be computed as follows:

$$L = \frac{1}{n}\sum_{k=1}^{n} l_k \quad (3)$$

The loss function is minimized to train the whole model to get a promising result.

4 Experiments

This section presents the experimental results of our approach.

4.1 Datasets

Two datasets are used in the experiments, MathConcepts and MOOCCube [15]. MathConcepts is a dataset built by ourselves, which contains concepts from mathematical textbooks for high schools. We manually annotated prerequisite relations between concepts in MathConcepts. MOOCCube is a recently published dataset built from Chinese MOOCs, it also contains manually annotated prerequisite relations between concepts in MOOCs. Detailed information on these two datasets is outlined in Table 1.

MathConcepts and MOOCCube contains 1147 and 1153 prerequisites respectively. The prerequisites in MathConcepts are from manually annotated prerequisite relationships of mathematical knowledge labeled by professionals, while those in MOOCCube are from the official dataset. Their definitions are acquired through Wikipedia.

We have 455 concepts for MathConcepts and 680 concepts for MOOCCube. For those concepts, we get their link relationships from Wikipedia to form a graph. The graph generated by MathConcepts has 455 nodes with 396,469 links, and the graph generated by MOOCCube has 680 nodes with 169,773 links.

For both datasets, the ratio between training set, test set, and evaluation set is 5:3:2.

Table 1. Details of the datasets

	MathConcepts	MOOCCube
#Concepts	455	680
#Prerequisites	1,147	1,153
#Prerequisites for Train	573	577
#Prerequisites for Test	344	346
#Prerequisites for Evaluate	230	230
#Links in the graph	396,469	169,773

4.2 Evaluation Settings

We implement our approach by using PyTorch, and run experiments on a workstation with Intel Xeon 2.30 GHz CPU, an NVIDIA Tesla P4 GPU with 8 GB memory. We use Precision, Recall and F1-score as the evaluation metrics, which are popular and widely used in the evaluation of classification models. Precision measures the percentage of true positive samples to the total predicted positive samples, calculated as follows:

$$\text{Precision} = \frac{\text{TP}}{\text{TP} + \text{FP}} \tag{4}$$

Recall is the percentage of true positive samples to the all samples in actual positive samples.

$$\text{Recall} = \frac{\text{TP}}{\text{TP} + \text{FN}} \tag{5}$$

F1-score is the weighted average of Precision and Recall.

$$\text{F1-score} = 2 \cdot \frac{\text{Precision} \cdot \text{Recall}}{\text{Precision} + \text{Recall}} \tag{6}$$

The higher these values are, the better the performance is. We consider that the learning rate in BERT is 0.00001, and the learning rate in the Fully Connected Network is 0.001. The threshold for classifying concept pairs in our approach is among $\{0.3, 0.4, 0.5, 0.6, 0.7\}$. Best configurations of the model in our approach are selected based on all of the Precision, Recall, and F1-score.

We perform several experiments to compare our approach with other prerequisite relation prediction methods, including MOOC-LR [12], MOOC-XG [12], PREREQ [13], PREREQ-S, PCNN, and PRNN. The latter three methods are designed in the paper of MOOCCube [15].

4.3 Results

Contributions of Component Models. In order to evaluate the effectiveness of each component module in our approach, we designed experiments on the MathConcept dataset using different modules. Table 2 outlines the results of BERT, $\text{BERT}_{fine-tuned}$, Node2Vec and our approach. Results of BERT are obtained by using pre-trained BERT as the encoder of descriptions of concept pairs, without fine-tuning process. Results of $\text{BERT}_{fine-tuned}$ are obtained by fine-tuning BERT for the task of prerequisite learning. Results of Node2Vec are obtained by only using latent representations of graph embedding for prerequisite prediction. It shows that two component modules of our approach are both effective for improving the performance of prerequisite learning. $\text{BERT}_{fine-tuned}$ performs better than pre-trained BERT. There are 10.4%, 36.2%, 25.6% improvements in Precision, Recall and F1-score after fine-tuning the pre-trained language model. Our proposed approach obtains 3.1%, 0.9% improvements in Precision, Recall over $\text{BERT}_{fine-tuned}$, when learning $\text{BERT}_{fine-tuned}$ and Node2Vec jointly.

Table 2. Results of ablation experiments

	Precision	Recall	F1-score
BERT	0.788	0.541	0.641
$BERT_{fine-tuned}$	0.892	0.903	0.897
Node2Vec	0.535	0.812	0.645
The proposed	**0.923**	**0.912**	**0.917**

Overall Comparisons. Table 3 shows the results of our approach and the compared ones on MOOCCube dataset. It shows that our approach achieves promising improvements compared with previous approaches. Our approach outperforms all the compared approaches on MOOCCube in terms of Precision, Recall and F1-score. Compared with PRNN, which is the best among the other six approaches in terms of Precision, our approach gets an improvement of 9.7%. PREREQ performs the best on Recall among the other six approaches, and our approach outperforms it by 0.9% on Recall. Our approach gets better results of F1-score than PREREQ-S, which performs best on F1-score among the other six approaches.

Table 3. Results of all approaches on MOOCCube

	Precision	Recall	F1-score
MOOC-LR	0.667	0.479	0.565
MOOC-XG	0.607	0.507	0.552
PREREQ	0.606	0.755	0.672
PREREQ-S	0.651	0.730	0.688
PCNN	0.629	0.636	0.630
PRNN	0.681	0.668	0.659
The proposed	**0.778**	**0.764**	**0.771**

5 Conclusion

In this paper, we propose a new prerequisite learning approach based on pre-trained language and graph embedding models. To fully utilize textual and structural information of concepts, our approach jointly learns two models to generate latent features of concepts for prerequisite prediction. Specifically, pre-trained BERT and Node2Vec are fine-tuned for prerequisite learning. Experiments on manually annotated datasets show that our approach can achieve better results than the state-of-the-art prerequisite learning approaches.

References

1. ALSaad, F., Boughoula, A., Geigle, C., Sundaram, H., Zhai, C.: Mining MOOC lecture transcripts to construct concept dependency graphs. International Educational Data Mining Society (2018)
2. Devlin, J., Chang, M.W., Lee, K., Toutanova, K.: BERT: pre-training of deep bidirectional transformers for language understanding. In: Proceedings of the 2019 Conference of the North American Chapter of the Association for Computational Linguistics: Human Language Technologies, Volume 1 (Long and Short Papers), Minneapolis, Minnesota, pp. 4171–4186. Association for Computational Linguistics (Junuary 2019). https://doi.org/10.18653/v1/N19-1423, https://www.aclweb.org/anthology/N19-1423
3. Fabbri, A., et al.: TutorialBank: a manually-collected corpus for prerequisite chains, survey extraction and resource recommendation. In: Proceedings of the 56th Annual Meeting of the Association for Computational Linguistics (Volume 1: Long Papers), Melbourne, Australia, pp. 611–620. Association for Computational Linguistics (July 2018). https://doi.org/10.18653/v1/P18-1057, https://www.aclweb.org/anthology/P18-1057
4. Gordon, J., Zhu, L., Galstyan, A., Natarajan, P., Burns, G.: Modeling concept dependencies in a scientific corpus. In: Proceedings of the 54th Annual Meeting of the Association for Computational Linguistics (Volume 1: Long Papers), Berlin, Germany, pp. 866–875. Association for Computational Linguistics (August 2016). https://doi.org/10.18653/v1/P16-1082, https://www.aclweb.org/anthology/P16-1082
5. Grover, A., Leskovec, J.: Node2vec: scalable feature learning for networks. In: Proceedings of the 22nd ACM SIGKDD International Conference on Knowledge Discovery and Data Mining, KDD 2016, pp. 855–864. Association for Computing Machinery, New York (2016). https://doi.org/10.1145/2939672.2939754, https://doi.org/10.1145/2939672.2939754
6. Huang, X., et al.: Constructing educational concept maps with multiple relationships from multi-source data. In: 2019 IEEE International Conference on Data Mining (ICDM), pp. 1108–1113 (2019). https://doi.org/10.1109/ICDM.2019.00132
7. Li, I., Fabbri, A., Hingmire, S., Radev, D.: R-VGAE: relational-variational graph autoencoder for unsupervised prerequisite chain learning. In: Proceedings of the 28th International Conference on Computational Linguistics, pp. 1147–1157. International Committee on Computational Linguistics (December 2020). https://doi.org/10.18653/v1/2020.coling-main.99, https://www.aclweb.org/anthology/2020.coling-main.99
8. Li, I., Fabbri, A.R., Tung, R.R., Radev, D.R.: What should i learn first: introducing lecturebank for NLP education and prerequisite chain learning. In: Proceedings of the AAAI Conference on Artificial Intelligence, vol. 33, no. 01, pp. 6674–6681 (2019)
9. Liang, C., Wu, Z., Huang, W., Giles, C.L.: Measuring prerequisite relations among concepts. In: Proceedings of the 2015 Conference on Empirical Methods in Natural Language Processing, Lisbon, Portugal, pp. 1668–1674. Association for Computational Linguistics (September 2015). https://doi.org/10.18653/v1/D15-1193, https://www.aclweb.org/anthology/D15-1193
10. Liang, C., Ye, J., Wang, S., Pursel, B., Giles, C.L.: Investigating active learning for concept prerequisite learning (2018). https://www.aaai.org/ocs/index.php/AAAI/AAAI18/paper/view/17265

11. Liang, C., Ye, J., Wu, Z., Pursel, B., Giles, C.: Recovering concept prerequisite relations from university course dependencies. Proceedings of the AAAI Conference on Artif. Intell. **31**(1) (2017). https://ojs.aaai.org/index.php/AAAI/article/view/10550
12. Pan, L., Li, C., Li, J., Tang, J.: Prerequisite relation learning for concepts in MOOCs. In: Proceedings of the 55th Annual Meeting of the Association for Computational Linguistics (Volume 1: Long Papers), Vancouver, Canada, pp. 1447–1456. Association for Computational Linguistics (July 2017). https://doi.org/10.18653/v1/P17-1133, https://www.aclweb.org/anthology/P17-1133
13. Roy, S., Madhyastha, M., Lawrence, S., Rajan, V.: Inferring concept prerequisite relations from online educational resources. In: Proceedings of the AAAI Conference on Artificial Intelligence, vol. 33, no. 01, pp. 9589–9594 (July 2019)
14. Talukdar, P., Cohen, W.: Crowdsourced comprehension: predicting prerequisite structure in Wikipedia. In: Proceedings of the 7th Workshop on Building Educational Applications Using NLP, Montréal, Canada, pp. 307–315. Association for Computational Linguistics (June 2012). https://www.aclweb.org/anthology/W12-2037
15. Yu, J., et al.: MOOCCube: a large-scale data repository for NLP applications in MOOCs. In: Proceedings of the 58th Annual Meeting of the Association for Computational Linguistics, pp. 3135–3142. Association for Computational Linguistics (July 2020). https://doi.org/10.18653/v1/2020.acl-main.285, https://www.aclweb.org/anthology/2020.acl-main.285

Summarization and Generation

Variational Autoencoder with Interactive Attention for Affective Text Generation

Ruijun Chen, Jin Wang(✉), and Xuejie Zhang

School of Information Science and Engineering, Yunnan University, Kunming, China
wangjin@ynu.edu.cn

Abstract. Human language has the ability to convey the speaker's emotions, such as happiness, sadness, or anger. Existing text generation methods mainly focus on the sequence-to-sequence (Seq2Seq) model that applied an encoder to transform the input text into latent representation and a decoder to generate texts from the latent representation. To control the sentiment of the generated text, these models usually concatenate a disentangled feature into the latent representation. However, such a method is only suitable for short texts, since the sentiment information may gradually dissipate as the text becomes longer. To address this issue, a variational autoencoder with interactive variation attention was proposed in this study. Unlike the previous method of directly connecting sentiment information with the latent variables to control text generation, the proposed model adds the sentiment information into variational attention with a dynamic update mechanism. At each timestep, the model leverage both the variational attention and hidden representation to decode and predict the target word and then uses the generated results to update the emotional information in attention. It can keep track of the attention history, which encourages the attention-based VAE to control better the sentiment and content of generating text. The empirical experiments were conducted using the SST dataset to evaluate the generation performance of the proposed model. The comparative results showed that the proposed method outperformed the other methods for affective text generation. In addition, it can still maintain accurate sentiment information and sentences smoothness even in the longer text.

Keywords: Affective text generation · Variational autoencoder · Variational attention · Sentiment information

1 Introduction

Language is used to communicate the content of language itself and the sentiment associated with the content. For instance, people's utterances can convey that they are very sad, slightly angry, absolutely elated, etc. Computers need to be able to understand human emotions [19] and be able to generate natural sentences. However, existing methods for language generation are mainly focused on

L. Wang et al. (Eds.): NLPCC 2021, LNAI 13029, pp. 111–123, 2021.
https://doi.org/10.1007/978-3-030-88483-3_9

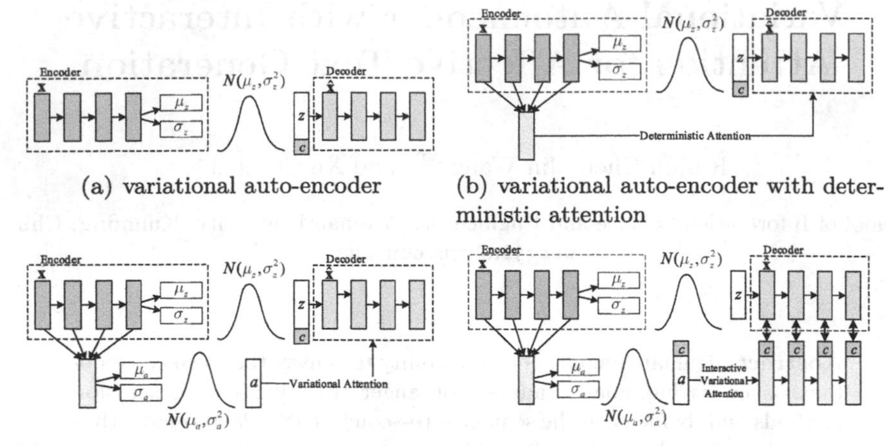

(a) variational auto-encoder

(b) variational auto-encoder with deterministic attention

(c) variational auto-encoder with variational attention

(d) variational auto-encoder with interactive variational attention

Fig. 1. Different variational auto-encoder for affective text generation.

contextual information while affective information is ignored. Using computers for generating text with a specified emotion is a major challenge.

For text generation, recent studies recommended using the sequence-to-sequence model (Seq2Seq) [14], which consists of an encoder and a decoder, and it turns one sequence of words into another. The encoder uses a stack of several recurrent units, e.g., long short-term memory (LSTM) [4] or recurrent gated units [6] (GRU) cells, to convert the words' source sequence into a hidden representation in latent space. The decoder, on the other hand, converts the representation into the target sequence of words. Similarly, transformer-based models, e.g., GPT-2 [11], can predict or generate the next tokens in a sequence based on preceding tokens. One issue with those Seq2Seq models is the lack of control in the content it generates. That is, an introduction or instruction can be fed into GPT-2; however, it is still hard to pick and arrange the texts from multiple outputs to something cohesive, not to mention a certain emotion.

The application of affective text generation required good control over the model output. The method proposed by Zhang et al. [18] can obtain sentiment representation from sentences. To further enhanced the performance of Seq2Seq, the models, such as the variational autoencoder (VAE) [8] and generative adversarial networks (GAN) [16,17] can be used for affective text generation. The VAE encodes the input as a distribution over the latent space, making it possible to generate diversified data from the latent space. The GAN uses adversarial learning between the generator and the discriminator to make the generated text more natural.

To control the VAE model to generate text with a specific sentiment, the most common method models the sentiment features as a structured or discrete representation, which was used to guide text generation. As shown in Fig. 1(a), these methods mainly focus on disentangling the sentiment and content representation in latent space, which is used as the initial hidden state to generate text by the decoder. Instead of only depending on the single vector from the encoder, the attention mechanism was introduced to focus on a few relevant words in the input to help the decoder predicting the target word, as shown in Fig. 1(b). Such an attention mechanism can achieve competitive results with far fewer parameters and training consumption. The superiority in efficiency mainly comes from the mechanism of dynamic alignment, helping the decoder to get more information from the input texts.

However, Bahuleyan et al. [1] reported that the traditional deterministic attention might bypass the latent space so that the generated texts were the same as the input texts. Since the variational latent space did not learn much, as long as the attention mechanism itself is too powerful to capture the information from the input texts. To address this issue, they proposed variational attention to enhance VAE, where the attention vector was also modeled as latent variables, as shown in Fig. 1(c), to balance the abilities of attention and latent variables and increase the diversity of generated texts.

The aforementioned methods for affective text generation have one thing in common that the desired sentiment information was directly concatenated into the latent variables to control the content of the generated text. However, these attention models are conducted only with the reading operation. This may lead the decoder to ignore past attention information. As the decoding progresses, the latent variables with sentiment information will dissipate. Consequently, the longer the training sentence used in VAE, the more uncontrolled the emotion and contents were generated.

In this paper, a variational auto-encoder with interactive attention was proposed to advance the idea of variational attention. Interactive attention was introduced in the decoding phase to keep track of the attention history, which encourages the attention-based VAE to better control the sentiment and content of the generated text. Instead of disentangling the sentiment information with latent variables, the proposed model added the sentiment information into attention with a dynamic update mechanism. At each timestep, the model leverage both the variational attention and hidden representation to decode and predict the target word, and then uses the generated results to update the emotional information in attention. When the texts are becoming longer, the proposed interactive attention can better preserve the sentiment and content information, resulting in better performance of generation.

The empirical experiments were conducted on the Stanford Sentiment TreeBank (SST) [12] corpus. To investigate the generation performance for long texts, we selected texts with different lengths, e.g., 16, 32, and 64, for training. The results show that the sentiment accuracy of the text generated by the proposed model outperformed the other models. Furthermore, the generated text also has a relatively high degree of smoothness, indicating that the proposed model is promising for affective text generation.

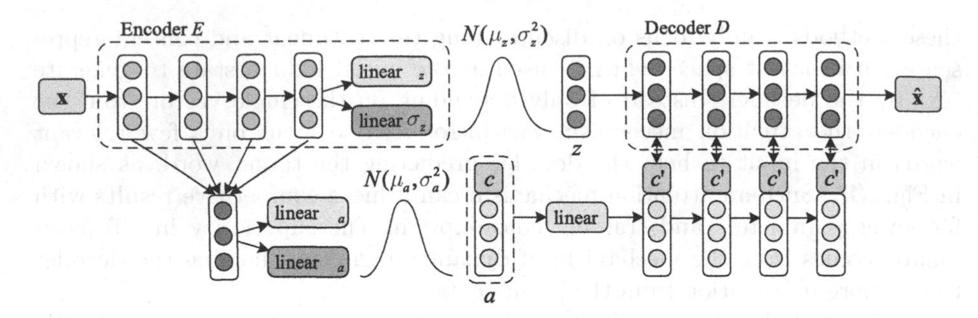

Fig. 2. Conceptual diagram of the proposed variational auto-encoder with interactive attention.

The rest of the paper is organized as follows. Section 2 presents the details of the proposed variational auto-encoder with interactive attention for affective text generation. Section 3 summarizes and reports the experimental results. Section 4 finally concluded the current study.

2 Variational Autoencoder with Interactive Attention

Figure 2 illustrates the overall architecture of the proposed variational auto-encoder with interactive attention to extend the Seq2Seq generation model. It consists of three parts: an encoder E, a decoder D, and a variational attention A. The traditional vanilla VAE [2] uses an unstructured variable $\mathbf{z}$, which encoding the useful information. Furthermore, we added another emotional code $\mathbf{c}$, to encode extra affective information, indicating salient affective features of texts. A detailed description of each component is present as follows.

2.1 Encoder

The encoder E is a single-layer GRU parameterized by θ_E, which takes the word embedding sequence $\mathbf{x} = \{x_1, \ldots, x_{|\mathbf{x}|}\}$ of a text as input, and outputs the encoded hidden representation of the text $h = \{h_1, \ldots, h_{|\mathbf{x}|}\}$, denoted as,

$$h_i = GRU(h_{i-1}, x_i; \theta_E) \tag{1}$$

where $|\mathbf{x}|$ is the length of the input sentence. Then, the variational autoencoder used two dense connection layers on hidden representation to learn the mean vector μ_z and the standard deviation vector σ_z. The latent variables $\mathbf{z}$ is generated by sampling from the Gaussian distribution $\mathcal{N}(\mu_z, diag(\sigma_z^2))$, denoted as,

$$\mathbf{z} \sim q_E^{\mathbf{x}}(\mathbf{z}|\mathbf{x}) = \mathcal{N}(\mu_z, diag(\sigma_z^2)) \tag{2}$$

2.2 Variational Attention

The variational attention A was used to dynamically align $\hat{\mathbf{x}} = \{\hat{x}_1, \ldots, \hat{x}_{|\hat{\mathbf{x}}|}\}$ and $\mathbf{x} = \{x_1, \ldots, x_{|\mathbf{x}|}\}$ during generation. The attention mechanism computes a probabilistic distribution at j-th timestep in the decoder,

$$\alpha_{ji} = \frac{\exp(e_{ji})}{\sum_{i'=1}^{|\mathbf{x}|} \exp(e_{ji'})} \tag{3}$$

where e_{ji} is the pre-normalized score, defined as $e_{ji} = h_j^{\hat{\mathbf{x}}} W^T h_i^{\mathbf{x}}$. Here, $h_j^{\hat{\mathbf{x}}}$ and $h_i^{\mathbf{x}}$ are the j-th and the i-th hidden representation in output and input sentence, respectively, and W is a bilinear term that interacts with these two vectors and captures a specific relation. Then, the attention vector is calculated by a weighted summation, defined as,

$$a_j = \sum_{i=1}^{|\mathbf{x}|} \alpha_{ji} h_i \tag{4}$$

In addition, the variational autoencoder is performed on the attention vector in j-th step to model the posterior of $q_E^a(a_j|\mathbf{x})$ as another Gaussian distribution $\mathcal{N}(\mu_{a_j}, diag(\sigma_{a_j}^2))$, denoted as

$$a_j \sim q_A^a(a_j|\mathbf{x}) = \mathcal{N}(\mu_{a_j}, diag(\sigma_{a_j}^2)) \tag{5}$$

For implementation, we applied an identity transformation $\mu_{a_j} \equiv a_j^{\text{det}}$ to preserve the information in variation attention. Then, a dense connection layer with tanh activation function was used to transform a_j^{det} and then followed by another dense connection layer with exp activation function (ensure that the values are positive) to get σ_{a_f}.

To initialize the interactive attention vector, the latent affective variable c_0' is encoded by a linear transformation of the concatenation of both the initial variational attention a and the scalar value $\mathbf{c}$, defined as,

$$c_0' = f(a_0 \oplus c) \tag{6}$$

where $\oplus$ is the concatenate operator, $c' \in \mathbb{R}^{d_c}$ is the latent affective variables, and f is a linear transformation.

2.3 Decoder

The decoder D consists of a decoder GRU and an update layer U. Considering that the affective information of each part in the sentence will continue to change during the process of sentence generation, the update layer U is designed to vigorously adjust the affective information variable c' at j-th timestep in decoder by,

$$c_j' = GRU(\hat{h}_j, c_{j-1}') \tag{7}$$

Then, decoder GRU adjusts the hidden representation $\hat{h}_j$ in the j-th step by inputting the word embeddings $\hat{x}_{y-1}$, which is the ground-truth in the training phase and the prediction from the last step in the testing phase. The decoder is also a single GRU with an affective information variable from the update layer and the attention vector from variational attention mechanism, denoted by,

$$\hat{h}_j = \text{GRU}(\hat{h}_{j-1}, [\hat{x}_{j-1} \oplus a_j \oplus c'_j]) \tag{8}$$

The process of generating text is transformed into a process that is similar to discrete sequential decision-making in the Seq2Seq framework. Then, a softmax function is used to predict the tokens $\hat{x}_j$ using the hidden representation $\hat{h}_j$, denote as,

$$p(\hat{x}_t|\hat{x}_{1:t-1}, \hat{h}_t) = \text{softmax}(W_h\hat{h}_t + b_h) \tag{9}$$

By inputting the affective information variable c', the model forces the decoder to produce realistic texts with affective constraints. θ_D and θ_E denote the parameters in decoder D and encoder E, respectively. The loss function of the decoder D in each step is,

$$\begin{aligned}\mathcal{L}_j(\theta_D, \theta_E; \mathbf{x}) &= -\text{KL}(q_E(\mathbf{z}, a|\mathbf{x})||p(\mathbf{z}, a)) + \mathbb{E}_{q_E(\mathbf{z},a|\mathbf{x})}[\log p_D(\hat{\mathbf{x}}|\mathbf{z}, a)] \\ &= -\text{KL}(q_E(\mathbf{z}|x_i)||p(\mathbf{z})) - \text{KL}(q_E(a|x_i)||p(a)) \\ &+ \mathbb{E}_{q_E(\mathbf{z}|x_i)q_E(a|x_i)}[\log p_D(\hat{x}_i|\mathbf{z}, a)]\end{aligned} \tag{10}$$

where $\text{KL}(\cdot||\cdot)$ is the Kullback-Leibler divergence.

The posterior $q_E(\mathbf{z}, a|\mathbf{x}) = q_E(\mathbf{z}|\mathbf{x})q_E(a|\mathbf{x})$ since $\mathbf{z}$ and a are conditionally independent given x, the sampling procedure can be performed separately and the KL divergence can be computed independently. The overall training objective of VAE with variational latent variables $\mathbf{z}$, and variational attention a is to minimize,

$$\begin{aligned}\mathcal{L}_D(\theta_D, \theta_E) = \mathcal{L}_{rec}(\theta_D, \theta_E, \hat{x}) + \lambda_{KL}\{&\text{KL}[q_E(\mathbf{z}|\mathbf{x})||p(\mathbf{z})] \\ &+\lambda_a \sum_{j}^{|\hat{\mathbf{x}}|} \text{KL}[q_E(a_j|\mathbf{x})||p(a_j)]\}\end{aligned} \tag{11}$$

where the hyperparameter λ_{KL} and λ_a were used to balance each part of the loss function and $\mathcal{L}_{rec}$ was the reconstruction loss.

3 Experimental Results

This section conducted comparative experiments to evaluate the performance of the proposed variational auto-encoder with interactive attention for affective text generation against several previous models.

3.1 Dataset

Following the previous works [5,7], we used Stanford Sentiment Treebank-2 (SST-2) to train the model for generating affective texts. SST-2 consists of

8544/1101/2210 movie review sentences for the train/dev/test sets, respectively. The polarity labels (i.e., positive or negative) were used as the affective labels $\mathbf{c}$ for experiments. Compared with the previous studies [5] that only used the sentence less than 16 in length as training samples, we selected sentences less than 16, 32 and 64 for conducting three experiments to explore further the performance of the proposed model in long sentences.

3.2 Evaluation Metrics

Both semantic and sentiment tasks were conducted to evaluate whether the proposed method can generate sentences with desired sentiment and smooth expression. The evaluation metrics used in this experiment are presented as follow,

Accuracy. To test whether the sentiment polarity of the generated sentence is correct, we use accuracy to illustrate the performance of the model. We trained a sentiment classifier BERT-large-uncased [3], which has 92% accuracy on the SST-2 test set to calculate the accuracy of the affective generation texts. While generating sentences given emotional code $\mathbf{c}$, both latent variable $\mathbf{z}$ and attention a are sampled from the prior.

Forward Perplexity. Perplexity (PPL) [20] is the generated sentences' perplexity score predicted by a pre-trained language model. We measure the fluency of the generated sentences by the PPL calculated with the language model trained on the SST train set. The language model is adapted from the SRILM [13].

Reverse Perplexity. Reverse Perplexity (Reverse PPL) [20] is the perplexity of language models trained on the sampled sentences and evaluated on the real sentences, which evaluates the diversity and fluency of sampled sentences from a text generation model, the lower, the better. We also use SRILM to calculate reverse PPL by training the language model on 30K generated sentences and test on the SST training set.

3.3 Implementation Details

Various variational auto-encoders with different attention mechanisms were used as baselines for comparison. The details of the baselines are presented as follow,

Vanilla Variational Auto-encoder (VAE) [2]**.** The variational autoencoder to text using single-layer LSTM RNNs for both the encoder and the decoder. To control the emotional attributes, we concatenate the emotional attribute $\mathbf{c}$ in the latent variables.

Semi-supervised Variational Autoencoder (SVAE) [7]**.** A semi-supervised variational autoencoder, which reconstructs the input sentence for a given attribute code without using any discriminator.

Conditional Variational Autoencoder (CVAE) [5]**.** The variational autoencoder, which augments the unstructured variables $\mathbf{z}$ with a set of structured variables $\mathbf{c}$, which targets a salient and independent semantic feature of sentences.

Table 1. Comparative results of the sentiment polarity accuracy of the generated texts.

Model	Max sentence length		
	16	32	64
SVAE	0.822	–	–
CVAE	0.851	–	–
VAE	0.826	0.776	0.759
VAE-VA	0.895	0.867	0.830
VAE-interVA	**0.940**	**0.890**	**0.857**

Variational Autoencoder with Variational Attention (VAE-VA) [1]. It models the attention vector as random variables by imposing a probabilistic distribution, which is then used to sample the prior of the attention vector by a Gaussian distribution following the traditional VAE.

Variation Autoencoder with Interactive Variation Attention (VAE-interVA). The proposed VAE-interVA was implemented for comparison. It extends VAE-VA with interactive variational attention, which was introduced in the decoding phase to keep track of the attention history, and encourages the attention-based VAE to better control the sentiment and content of the generated text.

For a fair comparison, the encoder and the decoder of the above-mentioned models are single-layer GRU. The dimensionality of the hidden dimension is set to 300. The word embeddings used in this experiment was pre-trained on Common Crawl 840B[1] by GloVe [10]. For baselines, the polarity label $\mathbf{c}$ was treated as a scalar (i.e., 0 or 1) and directly concatenated into the latent variables. For the proposed model, the label $\mathbf{c}$ was employed to learn the interactive affective variable c'. Its dimensionality is 400. To avoid KL vanishing, the KL weight λ_{KL} is annealing from 0 to max KL weight during training, where max KL weight λ_{KL} is from 0.3 to 1.0, and the balanced KL weight λ_a is set to 0.8, 1.5, and 2.0. We use AdamW [9] to optimize the parameters when training[2].

3.4 Comparative Results

Table 1 shows the accuracy of polarity classification measured by the pre-trained sentiment classifier for the generated texts. We generate 15K positive sentences and 15K negative sentences by controlling the disentangled features $\mathbf{c}$ when latent variables $\mathbf{z}$ and attention a are prior sampled. A pre-trained classifier is employed to sign the sentiment polarity of texts generated by both baseline models and the proposed model. We calculate the classification accuracy between predicted labels and the specified sentiment polarities. Compared with baseline models, the proposed model achieves the highest accuracy when text length is

[1] https://nlp.stanford.edu/projects/glove/.
[2] https://github.com/Chenrj233/VAE-interVA

Table 2. Ablation study results of the model on generated sentences.

Model	Max length	Forward PPL	Reverse PPL	Accuracy
VAE	16	66.51	266.81	0.826
	32	67.32	**100.20**	0.776
	64	56.62	**77.09**	0.759
VAE-VA	16	51.48	266.67	0.895
	32	51.06	123.87	0.867
	64	46.66	89.05	0.830
VAE-interVA	16	**45.41**	**226.06**	**0.940**
	32	**43.40**	110.76	**0.890**
	64	**45.74**	82.07	**0.857**

less than 16, indicating that the proposed VAE-interVA can transfer sentiment more accurately. When the text length increases to 32 and 64, both the VAE-VA and VAE-interVA model achieve lower accuracy, while the VAE-interVA model still performs much better than VAE-VA model. It means that the VAE-interVA model can persistently maintain sentiment information while producing long text.

3.5 Ablation Experiment

We use the above model to generate 30K sentences for evaluation. Table 2 shows the ablation experiments to further investigate how the VAE-interVA models can benefit from each component. The addition of variational attention greatly improves the emotional accuracy of the generated text, and the dynamic update mechanism further improves the emotional accuracy of the text. We found that the forward PPL has declined to varying degrees with variational attention and update layer in different lengths of texts, suggesting that these two mechanisms can effectively improve the fluency of the generated texts. Compared with other models, VAE-interVA has a lower reverse PPL in short texts, which indicates that the proposed method can improve the emotional accuracy of the generated texts and maintain the fluency of the texts while increasing the diversity of the generated texts. However, in medium and long sentences, VAE performs better in reverse PPL, indicating that the variational attention and the dynamic update mechanism pay more attention to the fluency of sentences and emotional information while ignoring the diversity of texts. This is because the proposed method pays too much attention to certain words with strong emotional polarity in the decoding process, resulting in insufficient text diversity. Experimental results show that dynamically adjusting affective information variable c' during decoding can control the formation of sentence emotions and maintain the smoothness of sentences.

Figure 3 shows the comparative results on the SST when the KL weight λ_{KL} varies from 0.3 to 1.0 and the KL weight λ_a is set to 0.8, 1.5 and 2.0. We found

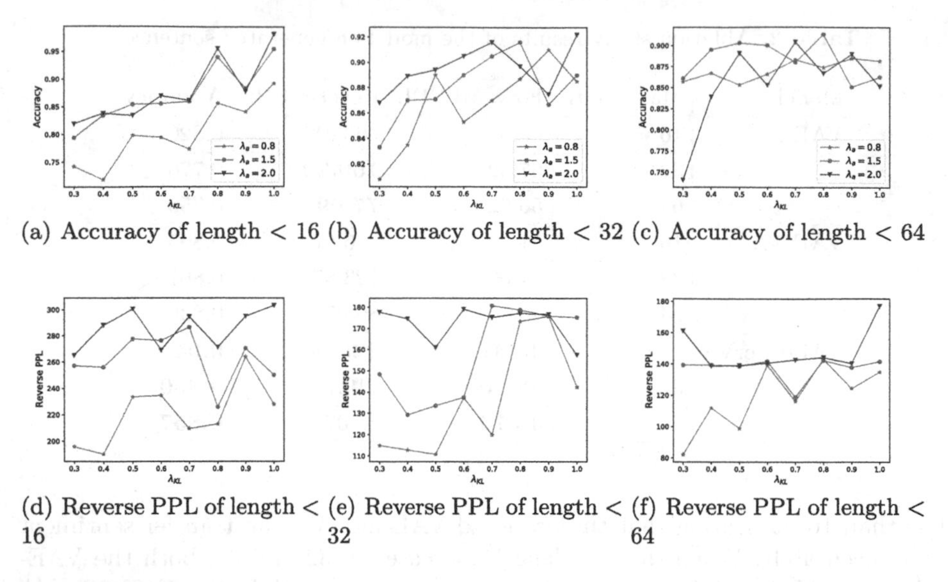

(a) Accuracy of length < 16 (b) Accuracy of length < 32 (c) Accuracy of length < 64

(d) Reverse PPL of length < 16 (e) Reverse PPL of length < 32 (f) Reverse PPL of length < 64

Fig. 3. Effect of the KL weight λ_a and λ_{KL} in the proposed model.

that as the text gets longer, the smaller the value of λ_a, making the model train better with a lower reverse PPL. We can select an appropriate value of λ_a to balance accuracy and reverse PPL. Meanwhile, a larger value of λ_{KL} will make the model performs worse on reverse PPL when the text gets longer.

3.6 Case Study

To further explore the effectiveness of the proposed VAE-interVA model, we selected four cases of sampled sentences from VAE, VAE-VA and VAE-interVA, as shown in Table 3. In VAE, the emotional feature **c** is simply concatenated into the latent variables, which can hardly affect the sentiment of the generated text. Meanwhile, the text generated by VAE is not natural enough. Compared with VAE, VAE-VA significantly improve the smoothness of sentences and emotional control. The text generated by VAE-VA is more natural and the generated emotion is more accurate. The proposed VAE-interVA model can capture the features of the sentence (e.g., sentiment information), and combine the structure knowledge of the sentence to generate a sentence with specific affective feature **c**. The texts generated by VAE-interVA are more fluent and natural.

Table 3. Sampled sentences from different models.

	VAE	VAE-VA	VAE-interVA
Positive	The film offers a number of holes, and a story of the end and shakes (**Wrong sentiment**)	A riveting story of determination and the human behavior	A moving documentary that provides a rounded and revealing overview of this ancient holistic healing system
	Return to never land is reliable, standard disney animated fare, with no crea-tive energy sparked by two actresses in their 50s (**Unreadable**)	This is a movie that refreshes the mind of spirit along with all of its audience – or indeed, and it is one of the great minds and the audience (**Unnatural**)	It's a fine, focused piece of work that reopens an interesting controversy, and thrill enough to watch
Negative	Attal pushes every sequel is more frus-trating than a mess (**Unreadable**)	A movie that lacks both a purpose and depth	I hate the feeling of having been slimed in the name of high art
	A muddled limp biscuit of a movie, using vampire soap opera ... doesn't make you to the 'laughing at' ve ever given by the time (**Unreadable**)	After all the time, the movie bogs down in insignificance, saying nothing about kennedy's performance (**Unnatural**)	The movie is so contrived, non-sensical and formulaic that it's not any of any depth

4 Conclusions

In this paper, variational attention models are proposed for affective text generation. Unlike previous work, we add explicit and implicit emotion information to the VAE during the decoding process, so that the emotion will not dissipate with the generation of the text, and then further improve the performance of the affective text generation. The comparative results showed that the proposed method outperformed the other methods for affective text generation. Another observation is that it can still maintain accurate sentiment information and smoothness of sentences even in the longer text.

In addition, based on the work of [15], this method should also have a certain effect on the generation of multi-dimensional emotional text, and we will investigate it in the future.

Acknowledgement. This work was supported by the National Natural Science Foundation of China (NSFC) under Grants Nos. 61702443, 61966038 and 61762091.

References

1. Bahuleyan, H., Mou, L., Vechtomova, O., Poupart, P.: Variational attention for sequence-to-sequence models. In: Proceedings of the 27th International Conference on Computational Linguistics, COLING, pp. 1672–1682 (2018)
2. Bowman, S.R., Vilnis, L., Vinyals, O., Dai, A.M., Józefowicz, R., Bengio, S.: Generating sentences from a continuous space. In: Proceedings of the 20th SIGNLL Conference on Computational Natural Language Learning, pp. 10–21 (2016)
3. Devlin, J., Chang, M.W., Lee, K., Toutanova, K.: Bert: pre-training of deep bidirectional transformers for language understanding. In: Proceedings of the 2019 Conference of the North American Chapter of the Association for Computational Linguistics, NAACL, pp. 4171–4186 (2019)
4. Hochreiter, S., Schmidhuber, J., Elvezia, C.: Long short-term memory. Neural Comput. **9**(8), 1735–1780 (1997)
5. Hu, Z., Yang, Z., Liang, X., Salakhutdinov, R., Xing, E.P.: Toward controlled generation of text. In: Proceedings of the International Conference on Machine Learning, ICML, pp. 1587–1596 (2017)
6. Jain, P., Agrawal, P., Mishra, A., Sukhwani, M., Laha, A., Sankaranarayanan, K.: Story generation from sequence of independent short descriptions (2017)
7. Kingma, D.P., Rezende, D.J., Mohamed, S., Welling, M.: Semi-supervised learning with deep generative models. In: Proceedings of the 27th International Conference on Neural Information Processing Systems, ICNIPS, pp. 3581–3589 (2014)
8. Kingma, D.P., Welling, M.: Auto-encoding variational Bayes. In: Proceedings of the 2nd International Conference on Learning Representations, ICLR (2014)
9. Loshchilov, I., Hutter, F.: Decoupled weight decay regularization. In: Proceedings of the 7th International Conference on Learning Representations, ICLR. OpenReview.net (2019)
10. Pennington, J., Socher, R., Manning, C.D.: GloVe: global vectors for word representation. In: Proceedings of the 2014 Conference on Empirical Methods in Natural Language Processing, EMNLP, pp. 1532–1543 (2014)
11. Radford, A., Wu, J., Child, R., Luan, D., Amodei, D., Sutskever, I.: Language models are unsupervised multitask learners. OpenAI Blog **1**(8), 9 (2019)
12. Socher, R., et al.: Recursive deep models for semantic compositionality over a sentiment treebank. In: Proceedings of the 2013 Conference on Empirical Methods in Natural Language Processing, EMNLP, pp. 1631–1642 (2013)
13. Stolcke, A.: SRILM - an extensible language modeling toolkit. In: Proceedings of the 7th International Conference on Spoken Language Processing, ICSLP (2002)
14. Sutskever, I., Vinyals, O., Le, Q.V.: Sequence to sequence learning with neural networks. In: Proceedings of Advances in Neural Information Processing Systems, NIPS, pp. 3104–3112 (2014)
15. Wang, J., Yu, L.C., Lai, K.R., Zhang, X.: Tree-structured regional CNN-LSTM model for dimensional sentiment analysis. IEEE/ACM Trans. Audio Speech Lang. Process. **28**, 581–591 (2019)
16. Wang, K., Wan, X.: Automatic generation of sentimental texts via mixture adversarial networks. Artif. Intell. **275**, 540–558 (2019)
17. Zhang, R., Wang, Z., Yin, K., Huang, Z.: Emotional text generation based on cross-domain sentiment transfer. IEEE Access **7**, 100081–100089 (2019)
18. Zhang, Y., Wang, J., Zhang, X.: Learning sentiment sentence representation with multiview attention model. Inf. Sci. **571**, 459–474 (2021)

19. Zhang, Y., Wang, J., Zhang, X.: Personalized sentiment classification of customer reviews via an interactive attributes attention model. Knowl. Based Syst. **226**, 107135 (2021)
20. Zhao, J.J., Kim, Y., Zhang, K., Rush, A.M., LeCun, Y.: Adversarially regularized autoencoders. In: Proceedings of the 35th International Conference on Machine Learning, ICML, vol. 80, pp. 5897–5906 (2018)

CUSTOM: Aspect-Oriented Product Summarization for E-Commerce

Jiahui Liang(✉), Junwei Bao, Yifan Wang, Youzheng Wu, Xiaodong He, and Bowen Zhou

JD AI Research, Beijing, China
{liangjiahui14,baojunwei,wangyifan15,wuyouzhen1,xiaodong.he, bowen.zhou}@jd.com

Abstract. Product summarization aims to automatically generate product descriptions, which is of great commercial potential. Considering the customer preferences on different product aspects, it would benefit from generating aspect-oriented customized summaries. However, conventional systems typically focus on providing general product summaries, which may miss the opportunity to match products with customer interests. To address the problem, we propose CUSTOM, aspect-oriented product summarization for e-commerce, which generates diverse and controllable summaries towards different product aspects. To support the study of CUSTOM and further this line of research, we construct two Chinese datasets, i.e., SMARTPHONE and COMPUTER, including 76,279/49,280 short summaries for 12,118/11,497 real-world commercial products, respectively. Furthermore, we introduce EXT, an extraction-enhanced generation framework for CUSTOM, where two famous sequence-to-sequence models are implemented in this paper. We conduct extensive experiments on the two proposed datasets for CUSTOM and show results of two famous baseline models and EXT, which indicates that EXT can generate diverse, high-quality, and consistent summaries (https://github.com/JD-AI-Research-NLP/CUSTOM).

1 Introduction

Product summarization aims to automatically generate product descriptions for e-commerce. It is of great commercial potential to write customer-interested product summaries. In recent years, a variety of researches focus on product summarization and have proposed practical approaches [11,13,21,23]. These models take product information as input and output a general summary describing a product. However, the general summaries usually face two problems: (1) they are singular and lack diversity, which may miss the opportunity to match products with customer interests; (2) it is not able to control the models to describe what aspects of the products in the general summaries, which hurts for personalized recommendations. Considering that different customers have preferences on different aspects of products on e-commerce platforms, e.g., the customers may care more about the "APPEARANCE" than the "PERFORMANCE" aspect

L. Wang et al. (Eds.): NLPCC 2021, LNAI 13029, pp. 124–136, 2021.
https://doi.org/10.1007/978-3-030-88483-3_10

of a smartphone, so generating aspect-oriented customized summaries will be beneficial to personalized recommendations.

Motivated by the above issues and observation, in this paper, we propose CUSTOM, aspect-oriented product summarization for e-commerce, which can provide diverse and controllable summaries towards different product aspects. In detail, given a product with substantial product information and a set of corresponding aspects for this category, e.g., smartphone and computer, the task of CUSTOM is to generate a set of summaries, each of which only describes a specified aspect of the product. Figure 1 shows an example of a smartphone where the product information includes a product title and substantial product details in a natural language recognized from abundant product images. Different colors in the product information represent different aspects of the product. When a certain aspect, i.e., CAMERA, APPEARANCE, or PERFORMANCE, is specified, the corresponding aspect-oriented summary is generated based on the product information. For example, the "green" summary describes the appearance of the smartphone. To the best of our knowledge, our task is related to the conditional generation [2,6,10,16]. Specifically, KOBE [2] is the most similar task to CUSTOM. KOBE focuses on writing expansion from short input, i.e., product title, where the generated text is likely to disrespect the truth of products, while CUSTOM concentrates on generating summary from long input, i.e., substantial product information, where the generated summary is consistent with the truth of products.

To support the study of CUSTOM and further this line of research, we construct two Chinese datasets, i.e., SMARTPHONE and COMPUTER, including 76,279/49,280 short summaries for 12,118/11,497 real-world commercial products, respectively. Furthermore, inspired by the content selection methods proposed in [3,7,9,24], we introduce EXT, an extraction-enhanced generation framework for CUSTOM, which equips the model with the ability to select aspect-related sentences from product information to enhance the correlation between the generated summary and the aspect. We conduct extensive experiments on the two proposed datasets and show results of Pointer-Generator, UniLM, and EXT, which indicate that EXT can generate more diverse, high-quality, and consistent summaries. Our contributions are as follows:

- We propose CUSTOM, aspect-oriented product summarization for e-commerce, to generate diverse and controllable summaries towards different product aspects.
- We construct two real-world Chinese commercial datasets, i.e., SMARTPHONE and COMPUTER, to support the study of CUSTOM and further this line of research.
- We introduce EXT, an extraction-enhanced generation framework. Experiment results on SMARTPHONE and COMPUTER show the effectiveness of the EXT.

2 Methodology

2.1 CUSTOM: Aspect-Oriented Product Summarization for E-Commerce

In this section, we formulate the CUSTOM task as follows. With respect to Fig. 1, the input of the model includes product information S and an aspect category a, where $S = \{s_i\}_{i=1}^{N}$ is a series of sentences by concatenating the title and product details, N represents the maximum number of sentences. The output is a short summary y that only describes the specified aspect a of the product.

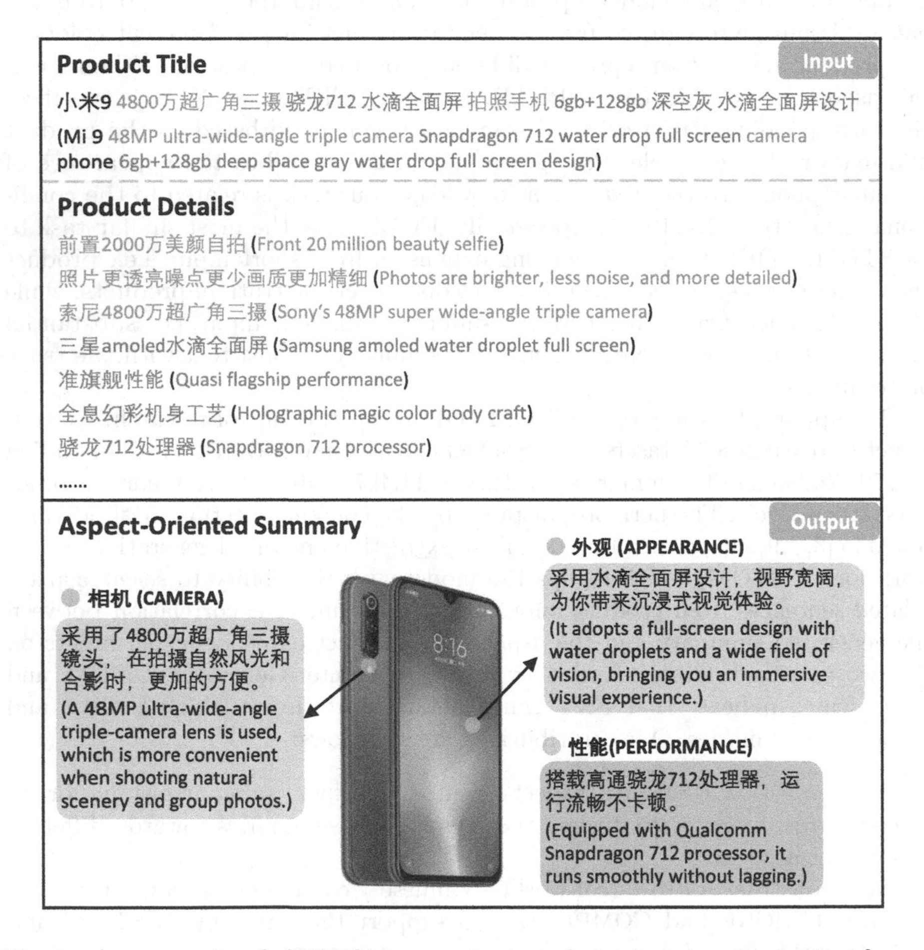

Fig. 1. An example of CUSTOM: aspect-oriented product summarization for e-commerce. (Color figure online)

2.2 SMARTPHONE and COMPUTER

To support the study of CUSTOM and further this line of research, we construct two Chinese datasets, i.e., SMARTPHONE and COMPUTER, including

76,279/49,280 short summaries for 12,118/11,497 real-world commercial products, respectively. We split the two datasets into training, development, and test for experiments. The number of instances and products of train, development, and test sets are shown in Table 1. As described in Sect. 2.1, each dataset is consist of a set of ⟨*product information*, *aspect*, *aspect summary*⟩ tuples. In the SMARTPHONE training set, the average number of different aspect summaries for each product is about 6.3, while it is 1.8 and 2.1 in development and test sets respectively. The reason is that we keep the pair of ⟨*product information*, *aspect*⟩ unique in the development and test sets in order to ensure that an input ⟨*product information*, *aspect*⟩ pair corresponds to an unique output summary. The same is true for COMPUTER dataset. For SMARTPHONE dataset, there are five kinds of aspects, i.e., *APPEARANCE*, *BATERY*, *CAMERA*, *PERFORMANCE*, and *FEATURE*. For COMPUTER dataset, there are three kinds of aspects, i.e., *FEATURE*, *PERFORMANCE*, *APPEARANCE*. Figure 2 shows the aspect distributions on training, development, and test set for SMARTPHONE and COMPUTER.

Table 1. Statistics of SMARTPHONE and COMPUTER. #sum: the number of aspect summaries. #prod: the number of products.

Category	Overall		Train		Dev		Test	
	#sum	#prod	#sum	#prod	#sum	#prod	#sum	#prod
SMARTPHONE	76,279	12,118	73,640	10,738	1,639	896	1,000	484
COMPUTER	49,280	11,497	47,284	10,268	996	615	1,000	614

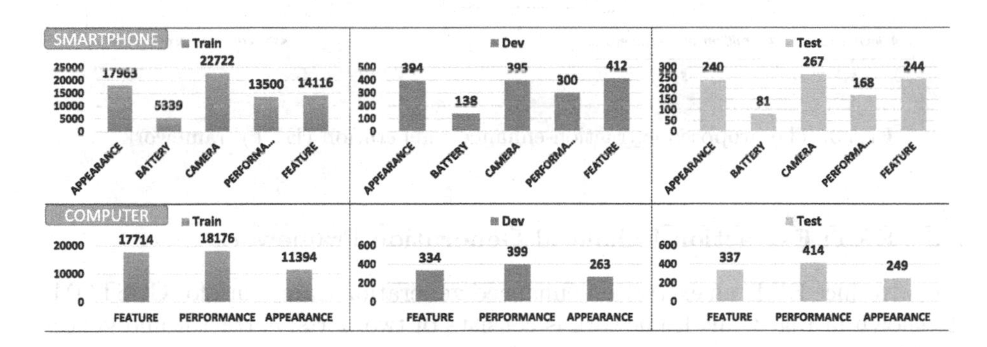

Fig. 2. Aspect Distributions for SMARTPHONE and COMPUTER.

The datasets are collected from a Chinese e-commerce platform and we have summaries written by professional writers that describe multiple aspects of the product. Since there are no ⟨*product information*, *aspect*, *aspect summary*⟩ tuples, we use a heuristic method to construct the datasets. First, we split the origin summary by periods, and only keep fragments with character lengths between

15 and 55. The shorter fragments may contain insufficient information while the longer fragments could contain more than one aspect description. We then use BERT [4] to obtain continuous vectors of each fragment and used them as input to the K-Means [17] clustering algorithm. The resulting clusters represent the description set of different aspects of the product, that is, the aspect summary set. The cluster label of each instance is the aspect category. We experiment with different numbers of clusters, and manually select 100 from each clustering result for evaluation, and select the result with the highest degree of separation of the clustering results as the final number of clusters. We have also tried some methods to automatically detect clustering results, such as the silhouette coefficient, but the results are not ideal. Finally, we find that the best number of clusters is 5 and 3 for the SMARTPHONE and COMPUTER, respectively.

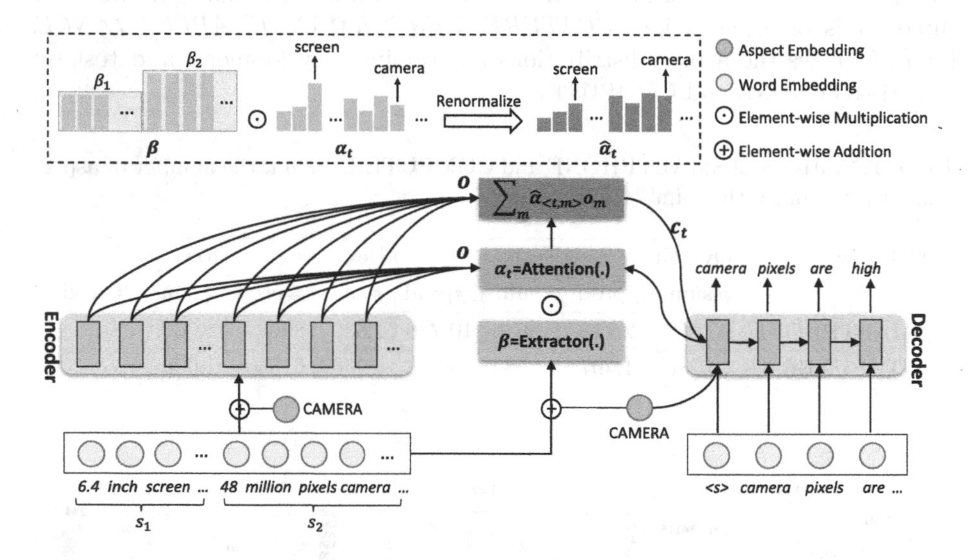

Fig. 3. The proposed extraction-enhanced generation (EXT) framework.

2.3 EXT: Extraction-Enhanced Generation Framework

We introduce EXT, an extraction-enhanced generation framework for CUSTOM. As shown in Fig. 3, the framework is consists of two parts: extractor and generator.

The extractor is responsible for scoring each sentence in the product information and outputs the sentence-level attention:

$$\beta_{\mathrm{i}} = \mathbf{Ext}(\mathbf{s}_i, \mathbf{a}) \tag{1}$$

where s_i is the ith sentence and a is the input aspect category, the greater the β_i, the more relevant this sentence is to the current aspect, and vice versa. The binary cross-entropy loss function is used to train the extractor:

$$L_{ext} = -\frac{1}{N}\sum_{i=1}^{N}(g_i \log \beta_i + (1 - g_i)\log(1 - \beta_i)) \tag{2}$$

where g_i is the aspect label for the ith sentence. g_i is 1 means that the description of the current sentence is consistent with the current aspect. We use a heuristic rule to obtain the aspect labels. Specifically, we calculate the overlap rate $r = \frac{\#overlap\_length}{\#sentence\_length}$ between each sentence in the product information and the aspect summary, where $\#overlap\_length$ is calculated by longest common subsequence (LCS) [8]. The aspect label is 1 if the overlap rate is above a certain threshold, which is 0.35 in our setting.

The generator is basically a sequence-to-sequence structure with attention mechanism [1]. Inspired by the study [9], at each decoding step t, we combine the sentence-level score β_i generated by the extractor and word-level attention α^t_m at time step t as follows:

$$\hat{\alpha}^t_m = \frac{\alpha^t_m \times \beta_{i(m)}}{\sum_m \alpha^t_m \times \beta_{i(m)}} \tag{3}$$

where m is the word index. The updated word-level attention weight $\hat{\alpha}^t_m$ is then used to compute context vector c_t as follows:

$$c_t = \sum_{m=1}^{|w|} \hat{\alpha}^t_m o_m \tag{4}$$

where o_m stands for the mth word representation.

We reuse the aspect information and fuse it to the generator to calculate the final word distribution $P(w) = \texttt{Gen}(y_{t-1}, \mathbf{c}_t, \mathbf{a})$ at step t. We train the generator with the average negative log-likelihood loss:

$$L_{gen} = -\frac{1}{T}\sum_{t=1}^{T} \log P(y_t) \tag{5}$$

where y_t is the ground-truth target word. The extractor and generator are trained end-to-end, the final loss is as below:

$$L = L_{ext} + L_{gen} \tag{6}$$

We apply this framework to two typical summarization models, one is an RNN-based end-to-end model called Pointer-Generator [20], and the other is a Transformer-based pretrained model called UniLM [5]. We noticed that UniLM, T5 [19], and BART [12] are all SOTAs of pretrained generative models. In this paper, we choose UniLM since it is representative and only UniLM has released the Chinese version pretrained parameters.

EXT on Pointer-Generator. First, we concatenate each sentence in S with a period to form a sequence of words $\{w_m\}_{m=1}^{|w|}$. Then each word w_m and the aspect category a are embedded to obtain continuous vector e_{w_m} and e_a respectively. The word embedding and aspect embedding are added by column to obtain the fused aspect-aware word embedding e_{f_m}. The fused embedding is used as initial input for both extractor and generator.

For extractor, we use a hierarchical bidirectional GRU as in [18] to extract sentence representations $\mathbf{H}$. Then we apply a matrix dot operation and uses the sigmoid function to predict the sentence-level attention β_i:

$$\beta_{\mathbf{i}} = \text{Sigmoid}(\mathbf{e_a}^{\mathrm{T}}\mathbf{h_i}) \tag{7}$$

where $\mathbf{h_i}$ stands for i^{th} sentence vector.

The generator is basically the pointer-generator network and we combine the sentence-level attention and word-level attention as described in equation (3). Besides being used to compute context vector c_t, the updated word attentions are also used as the new copy distribution. The aspect embedding is also added to the previous word embedding by column at each decoding step.

EXT on UniLM. We use the same fused embedding e_{f_m} as input, except e_{w_m} is the summation of 3 kinds of embeddings including word embedding, position embedding, and segment embedding. The input embeddings are encoded into contextual representations $\mathbf{O} = [\mathbf{o_1}, \ldots, \mathbf{o_{|w|}}]$ using Transformer. Since the sentences are separated by periods, we simply take out the hidden vector of each period to represent the sentence representation $\mathbf{H}$:

$$\mathbf{H} = \mathbf{O}[I_1, \ldots, I_M] \tag{8}$$

where $\mathbf{I_i}$ denotes the index of period of sentence $\mathbf{s_i}$. Then we apply a two-layer feedforward network to compute the sentence score:

$$\beta_{\mathbf{i}} = \text{Sigmoid}(\mathbf{FFN}(\mathbf{h_i})) \tag{9}$$

The sentence scores are used to update word-level attention as in Eq. (3). The generator shares the parameters of the transformer encoder with the extractor. Finally, we apply the same fine-tuning method as in UniLM [5].

3 Experiment

3.1 Comparison Methods

In this section, we note the baselines and our proposed methods. **PGen w/o Aspect** and **UniLM w/o Aspect** denotes the original Pointer-Generator Network and UniLM model without aspect information as input. **PGen w/ Aspect** and **UniLM w/ Aspect** means we fuse aspect information as describe in Sect. 2.3 but without incorporating **EXT**. **EXT-PGen** denotes our proposed PGen-based model. **EXT-UniLM** represents our proposed UniLM-based model.

3.2 Implementation Details

For PGNet-based models, the model parameters remain the same as work [9]. For UniLM-based models, we use the base version and load the pre-trained parameters published in work[1] for initializing. The maximum numbers of the characters in the input and target are 400 and 70, respectively. During training, the mini-batch size is set to 20. We choose the model with the smallest perplexity on the development dataset for evaluation. During inference, the beam size is set to 5, and the maximum decoding length is set to 80 for all models.

3.3 Diversity Evaluation for CUSTOM

We argue that CUSTOM generates more diverse summaries than conventional product summarization. To evaluate diversity, we follow [14] to use Dist-2/3/4. Since different instances of the same product have the same product information, conventional models without aspect information as input (PGen/UniLM w/o Aspect) generate the same summaries for different aspects of the same product (Top-1). To improve the diversity of the conventional models, we also report results of conventional models without aspect information (PGen/UniLM w/o Aspect) which keep the top K candidates in the beam as outputs for evaluation (Top-K), where K is the number of instances for the product in the test set. Experiment results are shown in Table 2. Our proposed CUSTOM achieves the highest performance on all diversity indicators on two datasets.

Table 2. Diversity evaluation results. Dist-2: Distinct-2. Dist-3: Distinct-3. Dist-4: Distinct-4.

Model	SMARTPHONE			COMPUTER		
	Dist-2	Dist-3	Dist-4	Dist-2	Dist-3	Dist-4
PGen w/o Aspect (Top-1)	0.126	0.187	0.226	0.034	0.047	0.056
PGen w/o Aspect (Top-K)	0.155	0.242	0.307	0.038	0.053	0.065
PGen w/ Aspect	**0.199**	**0.313**	**0.387**	**0.107**	**0.160**	**0.196**
UniLM w/o Aspect (Top-1)	0.127	0.186	0.222	0.070	0.094	0.107
UniLM w/o Aspect (Top-K)	0.140	0.216	0.268	0.077	0.107	0.126
UniLM w/ Aspect	**0.232**	**0.372**	**0.462**	**0.140**	**0.205**	**0.246**

3.4 Quality Evaluation for EXT

Following the traditional text generation tasks, we calculate character-based Rouge-1, Rouge-2 and Rouge-L [15] F1 scores to evaluate the text quality. We use the same ROUGE-1.5.5 toolkit as in UniLM [5]. We evaluate Rouge scores for overall-level and the aspect-level at the same time. Experiment results are

[1] https://github.com/YunwenTechnology/Unilm.

shown in Table 3 and Table 4. We can conclude that, compared with the original models, both the EXT-PGen and EXT-UniLM models can obtain better Rouge scores by incorporating with the EXT framework, which verifies that EXT can boost the quality of generated aspect summaries.

Table 3. Experiment results on SMARTPHONE. R-1: Rouge-1. R-2: Rouge-2. R-L: Rouge-L.

Model	Overall	APPEARANCE	BATTERY	CAMERA	PERFORMANCE	FEATURE
	R-1/R-2/R-L	R-1/R-2/R-L	R-1/R-2/R-L	R-1/R-2/R-L	R-1/R-2/R-L	R-1/R-2/R-L
PGen w/Aspect	36.3/20.4/33.4	37.0/20.3/33.8	37.7/20.0/33.9	39.7/23.0/37.2	40.7/26.9/38.3	28.5/13.3/25.6
EXT-PGen	**37.4/21.1/34.3**	**38.7/20.9/34.8**	**39.2/22.4/35.8**	**40.1/23.0/37.4**	**42.4/27.4/39.1**	**29.3/14.5/26.9**
UniLM w/Aspect	37.3/21.5/34.2	**38.8/22.7/35.5**	41.6/24.5/38.2	38.9/21.8/35.8	**43.0**/27.8/**39.5**	28.7/14.7/26.3
EXT-UniLM	**38.0/22.3/34.8**	38.3/22.3/34.7	**43.5/25.7/38.9**	**40.9/24.1/37.7**	42.9/**27.9**/39.4	**29.5/15.7/27.2**

Table 4. Experiment results on COMPUTER. R-1: Rouge-1. R-2: Rouge-2. R-L: Rouge-L.

Model	Overall	FEATURE	PERFORMANCE	APPEARANCE
	R-1/R-2/R-L	R-1/R-2/R-L	R-1/R-2/R-L	R-1/R-2/R-L
PGen w/ Aspect	31.4/14.9/28.1	**24.1**/9.6/21.7	35.5/17.7/31.7	34.6/17.3/30.9
EXT-PGen	**32.5/16.4/29.6**	23.9/**9.6/21.8**	**36.3/19.5/33.1**	**38.1/20.5/34.4**
UniLM w/ Aspect	33.1/16.6/30.1	26.2/11.9/24.3	35.9/18.2/32.2	38.0/20.4/34.5
EXT-UniLM	**33.9/17.7/31.0**	**26.8/12.6/25.0**	**36.1/19.2/32.6**	**39.8/22.3/36.4**

3.5 Human Evaluation

Table 5. Human evaluation results.

Datasets	Win	Lose	Tie
SMARTPHONE	27.00%	11.30%	61.70%
COMPUTER	24.70%	13.00%	62.30%

We conduct a human evaluation study where three participants were asked to compare summaries generated by EXT-PGen and PGen in terms of summary quality and aspect-summary consistency. We randomly sample 100 instances for each participant from the test set for evaluation. Table 5 shows the human evaluation results, where **Win** means the summary generated by EXT-PGen is better than PGen w/ Aspect, **Lose** means the summary generated by EXT-PGen is worse than PGen w/ Aspect and **Tie** means the summaries generated by EXT-PGen and PGen w/ Aspect obtain similar scores. We can see that EXT-PGen surpasses the PGen w/ Aspect model at a large margin (over 10%) on both datasets.

3.6 Extractor Analysis

We randomly select one instance from the SMARTPHONE test set and plot heat maps of the extractor module. As shown in Fig. 4, we find that the extractor module learned to capture sentences most relevant to the specified aspect from the product information.

Outdoor look more comfortable. Software algorithm upgrade to unlock faster, smarter. 639 AMOLED extreme full screen. Sony 48MP super wide-angle triple camera. Faster click response. Set 20 million panoramic selfies. 3D four-curved glass body. Front 20 million lift camera. The unlocking speed is one step faster, the seventh-generation screen fingerprint recognition technology. Press protection. The seventh-generation screen fingerprint unlock. Qualcomm Snapdragon 855Plus processor. Powerful camera combination. The larger the memory of the same processor, the stronger the performance, 12GB large memory. Elevating front camera achieves the 919's full screen. Super night view. 4000mAh long battery life. Large memory. Fast charging in 30 minutes 58, no longer waiting.

(a) "CAMERA" aspect

Outdoor look more comfortable. Software algorithm upgrade to unlock faster, smarter. 639 AMOLED extreme full screen. Sony 48MP super wide-angle triple camera. Faster click response. Set 20 million panoramic selfies. 3D four-curved glass body. Front 20 million lift camera. The unlocking speed is one step faster, the seventh-generation screen fingerprint recognition technology. Press protection. The seventh-generation screen fingerprint unlock. Qualcomm Snapdragon 855Plus processor. Powerful camera combination. The larger the memory of the same processor, the stronger the performance, 12GB large memory. Elevating front camera achieves the 919's full screen. Super night view. 4000mAh long battery life. Large memory. Fast charging in 30 minutes 58, no longer waiting.

(b) "APPEARANCE " aspect

Fig. 4. Heatmap of the extractor

3.7 Case Study

We perform a case study on the SMARTPHONE dataset and compare the generation results of EXT-UniLM and UniLM on different aspects. Table 6 shows the comparison results. We can see that when incorporated with the EXT framework, the summary is more informative and descriptive. For example, when describing the APPEARANCE aspect of the first instance, the EXT-UniLM generates more details, including the screen size is 6.4 in. and the screen-to-body ratio is 90.77%. This shows that our EXT framework can select key information related to the current aspect from the input to improve the quality of the generated summaries.

Table 6. Comparison of the generated aspect summaries of two models on SMARTPHONE dataset. Only English translation are shown in table due to space limitation.

Product information	Aspect	UniLM	EXT-UniLM
OPPO Reno 10x zoom version Qualcomm Snapdragon 855 48 MP ultra-clear triple camera 6 GB + 256 GB Extreme night black Full Netcom Full screen camera Smart game phone Smart protection The night scene is more pure Panoramic screen 48 million pixels Make beauty more natural ···	PERFORMANCE	Equipped with Qualcomm Snapdragon 855 processor, the performance is strong, and the operation is smooth and not stuck	Equipped with Qualcomm Snapdragon 855 processor, it has strong performance, greatly improves the running speed, and brings a smoother gaming experience
	CAMERA	The rear 48 million ultra-clear three-shot, the picture is clearer	The rear 48 MP ultra-clear three-shot, supports 10x hybrid optical zoom, and the photos are clearer
	APPEARANCE	Adopting a full-screen design to bring an immersive visual experience	This phone uses a 6.4-in. panoramic screen design with a screen-to-body ratio of 90.77%, bringing you an immersive visual experience
Honor 20 Youth Edition AMOLED screen fingerprint 4000mAh large battery 2 W fast charge 48 million mobile phone 6 GB + 128 GB blue water emerald full screen gestures The night is more moving Ultra-wide-angle photography has a wider field of view 209 slender design ···	BATTERY	Built-in 4000 mah high-capacity battery, long-lasting battery life	Built-in 4000 mah high-capacity battery, support 2 w fast charge, longer battery life
	FEATURE	Equipped with a full-scene acceleration engine, it brings a smooth experience without stuttering	Equipped with a full-scene acceleration engine and multiple core optimizations, it brings a smooth experience without stuttering
	APPEARANCE	With a 6.3-in. amoled pearl screen design, the screen display is clear and delicate	With a 6.3-in. amoled pearl screen design, the screen display is clear and vivid, bringing you an immersive viewing experience

4 Related Work

4.1 Product Summarization

Xiao et al. [22] present two extractive summarization approaches using a bi-directional LSTM encoder-decoder network with attention mechanism to summarizing titles of e-commerce products. Khatri et al. [11] propose a novel

Document-Context based Seq2Seq models for abstractive and extractive summarizations in e-commerce, which should be started with contextual information at the first time-step of the input to obtain better summaries. Shao et al. [21] propose a Planning-based Hierarchical Variational Model to generate long and diversified expressions. They decompose long text generation into dependent sentence generation sub-tasks and capture diversity with a global planning latent variable and a sequence of local latent variables. Li et al. [13] propose an Aspect-aware Multimodal Summarizer to improve the importance, non-redundancy, and readability of the product summarization.

4.2 Conditional Text Generation

Hu et al. [10] propose a generative model which combines variational auto-encoders (VAEs) and attribute discriminators to produces sentences with desired attributes. Fan et al. [6] present a neural summarization model to enable users to control the shape of the final summary in order to better suit their needs, such as selecting desired length or style. They introduce special marker tokens when training and testing. Chen et al. [2] propose KOBE which is the most similar task to CUSTOM. KOBE focuses on writing expansion from short input where the generated text is likely to disrespect the truth of products, while CUSTOM concentrates on generating consistent summary from long input.

5 Conclusion

In this paper, we propose CUSTOM, aspect-oriented product summarization for e-commerce, to generate diverse and controllable summaries towards different product aspects. To support the study of CUSTOM and further this line of research, we construct two real-world Chinese commercial datasets, i.e., SMARTPHONE and COMPUTER. Furthermore, we introduce EXT, an extraction-enhanced generation framework for CUSTOM. Experiment results on SMARTPHONE and COMPUTER show the effectiveness of the proposed EXT.

Acknowledgments. We are grateful to all the anonymous reviewers. This work is supported by the National Key Research and Development Program of China under Grant (No. 2018YFB2100802).

References

1. Bahdanau, D., Cho, K., Bengio, Y.: Neural machine translation by jointly learning to align and translate. In: ICLR (2015)
2. Chen, Q., Lin, J., Zhang, Y., Yang, H., Zhou, J., Tang, J.: Towards knowledge-based personalized product description generation in e-commerce. In: SIGKDD (2019)
3. Cho, J., Seo, M., Hajishirzi, H.: Mixture content selection for diverse sequence generation. In: EMNLP-IJCNLP, Hong Kong, China (2019)

4. Devlin, J., Chang, M.W., Lee, K., Toutanova, K.: BERT: pre-training of deep bidirectional transformers for language understanding. In: NAACL-HLT, Minneapolis, Minnesota (2019)
5. Dong, L., et al.: Unified language model pre-training for natural language understanding and generation. In: NeurIPS (2019)
6. Fan, A., Grangier, D., Auli, M.: Controllable abstractive summarization. In: NMT@ACL, Melbourne, Australia (2018)
7. Gehrmann, S., Deng, Y., Rush, A.: Bottom-up abstractive summarization. In: EMNLP, Brussels, Belgium (2018)
8. Hirschberg, D.S.: Algorithms for the longest common subsequence problem. J. ACM **24**, 664–675 (1977)
9. Hsu, W.T., Lin, C.K., Lee, M.Y., Min, K., Tang, J., Sun, M.: A unified model for extractive and abstractive summarization using inconsistency loss. In: ACL, Melbourne, Australia (2018)
10. Hu, Z., Yang, Z., Liang, X., Salakhutdinov, R., Xing, E.P.: Toward controlled generation of text. In: ICML (2017)
11. Khatri, C., Singh, G., Parikh, N.: Abstractive and extractive text summarization using document context vector and recurrent neural networks. arXiv (2018)
12. Lewis, M., et al.: BART: denoising sequence-to-sequence pre-training for natural language generation, translation, and comprehension. In: ACL (2020)
13. Li, H., Yuan, P., Xu, S., Wu, Y., He, X., Zhou, B.: Aspect-aware multimodal summarization for Chinese e-commerce products. In: AAAI (2020)
14. Li, J., Galley, M., Brockett, C., Gao, J., Dolan, B.: A diversity-promoting objective function for neural conversation models. In: NAACL, San Diego, California, pp. 110–119 (2016)
15. Lin, C.Y.: ROUGE: a package for automatic evaluation of summaries. In: ACL (2004)
16. Liu, D., et al.: Diverse, controllable, and keyphrase-aware: a corpus and method for news multi-headline generation. In: EMNLP (2020)
17. MacQueen, J., et al.: Some methods for classification and analysis of multivariate observations. In: Proceedings of the 5th Berkeley Symposium on Mathematical Statistics and Probability, pp. 281–297 (1967)
18. Nallapati, R., Zhai, F., Zhou, B.: SummaRuNNer: a recurrent neural network based sequence model for extractive summarization of documents. In: AAAI (2017)
19. Raffel, C., et al.: Exploring the limits of transfer learning with a unified text-to-text transformer. arXiv (2020)
20. See, A., Liu, P.J., Manning, C.D.: Get to the point: summarization with pointer-generator networks. In: ACL, Vancouver, Canada (2017)
21. Shao, Z., Huang, M., Wen, J., Xu, W., Zhu, X.: Long and diverse text generation with planning-based hierarchical variational model. In: EMNLP, Hong Kong, China (2019)
22. Xiao, J., Munro, R.: Text summarization of product titles. In: eCOM@SIGIR (2019)
23. Zhang, T., Zhang, J., Huo, C., Ren, W.: Automatic generation of pattern-controlled product description in e-commerce. In: WWW (2019)
24. Zhou, Q., Yang, N., Wei, F., Zhou, M.: Selective encoding for abstractive sentence summarization. In: ACL, Vancouver, Canada (2017)

Question Answering

FABERT: A Feature Aggregation BERT-Based Model for Document Reranking

Xiaozhi Zhu[1], Leung-Pun Wong[2], Lap-Kei Lee[2], Hai Liu[1](✉), and Tianyong Hao[1](✉)

[1] School of Computer Science, South China Normal University, Guangzhou, China
{2020022975,haoty}@m.scnu.edu.cn

[2] The Open University of Hong Kong, Hong Kong, China
{s1243151,lklee}@ouhk.edu.hk

Abstract. In a document reranking task, pre-trained language models such as BERT have been successfully applied due to their powerful capability in extracting informative features from queries and candidate answers. However, these language models always generate discriminative features and pay less attention to generalized features which contain shared information of query-answer pairs to assist question answering. In this paper, we propose a BERT-based model named FABERT by integrating both discriminative features and generalized features produced by a gradient reverse layer into one answer vector with an attention mechanism for document reranking. Extensive experiments on the MS MARCO passage ranking task and TREC Robust dataset show that FABERT outperforms baseline methods including a feature projection method which projects existing feature vectors into the orthogonal space of generalized feature vector to eliminate common information of generalized feature vectors.

Keywords: Document ranking · BERT · Feature aggregation

1 Introduction

Document reranking task aims to return the most relevant permutation of candidate answers to a given question posed in natural language, which has been a longstanding problem in natural language processing (NLP) and information retrieval (IR). In recent years, neural pre-trained language models such as BERT [1], Roberta [2] and OpenAI GPT [3], have achieved significant improvement on a series of NLP tasks and outperformed traditional document ranking methods [4–8].On the MS MARCO passage ranking task, fine-tuning BERT simply with an extra linear layer on the top of BERT has achieved state-of-the-art results by treating the ranking task as a binary classification problem [9]. Dual-encoder models based on BERT have been explored recently to learn the dense

L. Wang et al. (Eds.): NLPCC 2021, LNAI 13029, pp. 139–150, 2021.
https://doi.org/10.1007/978-3-030-88483-3_11

representation of question-answer pairs and then use this representation to predict the relevance scores in open-domain question answering [10,11]. Multi-stage reranking model based on BERT has outperformed single BERT by a large margin, which is consists of BERT trained with the pointwise loss function and another trained with pairwise loss function separately [12]. These methods use interactive-based discriminative feature vectors generated by BERT to evaluate the relevance of QA pairs. In cross-domain classification, generalized features produced through a gradient reverse layer (GRL) [13], serve as the domain-shared features between source domain and target domain [14,15]. To obtain a more discriminative representation, FP-Net [16] projects the existing feature vectors into the orthogonal space of generalized feature vectors to eliminate generalized information for text classification. As proposed by FP-Net, generalized features are class-shared and have little discriminative information for classification, which prevents the classifier from performing correct classification. Although the generalized feature vectors cannot be utilized as the answer vectors for document ranking, the way of adopting the information is worth exploring. As shown in Table 1, there is a question about *Badminton players who have won Olympic titles*, and two candidate passages about *Lin Dan* and *Lee Chong Wei.* The word "Olympic" and "badminton player" colored in blue, are common information on both candidate passages for guiding models on the object "Olympic" rather than other objects like "master", "Malaysian", etc., while the red words are the discriminative feature for answering the question.

Table 1. Example of generalized feature and discriminative feature. The word colored in blue represents generalized feature while the word colored in red represents discriminative feature.

Question: *Badminton players who have won Olympic titles*	
Passage 1 Lin Dan (born 14 October 1983) is a Chinese former professional badminton player and accept masters degree presented at Huaqiao University. He is a two-time Olympic champion, five-time World champion.	Passage 2 Lee Chong Wei (born 21 October 1982) is a former Malaysian badminton player with D-.C.S.M Dato W-illa. Lee is a triple silver medalist at the Olympic Games, and the sixth Malaysian to win an Olympic medal.

In this paper, a feature aggregation BERT-based model called FABERT is proposed, which uses the information of both generalized feature vectors and discriminative feature vectors for document ranking. FABERT is composed of two sub-networks, a generalized feature vector generation network referred to as G-net for extracting generalized information of QA pairs and an aggregation network referred to as A-net for combining generalized feature vectors and discriminative feature vectors into an answer vector. To obtain the generalized feature vector, G-net uses a BERT-base model as a feature extractor to encode the QA pairs and utilizes a Gradient Reverse Layer (GRL) after the encoder to reverse the gradient direction. A-net uses another BERT-based model as the feature extractor as well as the G-net, and proportionally integrates the semantic information of generalized feature vectors and exiting feature vectors into an answer vector with an attention mechanism. Our model is evaluated on two widely used document ranking datasets: the MS MARCO dataset and Robust 2004 dataset. Compared with a list of baseline methods including the Vanilla BERT-base model, FABERT obtains an improvement of 1.9% in MRR on the MS MARCO dataset and an improvement of 1.7% in NDCG@20 on the Robust04 dataset at least.

In summary, the major contribution of the paper lies on: 1) A feature aggregation method that incorporates generalized features at the document level is proposed for improving document ranking. 2) A novel FABERT model that utilizes an attention mechanism to extract essential information from both generalized feature and discriminative feature is proposed for enhancing representation learning. 3) Evaluation on standard MS MARCO and Robust04 datasets confirms that FABERT outperforms the baseline methods.

2 Related Work

Language model pre-trained on large-scale corpus achieved impressive improvement on information retrieval tasks. Nogueira et al. [9] first applied pre-trained language model BERT to predict relevance scores between passages and questions on MS MARCO and TREC-CAR datasets and substantially outperformed traditional neural ranking models like DUET [6], Co-PACRR [7] and Conv-KNRM [4]. CEDR [17] combined the pre-trained model BERT with existing IR models by feeding the output of BERT into the IR models for document ranking. Inspired by the capability of Transformer [18] for representation learning, PARADE [19] proposed a passage representation aggregation method that allows information interaction between paragraphs using Transformer. RankTxNet [20] employed a Bert-based model as a sentence encoder at the first stage and then used an order invariant Bert-based model for paragraph encoding. These models combined the discriminative feature vectors from BERT with different interactive architectures to improve the representation learning of the models, but ignored the generalized features of the text.

The gradient reverse layer (GRL) [13] was designed for learning cross-domain generalized features which contain common information between domains. IATN [14] proposed an interactive attention transfer mechanism that considered the information of both sentences and aspects, and employed GRL to obtain invariance features for cross-domain sentiment classification. Du et al. [15] designed a post-training method that enabled Bert-based models to extract specific features for different domains and to be domain-aware for domain transferring classification. Domain-adaption models utilized GRL to learn domain-shared feature, which represented data from both domains and was generalized at the domain level. Grand et al. [22] and Belinkov et al. [21] encouraged model to learn bias-free representations by means of adversarial approach for visual question answering task and natural language inference. To obtain more discriminative features in text classification, Qin et al. [16] first proposed a feature purification network (FP-Net) for improving representation learning, which projected existing feature vectors into the orthogonal space of generalized feature vectors to eliminate common information of generalized feature vectors. In the document ranking task, generalized feature learned by model were at passage-level, and eliminating the generalized information leaded to a reduction in representation capability of model. Therefore our work uses an attention mechanism to selectively aggregate generalized feature vectors with existing feature vectors.

3 Model

A new BERT-based model named FABERT is proposed to aggregate discriminative features vector and generalized features vector for document reranking. FABERT mainly contains two parts, i.e., one network produces the generalized feature vector referred to as G-net and other network aggregates feature referred to as A-net. A fine-tuned BERT model which is composed of an embedding layer and representation layer serves as an encoder for the G-Net and the A-Net separately. On the top of the BERT encoder, there is an attention layer in the A-net and a GRL in the G-net. Finally, the output of the A-net is sent to a passage ranker to produce a relevance score for each pair of query and candidate answers. The architecture of FABERT is shown in Fig. 1.

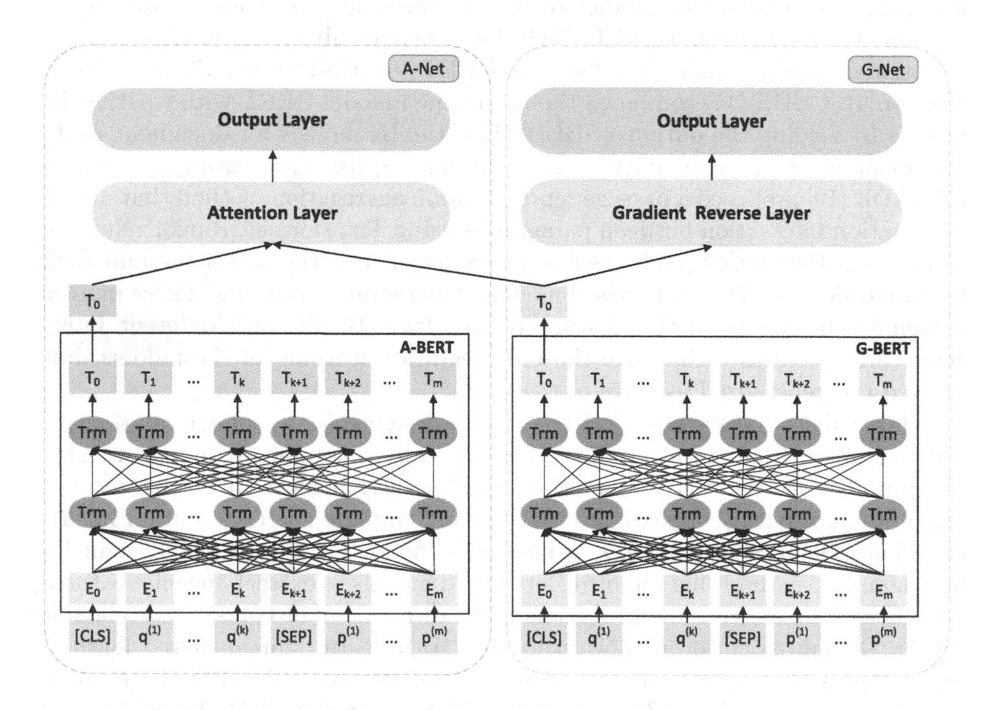

Fig. 1. The architecture of FABERT for document ranking

3.1 Problem Definition

In the task of open-domain QA, a natural language question and a list of candidate passages to given question are provided. The question q can be viewed as a k-length sequence of tokens $= <q_1, q_2, \cdots, q_k>$, while each candidate passage p can be denoted by an m-length sequence of tokens $= <p_1, p_2, \cdots, p_m>$. An open-domain QA task is to learn informative representation and a precise relevance ranker in the training process to return the best permutation of candidate passages.

3.2 QA Pairs Encoder

A BERT-base model was employed as document encoder for the A-Net and the G-Net respectively. For open-domain QA, the given question-answer pairs are preprocessed into a sequence $I = \{[CLS], q_1, \cdots, q_k, [SEP], p_1, \cdots p_m, [SEP]\}$, where special token $[CLS]$ is a symbol for indicating the start of a sentence and token $[SEP]$ for indicating the end of a sentence. In the token sequence, the $i-$th token is denoted as token embedding $T_i \in \mathbb{R}^d$ which is a d dimensional real-valued vector. We denote position embedding as $P_i \in \mathbb{R}^d$ and segment embedding as $S_i \in \mathbb{R}^d$ and obtain the input representation $E = (E_0, E_1, \cdots E_{n-1}, E_n)$ where n is the max-length of the sequence by summing the embedding vectors up as Eq. (1).

$$E_i = T_i + P_i + S_i \tag{1}$$

BERT is composed of a stack of L identical layers which consists of two sublayers with a self-attention mechanism. The first sublayer is a multi-head self-attention mechanism, and the second is a position-wise fully connect feed-forward neural network. The self-attention, called "Scaled Dot-Product Attention", aggregates the input consists of query matrix $Q \in \mathbb{R}^{n \times d}$, key matrix $K \in \mathbb{R}^{n \times d}$ and value matrix $V \in \mathbb{R}^{n \times d}$ in Eq. (2).

$$Attention(Q, K, V) = softmax(\frac{QK^{\mathrm{T}}}{\sqrt{d_k}})V \tag{2}$$

A multi-head attention mechanism can be viewed as ensembling h different self-attention simultaneously and then concatenates the output of each self-attention part into final values in Eq. (3).

$$MultiHead(Q, K, V) = Concat(head_1, \ldots, head_h)W^{\mathrm{O}} \tag{3}$$

$$head_i = Attention(QW_i^Q, KW_i^K, VW_i^V) \tag{4}$$

In the Eq. (4), $W_i^Q \in \mathbb{R}^{d_{model} \times d}$, $W_i^K \in \mathbb{R}^{d_{model} \times d}$ and $W_i^V \in \mathbb{R}^{d_{model} \times d}$ are learnable parameter matrices. Although the A-net and the G-net apply the same architecture for encoding the question-answer pairs, the output of the G-BERT is feed into the gradient reverse layer during the backpropagation to generate generalized feature vectors. The attention layer in the A-net aggregates generalized feature vectors and the outputs of the A-BERT into an indicative vector used to evaluate the relevance score of the QA pairs. Specifically, the representations T_{CLS} at the first token (e.g., $[CLS]$ symbol in BERT) is taken as the output for encoding.

3.3 Feature Aggregation

To generalize feature vectors, the G-net aims to extract generalized features from the input QA pairs, which contain necessary semantic information. In addition, our goal is to learn the generalized features which prevent the passage ranker of the G-net from providing distinct scores for two QA pairs effortlessly. To achieves that, the Gradient Reverse Layer (GRL) is added after the G-BERT to reverse the gradient direction in training process. Through this procedure, the generalized feature vectors are generated as passage shared to a given query.

With mathematical formulation, the gradient reverse layer is treated as a "pseudo-function" defined by two incompatible equations with a hyper-parameter λ describing its forward and backpropagation behaviors in Eq. (5) and (6).

$$GRL(x) = x \tag{5}$$

$$\frac{\partial GRL(x)}{\partial x} = -\lambda \mathrm{I} \tag{6}$$

We denote the QA pairs representation T_{CLS}^{G} through the GRL as $GRL(T_{CLS}^{G}) = \tilde{T}_{CLS}^{G}$ and then feed it to the Linear layer which produces a relevance score for each QA pair to quantify their relevance:

$$score_G = W_G \cdot \tilde{T}_{CLS}^{G} + b_G, \tag{7}$$

where W_G and b_G are the weight and bias of the Linear layer, which are learnable parameters. The pairwise margin ranking loss is adopted in the G-net:

$$Loss_G(socre_G^k, socre_G^{k+1}, y) = \max(0, y \cdot (socre_G^k - socre_G^{k+1}) + \gamma), \tag{8}$$

where $y = 1$ if the $socre_G^k$ is ranked higher than the $socre_G^{k+1}$, and vice-versa for $y = -1$. γ is the margin hyperparameter. The target is to minimize the pairwise ranking loss $Loss_G$ for all QA pairs in the G-net and to allow the model to learn discriminative features that distinguish QA pairs. As the gradient passes through thr GRL, the gradient is changed to the opposite direction, which enables the G-BERT to extract common features of passages to a given query.

To aggregate the generated features, the A-net focuses on extracting semantic information of an input example and returns a relevance score to a given QA pair. In the classification task, the generalized feature vectors in FP-Net [16], which hinder the classifier from making the correct judgment, are removed through an orthogonal projection layer. In the contrast, an extra attention layer is added on the top of the A-BERT to aggregate the generalized feature vectors T_{CLS}^{G} and the existing feature vectors T_{CLS}^{A} produced by the A-BERT into the final vectors T_{CLS}^{F}, which leverages the information in generalized feature vectors instead of eliminating them completely. The final vector T_{CLS}^{F} is calculated by the weighted sum of the generalized feature vectors T_{CLS}^{G} and the existing feature vectors T_{CLS}^{A} as follows:

$$T_{cls}^{F} = \alpha_{cls}^{A} \cdot T_{cls}^{A} + \alpha_{cls}^{G} \cdot T_{cls}^{G}, \tag{9}$$

where α_{cls} is defined as:

$$\alpha_{cls} = soft\max(W_a \cdot T_{cls} + b_a) \tag{10}$$

W_a and b_a are learnable parameters in the attention layer. The attention weight α_{cls} is served as a feature selector to determine how much information of the generalized feature vectors T_{CLS}^{G} and the existing feature vectors T_{CLS}^{A} are retained. To obtain the scores of QA pairs, an output layer is added after the attention layer to measure the relevance of the input pairs:

$$score_F = W_F \cdot T_{cls}^{F} + b_F \tag{11}$$

$$Loss_F(socre_F^{k}, socre_F^{k+1}, y) = \max(0, y \cdot (socre_F^{k} - socre_F^{k+1}) + \gamma) \tag{12}$$

The parameters of A-Net and G-Net are trained with pairwise loss functions $Loss_F$ and $Loss_G$ separately. In addition, the gradients of $Loss_F$ and $Loss_G$ are passed back to each network simultaneously in training process, while the gradients of $Loss_F$ do not update the parameters of the G-BERT.

4 Experiment and Results

4.1 Dataset and Baselines

The proposed method is evaluated on two widely used datasets: MS MARCO Passage Ranking [23] and TREC Robust 2004 [24]. The MS MARCO dataset is a large-scale dataset consist of anonymized questions sampled from Bing Search and human-generated answers. We evaluate our model on the passage ranking task which contains 8.8 million passages and the goal is to rank based on the relevance of QA pairs. Robust04 is a newswire collection used by TREC 2004 Robust track, which comprises 250 queries and 0.5 million documents (TREC Disks 4 and 5).

The proposed model FABERT is compared with the following methods:

1) **BM25** is a traditional method based on IDF-weighted counting for Information Retrieval (IR), which retrieves documents from the large collection. The BM25 baseline is following the implementation by Anserini [25].
2) **ConvKNRM** [4] is the n-grams version of KNRM, which employed a convolutional layer on the top of the word embedding layer to obtain n-gram representation.
3) **Vanilla BERT** [9] is a BERT-based relevance ranker, composed of a stack of 12 identical layers (BERT-Base, Uncased) with an extra linear layer at the top of the last layer.
4) **BERT+CNN** uses the BERT-based model as a sentence encoder and combines CNN as a feature extractor for MS MARCO Passage Ranking.
5) **BM25+BERT** [26] utilizes BM25 as a document retriever to produce a ranked list, and a fine-tuned BERT reranks the candidate documents based on the relevance between query and candidate answers in the training process.

4.2 Evaluation Metrics

Three widely used evaluation metrics are employed in experiments to measure the performance of FABERT, including MRR@10 on the MS MARCO dataset, P@20 and NDCG@20 on the Robust04 dataset.

MRR@10 (Mean Reciprocal Rank) focuses on whether the relevant document is in the top 10 candidate documents and its position is regarded as ranking. The metric takes the reciprocal rank of the first relevant document in the ranked list to a given query to measure its precision, as Eq. (13), where $|Q|$ represents the number of queries:

$$MRR = \frac{1}{|Q|} \sum_{i=1}^{|Q|} \frac{1}{rank_i} \tag{13}$$

P@20 (Precision) is precision at the fixed rank of 20 in the ranking, which calculates the proportion of relevant documents in the top 20 candidate documents.

NDCG@20 (Normalized Discounted Cumulative Gain) considers the relevance score and the position of a relevant document in the ranking, which is defined as follows:

$$CG_p = \sum_{i=1}^{p} score_i, \tag{14}$$

$$DCG_p = \sum_{i=1}^{p} \frac{2^{score_i} - 1}{\log_2(i+1)}, \tag{15}$$

$$NDCG_p = \frac{DCG_p}{IDCG_p}, \tag{16}$$

where $score_i$ is the predictive score of $i-$th candidate document produced by the model and p represents the number of candidate documents in the ranked list. Discounted cumulative gain (DCG) increases the impact of top-ranked documents on the results, which means that the bottom-ranked documents have little effect on the results. NDCG normalizes DCG by the ideal DCG(IDCG), which is the best-ranking result based on DCG.

4.3 Setting

For both datasets, the pre-trained BERT-base model is used as the encoder for the A-net and the G-net independently, which are fine-tuned with the default hyper-parameters such as embedding dimension, the number of multi-head attention and vocabulary size. Following the setting of BM25+BERT[26] on the MS MARCO dataset, BM25 is employed as the retriever at the first stage and train FABERT using the pairwise loss function. FABERT is optimized by Adam with a learning rate of 3e-6 and a batch size of 8. The maximum sequence length is limited to 256 tokens. On the Robust04 dataset, FABERT is optimized by Adam with a learning rate of 5e-6 for BERT layers and 1e-3 for the rest layers. The documents are truncated to 800 tokens and the model is trained for 100 epochs with a batch size of 16. In the GRL, the hyper-parameter λ is set to 1 in the training process. MRR@10 is adopted on the MS MARCO dataset for the reason that there is at most one relevant passage for a given query. However, there are multiple relevant documents for a given query on the Robust04 dataset, and thus P@20 and NDCG@20 are used to evaluate the performance of models on the Robust04 dataset.

4.4 Results

The ranking performance of all methods is evaluated on MS MARCO Passage Ranking and Robust04, which are shown in Table 2. The Vanilla BERT substantially outperforms the BM25 and Conv-KNRM methods on both datasets, which indicates the improvement in feature representation capability of the pre-trained language model. Compared with Vanilla BERT-base, BM25+BERT-base utilizes BM25 to retrieve top-k candidate answers and then their relevance scores are predicted by BERT-base. Training model with negative examples retrieved by BM25, the BM25+BERT method obtains improvements of 1% with NDCG@20 on Robust04 dataset and 0.9% on MS MARCO Passage Ranking Eval Set.

Table 2. Ranking performance of all methods on the MS MARCO Passage Ranking and Robust04 datasets.

Methods	MS MARCO passage ranking		Roubust04	
	MRR@10(Dev)	MRR@10(Eval)	P@20	NDCG@20
BM25	0.167	0.165	0.312	0.414
Conv-KNRM	0.290	0.271	0.335	0.381
Vanilla BERT-base	0.347	0.336	–	0.444
BERT-base+CNN	0.346	0.333	–	–
BM25+BERT-base	0.349	0.345	0.404	0.454
BERT-base+FP-Net	0.347	–	0.393	0.431
FABERT	**0.361**	**0.355**	**0.429**	**0.461**

The performance of BERT+CNN and BERT+FP-Net is comparable and worse than Vanilla BERT-base, suggesting that neither the models of using convolutional layers to locally aggregate feature nor the models of eliminating the information of generalized feature vectors are effective in improving the performance of the model. FABERT consistently outperforms the other baseline methods on the MS MARCO and Robust04 datasets, achieving a highest MRR@10 of 0.355 and NDCG@20 of 0.461. The results demonstrate that applying feature aggregation with the attention mechanism obtains more informative features and thus improves the performance of the passage ranking task.

The t-SNE visualization of the distribution of generalized feature vectors produced by G-Net and final answer vectors after feature aggregation on MS MARCO and Robust04 datasets is shown in Fig. 2. The gradient reverse layer and pairwise loss function encourage G-BERT to learn the generalized information between two candidate answers, as the relevant documents contain similar information for the same query. As illustrated in (a) and (b), the feature vectors learned by the G-NET automatically cluster and overlap towards a region, indicating that the feature vectors within that region contain similar information. The reason for multiple regions is that the text content of the candidate answers varies, and thus there are multiple generalized features learned.

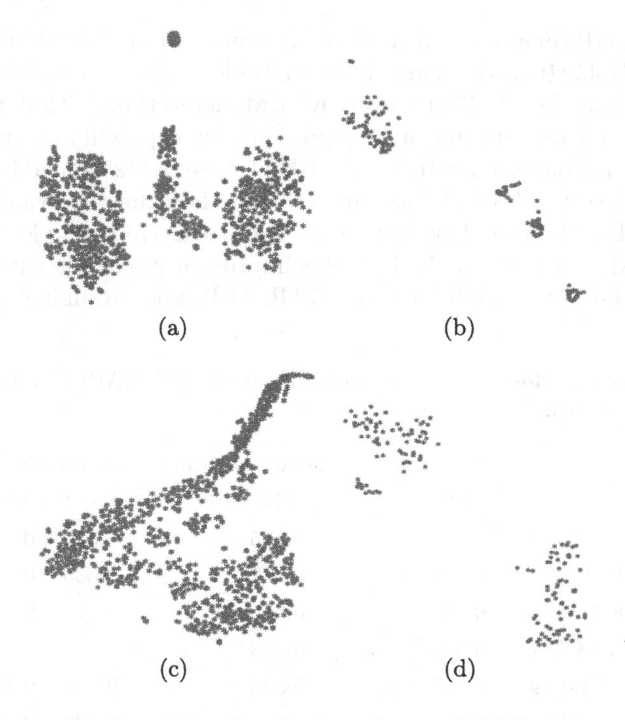

Fig. 2. t-SNE visualization of the distribution of generalized feature vectors and final answer vectors. (a) and (b) show the distribution of generalized feature vectors for the same query on the MS MARCO and Robust04 datasets respectively, while (c) and (d) show the distribution of answer vectors after feature aggregation

By selectively aggregating the generalized information and discriminative information through an attention layer, the feature vectors in (c) and (d) become discrete and not as densely distributed as the generalized feature vectors in (a) and (b), suggesting that FABERT can identify the differences in each candidate answer. After feature aggregation, the generalized information allows the feature vectors learned by FABERT during training process to be aware of other candidate documents, resulting in the distribution of answer vectors remaining regional.

5 Conclusion

This paper proposed a feature aggregation model FABERT to improve the performance of document ranking. The model FABERT learned the generalized information between candidate documents before aggregating the information of generalized feature vectors and discriminative feature vectors into answer vectors for predicting the relevant scores between documents and queries. Experiments on standard datasets demonstrated the effectiveness of the proposed FABERT for document ranking compared with a list of baseline methods.

References

1. Devlin, J., Chang, M.W., Lee, K., Toutanova, K.: BERT: pre-training of deep bidirectional transformers for language understanding. In: Proceedings of the 2019 Conference of the North American Chapter of the Association for Computational Linguistics: Human Language Technologies, vol. 1, pp. 4171–4186 (2019)
2. Liu, Y., et al.: RoBERTa: a robustly optimized BERT pretraining approach. arXiv preprint arXiv:1907.11692 (2019)
3. Radford, A., Narasimhan, K., Salimans, T., Sutskever, I.: Improving language understanding by generative pre-training (2018)
4. Dai, Z., Xiong, C., Callan, J., Liu, Z.: Convolutional neural networks for soft-matching n-grams in ad-hoc search. In: Proceedings of the Eleventh ACM International Conference on Web Search and Data Mining, pp. 126–134 (2018)
5. Robertson, S., Zaragoza, H., et al.: The probabilistic relevance framework: BM25 and beyond. Found. Trends® Inf. Retr. **3**(4), 333–389 (2009)
6. Guo, J., Fan, Y., Ai, Q., Croft, W.B.: A deep relevance matching model for ad-hoc retrieval. In: Proceedings of the 25th ACM International on Conference on Information and Knowledge Management, pp. 55–64 (2016)
7. Hui, K., Yates, A., Berberich, K., De Melo, G.: CO-PACRR: a context-aware neural IR model for ad-hoc retrieval. In: Proceedings of the Eleventh ACM International Conference on Web Search and Data Mining, pp. 279–287 (2018)
8. Mitra, B., Craswell, N.: An updated duet model for passage re-ranking. arXiv preprint arXiv:1903.07666 (2019)
9. Nogueira, R., Cho, K.: Passage re-ranking with BERT. arXiv preprint arXiv:1901.04085 (2019)
10. Karpukhin, V., et al.: Dense passage retrieval for open-domain question answering. arXiv preprint arXiv:2004.04906 (2020)
11. Ding, Y.Q.Y., et al.: RocketQA: an optimized training approach to dense passage retrieval for open-domain question answering. arXiv preprint arXiv:2010.08191 (2020)
12. Nogueira, R., Yang, W., Cho, K., Lin, J.: Multi-stage document ranking with BERT. arXiv preprint arXiv:1910.14424 (2019)
13. Ganin, Y., Ustinova, E., Ajakan, H., Germain, P., Larochelle, H., Laviolette, F., Marchand, M., Lempitsky, V.: Domain-adversarial training of neural networks. J. Mach. Learn. Res. **17**(1), 2096–2130 (2016)
14. Zhang, K., Zhang, H., Liu, Q., Zhao, H., Zhu, H., Chen, E.: Interactive attention transfer network for cross-domain sentiment classification. In: Proceedings of the AAAI Conference on Artificial Intelligence, vol. 33, pp. 5773–5780 (2019)

15. Du, C., Sun, H., Wang, J., Qi, Q., Liao, J.: Adversarial and domain-aware BERT for cross-domain sentiment analysis. In: Proceedings of the 58th Annual Meeting of the Association for Computational Linguistics. pp. 4019–4028 (2020)
16. Qin, Q., Hu, W., Liu, B.: Feature projection for improved text classification. In: Proceedings of the 58th Annual Meeting of the Association for Computational Linguistics, pp. 8161–8171 (2020)
17. MacAvaney, S., Yates, A., Cohan, A., Goharian, N.: CEDR: contextualized embeddings for document ranking. In: Proceedings of the 42nd International ACM SIGIR Conference on Research and Development in Information Retrieval, pp. 1101–1104 (2019)
18. Vaswani, A., et al.: Attention is all you need. In: Proceedings of the 31st International Conference on Neural Information Processing Systems, pp. 6000–6010 (2017)
19. Li, C., Yates, A., MacAvaney, S., He, B., Sun, Y.: PARADE: passage representation aggregation for document reranking. arXiv preprint arXiv:2008.09093 (2020)
20. Kumar, P., Brahma, D., Karnick, H., Rai, P.: Deep attentive ranking networks for learning to order sentences. In: Proceedings of the AAAI Conference on Artificial Intelligence, vol. 34, pp. 8115–8122 (2020)
21. Belinkov, Y., Poliak, A., Shieber, S.M., Van Durme, B., Rush, A.M.: On adversarial removal of hypothesis-only bias in natural language inference. In: Proceedings of the Eighth Joint Conference on Lexical and Computational Semantics (* SEM 2019), pp. 256–262 (2019)
22. Grand, G., Belinkov, Y.: Adversarial regularization for visual question answering: Strengths, shortcomings, and side effects. In: Proceedings of the Second Workshop on Shortcomings in Vision and Language, pp. 1–13 (2019)
23. Nguyen, T., et al.: MS MARCO: a human generated machine reading comprehension dataset. In: CoCo@ NIPS (2016)
24. Voorhees, E.M.: Overview of the TREC 2004 robust retrieval track (2004)
25. Yang, P., Fang, H., Lin, J.: Anserini: enabling the use of Lucene for information retrieval research. In: Proceedings of the 40th International ACM SIGIR Conference on Research and Development in Information Retrieval, pp. 1253–1256 (2017)
26. Xiong, C., et al.: CMT in TREC-COVID round 2: mitigating the generalization gaps from web to special domain search. arXiv preprint arXiv:2011.01580 (2020)

Generating Relevant, Correct and Fluent Answers in Natural Answer Generation

Yongjie Huang[1], Meng Yang[1,2](✉), and Ni Yang[1]

[1] School of Computer Science and Engineering, Sun Yat-sen University, Guangzhou, China
{huangyj229,yangn57}@mail2.sysu.edu.cn

[2] Key Laboratory of Machine Intelligence and Advanced Computing (SYSU), Ministry of Education, Guangzhou, China
yangm6@mail.sysu.edu.cn

Abstract. This study focuses on natural answer generation that generates a complete answer using given paragraphs. Existing methods of only extracting the answer span loses the naturalness while those of generating it word by word increases the learning difficulty. In this paper, we propose to split the answering process into two branches, which share the same encoder and deeply interact with each other. On one branch, we generate the answer template based on the question through text editing method. On the other branch, we extract the answer span based on the documents. Finally, the answer is composed of the generated answer template and the extracted answer span. Besides, we propose to select the span in candidates to better fit the template with more fluency. Experiments show that our method improves the performance on natural answer generation.

Keywords: Natural Answer Generation · Answer template · Text editing · Answer span selection

1 Introduction

Reading Comprehension (RC) task aims at exploring the answer through one or more paragraphs, and usually the answer is a span that can be extracted from the given paragraphs [4,16,21]. In the RC task, most existing models predict the start and end positions of the answer span and extract it as the final answer [5,17,24]. However, only directly extracting the span without polishing it leads to a blunt answer.

Natural Answer Generation (NAG) requires the model to generate natural and complete answers [1,12], where the abovementioned extracting method cannot work well, and the model need to generate the complete answer sentence in its own words. For example, the MS MARCO dataset [1] contains well-formed abstractive answers generated by human editors, increasing the difficulty of generating complete answers with the model, but closer to the real world.

L. Wang et al. (Eds.): NLPCC 2021, LNAI 13029, pp. 151–162, 2021.
https://doi.org/10.1007/978-3-030-88483-3_12

One approach for achieving NAG is to stubbornly adopt extracting methods to extract spans that are close to the ground-truth natural answers, which reduces the difficulty of generating answers [19,20]. However, this approach is limited by the original expression in the paragraph, leading to an unbreakable bottleneck. In other words, some ground-truth natural answers cannot be directly extracted from the given paragraphs, which are originally generated by summarizing and rewriting some sections in the paragraphs.

Another approach is to use sequence-to-sequence (Seq2seq) way to generate answers word by word, showing greater potential than extracting methods. Since longer text increases the difficulty of training Seq2seq models, a compromised way is to extract the preliminary evidences and then feed them to the Seq2seq model to generate the final answer [18], but it still lacks enough training data. Using multi-source training datasets can alleviate the dearth of training data to some extent [15]. In theory, since the Seq2seq methods can generate the answers word by word through a pre-defined vocabulary, they have the ability to break through the limitation of the original expression of the paragraphs. They can generate each word by selecting words in the vocabulary or copying words [10] from the paragraphs.

However, problems still associated to the existing Seq2seq methods. For example, they are hard to be trained and cannot generate long answers. The generated long answers always show poor quality with some repeated or meaningless words. Besides, although the lack of training data was alleviated [15], it has not been resolved fundamentally.

To solve the aforementioned problems, we provide an insight into the NAG task. Usually, a natural answer can be composed of two parts. On one hand, it covers the meanings of the question and gives explanation of it. On the other hand, it contains the specific answer that is obtained from documents or paragraphs. For example, given a question "What's the average years of dentist college?" and paragraph context "...8 years of the cost of dental school...", the natural answer can be "An average dentist college is of 8 years.", where "An average dentist college..." is the explanation of the question and "8 years" can be directly extracted from the given paragraph.

In this paper, we propose to split the answering process into answer template generation and span extraction, which share one context encoder that allows deep interaction between them. Since the answer template is modified from the question using Text Editing method [6,9,13,25], it is quite related to the question, contributing to a **relevant** answer. At the same time, using a powerful encoder, we extract the answer span and fill it into the template to generate and form the **correct** answer. In addition, we propose to produce more candidate spans and select one that is more **fluent** in connection with the answer template. Detail advantages are shown in Sect. 5.6. Experiments show that the proposed method does improve the performance on NAG dataset.

The main contributions of this paper are threefold: (1) We propose to split the answering process into two branches (i.e., template generation and span extraction), which can form natural answers without generating it word by word.

(2) We introduce text editing method to generate the answer template based on the question, most words are generated through tags instead of using a huge vocabulary. (3) We provide two ways to select better span that fluently connects the generated template.

2 Related Words

2.1 Natural Answer Generation

Some previous works studied NAG based on knowledge base [8,11,23], most of which follow the Encoder-Decoder framework [3]. Recently, multi-document NAG task is catching more and more attention. Some methods regarded the NAG task as an extracting task [19,20], limited by the original context expression. Therefore, a more reasonable way for achieving NAG is to generate the answer word by word. Instead of feeding the whole context into the model, SNET [18] used the extracted preliminary evidences for generation. To solve the lack of training data, Masque [15] proposed to use other source of training datasets to train the NAG model. However, problems still exist when training a NAG model, especially for long answers and insufficient training data.

2.2 Text Editing

An effective way for text editing is to execute simple operations on the input. Deletion is a simple method for text editing, where each token is predicted to zero or one, corresponding to deletion decisions [7]. Similar operation on sentence-level has also been studied [14]. Beyond deleting operation, Reference [13] regarded the text editing as text tagging task with three editing operation (i.e., keeping a token, deleting it, and adding a phrase before the token). Similar to it, Reference [6] realized the adding operation by generating words one by one, which costs more inference time but can generate more diverse output. Apart from sequential data, tree structure data has also been edited in [22]. The lack of training data is also a severe problem. More instances were generated with fluency scores during training, allowing the model to correct a sentence incrementally through multi-round inference [9]. Denoising auto-encoder was proposed and was pre-trained with unlabeled data [25]. However, text editing method has not been used to solve the question answering or NAG task.

3 Splitting Answering Process into Template Generation and Span Extraction

In this task, for each instance, the input contains a question and multiple documents, and the output is a sentence of natural answer. Given a question and multiple documents, our model extracts an answer span from the documents, generates a template by text editing from the question, and then fill the template with the answer span to generate the final natural answer. The overall framework of our model is shown in Fig. 1

Since the main part of the natural answer (i.e., answer span) is extracted from the documents, and other explanation words that are related to the question and is rewrote from the question (i.e., answer template), the model can learn to form the natural answer more easily than directly generating the whole natural answer word by word.

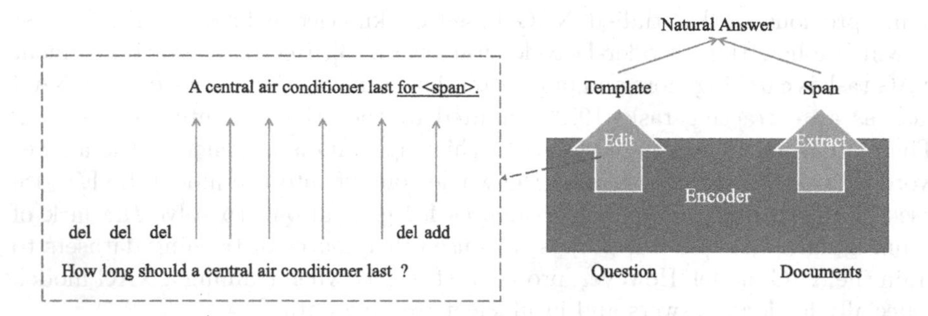

Fig. 1. The overview of our model. The right is framework that split the answering process into two branches, and the left is template generation based on the question.

3.1 Span Extraction

Given a question Q and multiple documents $\{P_i\}$, we concatenate them together and feed them into the encoder: $\mathbf{H} = \text{Encoder}(Q, \{P_i\})$. Since the documents are quite long (always $\geq$ 512 tokens), we use a Longformer model [2] for encoding.

The score of the j-th token in the encoded context being the start and the end of the answer span is calculated as:

$$p_{start}(y^j) = softmax(\text{MLP}_1(\mathbf{H}[j,:])),\ p_{end}(y^j) = softmax(\text{MLP}_2(\mathbf{H}[j,:])), \tag{1}$$

where MLP_1 and MLP_2 are two-layer Multi-Layer Perceptrons (MLP), and $[j,:]$ means picking all the hidden values for the j-th token.

During inference, we can select the answer span with the highest score. And another better way is to select multiple candidate spans to form a more fluent natural answer (illustrated in Sect. 4).

3.2 Template Generation with Editing

Since the natural answer is always relevant to the question and overlaps with it in many words, we generate the answer template based on the question in an editing way. In other words, the model receives the question and predicts a sequences of operation tags that are corresponding to the question words. Then we can use these tags to edit the question and generate the template.

We obtain the token-level question representation by extracting the representation from the encoder: $\mathbf{Q} = \mathbf{H}[0 : L_Q, :]$, where L_Q is the question length. Different from existing text editing methods, to reduce the vocabulary size, we design our template generation as Keeping, Deleting and Adding for each token in the question. Rather than predicting a sequence of tags one by one, we predict them in a parallel way with less time complexity.

Keeping or Deleting. In this part, we calculate the probability of each token to be kept or deleted. Each token is mapped into a binary classification. For the j-th token in the question:

$$\mathbf{b}_j = \text{softmax}(\text{MLP}_3(\mathbf{Q}[j, :])), \tag{2}$$

where $\mathbf{b}_j \in \mathbb{R}^2$ for a binary classification, one dimension is the probability of keeping the j-th token while the other is that of deleting it.

Adding. For each token, it's also required to select a word/phrase in a predefined vocabulary for generating new words/phrases that cannot be retrieved in the paragraphs.

The vocabulary V includes: (1) words and phrases with high frequencies that are obtained from the train-set; (2) a "*<null>*" symbol that means generating no word; (3) a "*<span>*" symbol that will be replaced with the extracted span (the replacement is illustrated in Sect. 3.3).

The probability of generating any new word in the position on j-th token is calculated by:

$$\mathbf{g}_j = \text{softmax}(\text{MLP}_4(\mathbf{Q}[j, :])), \tag{3}$$

where $\mathbf{g}_j \in \mathbb{R}^{|V|}$ for classification (i.e., selecting a word/phrase/symbol in V), $|V|$ is the total number of words/phrases/symbols in V.

If the model predicts Keeping and Adding at the same time, it adds the generated word/phrase after the original word.

Ground-Truth Template Tagging. Before training, an essential work is to convert the original answers into the tag sequences. For each training instance, the ground-truth template tagging is a three-step process:

(1) We match the span[1] in the natural answer.
(2) We tag the span in the natural answer with the symbol "*<span>*".
(3) Then similar to the algorithm in [13] that converts a target string to tags, we iterate over the words in the natural answer and greedily attempt to match them. The difference is that [13] predicts the keeping/deleting and adding together for each token, while we predict them separately to make the vocabulary smaller and the training easier.

[1] The span is provided in the dataset. If not, we can adopt Rouge-L or Edit Distance to match it in the documents with the natural answer.

3.3 Filling the Template

After extracting the span and generating the template, the next step is to simply fill the template with the span. If there are multiple "<*span*>" symbols in the template, we remove the latter ones to prevent duplication. Finally, we replace the "<*span*>" symbol with the extracted span that is done in Sect. 3.1, and obtain the final natural answer.

3.4 Training

Since multiple training objects are included in our model, the loss function is:

$$\begin{aligned} \mathcal{L} &= \mathcal{L}_{CE}(\hat{y}^{start}, y^{start}) + \mathcal{L}_{CE}(\hat{y}^{end}, y^{end}) \\ &+ \lambda_1 \cdot \mathcal{L}_{BCE}(\hat{z}^{kd}, z^{kd}) + \lambda_2 \cdot \mathcal{L}_{CE}(\hat{z}^{add}, z^{add}) \end{aligned} \tag{4}$$

where $\mathcal{L}_{CE}(\hat{y}^{start}, y^{start})$ and $\mathcal{L}_{CE}(\hat{y}^{end}, y^{end})$ are cross-entropy losses for span extraction. $\mathcal{L}_{BCE}(\hat{z}^{kd}, z^{kd})$ is the binary cross-entropy loss for Keeping/Deleting and $\mathcal{L}_{CE}(\hat{z}^{add}, z^{add})$ is the cross-entropy loss for Adding.

4 Selecting in Candidate Spans

During inference, to obtain the final answer with better fluency, it's essential to select a better span that fits the template (i.e., connect better with the template). Therefore, we suggest generating more candidate answers for selection.

Specifically, during span extraction (Sect. 3.1), rather than only extracting a single answer span, we extract more candidate spans with top-K scores. Then, for each candidate span, we calculate the connecting score with the template. The connecting score is composed of the forward score $p_{forward}(\cdot)$ and the backward score $p_{backward}(\cdot)$. One stands for the connecting fluency of the start of the candidate span and the other stands for that of the end of the candidate span. In this paper, we provide two different ways for calculation, shown in Fig. 2.

4.1 Using Statistic in Training Data

An easy way to obtain $p_{forward}(\cdot)$ and $p_{backward}(\cdot)$ is to summarize the train-set:

$$p_{forward}(start = y_i | T_{start=y_j}) = c(y_j, y_i)/c(y_j), \tag{5}$$

$$p_{backward}(end = y_i | T_{end=y_j}) = c(y_i, y_j)/c(y_j), \tag{6}$$

where $start = y_i$ means that the start token of the candidate span is y_i, and similarly, $end = y_i$ means that the end token of the candidate span is y_i. $T_{start=y_j}$ denotes the generated template where the token before the "<*span*>" symbol is y_j, and similarly, $T_{end=y_j}$ denotes that where the token after the "<*span*>" symbol is y_j. $c(y_i)$ is the appearance time of word "y_i", $c(y_i, y_j)$ is the appearance time of bi-gram "$y_i y_j$" that contains the two words.

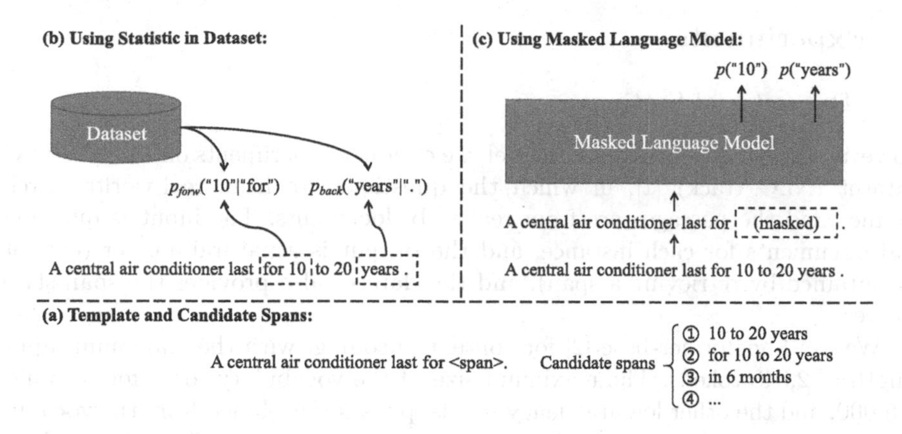

Fig. 2. Two ways for selection with candidate spans.

4.2 Using Masked Language Model

Another way is to introduce masked language model to predict the two scores. Since the masked language model have the ability of predicting masked tokens, this idea is reasonable. Specifically, we fill the template with each candidate span (Sect. 3.3) as a candidate answer, and feed it into a BERT with the span masked. The output of the BERT is denoted as $\mathbf{O} \in \mathbb{R}^{L \times |V'|}$, where L is the length of the candidate answer, $|V'|$ is the BERT's vocabulary size:

$$p_{forward}(start = y_i|T) = \mathbf{O}[s_0, i], \tag{7}$$

$$p_{backward}(end = y_i|T) = \mathbf{O}[s_{-1}, i], \tag{8}$$

where T denotes the generated template, s_0 and s_{-1} denote the start and end position of the candidate span in the candidate answer respectively, and y_i is the i-th token in the vocabulary V'.

4.3 Final Score

Finally, the overall score of a candidate span is calculated as:

$$\begin{aligned} \psi_(start = y^i, end = y^j) =& log(p_{start}(y^i)) + log(p_{end}(y^j)) \\ &+ \mu_1 \cdot log(p_{forward}(\cdot)) + \mu_2 \cdot log(p_{backward}(\cdot)) \end{aligned} \tag{9}$$

We select the span with highest score to form the final natural answer. However, in our experimental dataset, we find that only use the forward connecting score can lead to better performance. A probable reason may be that the spans are located in the ends of the sentences in many instances.

5 Experiments

5.1 Dataset and Settings

To verify the effectiveness of our model, we conduct experiments on MS-MARCO dataset (NLG track) [1], in which the questions are from real-world search engines and the passages are from real web documents. The input is question and documents for each instance, and the output is a natural answer (cannot be obtained by retrieving a span), and the dataset also provides the span-style answer.

We use Longformer-base [2] for context encoding, with the maximum input length of 2,048 tokens. The maximum size of the vocabulary (only for Adding) is 5,000, and the other low-frequency words/phases is excluded from the vocabulary. In span extraction, we choose top-5 spans as candidates. The weights of the loss function are $\lambda_1 = 5$, $\lambda_2 = 3$. When calculating final score, we set $\mu_1 = 0.02$, $\mu_2 = 0$. We use Adam with learning rate of 2e−5, learning rate warm-up over the first 1,000 steps and linear decay to 0.

5.2 Results

We conduct our experiments on MS MARCO dataset. The results are shown in Table 1. On the leaderboard, Cascade and VNET are extracting models, while SNET and Masque are generative models. Both extracting models and generative models show a gap to Human performance. Since the test set[2] is closed and not publicly available, we evaluate our model on the development set. Since Longformer[2] is an extracting model that cannot generate new words and the answers may be cut with fixed length, it cannot achieve high result. Our basic

Table 1. Results on MS MARCO dataset. Model annotated with "†" is on the development set.

Model	ROUGE-L	BLEU-1
Cascade [20]	35.1	37.4
SNET [18]	45.2	43.8
VNET [19]	48.4	46.8
Masque [15]	49.6	50.1
Human Performance	63.2	53.0
VNET† [18]	45.7	–
Longformer†	35.2	35.8
Our Model†	60.0	58.7
Our Model+sum†	**60.9**	**59.0**

[2] https://microsoft.github.io/msmarco/.

model achieves 60.0 ROUGE-L and 58.7 BLEU-1. Then, with the method proposed in Sect. 4.1, our model achieves 60.9 and 59.0 performance, which is even higher than our basic model.

5.3 Ablations

For ablation study, we remove the Deleting, Adding and the whole template (i.e., without Deleting and Adding, only with an encoder that extracts the answer span). The ablation result is shown in Table 2.

Table 2. Ablation results on MS MARCO development set.

Model	ROUGE-L	BLEU-1
Our Model	**60.0**	**58.7**
- Deleting	52.4	49.7
- Adding	54.9	54.2
- Template	35.2	35.8

When removing the Deleting module, the model can only add words in Sect. 3.2. The ROUGE-L drops at 7.6 and BLEU-1 drops at 9.0, showing that not all the words in the question are useful for generating the answer. When removing the adding module, the model can only delete the question words in Sect. 3.2. The performance drops at 5.1 ROUGE-L and 4.5 BLEU-1, showing that new words are required to form the natural answer. Finally, without the whole template (i.e., Longformer), the performance drops greatly, indicating that only adopt a extracting module is insufficient for a natural answer.

5.4 Discussion About Candidate Span Selection

In Sect. 4, we provide two ways for calculating the score to select a better candidate span that fits the generated template. In this sub-section, we discuss which way is better. The performance of the method of summarizing training data and using masked language model is shown in Table 3.

Table 3. Results with different ways of candidate span selection.

Model	ROUGE-L	BLEU-1
Our Model	60.0	58.7
Our Model+sum	**60.9**	**59.0**
Our Model+mlm	60.4	58.9

When summarizing the bi-gram posterior probability with training data, the ROUGE-L is improved at 0.9. And when use a masked language model to predict the posterior probability based on the template, the ROUGE-L is improved but less than that of the summarizing way. A possible reason is that the answers are different from our daily language to some extent. For example, the answers always have some fixed sentence patterns, as "The ... is ...". Differently, the dataset that is used to pre-train the language model may contain more diverse sentence pattern. For this reason, directly adopt a widely-trained masked language model may not be suitable for fixed sentence patterns.

Although the masked language model method does not show the best performance, it inspires future works. When the training data is noisy or lacking, the masked language model method may have better generalization ability.

5.5 Case Study

Figure 3 shows a case through our model. Given the question "When did Windows XP come out?", the model deletes four tokens "When", "did", "come" and "?", at the same time it adds two new ones "came" and "on *<span>*.". On the other model branch, it extracts the span "2021" from the paragraphs. Finally, by filling the span into the generated template, it forms the final natural answer.

Question:	When	did	Windows	XP	come	out	?
Keep/Del:	del	del	keep	keep	del	keep	del
Add:	\	\	\	\	came	\	on <span>.
Template:			Windows	XP	came	out	on <span>.
Final answer:			Windows	XP	came	out	on 2001.
Ground-truth:	Windows XP came out in August 2001.						

Fig. 3. A case generated by our model.

Although we didn't define a modifying operation, the model modifies the word "come" to a new word "came" using Deleting and Adding operation, showing the ability to generate diverse templates to form the answers. Besides, since the two branches are calculated in one encoder, they can interact with each other and cooperate well, so it generates new phrase "on *<span>*." in the template.

5.6 Advantages Compared to Extracting and Generative Models

Compared to extracting models, our model can match ground-truth answers to a higher degree. Specifically, our model can generate new words regardless of the paragraph original expression. In Fig. 4, with the MS MARCO dataset, we show the frequency histogram of the number of new words (cannot be extracted) in each instance. About 19% answers need no new word, they can be obtained by extracting one or multiple spans. On the other hand, more than 80% answers are relied on the new words and most of them require 6–12 new words.

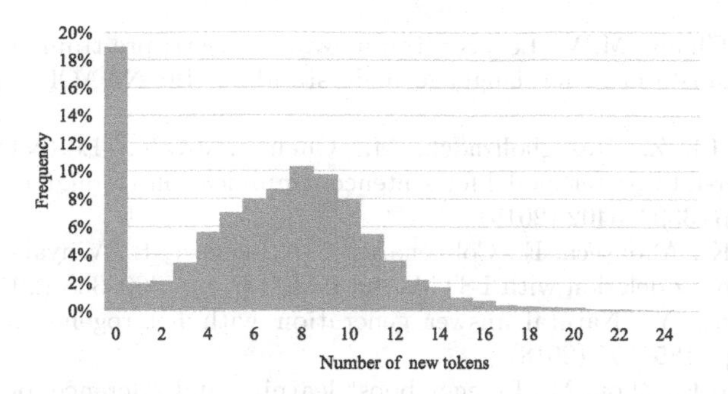

Fig. 4. Distribution of the number of words that cannot be extracted.

Compared to generative models, our model has the following advantages. Firstly, the answer span can be directly extracted, avoiding the trouble of generating or copying it word by word. Secondly, by designing the answer template that is modified from the question, most words are generated through the tags instead of using a huge vocabulary. Thirdly, the ways of predicting tags are parallel, which is less time-consuming than predicting them one by one. In short, our model decreases the learning difficulty and complexity.

6 Conclusion

In this paper, we proposed a NAG model that can generate relevant, correct and fluent answers. Since the template is edited from question, the answer is relevant to the question with more explanation of the question. And using extracting method with powerful encoder, our model has the ability to extract correct answer span. Finally, with more candidate spans that can be selected through summarizing training data or using masked language model, we can find a span that better fits the generated template and generate a fluent answer.

Acknowledgement. This work is partially supported by National Natural Science Foundation of China (Grants no. 61772568), Guangdong Basic and Applied Basic Research Foundation (Grant no. 2019A1515012029), and Youth science and technology innovation talent of Guangdong Special Support Program.

References

1. Bajaj, P., et al.: MS MARCO: a human generated machine reading comprehension dataset. arXiv preprint arXiv:1611.09268 (2016)
2. Beltagy, I., Peters, M.E., Cohan, A.: Longformer: the long-document transformer. arXiv preprint arXiv:2004.05150 (2020)
3. Cho, K., et al.: Learning phrase representations using RNN encoder-decoder for statistical machine translation. In: EMNLP, pp. 1724–1734 (2014)
4. Cui, Y., et al.: A span-extraction dataset for Chinese machine reading comprehension. In: EMNLP, pp. 5886–5891 (2019)

5. Devlin, J., Chang, M.W., Lee, K., Toutanova, K.: Bert: pre-training of deep bidirectional transformers for language understanding. In: NAACL, pp. 4171–4186 (2019)
6. Dong, Y., Li, Z., Rezagholizadeh, M., Cheung, J.C.K.: EditNTS: an neural programmer-interpreter model for sentence simplification through explicit editing. In: ACL, pp. 3393–3402 (2019)
7. Filippova, K., Alfonseca, E., Colmenares, C.A., Kaiser, Ł., Vinyals, O.: Sentence compression by deletion with LSTMs. In: EMNLP, pp. 360–368 (2015)
8. Fu, Y., Feng, Y.: Natural answer generation with heterogeneous memory. In: NAACL, pp. 185–195 (2018)
9. Ge, T., Wei, F., Zhou, M.: Fluency boost learning and inference for neural grammatical error correction. In: ACL, pp. 1055–1065 (2018)
10. Gu, J., Lu, Z., Li, H., Li, V.O.: Incorporating copying mechanism in sequence-to-sequence learning. In: ACL, pp. 1631–1640 (2016)
11. He, S., Liu, C., Liu, K., Zhao, J.: Generating natural answers by incorporating copying and retrieving mechanisms in sequence-to-sequence learning. In: ACL, pp. 199–208 (2017)
12. He, W., et al.: DuReader: a Chinese machine reading comprehension dataset from real-world applications. In: ACL, pp. 37–46 (2018)
13. Malmi, E., Krause, S., Rothe, S., Mirylenka, D., Severyn, A.: Encode, tag, realize: High-precision text editing. In: EMNLP, pp. 5054–5065 (2019)
14. Narayan, S., Cohen, S.B., Lapata, M.: Ranking sentences for extractive summarization with reinforcement learning. In: NAACL, pp. 1747–1759 (2018)
15. Nishida, K., et al.: Multi-style generative reading comprehension. In: ACL, pp. 2273–2284 (2019)
16. Rajpurkar, P., Zhang, J., Lopyrev, K., Liang, P.: Squad: 100,000+ questions for machine comprehension of text. In: EMNLP, pp. 2383–2392 (2016)
17. Seo, M., Kembhavi, A., Farhadi, A., Hajishirzi, H.: Bidirectional attention flow for machine comprehension. arXiv preprint arXiv:1611.01603 (2016)
18. Tan, C., Wei, F., Yang, N., Du, B., Lv, W., Zhou, M.: S-net: from answer extraction to answer generation for machine reading comprehension. arXiv preprint arXiv:1706.04815 (2017)
19. Wang, Y., et al.: Multi-passage machine reading comprehension with cross-passage answer verification. In: ACL, pp. 1918–1927 (2018)
20. Yan, M., et al.: A deep cascade model for multi-document reading comprehension. In: AAAI, pp. 7354–7361 (2019)
21. Yang, Z., et al.: HotpotQA: a dataset for diverse, explainable multi-hop question answering. In: EMNLP, pp. 2369–2380 (2018)
22. Yao, Z., Xu, F., Yin, P., Sun, H., Neubig, G.: Learning structural edits via incremental tree transformations. In: ICLR (2021)
23. Yin, J., Jiang, X., Lu, Z., Shang, L., Li, H., Li, X.: Neural generative question answering. In: IJCAI, pp. 2972–2978 (2016)
24. Yu, A.W., et al.: QANet: combining local convolution with global self-attention for reading comprehension. In: ICLR (2018)
25. Zhao, W., Wang, L., Shen, K., Jia, R., Liu, J.: Improving grammatical error correction via pre-training a copy-augmented architecture with unlabeled data. In: NAACL, pp. 156–165 (2019)

GeoCQA: A Large-Scale Geography-Domain Chinese Question Answering Dataset from Examination

Zhen Cui(✉), Bin Wang, and Jiangzhou Ju

National Key Laboratory for Novel Software Technology, Nanjing University, Nanjing 210023, China
{cuiz,wangbin,jujiangzhou}@smail.nju.edu.cn

Abstract. We present GeoCQA, the largest multiple-choice Chinese Question answering dataset in the geographic domain, evaluating the high-level reading ability of logic reasoning and prior geographic domain knowledge integration of a question answering (QA) model. GeoCQA contains 58,940 questions from real-world scenarios and has been collected from the high school geography examination which aims to evaluate students' mastery of the geographic concept and their ability to use geographic knowledge to solve problems. To investigate the challenges of GeoCQA to existing methods, we implement both rule-based and best neural methods and find that the current best method can achieve 71.90% of test accuracy, while unskilled humans and skilled humans can reach 80% and 96% accuracy respectively, which shows that GeoCQA is challenging to the current methods and the performance still has space to improve. We will release GeoCQA and our baselines to bring more data sources to the community and hope that it can help to promote much stronger Chinese QA models in the future (https://github.com/db12138/GeoCQA).

Keywords: Geography-domain question answering · OpenQA task · Retriever-reader methods

1 Introduction

With the advent of large-scale datasets [1–4,13,18], machine reading comprehension (MRC), a text-dependent Question Answering (QA) task where a model is required to comprehend a given text to answer questions, has achieved a great progress. However, in real-world QA scenarios, the passages or evidence associated with the question are not labeled as it is in the MRC task. In this case, models are required to find and extract relevant information to questions from large-scale text sources [14,19]. This type of task setting is called open-domain question answering (OpenQA), which has recently attracted lots of attention from the NLP community but remains far from being solved [19–24].

L. Wang et al. (Eds.): NLPCC 2021, LNAI 13029, pp. 163–175, 2021.
https://doi.org/10.1007/978-3-030-88483-3_13

Table 1. An example to illustrate the spatial location of city is crucial to answer the question.

Question: 下列地区中，使用太阳能热水器效果最好的是 Of the following cities, which city has the best effect of using solar water heater?
Option: ✗A.海口 ✗B.重庆 ✗C.上海 ✓D.吐鲁番 ✗A.HaiKou ✗B.ChongQing ✗C.ShangHai ✓D.Turpan
Explanation of human experts: 我国太阳能资源最丰富地区为:青藏高原;其次为内蒙古高原和西北地区。吐鲁番位于青藏高原上,所以太阳能热水器效果最好。 The most abundant solar energy resources in China are: Qinghai Tibet Plateau; Secondly, Inner Mongolia Plateau and northwest area. Turpan is located on the Qinghai Tibet Plateau, so the solar water heater has the best effect.

However, the answer format of most previous works [13,14,19] is span-based facts that are explicitly expressed in the text, which makes the models only have the ability of surface word matching [25]. In order to develop more advanced reading skills, another series of studies focus on tasks in the multiple-choice form such as ARC [15], OpenBookQA [16], and QASC [17]. These datasets bring more challenges, which require the model to retrieve relevant materials and to have high-level reading skills such as logical reasoning and prior knowledge integration, and to overcome the interference of distractor options. Unfortunately, these datasets consist of questions that require only elementary-level knowledge and are crowd-sourced, and the distractor options are automatically generated, bringing a significant amount of noise. So models trained on them may be unable to support more sophisticated real-world scenarios.

To address the aforementioned limitations and promote the development of QA techniques in real-world scenarios for Chinese, we introduce GeoCQA, a large-scale geographic-domain Chinese question answering dataset representing a demanding real-world scenario. The dataset contains 58,940 multiple-choice questions collected from geographic exams and mock tests for high-school Chinese students, which is the largest geography-domain Chinese question answering dataset. GeoCQA coverages a broad range of geographic topics including natural geography, population, urbanization, geographic information system, and so on. The questions were generated by human experts to evaluate the ability of students to comprehend geographic concepts and the ability to use geography knowledge to answer questions. The dataset is collected from digital textbooks and the website. We also provide a geographic knowledge corpus which contains the majority of text material to answer the questions.

The special features of questions in GeoCQA lie in it requires ability of deep understanding of the geographic concept and the ability to perform logic reasoning over prior geographic domain knowledge. Specifically, the ability to answer questions in GeoCQA can be summarized as follows: (1) For questions related to physical geography, the attribute information (such as spatial location, climate, topography) of the mentioned geographic entities is crucial to answering

Table 2. An example to illustrate that national policy is the main reason behind the phenomena.

Question: 广东省东莞大朗镇于1979年引进第一家毛织厂。上世纪90年代初，毛织产业已颇具规模。2002年该镇被中国纺织工业协会授予“中国羊毛衫名镇”称号。90年代初期大朗毛织业发展较快的主要原因是? Dalang Town, Dongguan, Guangdong Province introduced its first wool weaving factory in 1979. In the early 1990s, the wool weaving industry had a large scale. In 2002, the town was awarded the title of “famous town of Chinese woolen sweater” by China Textile Industry Association. What is the main reason for the rapid development of Dalang wool industry in the early 1990s?
Option: ✗A.当地劳动力丰富 ✗B.销售市场广阔 ✓C.国家政策支持 ✗D.科技水平高 ✗A.The local labor force is abundant ✗B.Broad sales market ✓C.National policy support ✗D.High level of science and technology
Explanation of human experts: 东莞改革开放政策实行的早，因此初期大朗毛织业发展较快的主要原因是国家政策支持。 The “Reform and opening-up policy” was implemented early in DongGuan, so the main reason for the rapid development of Dalang’s wool industry in the early stage was the support of national policies.

the questions. For example, in Table 1, we need to know the spatial location of each city in each option to figure out the solar energy resources of the city. (2) For questions related to cultural geography, The factors related to human activities (such as policy, population, urbanization, transportation, and so on) are necessary to analyze the reasons behind some phenomena. For example, in Table 2 we need to know that “China’s reform and opening-up policy in the 1990s” is the main reason for the phenomenon of “the rapid development of the woolen industry”. (3) The ability to perform logical reasoning over prior geographic knowledge is crucial to answering questions in GeoCQA. GeoCQA requires multiple reasoning abilities to answer the questions, including word matching, multi-hop reasoning, and numerical analysis. The detailed analysis can be found in the section of Reasoning Types.

To investigate the challenges of GeoCQA, We make a detailed statistical analysis of GeoCQA and implement both the rule-based method and several neural models for reading comprehension following the standard OpenQA framework [19], which is a two-stage pipeline and consists of an evidence retriever for finding relevant text material and a document reader that perform machine reading comprehension over the retrieved evidence. Experiment results have shown that even the model based on large pre-trained models can achieve 71.90% of test accuracy, while unskilled humans and skilled humans can reach 80% and 96% accuracy respectively, indicating the challenge of this dataset. Our error analysis shows that existing OpenQA methods still suffer from the inability of conducting multi-hop reasoning over the retrieving process.

We hope GeoCQA can help to develop more powerful Chinese OpenQA models which better support logical reasoning in both retrieval and reading comprehension stages.

2 Related Work

2.1 Machine Reading Comprehension

In recent years, numerous MRC datasets [1–4,9,12] have been proposed. MRC is a text-dependent QA task where a model is required to comprehend a given text to answer questions. The documents related to the questions in those datasets are annotated by humans. Datasets [10,13] contain multiple passages for each question. [25] provide a Chinese MRC dataset requiring prior knowledge integration. [11] studies the comprehension of Chinese idiom. Some neural models based on large pre-trained language models such as BERT [6], RoBERTa [7], and ALBERT [8] have made great progress in the MRC task, even achieved nearly saturated performance on some popular datasets [1,2].

2.2 Open-Domain Question Answering

In real-world scenarios, relevant text materials are expensive to obtain. Under the setting of OpenQA task [19], models need to retrieve relevant evidence and comprehend them for answering the questions. As a preliminary trial, [19] proposes the DrQA model, which is a two-stage pipeline consists a TF-IDF retriever to obtain relevant documents from Wikipedia and a trained reading comprehension model to extract the answer from the retrieved documents. And more sophisticated models have been introduced which either rerank passages retrieved before reading comprehension [20,21] or filter out those irrelevant retrieved texts [23,24]. The answer in the datasets mentioned above is span-based, extract from the document. Most questions' answer are facts that are explicitly expressed in the text, which makes the models only have the ability of surface word matching [25].

In order to develop more advanced reading skills. Another series of studies focus on multiple-choice QA datasets [15–17]. The performance of models on these datasets can be evaluated objectively because machine performance and human performance use the same evaluation metrics, accuracy. These datasets bring more challenges, which require the model to have high-level reading skills such as logical reasoning and prior knowledge integration and to overcome the interference of distractor options. However, these OpenQA datasets consist of questions that require only elementary-level knowledge. Options and questions of these datasets are either crowd-sourced or automatically generated, bringing a significant amount of noise. These datasets can not support more sophisticated real-world scenarios.

Table 3. Overall statistics of GeoCQA. Question/Option length and vocabulary size are calculated in Chinese characters. "Avg./Max. Question Len" represents "Average/Maximum option length".

	Train	Dev	Test
Question count	44192	7374	7374
Option count	176768	29496	29496
Avg./max. question len	28.51/375	28.80/354	28.89/325
Avg./max. option len	8.12/31	8.17/27	8.25/33

Table 4. The information of knowledge corpus

(a) An example of knowledge snippet

台湾西海岸盛产海盐成因：西海岸有广阔的淤泥质海滩地处台湾山脉的背风地带，气流下沉,燕发旺盛。

(b) The statistics information of the geographic knowledge corpus

The Number Of Snippets:	175084
Avg./Max. length:	110.40/250
Corpus Size:	53M

2.3 Comparison with Other Datasets

Although efforts have been made with a similar motivation, [26] introduced MEDQA for solving medical problems in professional medical board exams, and [27] introduce JEC-QA collected from the National Judicial Examination of China to promote legal-domain Question. GeoCQA focuses on the geographic domain and provides more large-scale questions. GeoCQA can help expand domain coverage of OpenQA models rather than biased to a single domain. In the geographic domain QA, GeoSQA [28] focuses on scenario-based question answering depending on diagrams (maps, charts) which have been manually annotated with natural language description. However, the size of the dataset is relatively small (4110 questions), which is not enough for training powerful deep neural networks whose success relies on the availability of relatively large training sets. GeoCQA only focuses on non-diagrams questions in which challenges are far from being solved.

3 Dataset Collection and Analysis

3.1 Dataset Collection

Questions and Answers. We collected the questions and their associated options and gold answer label from high-school-level geographic exams and mock tests. The sources come from digital textbooks via OCR and websites. All the texts obtained by OCR have been manually checked and corrected. These questions aim to evaluate students' ability to understand geographic concepts and integrate geographic knowledge to answer questions. There are four candidate options for each question and only one is correct. The correct answer for each question is labeled in the dataset. After removing duplicated questions there are 58,940 questions in GeoCQA. We randomly split the dataset with 75% training, 12.5% development, and 12.5% for the test set. The overall statistics of the dataset are summarized in Table 3. The percentage of each option as the correct answer in the whole dataset is nearly 25%, ensuring that models will not be biased to one option.

Document Collection. Geographic knowledge is needed to answer each question in GeoCQA, Students usually need to understanding and memorize a lot of geographic knowledge from a volume of geographic textbooks during years of training. Similarly, for QA models to be successful in this task, we need to grant models access to the same collection of text materials as humans have. Therefore, we collected text materials from geographic textbooks, teaching materials, and the Baidu encyclopedia to form a geographic knowledge corpus. The size of the corpus is 54M, which includes the majority of knowledge to answer questions. For the convenience of retrieval. We split the corpus into knowledge snippets. An example is shown in Table 4a. The total number of knowledge snippets is 175084. The statistics of the corpus are summarized in Table 4b.

3.2 Reasoning Types

Following the previous work [2,15], we summarize 3 different reasoning types required for answering questions in GeoCQA. We sampled 200 questions and got the estimation of the proportion of three types of questions. The statistical information is shown in Table 6a.

(1) **Word matching** is the simplest reasoning type which requires the model only to check which option is matched with the relevant evidence and the relevant evidence can be obtained through a simple retrieval strategy based on lexical-level semantic matching. Because the context of the question and the context of evidence is highly consistent. Questions in this reasoning type widely exist in traditional MRC tasks. The example is shown in Table 5a.
(2) **Numerical Analysis.** This type of reasoning requires models to have the ability of numerical comparison or even need to obtain a calculation formula and fit the correct value to the formula to get the correct answer in some cases. As shown in Table 5b. The model needs to compare the values mentioned in the options to select the largest plotting scale.
(3) **Multi-hop reasoning.** This type of question requires models to perform multi-hop reasoning over several evidence snippets to get the answer. Both retrieving relevant evidence and answering questions require the ability of logical reasoning. At least one of the relevant pieces of evidence is hard to retrieve based on lexical-level matching. The example is shown in Fig. 1. We can see that both evidence retrieving and reasoning over evidence is challenging.

Table 5. The examples of reasoning types

(a) An example reasoning type of "Word Matching". The context of the question and the relevant evidence is highly consistent. And Option C best matches the evidence.

Question: 一个国家或地区水资源的多少，通常用到哪一指标来衡量
Which indicator is used to measure the amount of water resources in a region ?
Option:
✗A.多年平均降水量 ✗B.多年平均蒸发量 ✓C.多年平均径流总量 ✗D.多年降水量与蒸发量之和
✗A.Average annual precipitation ✗B.Annual average evaporation ✓C.Annual average runoff ✗D.Sum of annual precipitation and evaporation
relevant evidence:
一个地区或一个国家水资源的丰歉程度，通常用**多年平均径流总量**来衡量。
The abundance of water resources in a region or a country is usually measured by the **average annual runoff**.

(b) An example reasoning type of "Numerical Analysis", the model needs to compare the values mentioned in the options to select the largest plotting scale.

Question: 在下列比例尺中，最大的是
Which of the following scales is the largest?
Option:
✗A:1：1500000 ✓B.五十万分之一 ✗C.1：5000000 ✗D.一百万分之一
✗A:1：1500000 ✓B.one part in 500,000 ✗C.1：5000000 ✗D.one part in 1000,000
relevant knowledge:
本题考查比例尺。分母越小，则数值越大，即比例尺越大
This question examines the plotting scale. The smaller the denominator is, the larger the numerical value is, that is, the larger the plotting scale is.

4 Experiments

We implement a classical rule-based method and representative state-of-the-art neural models. For evaluating human performance, skilled humans are college students who have participated in the college entrance examination and obtained high scores. In contrast, unskilled humans are college students who have not attended. We invited five people in each group to complete 50 questions sampled from the test set independently. Humans can read the retrieved paragraph mentioned in Sect. 4.1 as models do.

4.1 Rule-Based Method

Information Retrieval (IR). We apply the off-the-shelf text retrieval system built upon Apache Lucene, ElasticSearch,[1] to build a search engine indexing over the geographic knowledge corpus we collected. For each question q and each option o_i, we use $q + o_i$ as the search query and send it to the search

[1] https://www.elastic.co/.

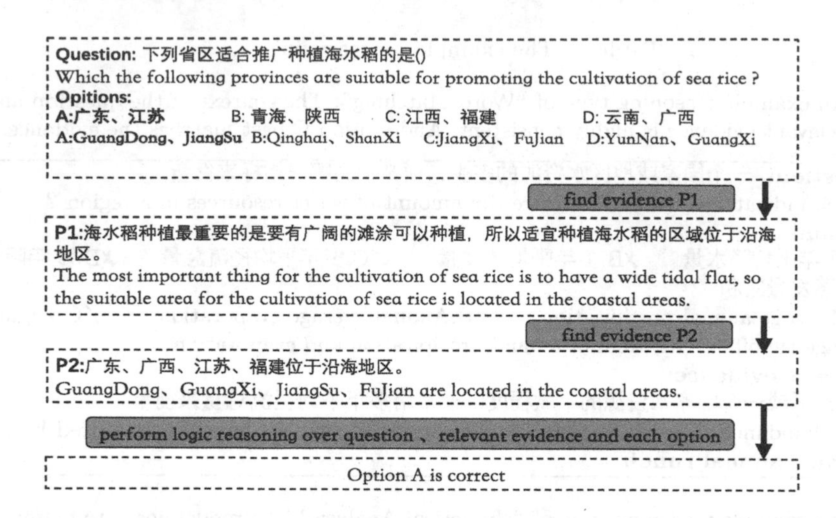

Fig. 1. An example of the reasoning type of multi-hop reasoning. The models must first find evidence **P1** shows that "suitable area for the cultivation of sea rice is located in the coastal areas", and then reason out the next evidence needed is about which province mentioned in the options is located in the coastal areas. Evidence **P2** is difficult to find through single-hop lexical-level semantic matching. Moreover, the models must have the ability to perform multi-hop reasoning over evidence snippets **P1** and **P2** to get the correct answer.

engine, and then we get top-N (N = 3) evidence snippets and corresponding score. And we calculate the average score of these N scores. This is repeated for all options and the option with the highest average score is selected as answer.

4.2 Neural Models

We follow the standard OpenQA framework which is a two-stage pipeline consisting of (1) a Document Retriever module for finding relevant evidence snippets (2) A Document Reader, a reading comprehension model to predict the correct answer over the question, options, and the evidence snippets retrieved in the first stage.

Retrieval Strategy. We use the same IR system described in Sect. 4.1. For each option, we use $q + o_i$ as the query to retrieve top-N (N = 3) relevant evident snippets and concatenate them into a long sequence c_i, and then pass them to the Document Reader for reading comprehension.

Neural Reader Models

Co-Matching: We implement Co-Matching [29], a Bi-LSTM-based model, which explicitly treats the question and the candidate answer as two sequences and jointly matches them to the given passage. The model is a single-paragraph reading comprehension model for multi-choice QA problems and achieves promising results on the RACE [2] dataset.

Table 6. The statistic information of reasoning types

(a) The Statistic information on the percentage of reasoning types

Type	Percentage
Word Matching	46.0%
Numerical Analysis	2.5%
Multi-hop reasoning	51.5%

(b) The percentage of reasoning types on 100 sampled questions from bad cases

Type	Percentage
Word Matching	18.0%
Numerical Analysis	7.0%
Multi-hop reasoning	75.0%

Table 7. The example shows a description of "the geographical characteristics of the middle and lower reaches of the Yangtze River Plain", We use the same whole word mask strategy to mask some geographic items for MLM.

长江中下游平原地理特征:地势低平、河湖众多、气候温暖、土壤肥沃,自古就是我国重要的种植业区和淡水鱼产区,为著名的"鱼米之乡"
[MASK][MASK]中下游[MASK][MASK]地理特征:地势低平、河湖众多、气候温暖、土壤肥沃,自古就是我国重要的种植业区和淡水鱼产区,为著名的"鱼米之乡"

Fine-Tuning Pre-trained Language Models: We noticed that on some popular MRC benchmarks, such as [1,2,5]. Most of the top-performing models are based on BERT and its variants, indicating that BERT has become new fundamental component in NLP. So, we adopt BERT and its variants pre-trained in Chinese corpus as our baseline reader models.

Our input sequence for BERT is constructed by concatenating [CLS], tokens in c_i, [SEP], tokens in $q + o_i$, [SEP], where [CLS] is a special token used for classification and [SEP] is sentence separator. We denote the hidden state output of [CLS] token as $\mathbf{H}_{cls} \in \mathbb{R}^{h\times 1}$ where h is the hidden size. We introduce a classification layer $W \in \mathbb{R}^{1\times h}$ and obtain the unnormalized probability $p(o_i|q, c_i) = W\mathbf{H}_{cls} \in \mathbb{R}^1$. We obtain the final prediction by applying a softmax layer over the unnormalized probabilities of all options associated with q and picking the option with the highest probability.

Infusing Geographic Knowledge into Pre-trained Language Modeling: Plenty of recent work [30–32] shows further pre-training on domain-specific corpus can help improve the performance of downstream tasks. [33] infuse disease knowledge into BERT through training BERT to infer the corresponding disease name and aspect from a disease-descriptive text. The knowledge infusion task is based on the BERT pre-training task, masked language modeling (MLM), and achieved performance improvement on downstream QA tasks. Inspired by this task, we performed the similar geographic knowledge infusion process before fine-tuning on QA tasks.

We masked geographic terms using the same mask strategy in [34] and train BERT to infer the masked term based on the corresponding description of geographical terms. As the example in Table 7 shown, BERT, infer masked term "长江" and "平原" via masked language modeling task.

Table 8. Performance of baselines in accuracy (%) on GeoCQA dataset. "KI" indicating knowledge infusion.

Methods	Dev	Test
Unskilled Humans	–	80% (40/50)
Skilled Humans	–	96% (48/50)
IR	41.09	39.09
Co-Matching	61.51	60.56
BERT-wwm-ext	66.63	67.13
RoBERTa-wwm-ext	68.51	68.33
RoBERTa-wwm-ext-large	**69.41**	**70.08**
BERT-wwm-ext +KI	67.90	67.44
RoBERTa-wwm-ext + KI	68.92	68.77
RoBERTa-wwm-ext-large + KI	**71.83**	**71.90**

4.3 Experiment Setting

Co-Matching: We replace the English tokenizer with a Chinese word segmenter in JieBa.[2] We use the 300-dimensional Chinese word embeddings released by [35] and the hidden size of model layer is 256.

Fine-Tuning Pre-trained Language Models: We use the pre-trained language models for Chinese released by [34]: Chinese BERT-base (denoted as **BERT-wwm-ext**), RoBERTa-Base (denoted as **ROBERTA-wwm-ext**), and RoBERTa-large (denoted as **ROBERTA-wwm-ext-large**) with whole word masking during pre-training over large corpus.

We set the learning rate and effective batch size to 2×10^{-5} and 18 for base model and 1×10^{-5} and 6 for large models. We truncate the longest sequence to 512 tokens (only truncate retrieved evidence). We fine-tune models for 16 epochs, set the warm-up steps to 2000, and keep the default values for the other hyper-parameters.

Geographic Knowledge Infusion: We randomly select a percentage of 30% words for whole word masking, and we set the learning rate as 1×10^{-5} for batch size as 16 and take 2 epochs to train MLM task.

4.4 Baseline Results

Table 8 summarizes the performance of all baselines on the GeoCQA. The result shows that the performance of pre-trained language models is better than that of the Co-Matching model based on Bi-LSTM. And the performance of pre-trained language models can be further enhanced by geographic knowledge infusion. However, even the best-performed model can only achieve an accuracy of 71.90%, indicating the great challenge of the dataset.

[2] https://github.com/fxsjy/jieba.

4.5 Error Analysis

We analyzed 100 bad cases that the model predicted wrong in the test set. The percentages of the three reasoning types on questions of these 100 bad cases are shown in Table 6b. Compared with the percentages of the reasoning types of all questions shown in Table 6a, The percentage of word matching decreased significantly, while the percentage of multi-hop reasoning and numerical analysis increased. It shows that current models do not perform well on the problems of multi-hop reasoning and numerical analysis. We also observed the retrieval results and found that it is difficult to retrieve all the evidence snippets needed to solve the problem using the existing BM25 based retrieval method. This is also the main bottleneck that leads to poor performance in multi-hop problems.

5 Conclusion

In this work, we present GeoCQA, a large-scale challenging Chinese QA dataset from real-world examinations. We implement several state-of-the-art neural reading comprehension models following the framework of OpenQA. The experimental result demonstrates that existing methods cannot perform well on GeoCQA. We hope GeoCQA can benefit researchers to develop more powerful OpenQA models that can be able to solve problems in real-world scenarios.

Acknowledgements. This work was supported by the National Key R&D Program of China (No. 2018YFB1005102) and the NSFC (No. 61936012,61976114).

References

1. Rajpurkar, P., Zhang, J., Lopyrev, K., Liang, P.: Squad: 100, 000+ questions for machine comprehension of text. In: EMNLP 2016 (2016)
2. Lai, G., Xie, Q., Liu, H., Yang, Y., Hovy, E.H.: RACE: large-scale reading comprehension dataset from examinations. In: EMNLP 2017 (2017)
3. Hermann, K.M., et al.: Teaching machines to read and comprehend. In: NIPS 2015 (2015)
4. Yang, Z., Qi, P., Zhang, P., Bengio, Y., et al.: HotpotQA: a dataset for diverse, explainable multi-hop question answering. In: EMNLP 2018 (2018)
5. Reddy, S., Chen, D., Manning, C.D.: CoQA: a conversational question answering challenge. Trans. Assoc. Comput. Linguist. **7**, 1–18 (2019)
6. Devlin, J., Chang, M.W., Lee, K., Toutanova, K.: BERT: pre-training of deep bidirectional transformers for language understanding. In: NAACL 2019 (2019)
7. Liu, Y., et al.: RoBERTa: a robustly optimized BERT pretraining approach. CoRR (2019)
8. Lan, Z., Chen, M., Goodman, S., Gimpel, K., Sharma, P., Soricut, R.: ALBERT: a lite BERT for self-supervised learning of language representations. In: ICLR 2020 (2020)
9. Cui, Y., Liu, T., Che, W., Xiao, L., Chen, Z., Ma, W., et al.: A span-extraction dataset for Chinese machine reading comprehension. In: EMNLP-IJCNLP 2019 (2019)

10. He, W., Liu, K., Liu, J., et al.: DuReader: a Chinese machine reading comprehension dataset from real-world applications. In: Workshop@ACL 2018 (2018)
11. Zheng, C., Huang, M., Sun, A.: ChID: a large-scale Chinese idiom dataset for cloze test. In: ACL 2019 (2019)
12. Trischler, A., et al.: NewsQA: a machine comprehension dataset. In: Rep4NLP@ACL 2017 (2017)
13. Joshi, M., Choi, E., Weld, D.S., Zettlemoyer, L.: TriviaQA: a large scale distantly supervised challenge dataset for reading comprehension. In: ACL 2017 (2017)
14. Dunn, M., Sagun, L., Higgins, M., Güney, V.U., Cirik, V., Cho, K.: SearchQA: a new q&a dataset augmented with context from a search engine. CoRR (2017)
15. Clark, P., Cowhey, I., Etzioni, O., Khot, T., Sabharwal, A., et al.: Think you have solved question answering? Try ARC, the AI2 reasoning challenge. CoRR (2018)
16. Mihaylov, T., Clark, P., Khot, T., Sabharwal, A.: Can a suit of armor conduct electricity? A new dataset for open book question answering. In: EMNLP 2018 (2018)
17. Khot, T., Clark, P., Guerquin, M., Jansen, P., Sabharwal, A.: QASC: a dataset for question answering via sentence composition. In: AAAI 2020 (2020)
18. Richardson, M., Burges, C.J.C., Renshaw, E.: MCTest: a challenge dataset for the open-domain machine comprehension of text. In: EMNLP 2013 (2013)
19. Chen, D., Fisch, A., Weston, J., Bordes, A.: Reading Wikipedia to answer open-domain questions. In: ACL 2017 (2017)
20. Wang, S., et al.: R^3: reinforced ranker-reader for open-domain question answering. In: AAAI 2018 (2018)
21. Wang, S., Yu, M., Jiang, J., Zhang, W., Guo, X., Chang, S., et al.: Evidence aggregation for answer re-ranking in open-domain question answering. In: ICLR 2018 (2018)
22. Clark, C., Gardner, M.: Simple and effective multi-paragraph reading comprehension. In: ACL 2018 (2018)
23. Lin, Y., Ji, H., Liu, Z., Sun, M.: Denoising distantly supervised open-domain question answering. In: ACL 2018 (2018)
24. Das, R., Dhuliawala, S., Zaheer, M., McCallum, A.: Multi-step retriever-reader interaction for scalable open-domain question answering. In: ICLR 2019 (2019)
25. Sun, K., Yu, D., Yu, D., Cardie, C.: Investigating prior knowledge for challenging Chinese machine reading comprehension. Trans. Assoc. Comput. Linguist. **8**, 141–155 (2020)
26. Jin, D., Pan, E., Oufattole, E., Weng, W., et al.: What disease does this patient have? A large-scale open domain question answering dataset from medical exams. CoRR (2020)
27. Zhong, H., Xiao, C., Tu, C., Zhang, T., Liu, Z., Sun, M.: JEC-QA: a legal-domain question answering dataset. In: AAAI 2020 (2020)
28. Huang, Z., Shen, Y., Li, X., Wei, Y., et al.: GeoSQA: a benchmark for scenario-based question answering in the geography domain at high school level. In: EMNLP 2019 (2019)
29. Wang, S., Yu, M., Jiang, J., Chang, S.: A co-matching model for multi-choice reading comprehension. In: ACL 2018 (2018)
30. Gururangan, S., Marasovic, A., Swayamdipta, S., Smith, N.A.: Don't stop pre-training: adapt language models to domains and tasks. In: ACL 2020 (2020)
31. Lee, J., et al.: BioBERT: a pre-trained biomedical language representation model for biomedical text mining. Bioinformatics **36**, 1234–1240 (2020)
32. Beltagy, I., Lo, I., Cohan, A.: SciBERT: a pretrained language model for scientific text. In: EMNLP 2019 (2019)

33. He, Y., Zhu, Z., et al.: Infusing disease knowledge into BERT for health question answering, medical inference and disease name recognition. In: EMNLP 2020 (2020)
34. Cui, Y., Che, W., Liu, T., Qin, B., Wang, S., Hu, G.: Revisiting pre-trained models for Chinese natural language processing. In: EMNLP 2020 (2020)
35. Li, S., Zhao, Z., Hu, R., Li, W., Liu, T., Du, X.: Analogical reasoning on Chinese morphological and semantic relations. In: ACL 2018 (2018)

Dialogue Systems

Generating Informative Dialogue Responses with Keywords-Guided Networks

Heng-Da Xu, Xian-Ling Mao(✉), Zewen Chi, Fanshu Sun, Jingjing Zhu, and Heyan Huang

School of Computer Science and Technology, Beijing Institute of Technology, Beijing, China
{xuhengda,maoxl,czw,fanshusun,zhujingjing,hhy63}@bit.edu.cn

Abstract. Recently, open-domain dialogue systems have attracted growing attention. Most of them use the sequence-to-sequence (Seq2Seq) architecture to generate dialogue responses. However, traditional Seq2Seq-based open-domain dialogue models tend to generate generic and safe responses, which are less informative, unlike human responses. In this paper, we propose a simple but effective Keywords-guided Sequence-to-sequence model (KW-Seq2Seq) which uses keywords information as guidance to generate open-domain dialogue responses. Specifically, given the dialogue context, KW-Seq2Seq first uses a keywords decoder to predict a sequence of topic keywords, and then generates the final response under the guidance of them. Extensive experiments demonstrate that the keywords information can facilitate the model to produce more informative, coherent, and fluent responses, yielding substantive gain in both automatic and human evaluation metrics.

Keywords: Dialogue system · Keywords-guided networks · Response generation

1 Introduction

Open-domain dialogue systems play an important role in the communication between humans and computers. However, it has always been a big challenge to build intelligent agents that can carry out fluent open-domain conversations with people [27]. Unlike the intelligence assistant agents that only focus on one or several predefined tasks, such as finding restaurants or booking tickets [9], open-domain dialogue agents should talk with people in broader topics [26]. People hope them to produce utterances not only fluent but also diverse, informative, and emotional [7,19,21]. In the early decades, people started to build open-domain chatbots with plenty of human-designed rules [24]. Recently, with the accumulation of data and the advancement of neural networks, more and more neural-based open-domain dialogue systems come into people's sight and achieve good results [7,27].

L. Wang et al. (Eds.): NLPCC 2021, LNAI 13029, pp. 179–192, 2021.
https://doi.org/10.1007/978-3-030-88483-3_14

Table 1. An example of a dialogue with keywords. The person would refuse the party invitation, so he first prepares the main idea in his mind (the keywords), and then express them to fluent sentences.

Context	Would you like to come to the party on Saturday night?
Keywords	sorry; want; but; finish; paper; weekend;
Response	I'm so **sorry**. I really **want** to go **but** I have to **finish** my **paper** on the **weekend**

The sequence-to-sequence (Seq2Seq) architecture has been empirically proven to be quite effective in building neural-based open-domain dialogue systems. Dialogue models based on Seq2Seq architecture don't require plenty of manually designed rules, but directly learn a mapping relation from the input to output utterances in a pure end-to-end manner [22]. However, Seq2Seq models tend to generate generic and less informative sentences such as *I'm not sure* and *I don't know*, which hinders the progress of conversations [26]. Many methods are proposed to alleviate the problem, such as equipping the maximum likelihood objective with the maximum mutual information (MMI) constraint [7], leveraging latent variables to model the one-to-many mapping from the input to output utterances [20,27], and using boosting to improve the response diversity [3]. All these methods generate dialogue responses in one step. However, the decoder has to complete the two tasks at the same time: predicting the main idea of the responses and organizing natural sentences. It's too hard for the decoder model to generate fluent and informative dialogue responses at the same time.

Intuitively, when someone prepares to say something in a conversation, he usually first conceives an outline or some keywords of what he wants to say in his mind, and then expands them into grammatical sentences. As Table 1 shows, the person wants to refuse the invitation to the party, so he first prepares the reason in his mind, which is represented by the keywords, and then organized the keywords to fluent and natural sentences. If a dialogue system explicitly models the two steps in human dialogues, the generated responses would be more specific and informative than the responses of traditional models.

There have been some works exploiting the usage of keywords to facilitate informative dialogue response generation. For example, Mou et al. [12] use the maximum pointwise mutual information (PMI) principle to pick up a word from the vocabulary as the keyword. The independent, untrainable keywords choosing method can only choose one noun as the response keyword. Another work is the MrRNN model [17], which models the keywords and response generation in a hierarchical recurrent encoder-decoder (HRED) framework [18]. Similarly, it generates a sequence of nouns or task-specific entities as keywords. Recently, Pandey et al. [15] propose the Mask & Focus model, which picks out keywords by masking context words of a trained conversation model and computing the pointwise mutual information (PMI) between the masked words and the response. However, it's computationally expensive to run an extra conversation model multiple times. The variational training procedure in it is also quite complicated. We propose that the main idea of utterances should consist of any kind of words (e.g. nouns, verbs, adjectives), and the model architecture and training procedure should be as simple as possible.

In this paper, we propose a novel Keywords-guided Sequence-to-sequence model (KW-Seq2Seq), which imposes more relaxed constraints on keywords than previous methods. Besides the standard encoder and decoder components, KW-Seq2Seq has an additional pair of encoder and decoder to deal with keywords information, i.e. the *keywords encoder* and *keywords decoder*. In our model, keywords are a sequence of any kind of words. After the dialogue context is mapped to its hidden representation, the keywords decoder first predicts some keywords from it, and the keywords encoder re-encodes the generated keywords to obtain the keywords hidden representation. The hidden representation of the dialogue context and the keywords are concatenated to decode the final response.

In order to obtain the ground truth keywords for training, we calculate the TF-IDF [16] value of each token in the response utterances. The tokens with high TF-IDF values are chosen as the keywords of the response. We use an additional keywords loss on the output of the keywords decoder so that the generated keywords can capture the main idea of the response to be generated. Moreover, we use a *cosine annealing mechanism* to make the response decoder better learn to leverage the keywords information. The inputs of the keywords encoder are switched gradually from the ground truth to the generated keywords, so the response decoder can learn to incorporate keywords information into responses and keep this ability in the testing stage.

We conduct experiments on a popular open-domain dialogue dataset DailyDialog [10]. The results on both automatic evaluation and human judgment show that KW-Seq2Seq can generate appropriate keywords that capture the main idea of the responses and leverage the keywords well to generate more informative, coherence, and topic-aware response sentences.

2 Related Work

Sequence-to-sequence (Seq2Seq) is the most commonly used framework for generative dialogue models. It can learn the mapping relationship between arbitrary variable-length sequences. However, the dialogue models based on the sequence-to-sequence framework suffer from the degeneration problem of tending to output generic responses like *I don't know* or *I'm not sure*. There have been many methods proposed to alleviate the generic response problem. Li et al. [7] use Maximum Mutual Information (MMI) as the training objective to strengthen the relevance of dialogue context and responses. Li et al. [8] propose a beam search decoding algorithm that encourages the model to choose hypotheses from diverse parent nodes in the search tree and penalizes the tokens from the same parent node. There is also some researches utilizing latent variables to improve response diversity. Shen et al. [20] build a conditional variational dialogue model that generates specific dialogue responses based on the dialog context and a stochastic latent variable. Zhao et al. [27] capture the dialogue discourse-level diversity by using latent variables to learn a distribution over potential conversational intents as well as integrating linguistic prior knowledge to the model.

Some researches try to leverage keywords to improve the quality of responses in generative dialogue systems. Xing et al. [26] propose a topic aware sequence-to-sequence (TA-Seq2Seq) model which uses an extra pre-trained LDA topic model to generate the topic keywords of the input utterances, and decode responses with a joint attention mechanism on the input messages and topic keywords. Mou et al. [12] use the maximum pointwise mutual information (PMI) principle to pick up just one noun word from the vocabulary, and use a backward and forward conditional language model [13] to generate the final response. Serban et al. [17] propose a hierarchical recurrent encoder-decoder (HRED) [18] based model MrRNN, which predicts a sequence of noun words or task-specific entities as the coarse representation of responses. However, the coarse representation is only conditioned on the keywords of the previous utterances, but not the whole dialogue context. Unlike the models mentioned above, our proposed KW-Seq2Seq model predicts any number of words of any kind as the keywords to capture the main idea of the dialogue response, and is trained in an end-to-end manner without any outside auxiliary resources. The keywords prediction is conditioned on the hidden states of the whole dialogue context, so all the information of the input can be fully leveraged. With these advantages equipped, our KW-Seq2Seq model is able to produce more diverse and informative dialogue responses.

3 Keywords-Guided Sequence-to-Sequence Model

The keywords-guided sequence-to-Sequence (KW-Seq2Seq) model consists of four components: *context encoder*, *response decoder*, *keywords decoder*, and *keywords encoder*. As in the conventional sequence-to-sequence (Seq2Seq) model, the context encoder transforms the dialogue context to its hidden representation, and the response decoder generates the response utterance conditioned on it. In KW-Seq2Seq, we additionally use the keywords decoder to generate a sequence of words as keywords, and use the keywords encoder to maps the generated keywords to their hidden representation. The response are generated conditioned on the dialogue context and the predicted keywords. The overall architecture of KW-Seq2Seq is shown in Fig. 1.

3.1 Context Encoder and Response Decoder

The context encoder takes the dialogue context as input, which consists of a sequence of utterances of two interlocutors. We follow the processing of the pretrained language model BERT [2] that the embedding of each token is the sum of the word embedding, type embedding, and position embedding. All the context utterances are concatenated to a whole sequence. We add the `[CLS]` token at the beginning of the sequence and the `[SEP]` token at the end of each utterance.

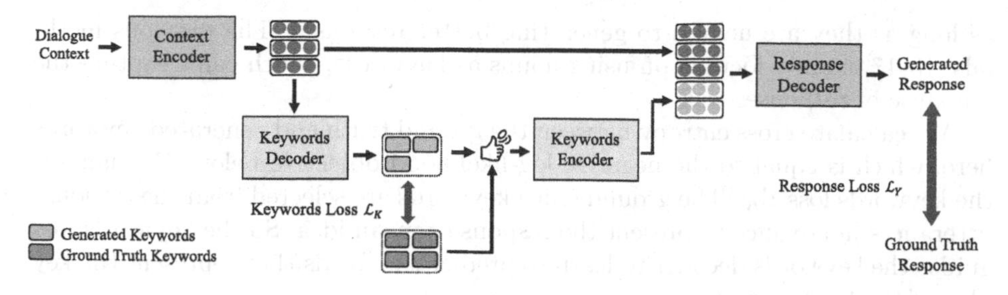

Fig. 1. The overall architecture of our KW-Seq2Seq model. In the training stage, the ground truth keywords (blue rectangles) are fed into the keywords encoder with the probability of p, and the generated keywords (green rectangles) with the probability of $1 - p$. (Color figure online)

The context encoder is a Transformer [23] encoder, which consists of N layers of residual multi-head self-attention layers with feed-forward connections. The i-th layer obtains its hidden states $\mathbf{H}^{(i)}$ by the following operations:

$$\tilde{\mathbf{H}}^{(i)} = \text{LayerNorm}(\mathbf{H}^{(i-1)} + \text{SelfAtten}(\mathbf{H}^{(i-1)})) \quad (1)$$

$$\mathbf{H}^{(i)} = \text{LayerNorm}(\tilde{\mathbf{H}}^{(i)} + \text{FC}(\tilde{\mathbf{H}}^{(i)})) \quad (2)$$

where LayerNorm is layer normalization, SelfAtten and FC are the self-attention and fully connected sub-layers. The output of the last layer is used as the hidden representation of the dialogue context $\mathbf{H}_X = \mathbf{H}^{(N)}$.

We use Transformer [23] decoder as the response decoder. In conventional Seq2Seq model, the response decoder generates each token of the output sequence conditioned on the context hidden status and the generated output tokens:

$$p(\mathbf{y}|\mathbf{X}) = \prod_{i=1}^{M} p(y_t|y_{<t}, \mathbf{H}_X) \quad (3)$$

In KW-Seq2Seq, we additionally provide keywords information to the response decoder. The details are described in the next subsection.

3.2 Keywords Decoder and Keywords Encoder

The architectures of the keywords decoder and keywords encoder are same as the response decoder and context encoder. With the dialogue context hidden states $\mathbf{H}_X$ as input, the keywords decoder generates a sequence of words $\mathbf{k}$ as the keywords of the response utterance:

$$p(\mathbf{k}|\mathbf{X}) = \prod_{t=1}^{S} p(k_t|k_{<t}, \mathbf{H}_X) \quad (4)$$

Note that the keywords decoder has the same vocabulary with the response decoder, so it can generate any words as keywords (e.g. nouns, verbs, adjectives) as long as they are helpful to generating better responses. The previous methods [13,17] mainly focused on using nouns as keywords, which can't capture the semantic of responses well.

We calculate cross entropy between the ground truth and generated keywords here, which is equal to the negative log-likelihood objective below. We name it the keywords loss $\mathcal{L}_K$. The ground truth keywords are selected from the response utterances in advance to present the response's mean idea. So the keywords loss guides the keywords decoder to learn to predict the words that represent the key idea of the final response.

$$\mathcal{L}_K = \sum_{i=1}^{N}\sum_{t=1}^{S} - \log p(k_t|k_{<t}, \mathbf{H}_X) \tag{5}$$

In order to sample out the predicted keywords from the decoder output distributions and still maintain the differentiability in the training procedure, we resort to Gumbel-Softmax [4], which is a differentiable surrogate to the argmax function. The probability distribution of the t-th keywords is:

$$\boldsymbol{m}(k_t) = \text{Gumbel-Softmax}\left(p(k_t|k_{<t}, \mathbf{H}_X)\right) \tag{6}$$

$$\text{Gumbel-Softmax}\left(\pi_i\right) = \frac{e^{(\log(\pi_i)+g_i)/\tau}}{\sum_{j=1}^{k} e^{(\log(\pi_j)+g_j)/\tau}} \tag{7}$$

where $(\pi_1, \pi_2, \ldots, \pi_k)$ represents the probabilities of the original categorical distribution, g_j are i.i.d samples drawn from the Gumbel distribution Gumbel$(0,1)$[1] and τ is a constant that controls the smoothness of the distribution. When $\tau \to 0$, Gumbel-Softmax performs like argmax, while if $\tau \to \infty$, Gumbel-Softmax performs like a uniform distribution.

The generated keywords are then encoded by keywords encoder to obtain their hidden representation $\mathbf{H}_K$. The context hidden states $\mathbf{H}_X$ and keywords hidden states $\mathbf{H}_K$ are concatenated together and feed into the response decoder to produce the final dialogue response $\mathbf{y}$:

$$p(\mathbf{y}|\mathbf{k}, \mathbf{X}) = \prod_{t=1}^{M} p(y_t|y_{<t}, \mathbf{H}_X \oplus \mathbf{H}_K) \tag{8}$$

where $\oplus$ donates the concatenation operation of two groups of vectors.

Finally, we calculate the negative log-likelihood (cross entropy) loss of the responses and sum the keywords loss and response loss weighted by α and β to obtain the final training loss:

$$\mathcal{L}_Y = \sum_{i=1}^{N}\sum_{t=1}^{M} - \log p(y_t|y_{<t}, \mathbf{H}_X \oplus \mathbf{H}_K) \tag{9}$$

$$\mathcal{L} = \alpha\mathcal{L}_K + \beta\mathcal{L}_Y \tag{10}$$

[1] If $u \sim$ Uniform$(0,1)$, then $g = -\log(\log(u)) \sim$ Gumbel$(0,1)$.

3.3 The Cosine Annealing Mechanism

Although we feed keywords hidden states to the response decoder, it's not guaranteed that the response decoder makes use of the keywords information well and generates responses related to the keywords. Inspired by the scheduled sampling technology [1], we propose the *cosine annealing mechanism* to guide the response decoder to leverage the keywords information better.

In the training stage, we feed ground truth keywords to the keyword encoder with probability p, and feed generated keywords with probability $1 - p$. The initial value of p is set to 1. As training progresses, p gradually drops to 0 by a cosine function. The relation between probability p and training progress x is formulated as following:

$$p = \begin{cases} 1, & 0 \leqslant x < x_1 \\ \frac{1}{2}\left(1 + \cos \pi x\right), & x_1 \leqslant x \leqslant x_2 \\ 0, & x_2 < x \leqslant 1 \end{cases} \tag{11}$$

At the beginning of the training procedure ($x \leqslant x_1$), we only feed the ground truth keywords to the keywords encoder. The true keywords make it easier for the response decoder to generate the responses, so the response decoder can learn to pay more attention to the keywords hidden states when decoding. As the training progresses ($x_1 \leqslant x \leqslant x_2$), we gradually decrease the probability p so the keywords encoder and response decoder have more opportunities to access the generated keywords. At last ($x > x_2$), we only use the generated keywords to train the model so the model can learn to do better when testing.

3.4 Keywords Acquisition

In order to obtain the ground truth keywords of each response utterance, we use the TF-IDF [16] value to indicate the importance of each word, which has been proven to be a simple but effective unsupervised method to extract keywords from large text corpus [6]. Specifically, we treat all the response utterances as a text corpus and calculate the TF-IDF value of each word in it. For every utterance, we choose the top 30% tokens in each response with the highest TF-IDF values as the keywords of it. We also try different keywords ratios to obtain the ratio value that produces the best dialogue responses. The experiment details are described in Sect. 4.5.

4 Experiments

4.1 Experiments Setting

The four components in KW-Seq2Seq are all 6-layer Transformer encoders or decoders. For hyper-parameters, we follow the settings of the $\text{BERT}_{\text{BASE}}$ model [2] (e.g. 768 hidden size and 12 attention heads). The sample temperature τ in Gumbel-Softmax is set to 1. We use Adam [5] to optimize the model

Table 2. The automatic metrics results. The best scores are in bold style. † means the model is not initialized with BERT's parameters.

	Overlay metric			Embedding metric			Keywords metric	
	BLEU-4	Rouge-L	Meteor	Average	Greedy	Extrema	KW-F1	KW-Recall
Seq2Seq†	8.8	20.5	9.8	86.4	68.9	47.3	–	–
Seq2Seq	18.5	30.1	14.7	88.9	73.4	54.3	–	–
Seq2Seq-12†	12.3	24.0	11.5	87.7	70.8	49.5	–	–
Seq2Seq-12	23.7	35.9	17.8	89.9	75.7	58.0	–	–
KW-Seq2Seq†	26.7	34.8	18.7	89.6	75.5	57.4	26.4	**87.6**
KW-Seq2Seq	**30.4**	**38.6**	**20.7**	**90.3**	**76.9**	**59.5**	**30.7**	86.6

parameters with learning rate of 1×10^{-5}. The weighting factors of the two loss items α and β are both set to 0.5. In the cosine annealing mechanism, we begin to decrease the probability p at the 50th epoch and after the 200th epoch p becomes 0. We don't use fixed batch size but set the max number of tokens to 2500 in a batch, which much improves the training efficiency [14]. We use the parameters of the first 6 layers of $\text{BERT}_{\text{BASE}}$ to initialize all the components in the model. There's no cross-attention module in BERT, so we copy the parameters of the self-attention module in each layer to the corresponding cross-attention module. We implement KW-Seq2Seq in PyTorch and use the pretrained BERT in the Transformers library [25].

4.2 Datasets

We train our model on a popular open-domain multi-turn dialogue dataset **DailyDialog** [10], which consists of 13K multi-turn conversations crawled from English practice websites. Each conversation is written by exactly two English learners and the content is mainly about people's daily life. To prepare the data for training and testing, we use a sliding window of size 6 to crop the conversations. The first 5 utterances in the window are used as the dialogue context and the last one as the response.

4.3 Automatic Evaluation

We train KW-Seq2Seq and two baseline Seq2Seq models: Seq2Seq-6 and Seq2Seq-12. Seq2Seq-6 has 6-layers encoder and decoder, which are the same as the context encoder and response decoder in KW-Seq2Seq. Seq2Seq-12 has 12-layers encoder and decoder so it has the same number of parameters as KW-Seq2Seq. We train all the models in both with and without BERT initialization settings. The results are shown in Table 2.

Overlap and Embedding Metrics We use three overlap-based metrics to evaluate generated dialogue responses: **BLEU**, **Rouge** and **Meteor**. They calculate the scores of two sentences by the number of co-occurring words or n-grams between them. Meanwhile, many papers point out that the overlay-based metrics cannot reflect the real quality of the responses in the dialogue task. So we also conduct three embedding-based evaluations: **Average**, **Greedy** and **Extrema** [11], which map sentences into embedding space and compute their cosine similarity. The embedding-based metrics can be used to measure the semantic similarity and test the ability of successfully generating a response sharing a similar topic with the golden answer. From Table 2 we can see that KW-Seq2Seq achieves higher scores than the Seq2Seq baselines on all the overlap and embedding-based metrics, which indicates the keywords information can help to generate more accurate and semantically relevant dialogue responses. It's worth noting that, when we train the models without BERT's pretrained parameters, KW-Seq2Seq still achieves good results while the performance of Seq2Seq drops sharply.

Keywords Metrics In order to check the performance of the keywords decoder and keywords encoder, we design two keywords-related metrics. First, we calculate the F1 score of the generated keywords and the ground truth keywords (**KW-F1**), which indicates the ability of the keywords decoder to capture the key idea of the response. Second, we count the number of generated keywords appearing in the final response sentence and calculate the recall score (**KW-Recall**). It reflects how well the keywords encoder and response decoder captures the meaning of the keywords and leverage them in the response sentence. As Table 2 shows, KW-Seq2Seq obtains about 30% KW-F1 score and nearly 90% KW-Recall score, that verifies the keywords decoder can predict keywords with reasonable accuracy and the keywords encoder and response decoder can effectively leverage the keywords information to guide the generation of the dialogue responses.

4.4 Human Evaluation

Accurate automatic evaluation of dialogue generation is still a big challenge [11]. We conduct human evaluation on KW-Seq2Seq and the baseline Seq2Seq model. We randomly sampled 300 dialogues from the evaluation results of the KW-Seq2Seq and Seq2Seq-6 models respectively and mix them together. We hire 3 undergraduate students majoring in English to score the dialogue responses. They are asked to give each response a score from 1 to 5 points according to the grammar, fluency, coherence, and informativeness of the sentences. Finally, Seq2Seq received the average score of **2.92** and KW-Seq2Seq got **3.16**. The ratio of each score is shown in Fig. 2. As the figure shows, more responses generated from the KW-Seq2Seq gain higher scores than Seq2Seq, which verifies KW-Seq2Seq can generate dialogue responses of higher quality and more informative.

4.5 The Keywords Ratio

To observe the effect of the keywords ratio on the quality of the generated responses, we choose the top 10%–50% of words with the largest TF-IDF values

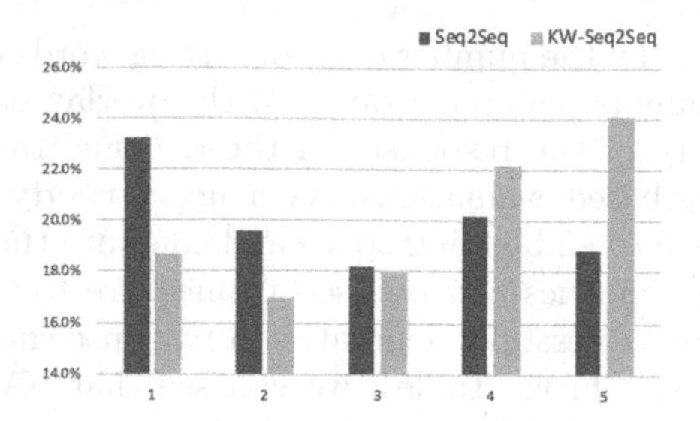

Fig. 2. Ratio of each score in the human evaluation results

as keywords to train the KW-Seq2Seq model separately. The results are shown in Table 3. We can see that the model trained with 30% keywords achieves the best scores on almost all the metrics, while the models trained with more or fewer keywords cannot outperform it. When training with fewer keywords, the keywords decoder cannot receive enough supervision information to learn the main idea of the responses. In turn, too many keywords bring more noise to the model and make the keywords decoder confuse to find key points of the dialogue. Therefore, we choose the keywords ratio of 30% to train the model, which gives the responses with the best quality.

Table 3. The results of training KW-Seq2Seq with different keywords ratio.

Ratio	10%	20%	30%	40%	50%
BLEU-4	20.8	26.7	**30.4**	27.1	25.0
Rouge-L	31.5	34.6	**38.6**	36.0	35.2
Meteor	15.5	18.4	**20.7**	18.8	18.8
Average	89.2	89.7	**90.3**	90.0	**90.3**
Greedy	74.1	75.4	**76.9**	75.7	75.8
Extrema	57.4	57.2	**59.5**	57.5	57.3
KW-F1	16.2	15.8	**30.7**	11.3	9.9
KW-Recall	62.9	64.2	86.6	86.8	**96.3**

5 Conclusion

We propose a keywords-guided sequence-to-sequence (KW-Seq2Seq) open-domain dialogue model, which predicts keywords from the dialogue context hidden states, and uses keywords information as guidance to generate final responses. Empirical experiments demonstrate that KW-Seq2Seq produces more informative, coherent, and fluent responses, yielding substantive gain in both automatic and human evaluation metrics.

Acknowledgments. The work is supported by National Key R&D Plan (No. 2020AAA0106600), NSFC (No. 61772076, 61751201 and 61602197), NSFB (No. Z181100008918002), and the funds of Beijing Advanced Innovation Center for Language Resources (No. TYZ19005). We would like to thank anonymous reviewers for their valuable suggestions and comments.

A Appendix

A.1 The Cosine Annealing Mechanism

The cosine annealing mechanism makes the model learn to leverage keywords information better in generating dialogue responses. Besides cosine annealing, we also try different training strategies for comparing. As shown in Table 4, we train the KW-Seq2Seq model with only feeding generated keywords to the keywords encoder (Predicted), and only feeding ground truth keywords (Ground Truth). As we can see, although the model with only ground truth keywords (Ground Truth) gets a high KW-Recall score (95.5), the model trained with cosine annealing mechanism (Cosine) gets the best results on all the other metrics. In the "Ground Truth" setting, the model can always access the ground truth keywords, which gives the model very strong guidance to leverage keywords information, so it's not surprising that it has an extremely high keywords recall score. But the score on the other metrics is not satisfactory. The experiments verify the importance of the cosine annealing mechanism in training KW-Seq2Seq. And also, the training method can be adapted to other models with middle supervision signals, which is left to people to explore.

Table 4. The results of training KW-Seq2Seq with cosine annealing mechanism (Cosine), only using predicted keywords (Predicted), and only using ground truth keywords (Ground Truth).

	BLEU-4	Average	KW-F1	KW-Recall
Predicted	27.4	89.9	15.1	40.2
Ground Truth	12.3	90.2	15.0	**95.5**
Cosine	**30.4**	**90.3**	**30.7**	86.6

A.2 Case Study

Table 5 shows two dialogue examples of KW-Seq2Seq and Seq2Seq baseline. Seq2Seq's responses are short and simple, while KW-Seq2Seq's responses are longer and more informative. We also find that the predicted keywords not only capture the topic idea of the dialogue but also bring new conceptions to the response. For example, in the first dialogue, KW-Seq2Seq generates new words "forest" and "suburb" in its keywords sequence, which brings more information

Table 5. Case Study. The Seq2Seq model directly generates the response utterance and our KW-Seq2Seq model first predicts a sequence of keywords and then generates the final response.

Case 1	Case 2
A: Lily, can you take part in our picnic this weekend? **B:** That sounds great. Where are you going?	**A:** Hello! Tomorrow I'm going to need a wake-up call. **B:** Not a problem. What time shall we call you? **A:** I always hit the snooze button, so give me two calls, one at 7 and another at 7 fifteen.
Seq2Seq Response: Going for a picnic with me. We can go boating and fishing.	**Seq2Seq Response:** Do you know when you can get up?
KW-Seq2Seq Keywords: we; plan; forest; suburb; there; river; fishing; **KW-Seq2Seq Response:** We plan to go to a forest in the suburb. There is a beautiful river there. We can do some fishing.	**KW-Seq2Seq Keywords:** command; expect; at; 7; fifteen; **KW-Seq2Seq Response:** Your wish is our command. Expect a call at 7:00 pm and again, sir.
Custom Keywords: go; park; flowers; beautiful; **KW-Seq2Seq Response:** We can go to the park with me and enjoy the flowers. It's very beautiful.	**Custom Keywords:** ok; no; problem; on; time; good; night; **KW-Seq2Seq Response:** Ok, no problem. We'll call you on time. Good-bye, sir. Good night.

to the final response. We also input some custom keywords to KW-Seq2Seq (the last row in Table 5). It generates the response formed by the new keywords, which indicates KW-Seq2Seq can not only generate meaningful and informative sentences but also gives people the opportunity to control the content and direction of the dialogues.

References

1. Bengio, S., Vinyals, O., Jaitly, N., Shazeer, N.M.: Scheduled sampling for sequence prediction with recurrent neural networks. In: NIPS (2015)
2. Devlin, J., Chang, M., Lee, K., Toutanova, K.: BERT: pre-training of deep bidirectional transformers for language understanding. In: Proceedings of the 2019 Conference of the North American Chapter of the Association for Computational Linguistics: Human Language Technologies, Volume 1 (Long and Short Papers), NAACL-HLT 2019, Minneapolis, MN, USA, 2–7 June 2019, pp. 4171–4186 (2019)
3. Du, W., Black, A.W.: Boosting dialog response generation. In: ACL (1), pp. 38–43. Association for Computational Linguistics (2019)

4. Jang, E., Gu, S., Poole, B.: Categorical reparameterization with Gumbel-Softmax. In: 5th International Conference on Learning Representations, ICLR 2017, Toulon, France, 24–26 April 2017. Conference Track Proceedings (2017)
5. Kingma, D.P., Ba, J.: Adam: a method for stochastic optimization. In: 3rd International Conference on Learning Representations, ICLR 2015, San Diego, CA, USA, 7–9 May 2015. Conference Track Proceedings (2015)
6. Lee, S.J., Kim, H.J.: Keyword extraction from news corpus using modified TF-IDF. J. Soc. e-Bus. Stud. **14**(4), 59–73 (2009)
7. Li, J., Galley, M., Brockett, C., Gao, J., Dolan, B.: A diversity-promoting objective function for neural conversation models. In: HLT-NAACL, pp. 110–119. The Association for Computational Linguistics (2016)
8. Li, J., Monroe, W., Jurafsky, D.: A simple, fast diverse decoding algorithm for neural generation. CoRR abs/1611.08562 (2016)
9. Li, X., Chen, Y., Li, L., Gao, J., Çelikyilmaz, A.: End-to-end task-completion neural dialogue systems. In: IJCNLP(1), pp. 733–743. Asian Federation of Natural Language Processing (2017)
10. Li, Y., Su, H., Shen, X., Li, W., Cao, Z., Niu, S.: DailyDialog: a manually labelled multi-turn dialogue dataset. In: IJCNLP(1), pp. 986–995. Asian Federation of Natural Language Processing (2017)
11. Liu, C., Lowe, R., Serban, I., Noseworthy, M., Charlin, L., Pineau, J.: How NOT to evaluate your dialogue system: An empirical study of unsupervised evaluation metrics for dialogue response generation. In: Proceedings of the 2016 Conference on Empirical Methods in Natural Language Processing, EMNLP 2016, Austin, Texas, USA, 1–4 November 2016, pp. 2122–2132 (2016)
12. Mou, L., Song, Y., Yan, R., Li, G., Zhang, L., Jin, Z.: Sequence to backward and forward sequences: a content-introducing approach to generative short-text conversation. In: COLING, pp. 3349–3358. ACL (2016)
13. Mou, L., Yan, R., Li, G., Zhang, L., Jin, Z.: Backward and forward language modeling for constrained sentence generation (2015)
14. Ott, M., et al.: fairseq: a fast, extensible toolkit for sequence modeling. In: Proceedings of NAACL-HLT 2019: Demonstrations (2019)
15. Pandey, G., Raghu, D., Joshi, S.: Mask & focus: conversation modelling by learning concepts. In: Proceedings of the AAAI Conference on Artificial Intelligence, vol. 34, pp. 8584–8591 (2020)
16. Salton, G., Buckley, C.: Term-weighting approaches in automatic text retrieval. Inf. Process. Manage. **24**(5), 513–523 (1988). https://doi.org/10.1016/0306-4573(88)90021-0
17. Serban, I.V., et al.: Multiresolution recurrent neural networks: an application to dialogue response generation. In: AAAI, pp. 3288–3294. AAAI Press (2017)
18. Serban, I.V., Sordoni, A., Bengio, Y., Courville, A.C., Pineau, J.: Building end-to-end dialogue systems using generative hierarchical neural network models. In: AAAI, pp. 3776–3784. AAAI Press (2016)
19. Shao, Y., Gouws, S., Britz, D., Goldie, A., Strope, B., Kurzweil, R.: Generating high-quality and informative conversation responses with sequence-to-sequence models. In: EMNLP, pp. 2210–2219. Association for Computational Linguistics (2017)
20. Shen, X., et al.: A conditional variational framework for dialog generation. In: Proceedings of the 55th Annual Meeting of the Association for Computational Linguistics, Volume 2: Short Papers, ACL 2017, Vancouver, Canada, 30 July–4 August 2017, pp. 504–509 (2017)

21. Song, Z., Zheng, X., Liu, L., Xu, M., Huang, X.: Generating responses with a specific emotion in dialog. In: ACL (1), pp. 3685–3695. Association for Computational Linguistics (2019)
22. Sordoni, A., et al.: A neural network approach to context-sensitive generation of conversational responses. In: HLT-NAACL, pp. 196–205. The Association for Computational Linguistics (2015)
23. Vaswani, A., et al.: Attention is all you need. In: Advances in Neural Information Processing Systems 30: Annual Conference on Neural Information Processing Systems 2017, Long Beach, CA, USA, 4–9 December 2017, pp. 5998–6008 (2017)
24. Weizenbaum, J.: ELIZA - a computer program for the study of natural language communication between man and machine. Commun. ACM **9**(1), 36–45 (1966)
25. Wolf, T., et al.: Transformers: state-of-the-art natural language processing. In: Proceedings of the 2020 Conference on Empirical Methods in Natural Language Processing: System Demonstrations, pp. 38–45. Association for Computational Linguistics, October 2020
26. Xing, C., et al.: Topic aware neural response generation. In: Proceedings of the Thirty-First AAAI Conference on Artificial Intelligence, San Francisco, California, USA, 4–9 February 2017, pp. 3351–3357 (2017)
27. Zhao, T., Zhao, R., Eskénazi, M.: Learning discourse-level diversity for neural dialog models using conditional variational autoencoders. In: ACL (1), pp. 654–664. Association for Computational Linguistics (2017)

Zero-Shot Deployment for Cross-Lingual Dialogue System

Lu Xiang[1,2], Yang Zhao[1,2], Junnan Zhu[1,2], Yu Zhou[1,2,3](✉), and Chengqing Zong[1,2]

[1] National Laboratory of Pattern Recognition, Institute of Automation, CAS, Beijing, China
{lu.xiang,yang.zhao,junnan.zhu,cqzong,yzhou}@nlpr.ia.ac.cn
[2] School of Artificial Intelligence, University of Chinese Academy of Sciences, Beijing, China
[3] Fanyu AI Laboratory, Zhongke Fanyu Technology Co., Ltd., Beijing, China

Abstract. The dialogue system is widely used in many application scenarios, while the construction of the dialogue system always faces the difficulty of zero-resource training data. To alleviate that, we propose a knowledge transfer framework to build a dialogue system based on existing machine translators and training data in data-rich language. Specifically, we first generate various kinds of pseudo data with cyclic translation procedure and different data combinations. Then we propose a noise injection method and a multi-task training method for the pipeline system and end-to-end system, respectively. The noise injection method optimizes each module by incorporating machine translation noises into the pipeline process to handle the error propagation problem, thus improving the whole system's robustness. The multi-task training method combines cross-lingual dialogue, monolingual dialogue, and machine translation into the end-to-end dialogue system's training process, thus reducing the impact of noises in pseudo data. The extensive experiments on a real-world e-commerce dataset demonstrate that our methods can achieve remarkable improvements over strong baselines.

Keywords: Cross-lingual dialogue system · Noise injection · Multi-task

1 Introduction

Dialogue systems have stimulated great interest from both academia and industry [16,22,25]. However, most existing dialogue systems are developed based on monolingual training data, making the dialogue service only available in the corresponding language. Along with globalization, there is an increasing need for commercial dialogue systems to handle different languages. However, collecting high-quality dialogue data for a new language is quite expensive, leading to the development of a dialogue system face the challenge of few-shot or even

L. Wang et al. (Eds.): NLPCC 2021, LNAI 13029, pp. 193–205, 2021.
https://doi.org/10.1007/978-3-030-88483-3_15

zero-resource training data. Therefore, in this paper, we focus on building a cross-lingual dialogue system based on the existing monolingual dialogue system.

Despite the attractive progress in cross-lingual dialogue systems [3,9,14,15], researchers mainly focus on the sub-modules in a dialogue system. To the best of our knowledge, there is none work trying to build a complete cross-lingual dialogue system under the zero-shot setting, which is the focus of this paper.

Benefiting from the excellent performance of machine translation (MT), we adopt MT systems as the language bridge, and two basic methods can be adopted to deploy a dialogue system for the zero-resource language:

i) **MT-based pipeline dialogue system.** It consists of three steps: translation step, dialogue step, and back-translation step. A machine translator translates a user's utterance into a language consistent with the dialogue system. Then the dialogue system generates a response based on the translated utterance. Finally, the machine translator translates the response back into the user's language. This method is easy to implement. However, this method's core challenge is the amplification of translation errors.
ii) **End-to-end dialogue system.** The machine translator is used to translate the dialogue training data in data-rich language into zero-resource language. Then an end-to-end dialogue system can be directly trained from the translated data. However, there are still many noises and errors in the translated data, which will seriously affect the dialogue system's performance.

In this paper, we propose two possible workarounds to deploy a dialogue system for the zero-resource language without any dialogue data in that language under the guidance of the MT systems and the dialogue dataset in data-rich language. Specifically, we first generate various pseudo data that contain the dialogue knowledge of the data-rich language and translation knowledge between data-rich language and zero-resource language through cyclic translation procedure. Based on generated pseudo data, we propose two methods to enhance the performance of the MT-based pipeline system and end-to-end system, respectively. For the MT-based pipeline system, a **noise injection** method is proposed to optimize each module in the pipeline paradigm. This method injects noises into both the MT systems and the dialogue system with generated pseudo data, making the MT systems more relevant to the dialogue and the dialogue system more robust to the noise input. For the end-to-end model, a **multi-task training** method is designed to augment the performance by combining the training process of three tasks: cross-lingual dialogue system, monolingual dialogue system, and machine translation task. This kind of synchronous learning can optimize the encoder and reduce the impact of noises in the pseudo data. The main contributions of this paper are as follows:

(1) To the best of our knowledge, we are the first to make a full investigation about how to deploy a dialogue system to a zero-resource language, which only uses the dialogue data in data-rich language and machine translators.
(2) Noise injection method and multi-task training method are proposed to boost the performance of the pipeline model and end-to-end model, respectively.

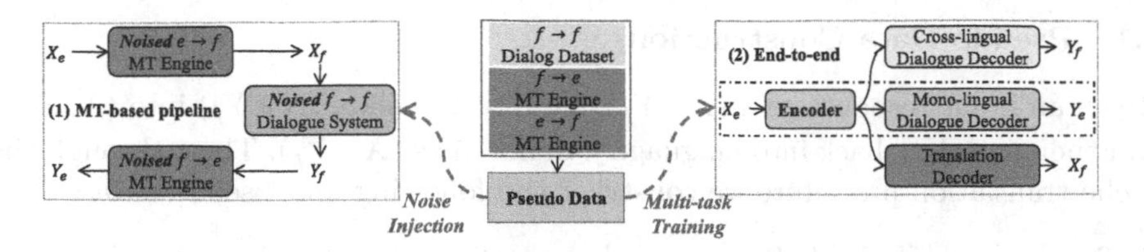

Fig. 1. The framework of our proposed methods.

(3) The proposed methods have been evaluated on the transfer of three language pairs. The results have shown the effectiveness of our methods.

2 Problem Definition and Background

Our goal is to construct a dialogue system for the new language e, which takes the contextual utterance X_e as input, and generates response Y_e. Due to the zero-resource training data, we make full use of the following two resources to transfer the dialogue system from data-rich language f to language e:

1) **Dialogue dataset in f**: $\mathcal{D}_{X_f \Rightarrow Y_f} = \{(X_f, Y_f)\}$, where X_f denotes the dialogue context and Y_f denotes the response.
2) **MT engines**, which can translate sentence from e to f (denoted as $MT_{e \Rightarrow f}$) and back-translate from f to e (denoted as $MT_{f \Rightarrow e}$). Hence, the dialogue system for e is formalized as follows:

$$Y_e = g(X_e | \mathcal{D}_{X_f \Rightarrow Y_f}, MT_{e \Rightarrow f}, MT_{f \Rightarrow e}) \tag{1}$$

We briefly describe the conversational model used in this paper. Considering the excellent text generation performance of the Transformer encoder-decoder network [21], we implement our neural conversational model entirely based on this framework. Given a dialogue data set $\mathcal{D} = \{(X, Y)\}$, where Y is a response of a dialogue context X. The encoder and decoder are trained jointly to maximize the conditional probability of response sequence given an input sequence:

$$L(\mathcal{D}; \theta) = \sum_{(X,Y) \in \mathcal{D}} \log p(Y|X; \theta) \tag{2}$$

3 Approach

To deploy the dialogue system for e, we first use the MT engines and dialogue dataset in f to construct various pseudo data. Then, we put forward a noise injection method for the pipeline system to alleviate the error propagation problem and a multi-task training method for the end-to-end model to reduce the influence of errors and noises in the pseudo data, as illustrated in Fig. 1.

3.1 Pseudo Data Construction

Given an input-response pair (X_f, Y_f), we first translate (X_f, Y_f) into (X_e, Y_e), and then translate back into language *f*, denoted as $(X_f^{'}, Y_f^{'})$. Thus, through the cyclic translation procedure we construct the following four pseudo datasets.

1) $\mathcal{D}_{X_e \Rightarrow Y_e} = \{(X_e, Y_e)\}$. It is a pseudo monolingual dialog dataset in language e consisting of the input-response pair (X_e, Y_e).
2) $\mathcal{D}_{X_e \Rightarrow Y_f} = \{(X_e, Y_f)\}$. It is a pseudo cross-lingual dialog dataset consisting of input-response pair (X_e, Y_f).
3) $\mathcal{D}_{X_f^{'} \Rightarrow Y_f^{'}} = \{(X_f^{'}, Y_f^{'})\}$. It is a pseudo monolingual dialog dataset in language f and contains input-response pair $(X_f^{'}, Y_f^{'})$.
4) $\mathcal{D}_{f \Rightarrow e} = \{(X_f, X_e) \cup (Y_f, Y_e)\}$. It is a pseudo parallel corpus consisting of each message including input and response and its translated message.

These four pseudo datasets contain dialogue knowledge from data-rich language and translation knowledge from MT engines. Then we use the datasets to optimize the MT-based pipeline system and the end-to-end system.

3.2 Noise Injection Method

For the MT-based pipeline system, the domain of the online MT engine is different from the dialogue scenario, which will introduce many errors and noises. Besides, the original dialogue system is trained on the clean dataset, making it impossible to work properly when given the translated utterances. The constructed pseudo datasets contain much knowledge from both the MT engine and dialogue. Therefore, we consider using the pseudo datasets to optimize each module in the pipeline system, as shown in Fig. 1. Since the modules are learning from noise data, they can better handle noise input. We name this method the **noise injection** method. It can be divided into two steps:

Noised NMT System. We use the generated pseudo dataset $\mathcal{D}_{f \Rightarrow e}$ to train Transformer-based neural machine translation (NMT) systems from both directions ($f \Rightarrow e$ and $e \Rightarrow f$). These two NMT systems are denoted as *Noised NMT Systems*. The two systems are more relevant to the dialogue task than the online MT engine since the pseudo parallel dataset is constructed from the original clean dialogue dataset.

The Transformer-based NMT also consists of an encoder and decoder. Given a parallel dataset $\mathcal{P} = \{(F, E)\}$, the loss function can be calculated as:

$$L(\mathcal{P}; \theta) = \sum_{(F,E) \in \mathcal{P}} \log p(E|F; \theta) \tag{3}$$

Hence, given the pseudo parallel dataset $\mathcal{D}_{f \Rightarrow e}$, the noised NMT systems can be trained by optimizing the loss function in Eq. 3.

Noised Dialogue System. To make the dialogue system better handle the noise input, we need to update the original dialogue system and let it experience more noise data. To achieve this, we merge the dataset $\mathcal{D}_{X'_f \Rightarrow Y'_f}$ with the original clean dataset $\mathcal{D}_{X_f \Rightarrow Y_f}$ and retrain the dialogue system. This system is denoted as *Noised Dialogue System.* Given the pseudo data $\mathcal{C} = \{(X', Y')\}$ and the original clean training data $\mathcal{D} = \{(X, Y)\}$, the loss function is calculated as follows:

$$L(\mathcal{D}, \mathcal{C}; \theta) = \sum_{n=1}^{|\mathcal{D}|+|\mathcal{C}|} \left\{ \underbrace{\log p(Y_{\mathcal{D}}^n | X_{\mathcal{D}}^n; \theta)}_{\text{Lossfromcleandata}} + \underbrace{\log p(Y_{\mathcal{C}}^n | X_{\mathcal{C}}^n; \theta)}_{\text{Lossfrompseudodata}} \right\} \tag{4}$$

3.3 Multi-task Training and Adaptation

In the noise injection method, the NMT systems and the dialogue system are optimized separately, indicating that the error propagation still exists. Thus, we would like to know whether we can directly train an end-to-end dialogue system for language e using the generated pseudo dataset. However, due to the translation errors and noises in the dataset, it is not enough to use $\mathcal{D}_{X_e \Rightarrow Y_e}$ to train the end-to-end dialogue system. Notice that the clean data can be used to enhance the noise data, we consider using multi-task learning to integrate different tasks to improve the end-to-end dialogue system in language e.

We employ the one-to-many scheme [10,26] to incorporate the training process of several tasks. As shown in Fig. 1, the scheme involves one shared encoder and multiple task-specific decoders for three language generation tasks: cross-lingual dialogue system, monolingual dialogue system, and MT. Here, cross-lingual dialogue system refers to the system, of which the input and response are in different languages.

Three pseudo datasets are used for the training procedure, including $\mathcal{D}_{X_e \Rightarrow Y_f}$, $\mathcal{D}_{X_e \Rightarrow Y_e}$ and $\mathcal{D}_{f \Rightarrow e}$. These datasets contain both clean data and pseudo data, and the clean data can help to improve the response generation for language e. Furthermore, the multi-task training procedure can enhance the encoder, thus minimizing the impact of the noise data. The loss function is as follows:

$$\begin{aligned} L(\theta_e, \theta_d^{ml}, \theta_d^{cl}, \theta_d^{mt}) = & \underbrace{\log p(Y_f | X_e; \theta_e, \theta_d^{cl})}_{\text{cross-lingual dialogue task}} + \underbrace{\log p(Y_e | X_e; \theta_e, \theta_d^{ml})}_{\text{monolingualdialoguetask}} \\ & + \underbrace{\log p(X_f | X_e; \theta_e, \theta_d^{mt})}_{\text{MTtask}} \end{aligned} \tag{5}$$

where θ_e denotes the shared encoder. θ_d^{ml}, θ_d^{cl}, and θ_d^{mt} are the decoder for monolingual dialogue task, cross-lingual dialogue task, and MT task, respectively.

4 Experiments

4.1 Experimental Settings

Dataset. We adopt a publicly available Chinese e-commerce dialogue corpus[1] [24] collected from Taobao[2] to conduct experiments. Chinese is the

[1] https://github.com/cooelf/DeepUtteranceAggregation.
[2] https://www.taobao.com.

high-resource language. We transfer the Chinese e-commerce dialogue service into English, Spanish and Korean under zero-shot setting. To verify our method, we manually translate the Chinese test set into the other three languages. More details about the dataset are given in Table 1.

Table 1. Dialog dataset statistics.

	Number of input-response pairs	Average of words	
		Input	Response
Train	517,525	31.28	11.77
Valid	4,402	31.64	11.65
Chinese-Test	5,204	32.63	11.73
English-Test	5,204	36.19	13.70
Spanish-Test	5,204	31.72	12.06
Korean-Test	5,204	21.40	7.93

Evaluation Metrics. We conduct evaluation with both automatic metrics and human evaluation. For automatic evaluation, We adopt several widely used metrics [6,7,11,19] to measure the performance of our proposed method, including word overlap metrics (BLEU-4, METEOR, ROUGE-L), distinct metrics (Dist-1/2), and normalized average sequence length (NASL). We also carry out a human evaluation for a more realistic comparison of our proposed methods to the baselines. We focus on evaluating the generated responses from three aspects: (1) **Relevance**: if the response is relevant to the given history; (2) **Informative**: if the response contains informative and interesting content; and (3) **Fluency**: whether the response is fluent without grammatical error. The details of human evaluation will be described in the corresponding part.
Implementation Details. We use Byte-Pair Encoding (BPE) with 30K merge operations to segment Chinese, English, Spanish, and Korean into subword granularities. For the Transformer-based dialogue system, the vocabulary size of the source and target words are both 30K. We train our models using configuration *transformer_base* [21], which contains a 6-layer encoder and a 6-layer decoder with 512 dimension hidden representations. During training, we apply Adam optimizer with $\beta_1 = 0.9$, $\beta_2 = 0.998$, and $\epsilon = 10^{-9}$. In the cyclic translation procedure, we adopt the Google translator[3] to generate the pseudo dataset. For the self-trained NMT systems in the noise injection method, we also use configuration *transformer_base* to train the NMT systems.

4.2 Experimental Results and Analysis

Results of Noise Injection Method. Table 2 shows the experimental results of the noise injection method. We can reach the following conclusions:

[3] https://translate.google.com/.

Table 2. Experimental results of the *noise injection method.* In the noise injection method, modules in the parentheses mean which modules use the noised models while the rest use the original. *Dial* denotes dialogue.

#	System	BLEU-4	METEOR	ROUGE-L	Dist-1/2	NASL
Upper bound (Input = ZH)						
1	Dial	**9.54**	**11.34**	**18.38**	**0.0410/0.2082**	1.0589
Baseline (Input = EN; $MT_1 = MT_2$ = Google Translator)						
2	MT_1+Dial	3.93	7.56	12.53	0.0384/0.1908	1.1377
3	MT_1+Dial+MT_2	2.55	5.26	9.31	0.0323/0.1987	1.1345
4	End-to-end	2.90	6.06	10.50	0.0307/0.1830	1.0539
Noise Injection method (Input = EN)						
5	*Noise*(MT_1+Dial+MT_2)	**3.57**	**6.56**	**11.13**	0.0346/0.2107	0.9799
6	*Noise*(MT_1+Dial)	3.35	6.48	11.00	0.0362/0.2154	0.9687
7	*Noise*(Dial+MT_2)	2.90	5.76	10.08	0.0349/0.2162	0.9873
8	*Noise*(MT_1+MT_2)	3.41	6.28	10.72	0.0340/0.1975	0.9725
9	*Noise*(MT_1)	3.24	6.29	10.69	**0.0366**/0.2088	0.9716
10	*Noise*(Dial)	2.76	5.80	10.03	0.0362/**0.2194**	0.9890

Table 3. Experimental results of *multi-task training method. MonoDial* denotes the monolingual dialogue system and *CrossDial* denotes the cross-lingual dialogue system.

#	System	BLEU-4	METEOR	ROUGE-L	Dist-1/2	NASL
1	Dial	**9.54**	**11.34**	**18.38**	**0.0410/0.2082**	1.0589
2	End-to-end	2.90	6.06	10.50	0.0307/0.1830	1.0539
3	*Noise*(MT_1+Dial+MT_2)	3.57	6.56	11.13	**0.0346/0.2107**	0.9799
Multi-task training method						
4	MonoDial+CrossDial+MT	**3.81**	**6.59**	**11.51**	0.0269/0.1447	0.9783
5	MonoDial+CrossDial	3.32	6.15	10.87	0.0335/0.1981	0.9832
6	MonoDial+MT	3.57	6.28	11.14	0.0280/0.1465	0.9371

i) The MT-based pipeline dialogue system suffers heavily from error propagation. Compared to the dialogue model given the clean Chinese test data (line 1), the system's performance degrades drastically if the input is a noise input translated from English (line 2). After Google translator translates the Chinese response back into English, the performance continues to decline (line 3). The performance of the end-to-end model (line 4) is better than the pipeline system (line 3), which proves the end-to-end model can avoid the problem of error propagation to a certain extent. However, the performance of the end-to-end model is still seriously harmed by the noises and errors in the translated pseudo data.

ii) The proposed noise injection method can boost the performance of the pipeline system (line 5 and line 3). After using the noised MT (including MT_1 and MT_2) and noised dialogue, the performance has gained impressively.

iii) We also investigate the effect of each noised model in the pipeline system (line 6 to line 10). We can see that each of the noised models can improve performance. Meanwhile, *noising* the first two models is more critical for improvement (line 6). This is because retraining the first two models can

Table 4. Experimental results of language transfer to other languages.

#	System	BLEU-4	METEOR	ROUGE-L	Dist-1/2	NASL
Chinese ⇒ Spanish						
1	MT_1+Dial+MT_2	2.37	4.20	6.07	**0.0620/0.2863**	1.0040
2	End-to-end	4.34	5.58	8.64	0.0493/0.2483	1.1579
3	*Noise*(MT_1+Dial+MT_2)	6.58	**6.88**	10.31	0.0539/0.2616	1.0934
4	Multi-task training	**6.66**	6.74	**10.49**	0.0434/0.2027	1.0594
Chinese ⇒ Korean						
5	MT_1+Dial+MT_2	1.42	7.46	2.68	**0.1411/0.4394**	1.0389
6	End-to-end	3.39	9.70	4.86	0.1247/0.3877	1.2519
7	*Noise*(MT_1+Dial+MT_2)	**5.46**	10.34	6.38	0.1362/0.4141	1.0943
8	Multi-task training	5.31	**10.64**	**6.73**	0.1119/0.3323	1.2291

make the dialogue system act more appropriately when given the translated utterances.

Besides, the performance of the Dist-1/2 is different from the word overlap metrics. The end-to-end model (line 4) achieves higher word overlap metrics while the MT-based pipeline dialogue system (line 3) obtains higher Dist-1/2. Our proposed method can improve both the word overlap metrics and diversity of the responses, demonstrating our proposed noise injection method's effectiveness compared with the MT-based pipeline dialogue system.

Results of Multi-task Training Method. The experimental results are shown in Table 3. Compared to the end-to-end model (line 2), training monolingual dialogue system and cross-lingual dialogue system simultaneously improves both the word overlap metrics and diversity of the generated responses (line 5). MT task is helpful for the word overlap metrics but harmful for the diversity. When combining the monolingual dialogue task and MT task (line 6), the word overlap metrics are higher than those of the end-to-end model, but the diversity is lower. *MonoDial+CrossDial+MT* (line 4) outperforms the other two (line 5–6) in word overlap metrics since it uses both the history and response in the Chinese dialogue dataset. However, the diversity reaches the lowest. This illustrates that the Chinese dialogue dataset can boost the performance of the end-to-end English dialogue system when only pseudo data is available.

Results of Transfer to Other Languages. We further conduct experiments on the transfer from Chinese to Spanish and Korean. The settings are the same as Chinese to English transfer. The experimental results are shown in Table 4. Our two methods outperform the baseline systems by a big margin from the word overlap perspective, demonstrating that our proposed two methods effectively transfer the dialogue system to a new language by only using knowledge in data-rich language and MT. The word overlap metrics except for METEOR on Korean are much lower than those in English and Spanish. This may be because the Chinese to Korean translation performance is not that ideal, and the generated Korean pseudo data may contain much more noise than that of English and Spanish. Besides, the diversity score of the multi-task training method is lower than the noise injection method both in Spanish and Korean.

Table 5. The effect of machine translation performance.

#	System	BLEU-4	METEOR	ROUGE-L	Dist-1/2	NASL
The size of MT training corpus: 500K						
1	MT_1+Dial+MT_2	0.25	2.71	5.82	**0.0512/0.3027**	0.8034
2	End-to-end	0.46	3.17	6.79	0.0409/0.2379	0.8414
3	*Noise*(MT_1+Dial+MT_2)	**0.63**	**3.85**	**7.38**	0.0435/0.2536	0.9140
4	Multi-task training	0.57	3.40	7.15	0.0340/0.1858	0.7946
The size of MT training corpus: 1M						
5	MT_1+Dial+MT_2	0.26	2.64	5.84	**0.0508/0.3181**	0.7656
6	End-to-end	0.51	3.38	6.76	0.0406/0.2369	0.8661
7	*Noise*(MT_1+Dial+MT_2)	**0.75**	**3.93**	**7.65**	0.0432/0.2590	0.8964
8	Multi-task training	0.63	3.75	7.26	0.0332/0.1873	0.8447
The size of MT training corpus: 2M						
9	MT_1+Dial+MT_2	0.33	2.77	5.97	**0.0450/0.2708**	0.7620
10	End-to-end	0.70	3.57	7.34	0.0387/0.2303	0.7985
11	*Noise*(MT_1+Dial+MT_2)	**1.03**	**4.35**	**8.37**	0.0401/0.2457	0.8790
12	Multi-task training	0.80	3.88	8.12	0.0337/0.1894	0.7571
MT_1=MT_2 = Google Translator						
13	MT_1+Dial+MT_2	2.55	5.26	9.31	0.0323/0.1987	1.1345
14	End-to-end	2.90	6.06	10.50	0.0307/0.1830	1.0539
15	*Noise*(MT_1+Dial+MT_2)	3.57	6.56	11.13	**0.0346/0.2107**	0.9799
16	Multi-task training	**3.81**	**6.59**	**11.51**	0.0269/0.1447	0.9783

Table 6. Human evaluation results.

System	Relevance	Informative	Fluency
MT_1+Dial+MT_2	2.26	2.90	3.36
End-to-end	2.71	2.91	3.48
Noise(MT_1+Dial+MT_2)	2.95	3.11	3.43
Multi-task training	**3.17**	**3.22**	**3.66**

The Effect of MT Performance. As introduced before, the MT engines play an essential role, since our proposed two methods are based on the pseudo data generated by the MT engines. The above experiments show the effectiveness of the two methods when adopting Google translator. However, MT performance is not always satisfied. *How will our methods help if we do not have a good MT?*

To investigate the effect of MT, we trained another three ZH → EN (EN → ZH) translation systems with *transformer_base* configuration using 500K, 1M, and 2M sentence pairs extracted from the English-Chinese Machine Translation track corpus[4]. We use these three translation systems to replace Google translator and simulate low resources and poor translation performance situations. The results are presented in Table 5.

Although the overall performance drops sharply when using the pseudo data generated by self-trained MT systems, our proposed two methods can help boost the performance to a certain extent. From Table 5, we can find an interesting

[4] https://challenger.ai/datasets/translation.

Input	C: What do you suggest to buy? S: Personally i like tea needles which still depends on hobbies C: I don't understand, just a brick tea, I don't know how to open it and what tools are needed to open it ?
Gold Response	Just choose a *tea knife* or a *tea needle*
Pipeline	Yes dear, what you shot was a three-piece set
End-to-End	I' m really sorry because the big warehouse in the order before the holiday has already been packed overtime. Oh usually it will be shipped to you within 48 hours and you wait patiently here.
Noise Injection	The effect is similar to the kisses this person sees that they like me personally they like *tea knives* most of them can be considered.
Multi-task	I suggest you use a large *tea knife*.

Fig. 2. Response examples with various systems.

phenomenon: When using Google translator to generate pseudo data, the *multi-task training method* achieves the best performance, while the *noise injection method* acquires the best when using self-trained MT systems. The reasons are two-fold: (1) When using Google translator, the quality of pseudo data is relatively high, and the multi-task training procedure can enhance the encoder and learn better representations for the noise data. (2) Due to the training corpus size, the self-trained MT's performance is much more unsatisfactory, resulting in more noises in the pseudo data. In this situation, the noise injection method is more helpful since it optimizes each module and injects the same noise into the MT systems and dialogue system, making the system more robust. All in all, our proposed two methods can be beneficial, even if only a weak MT system is available. Furthermore, the experimental results also indicate that the *multi-task training method* will be more useful when the quality of MT is good. Otherwise, the *noise injection method* will be more helpful.

Human Evaluation. We conduct the human evaluation on 150 random samples from the English test set, and these responses are based on distinct dialogue history. We compare responses generated by our methods with the responses generated by baselines. Three graduate students are asked to judge the quality of the responses according to relevance, informative, and fluency with a score from 1 (worst) to 5 (best). The student is presented with a dialogue history and four outputs with the name anonymized in each judgment. The average scores are presented in Table 6.

Compared to the baseline pipeline system, the end-to-end model generates better responses. Moreover, the fluency score is even a bit higher than that of the noise injection method. Our noise injection method significantly improves all three scores compared with the baseline pipeline system. This is mainly because the noised MT system is more relevant to the dialogue task. More importantly, the noised dialogue system has experienced more noise data from MT and better handles noise utterance. The multi-task training method outperforms the end-to-end by an impressive margin. Overall, the results suggest that our proposed methods can effectively improve dialogue systems' ability to generate more appropriate responses when transferring the dialogue system to a new language.

Case Study. The above results show that our proposed two methods can deploy and enhance a dialogue system for the new language. To further verify our methods, we show an example of response generation with various systems in Fig. 2. We can see that the responses generated by the noise injection method and multi-task training are better than the two baseline systems. The two responses generated by the two baseline systems are irrelevant to the dialogue context. The response generated by the noise injection method mentions *tea knives*, while the response generated by multi-task training can be regarded as a proper response. We can also find that some of the generated responses are not fluent. Nevertheless, it does not hinder the real application since people can understand as long as the system expresses critical information.

5 Related Work

The study of cross-lingual dialogue systems has gained much attention, and it studies how to adapt a dialogue system into the target language. The current work can be divided into three categories: cross-lingual NLU [1,8,9,13,15], cross-lingual DST [3,9,13] and cross-lingual response selection [14]. [2] proposed a multi-task learning architecture with share-private memory for multilingual open-domain dialogue generation, which is different from ours since they aimed at learning the common features among languages to boost dialogue systems.

Existing Cross-lingual transfer learning methods can be divided into two categories: transfer through cross-lingual representations [4,12,20] and transfer through MT [5,15,27]. In this paper, we focus on using MT to bridge the language gap between data-rich and zero-resource languages.

The differences between our work and the above work are two-fold: (1) There is no work in cross-lingual dialogue systems focusing on building a complete dialogue system for a new language under the zero-resource setting, which is the focus of this paper. (2) To the best of our knowledge, none of the work has explored how to use MT to transfer a generative dialogue system to a new language. This paper will study how to deploy a dialogue system for a new language by transferring knowledge from data-rich language and MT.

6 Conclusion

In this paper, we present cross-lingual transfer for dialogue systems under the zero-resource scenario. To alleviate this problem, we propose two methods to boost the pipeline system and the end-to-end system with the help of existing MT engines and training data in data-rich language. We first use MT and dialogue training data to generate various pseudo data. Then, the noise injection method is proposed to improve the pipeline system by injecting MT noises into the pipeline process, and the multi-task training method is proposed to enhance the end-to-end system. Experimental results have shown that our proposed methods can improve the dialogue system's performance for the new language. Furthermore, extended experiments demonstrate that our proposed methods are still useful even if only MT systems with poor performance are available.

Acknowledgments. This work was supported by the National Key R&D Program of China under Grant No. 2020AAA0108600.

References

1. Bai, H., Zhou, Y., Zhang, J., Zhao, L., Hwang, M.-Y., Zong, C.: Source critical reinforcement learning for transferring spoken language understanding to a new language. In: Proceedings of COLING (2018)
2. Chen, C., Qiu, L., Fu, Z., Liu, J., Yan, R.: Multilingual dialogue generation with shared-private memory. In: Proceedings of NLPCC (2019)
3. Chen, W., Chen, J., Su, Y., Wang, X., Yu, D., Yan, X., et al.: Xl-nbt: a cross-lingual neural belief tracking framework. In: Proceedings of EMNLP (2018)
4. Conneau, A., Lample, G.: Cross-lingual language model pretraining. In: Proceedings of NeurIPS (2019)
5. Jain, A., Paranjape, B., Lipton, Z. C.: Entity projection via machine translation for cross-lingual ner. In: Proceedings of EMNLP-IJCNLP (2019)
6. Liu, C. W., Lowe, R., Serban, I., Noseworthy, M., Charlin, L., Pineau, J.: How not to evaluate your dialogue system: An empirical study of unsupervised evaluation metrics for dialogue response generation. In: Proceedings of EMNLP (2016)
7. Li, J., Galley, M., Brockett, C., Gao, J., Dolan, B.: A diversity-promoting objective function for neural conversation models. In: Proceedings of NAACL-HLT (2016)
8. Liu, Z., Shin, J., Xu, Y., Winata, G. I., Xu, P., Madotto, A., Fung P.: Zero-shot cross-lingual dialogue systems with transferable latent variables. In: Proceedings of EMNLP- IJCNLP (2019)
9. Liu, Z., Winata, G. I., Lin, Z., Xu, P., Fung, P.: Attention-informed mixed-language training for zero-shot cross-lingual task-oriented dialogue systems. In: Proceedings of AAAI (2020)
10. Luong, M., Le, Q.V., Sutskever, I., Vinyals, O., Kaiser, L.: Multi-task sequence to sequence learning. In: Proceedings of ICLR (2016)
11. Olabiyi, O., Salimov, A.O., Khazane, A., Mueller, E.: Multi-turn dialogue response generation in an adversarial learning framework. In: Proceedings of the First Workshop on NLP for Conversational AI (2019)
12. Pires, T., Schlinger, E., Garrette, D.: How multilingual is multilingual bert? In: Proceedings of ACL (2019)
13. Qin, L., Ni, M., Zhang, Y., Che, W.: Cosda-ml: multi-lingual code-switching data augmentation for zero-shot cross-lingual nlp. In: Proceedings of IJCAI (2020)
14. Sato, M., Ouch, H., Tsuboi, Y.: Addressee and response selection for multilingual conversation. In: Proceedings of COLING (2018)
15. Schuster, S., Gupta, S., Shah, R., Lewis, M.: Cross-lingual transfer learning for multilingual task oriented dialog. In: Proceedings of NAACL (2019)
16. Serban, I., Sordoni, A., Bengio, Y., Courville, A., Pineau, J.: Building end-to-end dialogue systems using generative hierarchical neural network models. In: Proceedings of AAAI (2016)
17. Shang, L., Lu, Z., Li, H.: Neural responding machine for short-text conversation. In: Proceedings of ACL-IJCNLP (2015)
18. Shao, Y., Gouws, S., Britz, D., Goldie, A., Strope, B., Kurzweil, R.: Generating high-quality and informative conversation responses with sequence-to-sequence models. In: Proceedings of EMNLP (2017)

19. Sharma, S., El Asri, L., Schulz, H., Zumer, J.: Relevance of unsupervised metrics in task-oriented dialogue for evaluating natural language generation. arXiv preprint arXiv:1706.09799 (2017)
20. Sun, J., Zhou, Y., Zong, C.: Dual attention network for cross-lingual entity alignment. In: Proceedings of COLING (2020)
21. Vaswani, A., Shazeer, N., Parmar, N., Uszkoreit, J., Jones, L., Gomez, A.N., et al.: Attention is all you need. In: Proceedings of NeurIPS (2017)
22. Vinyals, O., Le, Q.: A neural conversational model. arXiv preprint arXiv:1506.05869 (2015)
23. Wu, Y., Wu, W., Yang, D., Xu, C., Li, Z: Neural response generation with dynamic vocabularies. In: Proceedings of AAAI (2018)
24. Zhang, Z., Li, J., Zhu, P., Zhao, H., Liu, G. Modeling multi-turn conversation with deep utterance aggregation. In: Proceedings of COLING (2018)
25. Zhao, T., Lee, K., Eskenazi, M.: Unsupervised discrete sentence representation learning for interpretable neural dialog generation. In: Proceedings of ACL (2018)
26. Zhu, J., et al.: Ncls: neural cross-lingual summarization. In: Proceedings of EMNLP- IJCNLP (2019)
27. Zhu, J., Zhou, Y., Zhang, J., Zong, C.: Attend, translate and summarize: an efficient method for neural cross-lingual summarization. In: Proceedings of ACL (2020)

MultiWOZ 2.3: A Multi-domain Task-Oriented Dialogue Dataset Enhanced with Annotation Corrections and Co-Reference Annotation

Ting Han[1], Ximing Liu[2], Ryuichi Takanabu[3], Yixin Lian[2], Chongxuan Huang[2], Dazhen Wan[3], Wei Peng[2(✉)], and Minlie Huang[3(✉)]

[1] University of Illinois at Chicago, Chicago, USA
than24@uic.edu
[2] Artificial Intelligence Application Research Center, AARC, Huawei Technologies, Shenzhen, China
{liuximing1,lianyixin1,huang.chongxuan,peng.wei1}@huawei.com
[3] Tsinghua University, Beijing, China
{gxly19,wandz19}@mails.tsinghua.edu.cn, aihuang@tsinghua.edu.cn

Abstract. Task-oriented dialogue systems have made unprecedented progress with multiple state-of-the-art (SOTA) models underpinned by a number of publicly available MultiWOZ datasets. Dialogue state annotations are error-prone, leading to sub-optimal performance. Various efforts have been put in rectifying the annotation errors presented in the original MultiWOZ dataset. In this paper, we introduce MultiWOZ 2.3, in which we differentiate incorrect annotations in dialogue acts from dialogue states, identifying a lack of co-reference when publishing the updated dataset. To ensure consistency between dialogue acts and dialogue states, we implement co-reference features and unify annotations of dialogue acts and dialogue states. We update the state of the art performance of natural language understanding and dialogue state tracking on MultiWOZ 2.3, where the results show significant improvements than on previous versions of MultiWOZ datasets (2.0–2.2).

1 Introduction

Task-oriented dialogue systems have made unprecedented progress with multiple state-of-the-art (SOTA) models underpinned by a number of publicly available datasets [1,12–15,17].

As the first publicly released dataset, MultiWOZ hosts more than 10K dialogues across eight different domains covering "Train", "Taxi", "Hotel", "Restaurant", "Attraction", "Hospital", "Bus" and "Police". MultiWOZ has been widely

T. Han and X. L—Both authors contributed equally to the work. The work was conducted when Ting Han interned at AARC.

L. Wang et al. (Eds.): NLPCC 2021, LNAI 13029, pp. 206–218, 2021.
https://doi.org/10.1007/978-3-030-88483-3_16

adopted by researchers in dialogue policy [6,11], dialogue generation [9] and dialogue state tracking [10,21–24] as it provides a means for modeling the changing states of dialogue goals in multi-domain interactions.

Dialogue state annotations are error-prone, leading to sub-optimal performance. For example, the SOTA joint accuracy for dialogue state tracking (DST) is still below or around 60%[1]. MultiWOZ 2.1 [16] was released to rectify annotation errors presented in the original MultiWOZ dataset. MultiWOZ 2.1 introduced additional features such as slot descriptions and dialogue act annotations for both systems and users via ConvLab [29]. Further efforts have been put into MultiWOZ 2.2 [18] to improve annotation quality. This schema-based dataset contains annotations allowing for directly retrieving slot values from a given dialogue context [10,20,24]. Despite achieving a noticeable annotation quality uplift compared to that for the original MultiWOZ, there is still room to improve. The focus of the corrections is on dialogue state annotations leaving the problematic dialogue act annotations untouched. Furthermore, the critical co-reference and ellipsis feature prevalent in the human utterance is not in presence.

To address the limitations above, we introduce an updated version, MultiWOZ 2.3[2]. Our contributions are as follow:

- We differentiate incorrect annotations in dialogue acts from those in dialogue states, and unify annotations of dialogue acts and dialogue states to ensure their consistency when publishing the updated dataset, MultiWOZ 2.3.
- We introduce co-reference features to annotations of the dialogue dataset to enhance the performances of dialogue systems in the new version.
- We re-benchmark a few SOTA models for dialogue state tracking (DST) and natural language understanding (NLU) tasks and provide a fair comparison using the updated dataset.

2 Annotation Corrections

The inconsistent annotations in the MultiWOZ dataset were caused by disparate interpretations from involved annotators during a crowdsourcing process. These errors can occur even when annotators attempt to apply unified rules. After analyzing annotation errors in both dialogue acts and dialogue states, we perform the following two data corrections.

2.1 Dialogue Act Corrections

The annotations for user dialogue acts were originally introduced by [29]. Following the pipeline provided in ConvLab, [16] re-annotated dialogue acts for both systems and users in MultiWOZ 2.1. We broadly categorize the incorrect annotations into three types (Table 1) based on our observations:

[1] https://github.com/budzianowski/multiwoz. Marked date: 6/1/2021.

[2] https://github.com/lexmen318/MultiWOZ-coref. Please be aware that all associated appendices are separately presented in the github link due to the limitataion of page numbers.

- **Under-annotated:** Annotation errors under this category are due to insufficient annotation even when exact information is available in the given dialogue utterances. The missing annotations should be added to the corresponding slots.
- **Over-annotated:** Sometimes, incorrect annotations are put down even when no corresponding information can be identified in the utterances. The over-annotated values should be removed to avoid confusion.
- **Wrongly-annotated:** This category refers to slots with incorrect values (or span information) and should be fixed.

Table 1. Example of different error types of dialogue acts. The red color in the table highlights incorrect annotations and corresponding repaired results. Note that MultiWOZ 2.2 is excluded from the table because it added missing dialogue act annotations and the remainings are the same as MultiWOZ 2.1.

Error Type	Dialogue ID	Utterance	2.1 Dialog_act	2.3 Dialog_act
Under-annotated	SSNG0348.json	For 3 people starting on Wednesday and staying 2 nights .	Hotel-Inform.Stay: 2	Hotel-Inform.Stay: 2 Hotel-Inform.Day: Wednesday Hotel-Inform.People: 3
	PMUL1170.json	Yes , one ticket please , can I also get the reference number ?	Train-Inform.People: 1	Train-Inform.People: one Train-Request.Ref: ?
	SNG01856.json	no, i just need to make sure it's cheap. oh, and i need parking	Hotel-Inform.Parking: yes	Hotel-Inform.Parking: yes Hotel-Inform.Price: cheap
Wrongly-annotated	PMUL2596.json	I will need to be picked up at the hotel by 4:45 to arrive at the college on tuesday .	Taxi-Inform.Leave:04:45 Taxi-Inform.Depart: arbury lodge guesthouse Hotel-Inform.Day: tuesday	Taxi-Inform.Leave: 4:45 Taxi-Inform.Dest: the college Taxi-Inform.Depart: the hotel Hotel-Inform.Day: tuesday
	PMUL3296.json	Yeah , could you book me a room for 2 people for 4 nights starting Tuesday ?	Hotel-Inform.Stay: 2 Hotel-Inform.Day: Tuesday Hotel-Inform.People:4	Hotel-Inform.Stay: 4 Hotel-Inform.Day: Tuesday Hotel-Inform.People:2
	PMUL4899.json	How about funkyu fun house , the are located at 8 mercers row , mercers ro industrial estate .	Attraction-Recommend.Name: funky fun house Attraction-Recommend.Addr: 8 mercers row Attraction-Recommend.Addr: mercers row industrial estate	Attraction-Recommend.Name: funky fun house Attraction-Recommend.Addr: 8 mercers row , mercers row industrial estate
Over-annotated	PMUL3250.json	No , I apoligize there are no Australian restaurants in Cambridege . Would you like to try another type of cuisine ?	Restaurant-Request.Food: ? Restaurant-NoOffer.Food: Australina Restaurant-NoOffer.Area: Cambridge	Restaurant-Request.Food: ? Restaurant-NoOffer.Food: Australian
	MUL1118.json	If there is no hotel availability , I will accept a guesthouse. Is one availabel ?	Hotel-Inform.Type: guesthouse Hotel-Inform.Stars: 4	Hotel-Inform.Type: guesthouse
	MUL0666.json	Just please book for that room for 2 nights .	Hotel-Inform.Price: cheap Hotel-Inform.Stay: 2	Hotel-Inform.Stay: 2

We apply two rules to sequentially correct "dialog_act" annotations: a) we use customized filters to select credible predictions generated from a MultiWOZ 2.1 pre-trained BERTNLU model [19] and merge them with original "dialog_act" annotations; b) we use assorted regular expressions to further clean "dialog_act" annotations from the previous step.

To fairly evaluate the quality of modified annotations, we sampled 100 dialogues from the test set and manually re-annotated the dialogue acts. Table 3 exhibits the ratios of "dialog_act" annotations of different datasets in terms of

Table 2. Example of updates on dialogue states. The red color in the figure highlights incorrect dialogue states and corresponding updated results. Note that MultiWOZ 2.2 is excluded from the figure because it is the same to MultiWOZ 2.1 in terms of inconsistent tracking. "a" and "r" used as slot names in the right two columns are abbreviations for "attraction" and "restaurant" respectively.

Dialogue ID	Utterance	MultiWOZ 2.1	MultiWOZ 2.3
MUL2602.json	*User*: Can you recommend me a nightclub where I can get jiggy with it? *Sys*: Well, I think the jiggiest nightclub in town is the Soul Tree Nightclub, right in centre city! Plis the entrance fee isonly 4 pounds	a-type=night club a-name=not mentioned a-area=not mentioned	a-type=nightclub a-name=not mentioned a-area=not mentioned
	User: That is perfect can I have the postcode please? *Sys*: Sure! The postcode is cb23qf	a-type=night club a-name=not mentioned a-area=not mentioned	a-type=nightclub a-name=soul tree nightclub a-area=not mentioned
MUL1455.json	*User*: I am also looking for a moderately priced chinses restaurant located in the north *Sys*: Golden wok is the moderate price range and in the north area would you like me to book it for you?	r-food=chinese r-pricerange=moderate r-name=not mentioned r-area=north	r-food=chinese r-pricerange=moderate r-name=not mentioned r-area=north
	User: Can I get the address and phone number please? *Sys*: Of course - the address is 191 Histon Road Chesterton cb43hl and the phone number is 01223350688	r-food=chinese r-pricerange=moderate r-name=not mentioned r-area=north	r-food=chinese r-pricerange=moderate r-name=golden wok r-area=north

Table 3. A comparison of annotation correctness ratios of "dialog_act" for MultiWOZ 2.1/2.2 and coref. The "Relax" rule indicates that the values of insignificant slots like "general-xxx" and "none" are removed.

Version	Rule	Slot level	Turn level
2.1/2.2	Strict	77.59%	68.83%
	Relax*	82.94%	77.19%
2.3	Strict	84.12%	76.09%
	Relax*	90.74%	86.83%

slot level and turn level using the manually-annotated 100 dialogues as golden annotations.

We added 24,405 slots and removed 4,061 slots in the "dialog_act" annotations. Roughly 16,800 slots are modified according to our estimation. Also note that in Table 1, boundaries for the three types are not strictly drawn. *PMUL2596.json* under wrongly-annotated type can also be treated as an under-annotated error when slot *Taxi-Inform.Dest* is missing.

Adding and removing operations for "dialog_act" annotations cause mismatches in paired span indices. When aligning span information with the modified dialogue acts, we note that original span information also contains incorrect annotations, such as abnormal span with ending index ahead of the starting index, incorrect span, and drifted span. The errors are all corrected, along with those for dialogue acts.

2.2 Dialogue State Corrections

The fixed "dialog_act" and the "span_info" annotations are propagated into the dialogue state annotations (i.e., "metadata" annotations), because we need to maintain the consistency among them.

Since the repairing for dialogue states is based on cleaned dialogue acts, we use the following rules to guide updating dialogue state annotations[3] (Table 2):

- **Slot Value Normalization**: Multiple slots values exist in MultiWOZ 2.2 due to a mismatch between given utterances and ontology, for example, "16:00" and "4 PM". This potentially leads to incomplete matching, as the values are not normalized. To this end, we follow the way how MultiWOZ 2.1 normalizes slot values based on utterances.
- **Consistent Tracking Strategy**: The inconsistent tracking strategy[4] was initially discussed (but not solved) in MulitWOZ 2.2. We track the user's requirements from slot values informed by the user, recommended by the system, and implicitly agreed by the user. We apply two sub-rules to resolve the implicit agreements: a) if an informing action is from the user to the system, the informed values are propagated to the next turn of dialogue states; b) if an informing/recommending action is from the system to the user, the informed or recommended values are propagated to the next turn of dialogues states if and only if one item is included. Multiple items are not considered to be valid in the implicit agreement settings.

3 Enhance Dataset with Co-Referencing

MultiWOZ contains a considerable amount of co-reference and ellipsis. As shown in Tabel 4, co-referencing frequently occurs in the cross-domain dialogues, especially when aligning the value of "Name" slot from a hotel (or restaurant) domain with those of "Departure/Destination" slots for taxi/train domains. The lack of co-reference annotations leads to poor performances presented in existing DST models.

A number of task-oriented dialogue models leveraged datasets enhanced with co-referencing features to achieve SOTA results [28]. By including co-reference in CamRest676 [13], GECOR [26] showed significant performance improvement compared to the baseline models. Through restoring incomplete utterances by annotating the dataset with co-reference labels, [25] boosted response quality of dialogue systems. [27] re-wrote utterances to cover co-referred and omitted information to realize notable success on their proposed model.

In MultiWOZ 2.1, the distributions of co-referencing among different slots are presented in Appendix C. In total, 20.16% dialogues are annotated with co-reference in the dataset, indicating the importance of co-referencing annotation.

[3] Statistics on the type of corrections on the "metadata" annotations is presented in Appendix A.

[4] Examples of inconsistent tracking are presented in Appendix B.

Table 4. Examples of co-reference annotations. Co-reference values are added to the original utterances and marked as light orange italic inside the brackets.

Dialogue ID	Utterance
PMUL1815.json	I'm traveling to Cambridge from lond liverpool street arriving by 11:45 the day *(saturday)* of my hotel booking. of my hotel booking.
PMUL2049.json	Thank you, can you also help me find a restaurant that is in the the same area*(centre)* as the Parkside pools?
PMUL2512.json	Thanks! I'm going to hanging out at the college (christ college) late tonight, could you get me a taxi back to the hotel*(the express by holiday inn cambridge)* at 2:45?

Table 5. Examples of co-reference annotations in the user goal. The red color highlights the difference between the original and new annotations

Dialogue ID	Goal description	Original annotation	New annotation
PMUL4372.json	You are slo looking for a *place to stay.* The hotel should *include free parking* and should be in the *same price range as the restaurant.* The hotel should *include free wifi.* Once you find the *hotel*, you want to book it for *the same group of people* and *3 nights* starting from *the same day.* If the booking fails how about *1 nights.* Make sure you get the *reference number.*	**Constraint** hotel.parking=yes hotel.pricerange=expensive hotel.internet=yes hotel.people=3 hotel.day=wednesday hotel.stay=3	**Constraint** hotel.parking=yes hotel.pricerange=[restaurant, pricerange] hotel.internet=yes hotel.people=[restaurant, people] hotel.day=[restaurant, day] hotel.stay=3→1 **Request** hotel.Ref=?
PMUL2512.json	You also want to book a *taxi* to *commute between the two places.* You want to leave the *attraction* by *02:45.* Make sure you get *contact number* and *car type.*	**Constraint** taxi.leaveAt=02:45 **Request** taxi.phone=? taxi.car type=?	**Constraint** taxi.departure=[attraction, None] taxi.destination=[hotel, None] taxi.leaveAt=02:45 **Request** taxi.phone=? taxi.car type=?

3.1 Annotation for Co-reference in Dialogue

We apply co-referencing annotations to problematic slots when necessary, for example, "Area/Price/People/Day/Depart/Dest/Arrive"[5]. The co-referencing annotations are added sequentially:

- We use first regular expressions to locate co-reference slots;
- Based on the current dialogue states, we trace back to the history utterances where the co-referred slots are first encountered;
- We use the corresponding dialogue acts with paired span information to retrieve co-referred values.

[5] Statistics of the amount of coreference annotation for each slot is presented in Appendix C.

The "coreference" annotations are applied to all "dialog_act" slots having co-referencing relationships with other slots. In total, we added 3,340 co-referencing annotations for "dialog_act"[6].

3.2 Annotation for Co-reference in User Goal

During the data collection process, the user converses with the system, following a given goal description [1]. Co-reference in the user utterances is derived from co-reference in user goals. However, the goal annotation, represented as several constraints and requests, is not consistent with the goal description and does not implement co-reference features. Table 5 shows two examples of user goals with co-reference. The original goal annotation misses a request, three constraints and all co-reference relations. The right arrow (hotel.stay = 3→1) indicates a possible goal change during a dialogue. The co-referencing relations are represented as referenced domains and slots. Note that the referenced slot of "taxi.departure/taxi.destination" is uncertain because the departure may be a name, an address, or "the attraction". *PMUL2512.json* in Table 4 shows the relation between the goal and utterance: the co-reference annotations of "the college" and "the hotel" realize the the referenced slot of "taxi.departure/taxi.destination" in the new annotations of user goal.

To introduce co-referencing annotation into user goals, we use regular expressions to extract all slot-value pairs and co-referencing relations from the goal descriptions. We manually check 150 random samples and confirm the correctness of the new goal annotations. The new goal annotations may contribute to better user simulators [7,8], which generate user responses or evaluate system performances based on user goals.

4 Benchmarks and Experimental Results

The updated dataset is evaluated for a natural language understanding task and a DST task. Experiment results are produced to re-benchmark a few SOTA models.

4.1 Dialogue Actions with Natural Language Understanding Benchmarks

BERTNLU [19] is introduced for dialogue natural language understanding. It tops extra two multilayer perceptron (MLP) layers on BERT [30] for slot recognition and intent classification [31], respectively. In practice, BERTNLU achieves better performance on classification and tagging tasks by including historical context and finetuning all parameters. We implement BERTNLU with inputs of current utterance plus the previous three history turns and finetune it based on the dialogue act annotations. The model's performance is evaluated by calculating F1 scores for intents, slots, and for both. Additionally, we use utterance

[6] Sample co-reference annotation is presented in Appendix D.

Table 6. Performance of BERTNLU on different datasets based on F1 score and utterance accuracy for slots, intents and both, respectively. Utterance accuracy is defined as the average accuracy of predicting all the slots, intents or both in an utterance correctly.

Dataset	F1(Slot/Intent/Both)	Utter. Acc.(Slot/Intent/Both)
MultiWOZ 2.1	81.18/88.34/83.77	81.89/86.23/71.68
MultiWOZ 2.2	80.61/88.34/83.41	81.94/86.41/71.85
MultiWOZ 2.3	**89.03/90.73/89.65**	**87.33/88.56/78.33**

Table 7. Joint goal accuracy of SUMBT and TRADE over different versions of dataset. MultiWOZ-coref refers to the dataset with co-reference applied. ♦ means the accuracy scores are adopted from the published papers.

Dataset	Sumbt	Trade
MultiWOZ 2.0	46.6%♦	48.6%♦
MultiWOZ 2.1	49.2%	45.6%
MultiWOZ 2.2	49.7%	46.6%
MultiWOZ 2.3	**52.9%**	**49.2%**
MultiWOZ-coref	**54.6%**	**49.9%**

accuracy as another metric to assess the model's effectiveness in understanding what the user expresses in an utterance. We score each utterance either 0 or 1 according to whether the predictions of all the slots, intents, or both in an utterance match the correct labels. The utterance accuracy is characterized as the average of this score across all utterances. Table 6 shows the performance of BERTNLU on different datasets (including dialogue utterances from both user and system sides) based on the above evaluation metrics.

4.2 Dialogue State Tracking Benchmarks

Multiple neural network-based models have been proposed to improve joint goal accuracy of dialogue state tracking tasks[7]. Existing belief state trackers could be roughly divided into two classes: span-based and candidate-based. The former approach [10,22,24] directly extracts slot values from dialogue history, while the latter approach [21] is to perform classification on candidate values, assuming all candidate values are included in the predefined ontology. To evaluate our updated dataset for DST task, we run experiments on TRADE [21] and SUMBT [22].

SUMBT uses a multi-head attention mechanism to capture relations between domain-slot types and slot values presented in the utterances. The attended context words are collected as slot values for corresponding slots. TRADE uses a pointer to differentiate, for a particular domain-slot, whether the slot value

[7] Full benchmarks with various models are available in Appendix E.

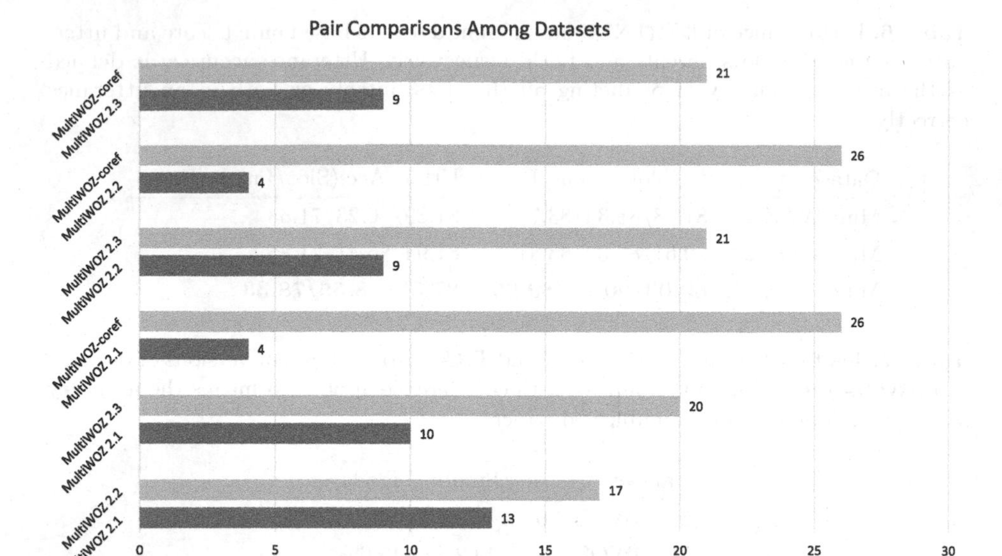

Fig. 1. Pairwise comparison between two datasets in terms of the number of higher accuracy slots. In total, there are 30 valid slots in the DST task. The number on top of each bar indicates the number of winning slots in comparison.

Table 8. Classification on slot gate for TRADE using different datasets. "Pointer", "dontcare" and "none" are three different slot gate classes. Precision, recall, and F1-score are used as metrics to evaluate among all datasets.

Dataset	Pointer (P/R/F1)	Dontcare (P/R/F1)	None (P/R/F1)
MultiWOZ 2.1	94.97/93.75/94.35	58.73/32.51/41.85	98.25/98.82/98.53
MultiWOZ 2.2	94.22/94.42/94.32	60.21/34.60/43.91	98.42/98.64/98.53
MultiWOZ 2.3	**96.41/96.15/96.28**	**67.80/41.62/51.58**	**98.79/99.11/98.95**

is from the given utterance or the predefined vocabulary. Both models perform predictions slot by slot and treat all slots equally.

Following the convention in dialogue state tracking task, joint goal accuracy is used to evaluate the models' performances for different datasets. The models also experiment with co-referencing enhanced datasets. Table 7 summarizes the joint goal accuracy of the two models using different datasets.

4.3 Experimental Analysis

As shown in Tables 6 and 7, substantial performance increases are achieved with the enhanced datasets compared to the previous datasets. BERTNLU trained using our dataset outperforms others with a margin of 5% improvement on both metric of F1-score and utterance accuracy. In the task of DST, models trained using our datasets also show superiority to those trained with the previous version MultiWOZ. By applying co-referencing features to dialogue state tracking, the joint goal accuracy is improved to approximately 55% using SUMBT.

5 Discussion

Note that SUMBT initially focused on MultiWOZ 2.0. Fixing dialogue states leads to enhanced data quality in MultiWOZ 2.1. This study adopts a rule-based method to correct the identified errors in MultiWOZ 2.1 further. With a customized pre-process script[8] for SUMBT, the joint goal accuracy can reach 54.54% for MultiWOZ 2.3 and 56.09% for MultiWOZ-coref, respectively. Since multiple slot values are allowed for MultiWOZ 2.2, it is not practical to identify errors in the dialogue states. We do not base this study on MultiWOZ 2.2 at this stage. Figure 1 shows pairwise comparisons between two datasets on the benchmarked scores. Our dataset (MultiWOZ 2.3) tops all the scores compared to previously updated datasets in all MultiWOZ specified slots. Details of slot accuracies are presented in Appendix G. Our dataset achieves the best performance for 17 out of all 30 slots. The performance is further enhanced with the co-reference version (24 out of all 30).

Table 8 shows precision, recall, and F1-score of slot gate classifications in the TRADE model across different datasets. For the three different classes, our dataset achieves top performances. As a result of the carefully designed error correction[9], our dataset outperforms others by at least 9% in all metrics for the "dontcare" gate.

Based on the contexts presented in utterances, we have fixed the dialogue acts and removed the inconsistency between dialogue acts and states. Span indices in the dialogue acts are further fixed with co-reference information introduced. By closely aligning the annotations to corresponding utterances mentioned above, we remove the inconsistency introduced by annotating a Wizard-Of-Oz dataset. Human evaluations by three volunteers on randomly sampled ten dialogues from both datasets are summarized in Table 9. The reported evaluation results further verify the qualify of MultiWOZ 2.3.

[8] Scores shown in Table 7 are achieved by using pre-process scripts provided by SUMBT and TRADE.

[9] Details of correction are shown in Appendix F.

Table 9. Scores of human evaluation on ten dialogues of the two datasets. 10 is given as an initial score for each turn. Any unnecessary, incorrect or missing annotations of dialogue acts or dialogue states lead to loss of 1 point. The final score of a dialogue is averaged across all turns.

Dialogue ID	# of Turns	MultiWOZ 2.1	MultiWOZ 2.3
MUL2261	16	9.23	9.96
MUL0784	24	9.35	9.94
PMUL2512	20	9.23	10.00
MUL2694	20	8.85	9.83
PMUL4510	26	9.48	9.92
MUL1313	16	9.47	10.00
MUL0859	12	9.23	9.83
MUL0195	20	9.16	9.87
PMUL0449	18	9.44	9.85
MUL0966	22	9.28	9.82

6 Conclusion

MultiWOZ datasets (2.0–2.2) are widely used in dialogue state tracking and other dialogue related subtasks. Mainly based on MultiWOZ 2.1, we publish a refined version, named MultiWOZ 2.3. After correcting annotations for dialogue acts and dialogue states, we introduce co-reference annotations, which supports future research to consider discourse analysis in building task-oriented dialogue systems. We re-benchmark the refined dataset using some competitive models. The experimental results show significant improvements for the associated scores, verifying the utility of this dataset. We hope to attract more alike research works to improve the quality of MultiWOZ datasets further.

References

1. Budzianowski, P., Wen, T.H., Tseng, B.H., Casanueva, I., Ultes, S., Ramadan, O., & Gašić, M.: MultiWOZ-a large-scale multi-domain wizard-of-oz dataset for task-oriented dialogue modelling. In: EMNLP, Brussels, pp. 5016–5026 (2018)
2. Mehri, S., Eric, M., Hakkani-Tur, D.: DialoGLUE: a natural language understanding benchmark for task-oriented dialogue. arXiv preprint arXiv:2009.13570 (2020)
3. Wang, Y., Guo, Y., Zhu, S.: Slot attention with value normalization for multi-domain dialogue state tracking. In: EMNLP, pp. 3019–3028, November 2020
4. Kim, S., Yang, S., Kim, G., Lee, S. W.: Efficient dialogue state tracking by selectively overwriting memory. In: ACL, pp. 567–582, July 2020
5. Ren, L., Ni, J., McAuley, J.: Scalable and accurate dialogue state tracking via hierarchical sequence generation. In: EMNLP-IJCNLP, Hong Kong, pp. 1876–1885, November 2019

6. Takanobu, R., Zhu, H., Huang, M.: Guided dialog policy learning: Reward estimation for multi-domain task-oriented dialog. In: EMNLP-IJCNLP, Hong Kong, pp. 100–110, November 2019
7. Schatzmann, J., Thomson, B., Weilhammer, K., Ye, H., Young, S.: Agenda-based user simulation for bootstrapping a POMDP dialogue system. In: NAACL-HLT, Companion Volume, pp. 149–152. Rochester, April 2007
8. Gür, I., Hakkani-Tür, D., Tür, G., Shah, P.: User modeling for task oriented dialogues. In: IEEE-SLT, Athens, pp. 900–906, December 2018
9. Chen, W., Chen, J., Qin, P., Yan, X., Wang, W.Y.: Semantically Conditioned Dialog Response Generation via Hierarchical Disentangled Self-Attention. In: ACL, Florence, pp. 3696–3709, July 2019
10. Zhang, J.G., Hashimoto, K., Wu, C.S., Wan, Y., Yu, P.S., Socher, R., Xiong, C.: Find or classify? dual strategy for slot-value predictions on multi-domain dialog state tracking. arXiv preprint arXiv:1910.03544 (2019)
11. Zhao, T., Xie, K., Eskenazi, M.: Rethinking action spaces for reinforcement learning in end-to-end dialog agents with latent variable models. In: NAACL-HLT, Volume 1 (Long and Short Papers), Minneapolis, pp. 1208–1218, June 2019
12. Rastogi, A., Zang, X., Sunkara, S., Gupta, R., Khaitan, P.: Towards scalable multi-domain conversational agents: the schema-guided dialogue dataset. In: AAAI, New York, pp. 8689–8696, April 2020
13. Wen, T.H., et al.: A network-based end-to-end trainable task-oriented dialogue system. In: EACL, Valencia, pp. 438–449, January 2017
14. Williams, J., Raux, A., Ramachandran, D., Black, A.: The dialog state tracking challenge. In: SIGDIAL, Metz, pp. 404–413 (2013)
15. Henderson, M., Thomson, B., Williams, J.D.: The second dialog state tracking challenge. In: SIGDIAL, Philadelphia, pp. 263–272 (2014)
16. Eric, M., et al.: MultiWOZ 2.1: a consolidated multi-domain dialogue dataset with state corrections and state tracking baselines. In: LREC, Marseille, pp. 422–428 (2020)
17. Zhu, Q., Huang, K., Zhang, Z., Zhu, X., Huang, M.: CrossWOZ: a large-scale chinese cross-domain task-oriented dialogue dataset. In: TACL, 8, pp. 281–295 (2020)
18. Zang, X., Rastogi, A., Sunkara, S., Gupta, R., Zhang, J., Chen, J.: MultiWOZ 2.2: a dialogue dataset with additional annotation corrections and state tracking baselines. In: ACL, pp. 109–117 (2020)
19. Zhu, Q., et al.: ConvLab-2: an open-source toolkit for building, evaluating, and diagnosing dialogue systems. In: ACL, System Demonstrations, pp. 142–149, July 2020
20. Gao, S., Sethi, A., Agarwal, S., Chung, T., Hakkani-Tur, D., AI, A.A.: Dialog state tracking: a neural reading comprehension approach. In: SIGDIAL, Stockholm, pp. 264–273 (2019)
21. Wu, C.S., Madotto, A., Hosseini-Asl, E., Xiong, C., Socher, R., Fung, P.: Transferable multi-domain state generator for task-oriented dialogue systems. In: ACL, Florence, pp. 808–819, July 2019
22. Lee, H., Lee, J., Kim, T.Y.: SUMBT: slot-utterance matching for universal and scalable belief tracking. In: ACL, Florence, pp. 5478–5483, July 2019
23. Zhou, L., Small, K.: Multi-domain dialogue state tracking as dynamic knowledge graph enhanced question answering. arXiv preprint arXiv:1911.06192 (2019)
24. Heck, M., et al.: TripPy: a triple copy strategy for value independent neural dialog state tracking. In: SIGDIAL, pp. 35–44, July 2020

25. Pan, Z., Bai, K., Wang, Y., Zhou, L., Liu, X.: Improving open-domain dialogue systems via multi-turn incomplete utterance restoration. In: EMNLP-IJCNLP, Hong Kong, pp. 1824–1833, November 2019
26. Quan, J., Xiong, D., Webber, B., Hu, C.: GECOR: an end-to-end generative ellipsis and co-reference resolution model for task-oriented dialogue. In: EMNLP-IJCNLP, Hong Kong, pp. 4539–4549, November 2019
27. Su, H., et al.: Improving multi-turn dialogue modelling with utterance ReWriter. In: ACL, Florence, pp. 22–31, July 2019
28. Ferreira Cruz, A., Rocha, G., Lopes Cardoso, H.: Coreference resolution: toward end-to-end and cross-lingual systems. Information **11**(2), 2078–2489 (2020)
29. Lee, S., et al.: ConvLab: multi-domain end-to-end dialog system platform. In: ACL, Florence, pp. 64–69, July 2019
30. Devlin, J., Chang, M.W., Lee, K., Toutanova, K.: BERT: pre-training of deep bidirectional transformers for language understanding. In: NAACL-HLT, Volume 1 (Long and Short Papers), pp. 4171–4186. Minneapolis, June 2019
31. Chen, Q., Zhuo, Z., Wang, W.: Bert for joint intent classification and slot filling. arXiv preprint arXiv:1902.10909 (2019)

EmoDialoGPT: Enhancing DialoGPT with Emotion

Yuxiang Jia[1(✉)], Shuai Cao[1,2], Changyong Niu[1], Yutuan Ma[1], Hongying Zan[1], Rui Chao[1], and Weicong Zhang[1]

[1] School of Information Engineering, Zhengzhou University, Zhengzhou, China
{ieyxjia,iehyzan}@zzu.edu.cn

[2] Zhengzhou Zoneyet Technology Co., Ltd., Zhengzhou, China

Abstract. Perceiving user emotions and generating responses with specific emotions are of great significance to a social dialogue system. Much previous work still utilizes seq2seq models as the backbone to generate emotional responses. With the popularity of the generative pre-training models, we propose an emotional response generation model EmoDialoGPT based on DialoGPT by introducing emotion embedding and emotion prediction loss. In order to obtain a large-scale dialogue dataset with emotion labels, we train an emotion classifier and automatically annotate the emotion labels for the dialogue data. We evaluate our models from three aspects, including emotion expression, response quality and human evaluation. The experimental results show that, the proposed models outperform seq2seq models and baseline DialoGPT models on most metrics.

Keywords: Emotion classification · Response generation · Dialogue system · Transformer

1 Introduction

Dialogue system has a wide application prospect and arouses great attention in NLP research. For a social dialogue system, it is crucial to detect the user emotion and respond with specific emotions. Studies have shown that injecting emotions in dialogue systems can significantly improve user satisfaction. The great amount of conversation data accumulated from online social media provide a good chance for open-domain dialogue system research based on deep learning. Especially, the use of large-scale pre-trained language models has yielded significant results for different tasks in NLP, including dialogue system. For example, OpenAI's GPT-2 model [12] is trained on large-scale datasets based on the Transformer architecture and can generate fluent, lexically diverse text. Microsoft's DialoGPT [18] uses 147M Reddit conversation data to fine-tune the GPT-2 model and is able to generate content-rich, contextually relevant responses.

However, emotional dialogue system still faces great challenges. The most significant challenge is the lack of large-scale dialogue corpora with human-labeled

L. Wang et al. (Eds.): NLPCC 2021, LNAI 13029, pp. 219–231, 2021.
https://doi.org/10.1007/978-3-030-88483-3_17

emotion tags. As an alternative, an emotion classifier could be trained and then labels the dialogue emotion automatically. But most emotion classification corpora and dialogue corpora are from different domains and results in low quality of emotion labels. Another challenge comes from the way of injecting emotion into existing dialogue system. Seq2seq models are explored to fuse emotion to generate emotional responses while relatively less attention has been paid to generative pre-trained models for emotional response generation.

Thus, we propose the EmoDialoGPT model by integrating emotion into DialoGPT, one of the state-of-the-art open-domain dialogue systems. We design three different approaches to embedding emotion, including a separate embedding layer, a special beginning token, and a special ending token. To obtain a large-scale dialogue corpus with emotion tags, we train an emotion classifier to automatically label dialogues with one of nine emotions. From three aspects, we compare our models with three seq2seq-based models and two baseline DialoGPT models.

The contributions of this work include:

- We propose an emotional response generation model based on DialoGPT by introducing emotion embedding and emotion prediction loss. We compare three different ways of emotion embedding.
- We evaluate the models from three aspects, and the experimental results show that the proposed models outperform seq2seq models and baseline DialoGPT models on most metrics.

2 Related Work

Emotion classification serves as one of the basics of this work. Ekman [3], an expert in emotion theory, first proposed six basic emotions, namely *anger*, *fear*, *joy*, *sadness*, *surprise* and *disgust*, in 1972. Based on the emotion data set collected from Twitter, Shahraki et al. [13] added other three emotions, *love*, *thankfulness* and *guilt*. We use this dataset to train the emotion classifier, which can identify the above nine emotions.

Inspired by phrase-based statistical machine translation techniques, Shang et al. [14] applied the seq2seq model to response generation and achieved good results. Despite the success of the seq2seq model in response generation, there is a tendency to generate generic responses like "i don't know" and a lack of consistency in response content and language style. The proposal of the Transformer model [15] greatly facilitated the development of large-scale pre-trained language generation models, and Radford et al. [11] proposed a large-scale generative pre-training model GPT based on the Transformer model. Then Radford et al. [12] trained a much larger GPT-2 model that can better mimic human-generated text. Zhang et al. [18] proposed the DialoGPT model, which fine-tuned the GPT-2 model based on 147 million conversations from the Reddit comment chain and achieved near-human response quality. The DialoGPT model shows that the Transformer model trained on large-scale datasets can capture long-term dependencies in text data and generate fluent, lexically diverse, and content-rich responses. However, previous models mainly focus on improving the content quality of the generated responses while neglect emotion information.

Li et al. [6] were able to generate personalized responses given a speaker, which is considered to be the first attempt to constrain the seq2seq model for response generation. Zhou et al. [20] proposed the Emotional Chatting Machine, which embeds emotion and words into the seq2seq model and uses internal and external emotion words to generate an emotional response. Based on the seq2seq model, Huang et al. [4] proposed three methods for incorporating emotions, including emotion label as the first special token of the input utterance, emotion label as the last special token of the input utterance, and embedding emotions into the decoding block. Nevertheless, in general, the seq2seq-based response generation models tend to generate generic responses and cannot capture emotions well.

3 Methodology

3.1 Model Architecture

Transformer [15] models are widely used for a variety of tasks in NLP, and research has shown that, compared with training models from scratch, better performance can be achieved by fine-tuning pre-trained models [17], such as GPT-2 and DialoGPT, which are very effective in language generation tasks. We propose an emotional response generation model, EmoDialoGPT, which has the same architecture as the DialoGPT model and is also a generative pre-trained model based on the Transformer decoder.

The overall architecture of the proposed model is shown in Fig. 1. It consists of L stacked layers of Transformer blocks, with each layer using multi-head masked attention to enable the model to perform autoregressive response generation. Unlike the DialoGPT model, we use the last hidden layer to compute both the loss of response generation and the loss of emotion prediction, and then apply weighted summation to obtain the final loss for backpropagation.

3.2 Input Representation

Suppose that the source sentence and golden response pair of a dialogue is denoted by (X, Y), where $X = \{x_1, x_2, ..., x_m\}$ and $Y = \{y_1, y_2, ..., y_n\}$ are sentences consisting of sequences of words. We tokenize each word using the same BPE (Byte Pair Encoding) algorithm as in GPT-2 [12] to obtain sub-words as the basic semantic units of the input text. Meanwhile, since words in different positions may yield different semantic information, we encode the position information of the input sequence as position embedding. Formally, we define the token embedding $T \in \mathbb{R}^{|V| \times d}$($|V|$ is the vocabulary size, and d is the size of the hidden layer), the position embedding $P \in \mathbb{R}^{1024 \times d}$ (the maximum input length is 1024), and the representation of each token x as the summation of the above two embeddings:

$$Emb(x) = T(x) + P(x) \tag{1}$$

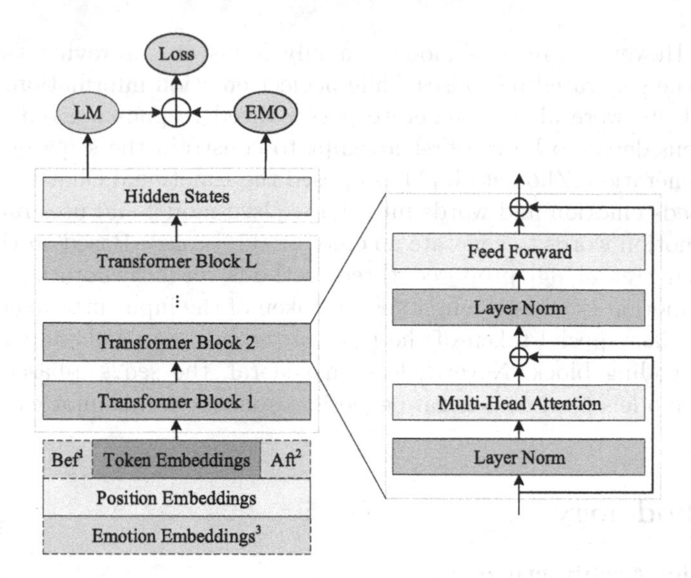

Fig. 1. The overall architecture of our EmoDialoGPT model. The three purple dashed blocks indicate different methods of emotion embedding.

3.3 Emotion Injection

Due to the lack of emotional information, it is difficult for DialoGPT models to generate responses that express specific emotions. To build EmoDialoGPT models that express specific emotions in responses, we propose three methods to incorporate emotions into DialoGPT models. In the first two methods, we add each emotion as a special token before and after the source sentence inspired by Huang et al. [4]. These methods reduce the two individual inputs into one so that they can be trained directly on the normal DialoGPT model. We name them as *EmoDialoGPT-bef* and *EmoDialoGPT-aft* respectively. As shown in the following equations:

$$X_1 = \{e, x_1, x_2, ..., x_m\} \tag{2}$$

$$X_2 = \{x_1, x_2, ..., x_m, e\} \tag{3}$$

X_1 is the input of *EmoDialoGPT-bef*, X_2 is the input of *EmoDialoGPT-aft*, and e is the emotion of the response.

Both of the above methods require to concatenate emotions in the source sentences. Another method is to convert the emotion into an emotion embedding vector using an emotion weight matrix $E \in \mathbb{R}^{|C| \times d}$ ($|C|$ is the number of emotion categories and d is the size of the hidden layer), thus enabling the model to learn emotion embeddings [19]. We name this method *EmoDialoGPT-emb*. For each token x and the emotion e of the response, the final input embedding becomes:

$$Emb(x, e) = T(x) + P(x) + E(e) \tag{4}$$

The computational process is abstracted as follows:

$$H^0 = Emb \tag{5}$$

$$H^l = BLOCK(H^{l-1}), 1 \leq l \leq L \tag{6}$$

$$P(y_i|X; y_{\leq i-1}, e) = softmax(W_V H_i^L)|_{y_i} \tag{7}$$

$P(y_i|X; y_{\leq i-1}, e)$ is modeled by multi-layer stacked Transformer blocks and $W_V \in \mathbb{R}^{|V| \times d}$ ($|V|$ is the vocabulary size, and d is the size of the hidden layer) is the weight matrix of the output layer. $BLOCK$ is the Transformer block that models the text sequence using multi-head masked attention.

$$P(c_i|X; Y, e) = softmax(W_C H_N^L)|_{c_i} \tag{8}$$

To ensure that our model can express specified emotion, we use a linear layer to transform the hidden vector at the last position of the hidden layer, H_N^L, into the distribution of emotion categories. As in Eq. 8, c_i is the category of emotion, and $W_C \in \mathbb{R}^{|C| \times d}$ ($|C|$ is the number of emotion categories and d is the size of the hidden layer) is the weight matrix of the output layer.

3.4 Optimization

The optimization object contains two parts. One part is the maximum likelihood estimation which is widely used in language models. We adopt NLL (Negative Log Likelihood) loss as follows:

$$\mathcal{L}_{NLL} = -\frac{1}{N} \sum_{i=1}^{N} \log P(y_i|X; y_{\leq i-1}, e) \tag{9}$$

where N is the length of the golden response, and e is its emotion.

The other part is the prediction loss of emotion:

$$\mathcal{L}_{EMO} = -\log P(c_i|X; Y, e) \tag{10}$$

So the final optimization object is the weighted summation of the above losses:

$$\mathcal{L} = \mathcal{L}_{NLL} + \lambda * \mathcal{L}_{EMO} \tag{11}$$

where λ is the weight of the emotion prediction loss. We set λ to 1.0 in our experiments.

4 Dataset

4.1 Emotion Classifier

Because of the lack of large-scale conversation datasets containing emotion labels, we first train an emotion classifier based on the CBET (Cleaned Balanced Emotional Tweets) dataset collected by Shahraki et al. [13] from Twitter and then use it to perform emotion prediction on large-scale conversation data. The CBET dataset contains nine emotion categories, namely *anger*, *fear*, *joy*, *love*, *sadness*, *surprise*, *thankfulness*, *disgust* and *guilt*. There are 81,163 tweets with emotion labels, of which 76,860 tweets with only one emotion label, of which each emotion category contains equal number of tweets. We choose tweets with single emotion label to train an emotion classifier, and divide the data of each emotion category into the training set, validation set, and test set by 8:1:1. We use the CBET dataset to fine-tune the BERT-Base, BERT-Large, RoBERTa-Base, and RoBERTa-Large models [2,8], respectively, with a learning rate that corresponds to the best performance of each model. The emotion classification results on the test set are shown in Table 1. RoBERTa-Large model has the best performance on this dataset with an F1-macro value (average of the F1 values for each emotion category) of 69.54%. Thus we use the RoBERTa-Large model as the emotion classifier for emotion prediction on the conversation dataset.

Table 1. Results of emotion classification (%)

Model	Accuracy	Precision	Recall	F1-macro	Learning rate
BERT-Base	63.85	64.05	63.86	63.91	2e−5
BERT-Large	65.91	66.16	65.91	66.01	1e−5
RoBERTa-Base	66.22	66.21	66.22	66.21	5e−5
RoBERTa-Large	**69.52**	**69.59**	**69.52**	**69.54**	1e−5

4.2 Dialogue Dataset with Emotion Labels

The OpenSubtitles dataset [7] is a large-scale noisy open-domain dialogue dataset containing about 60-70M movie subtitles. Li et al. [6] first preprocessed the OpenSubtitles dataset and extracted more than 11.3M single round dialogues. Then, we further remove duplicate data and data containing 'unk' to reduce noise, resulting in about 3.1M single round dialogues. Next, we use the trained emotion classifier to predict the emotion of the target utterance or response of each dialogue. We output the probability through softmax function and label the target utterance with a probability less than a threshold (we set the threshold to 0.9) as *None-emotion*. *None-emotion* is considered as a special emotion category, which is used to train the response generation model but not considered in the evaluation phase. Finally, 18.03% of the dialogues are labeled as *None-emotion*. The detailed distribution of the predicted emotions is shown in Table 2 (column "Ratio"). The validation set and test set each contains 50,000 dialogues, and those remaining are the training set. The dialogues in the validation set and test set are randomly selected according to the overall distribution of emotions in the whole dataset.

Table 2. Distribution of OpenSubtitles dataset with emotion labels

Emotion	Training	Validation	Test	Total	Ratio
anger	305,644	5,023	5,023	315,690	10.05%
fear	149,982	2,465	2,465	154,912	4.93%
joy	67,604	1,111	1,111	69,826	2.22%
love	71,334	1,172	1,172	73,678	2.35%
sadness	373,061	6,131	6,131	385,323	12.26%
surprise	285,620	4,694	4,694	295,008	9.39%
thankfulness	196,542	3,230	3,230	203,002	6.46%
disgust	321,426	5,282	5,282	331,990	10.56%
guilt	722,632	11,876	11,876	746,384	23.75%
None-emotion	548,588	9,016	9,016	566,620	18.03%
Total	3,042,433	50,000	50,000	3,142,433	100%

5 Experiments

5.1 Experimental Settings

All the models are implemented with PyTorch[1] and Transformers library[2]. For training the EmoDialoGPT models, we fine-tune the DialoGPT-medium model [16,18] with a parameter size of 345M, hidden size of 1024, the maximum sequence length of 1024, and the number of Transformer layer of 24. All the EmoDialoGPT models are fine-tuned for 5 epochs with a batch size of 8 and learning rate of 1e−5 on one NVIDIA RTX 2080 Ti GPU. We use Adam optimizer [5] and choose the model with the minimum loss on the validation set as the final model for evaluation. We use Beam search [10] to decode the model with a beam size of 5.

5.2 Baselines

To comprehensively verify the effectiveness and advantages of our proposed models, we make comparison with the previous state-of-the-art seq2seq models [4] using the same corpus and the same emotion classifier. The three baseline models are listed as follows:

- *Seq2Seq-bef* The emotion is added in the encoder as a special token before the input and then trained on the normal seq2seq model.
- *Seq2Seq-aft* The emotion is added in the encoder as a special token after the input and then trained on the normal seq2seq model.
- *Seq2Seq-dec* The emotion is injected into the decoder of the seq2seq model as an emotion vector.

[1] https://pytorch.org/.
[2] https://github.com/huggingface/transformers/.

5.3 Automatic Evaluation of Emotion Expression

In order to test whether the generated responses can accurately express specific emotions, we use 40,984 dialogues with emotion labels as the test set (ignoring "None-emotion"), as shown in Table 2. Inputing the source sentence and the emotion label of the response into the emotional response generation model results in an emotional response. The emotion classifier is used to predict the emotion of the generated response. Assuming that the input emotion label is the ground truth, if the emotion label predicted by the emotion classifier is consistent with the input emotion label, the emotion expression is correct. The emotion expression evaluation results are shown in Table 3. The DialoGPT-FT model uses the training set to fine-tune the DialoGPT model, without introducing emotion information. As can be seen, our three proposed models outperform all other models, including the three baseline seq2seq models, DialoGPT, and DialoGPT-FT, in all emotion expression evaluation metrics.

Table 3. Results of emotion expression in response generation (%)

Model	Accuracy	Precision	Recall	F1-macro
Seq2Seq-bef	70.46	65.44	63.59	63.99
Seq2Seq-aft	71.7	67.32	66.13	65.77
Seq2Seq-dec	74.18	68.03	70.45	68.62
DialoGPT	18.62	12.7	12.68	12.56
DialoGPT-FT	19.13	12.49	12.54	12.36
EmoDialoGPT-bef	78.6	74.33	73.01	73.30
EmoDialoGPT-aft	78.58	**74.62**	73.85	73.86
EmoDialoGPT-emb	**79.24**	74.29	**75.55**	**74.6**

5.4 Automatic Evaluation of Response Quality

We use 17 automatic metrics [1] to evaluate the effectiveness of our model, including: average response length $|U|$, per-word and per-utterance unigram (H_w^u, H_u^u) and bigram (H_w^b, H_u^b) entropy measuring diversity of responses, unigram and bigram-level KL divergence (D_{kl}^u, D_{kl}^b) between model and ground truth response sets measuring how well a model can approximate the ground truth distribution, embedding metrics average (AVG), extrema (EXT), and greedy (GRE) measuring similarity between response and ground truth embedding, the cosine similarity between pairs of input and response denoted by coherence (COH), distinct-1 and distinct-2 ($d1$, $d2$) for evaluating diversity of responses, and BLEU ($b1$, $b2$, $b3$, $b4$) measuring n-gram overlap between response and ground truth.

The results of the automatic response quality evaluation are shown in Table 4. We can find that, on average, responses generated by DialoGPT-style models

Table 4. Results of the automatic quality evaluation of response generation. "Human" represents the ground truth reference.

	$\|U\|$	$H_w^u \uparrow$	$H_w^b \uparrow$	$H_u^u \uparrow$	$H_u^b \uparrow$	$D_{kl}^u \downarrow$	$D_{kl}^b \downarrow$	$AVG \uparrow$	$EXT \uparrow$	$GRE \uparrow$	$COH \uparrow$	$d1 \uparrow$	$d2 \uparrow$	$b1 \uparrow$	$b2 \uparrow$	$b3 \uparrow$	$b4 \uparrow$
Human	10.2	9.29	15.2	94.6	136	0	0	100	100	100	.482	.032	.328	100	100	100	100
Seq2Seq-bef	9.4	8.06	11.81	**75.62**	94.73	1.403	1.962	35.62	46.92	46.17	.361	.005	.024	**10.42**	**12.06**	**12.35**	**11.71**
Seq2Seq-aft	**9.5**	7.95	11.60	75.24	**95.11**	1.607	2.297	35.94	46.79	46.23	.363	.002	.013	10.29	11.98	12.29	11.66
Seq2Seq-dec	9.3	8.07	11.79	74.88	93.15	1.448	1.976	35.56	46.77	46.08	.359	.005	.025	10.31	11.89	12.16	11.52
DialoGPT	7.4	**10.6**	**15.3**	55.9	54.0	1.351	**1.374**	45.62	34.57	39.71	.475	**.027**	**.168**	2.9	4.01	4.28	4.11
DialoGPT-FT	8.3	7.86	10.94	64.4	73.3	1.60	2.49	48.41	47.38	49.53	.485	.008	.040	9.54	10.59	10.71	10.07
EmoDialoGPT-bef	7.9	8.31	11.47	64.3	69.7	1.28	2.01	50.33	47.35	49.81	.486	.010	.051	9.41	10.26	10.28	9.62
EmoDialoGPT-aft	7.8	8.36	11.47	64.0	68.5	**1.21**	2.01	50.36	47.37	**49.83**	.486	.010	.051	9.37	10.18	10.18	9.52
EmoDialoGPT-emb	7.9	8.29	11.46	64.4	70.0	1.29	2.03	**50.38**	**47.4**	49.80	**.487**	.010	.051	9.44	10.31	10.33	9.67

are shorter than those generated by seq2seq models. Although the BLEU scores, per-utterance entropy of the seq2seq models are slightly higher than our models, the results of the remaining metrics are substantially improved. Both fine-tuned on the OpenSubtitles dataset, 10 metrics of our models are improved compared with the DialoGPT-FT model, indicating that injecting emotions can improve the quality of response generation to some extent. Furthermore, 11 metrics of our models have substantial performance improvement compared with the DialoGPT model. However, due to a large amount of noise in the OpenSubtitles dataset, H_w^u, H_w^b, and *distinct* of the fine-tuned models are still inferior to the DialoGPT model, even though we have removed some noisy data, which indicates that the quality of the dataset is crucial for the response generation model.

5.5 Human Evaluation

Human evaluation is carried out from three aspects, including relevance, fluency and emotion. Following [9], we randomly sample 100 utterances and get corresponding responses from each model. Then three annotators score each response from the above three aspects. The score ranges from 1 to 5 (1: not at all, 3: somewhat, 5: very much). A higher score means a better result. The average human rating scores are listed in Table 5. We can see that, our models outperform seq2seq models in terms of all the three metrics, among which relevance gets improved the most, followed by emotion. However, the score of fluency is always higher than those of the other two metrics for all models.

Table 5. Results of human evaluation

Model	Relevance	Fluency	Emotion
Seq2Seq-bef	3.2	3.79	3.12
Seq2Seq-aft	3.0	3.67	3.19
Seq2Seq-dec	3.13	3.79	3.14
EmoDialoGPT-bef	3.81	**3.92**	3.34
EmoDialoGPT-aft	3.8	3.89	3.37
EmoDialoGPT-emb	**3.83**	3.91	**3.56**

Table 6. An example of response generation

Source		i'm gonna get married
Target		you can't leave me like this
Seq2Seq-bef	*love*	i know that you'r e the most beautiful woman in the world
	sadness	i don't know anything about it
Seq2Seq-aft	*love*	i want you to come with me
	sadness	i will never see you again
Seq2Seq-dec	*love*	i want you to be my wife
	sadness	i know that you'r e not going to do anything for me
EmoDialoGPT-bef	*love*	i want to spend the rest of my life with you
	sadness	i just want to be with you
EmoDialoGPT-aft	*love*	it's a beautiful wedding ring
	sadness	i wanted to marry you but i couldn't
EmoDialoGPT-emb	*love*	you must come with me to the mountain and marry me there
	sadness	i won't gonna get happiness anymore

5.6 Case Study

Table 6 shows an example comparing responses generated by our models and three seq2seq models. Each model generates two responses corresponding to emotion *love* and *sadness* respectively. As can be seen, responses generated by our models are more relevant, diverse while those generated by the seq2seq models contain many frequently used language patterns, like "i want you to", "i know that", etc., which are safe, grammatically correct, but lack of useful information.

Table 7 shows an example of responses generated by the *EmoDialoGPT-emb* model corresponding to different emotions. "Source" is the input utterance, "Target" is the reference response, and "Predicted" is the emotion predicted by the emotion classifier based on the generated response. It can be seen that the generated responses have good relevance and fluency and are more evident in terms of emotion. But the emotion of the response generated according to *surprise* is predicted as *joy*, which may be an indication of emotion classification as a multi-label problem.

Table 7. An example of response generation by EmoDialoGPT-emb model

Source		shall i tell you something taplow?	Predicted
Target		you know i rather hope you don't get your promotion	
EmoDialoGPT-emb	*anger*	you can't do this to me !	*anger*
	fear	there's a big man in town	*fear*
	joy	when you see a good thing you will say	*joy*
	love	the most important thing in the world is love you	*love*
	sadness	i was on my way to the office	*sadness*
	surprise	you will got a present for me	*joy**
	thankfulness	i have a wife and a child	*thankfulness*
	disgust	i can't stand that man	*disgust*
	guilt	you know i am not supposed to talk	*guilt*

6 Conclusion and Future Work

In this paper, we focus on enhancing response generation with emotions and propose the EmoDialoGPT model by injecting emotions into DialoGPT with three different approaches. We compare the proposed models with three seq2seq-based models and the baseline DialoGPT models from three aspects and achieve the best results on most metrics. To build a large-scale dialogue corpus with emotion labels, we train an emotion classifier to automatically label dialogues with emotion tags.

The effect of emotion classification has a significant impact on the generation of emotional responses. Currently, emotion classification performance is not satisfactory due to the limited size of the emotion classification dataset and the difference of the domain with the dialogue dataset. In the next step, we will consider constructing an emotion classification dataset based on conversation data to achieve better emotion classification and improve emotional responses. We take English as the starting point, and then we will shift our focus to Chinese to study how to inject emotions into Chinese response generation better.

Acknowledgements. We would like to thank the anonymous reviewers for their insightful and valuable comments. This work was supported in part by National Key Research and Development Project (Grant No. 2017YFB1002101), Major Program of National Social Science Foundation of China (Grant No. 17ZDA138, 18ZDA295), National Natural Science Foundation of China (Grant No. 62006211), and China Post-doctoral Science Foundation (Grant No. 2019TQ0286, 2020M682349).

References

1. Csáky, R., Purgai, P., Recski, G.: Improving neural conversational models with entropy-based data filtering. In: Proceedings of the 57th Annual Meeting of the Association for Computational Linguistics, pp. 5650–5669 (2019)
2. Devlin, J., Chang, M.W., Lee, K., Toutanova, K.: Bert: pre-training of deep bidirectional transformers for language understanding. In: Proceedings of the 2019 Conference of the North American Chapter of the Association for Computational Linguistics: Human Language Technologies, Volume 1 (Long and Short Papers), pp. 4171–4186 (2019)
3. Ekman, P., Friesen, W.V., Ellsworth, P.: Emotion in the human face: Guidelines for research and an integration of findings (1972)
4. Huang, C., Zaiane, O.R., Trabelsi, A., Dziri, N.: Automatic dialogue generation with expressed emotions. In: Proceedings of the 2018 Conference of the North American Chapter of the Association for Computational Linguistics: Human Language Technologies, Volume 2 (Short Papers), pp. 49–54 (2018)
5. Kingma, D.P., Ba, J.: Adam: a method for stochastic optimization. arXiv preprint arXiv:1412.6980 (2014)
6. Li, J., Galley, M., Brockett, C., Spithourakis, G., Gao, J., Dolan, W.B.: A persona-based neural conversation model. In: Proceedings of the 54th Annual Meeting of the Association for Computational Linguistics (Volume 1: Long Papers), pp. 994–1003 (2016)
7. Lison, P., Tiedemann, J.: Opensubtitles 2016: Extracting large parallel corpora from movie and tv subtitles (2016)
8. Liu, Y., et al.: Roberta: a robustly optimized bert pretraining approach. arXiv preprint arXiv:1907.11692 (2019)
9. Liu, Y., Du, J., Li, X., Xu, R.: Generating empathetic responses by injecting anticipated emotion. In: ICASSP 2021–2021 IEEE International Conference on Acoustics, Speech and Signal Processing (ICASSP), pp. 7403–7407. IEEE (2021)
10. Paulus, R., Xiong, C., Socher, R.: A deep reinforced model for abstractive summarization. In: International Conference on Learning Representations (2018)
11. Radford, A., Narasimhan, K., Salimans, T., Sutskever, I.: Improving language understanding by generative pre-training (2018)
12. Radford, A., Wu, J., Child, R., Luan, D., Amodei, D., Sutskever, I.: Language models are unsupervised multitask learners. OpenAI blog **1**(8), 9 (2019)
13. Shahraki, A.G., Zaiane, O.R.: Lexical and learning-based emotion mining from text. In: Proceedings of the International Conference on Computational Linguistics and Intelligent Text Processing, vol. 9, pp. 24–55 (2017)
14. Shang, L., Lu, Z., Li, H.: Neural responding machine for short-text conversation. In: Proceedings of the 53rd Annual Meeting of the Association for Computational Linguistics and the 7th International Joint Conference on Natural Language Processing (Volume 1: Long Papers), pp. 1577–1586 (2015)
15. Vaswani, A., et al.: Attention is all you need. In: Proceedings of the 31st International Conference on Neural Information Processing Systems, pp. 6000–6010 (2017)
16. Wolf, T., et al.: Transformers: State-of-the-art natural language processing. In: Proceedings of the 2020 Conference on Empirical Methods in Natural Language Processing: System Demonstrations, pp. 38–45 (2020)
17. Zandie, R., Mahoor, M.H.: Emptransfo: a multi-head transformer architecture for creating empathetic dialog systems. arXiv preprint arXiv:2003.02958 (2020)

18. Zhang, Y., et al.: Dialogpt: large-scale generative pre-training for conversational response generation. In: Proceedings of the 58th Annual Meeting of the Association for Computational Linguistics: System Demonstrations. pp. 270–278 (2020)
19. Zheng, C., Liu, Y., Chen, W., Leng, Y., Huang, M.: Comae: a multi-factor hierarchical framework for empathetic response generation. arXiv preprint arXiv:2105.08316 (2021)
20. Zhou, H., Huang, M., Zhang, T., Zhu, X., Liu, B.: Emotional chatting machine: emotional conversation generation with internal and external memory. In: Proceedings of the AAAI Conference on Artificial Intelligence, vol. 32 (2018)

Social Media and Sentiment Analysis

BERT-Based Meta-Learning Approach with Looking Back for Sentiment Analysis of Literary Book Reviews

Hui Bao[1], Kai He[2], Xuemeng Yin[4], Xuanyu Li[3], Xinrui Bao[2], Haichuan Zhang[2], Jialun Wu[2], and Zeyu Gao[2](✉)

[1] Faculty of Liberal Arts, Northwest University, Xi'an 710127, Shaanxi, China
[2] School of Computer Science and Technology, Xi'an Jiaotong University, Xi'an 710049, Shaanxi, China
[3] Faculty of Humanities and Foreign Languages, Xi'an University of Technology, Xi'an 710048, Shaanxi, China
[4] Department of Economics, University of Southern California, Los Angeles 90089, CA, USA

Abstract. Sentiment analysis of literary book reviews is crucial for authors to understand the voice of readers and for humanities researchers to investigate the reader reception of the literary book. Constructing an automated sentiment analysis system can help us analyze literary reviews in a big-data way, achieving a complete view of the full understanding. The main challenge in developing such a system is the lack of large-scale datasets with precise annotations. Some related researches collect the star rates from social networks and convert them into corresponding sentiment annotations for data labeling. Such automatically generated annotations can greatly reduce the human labeling efforts but with two defects. The first is that such annotations are filling with a vast amount of noise. For example, different people have significantly different perceptions for sentiment markers (positive, neutral, and negative) of a five-star comment system. Besides, literary book reviews contain more positive emotions than neutral and negative. Namely, this task is also challenged by imbalanced data. This paper introduces an automatically generated sentiment analysis dataset for Chinese literary book reviews with parts of manual verification, containing 187 literary books with 109,286 reviews. Furthermore, we propose a novel meta-learning approach with *Looking Back* mechanism to build a robust BERT-based sentiment classification model. Specifically, we design an extra meta-model for learning a sample weighting function from the historical sample information, which mitigates the influence of noisy labels and class imbalance problems. The parameters of this meta-model are finely updated with the whole training process by the guidance of a small amount of unbiased meta data. The results show the F1 score raise of 13.51% with the *Looking Back* mechanism, showing that our method can significantly promote sentiment analysis performance with noisy labels and class imbalance data.

Keywords: Meta-learning · Sentiment analysis · Literary book review

L. Wang et al. (Eds.): NLPCC 2021, LNAI 13029, pp. 235–247, 2021.
https://doi.org/10.1007/978-3-030-88483-3_18

1 Introduction

Literary book reviews contain the readers' feelings and understandings, and they are crucial for evaluating the value and reception of literary books. Such evaluations play an essential role in humanities researches. With the rise of digital humanities, many digital and information technologies have been introduced to humanities researches [1,2], and sentiment analysis (SA) is one of the most important aspects [3]. SA has been proved to be an effective technique for companies to pay close attention to the voice of customers (VoC). Thus, it is potential in investigating the voice of readers (VoR) [4,5]. For a simple example, after extracting main topics from positive and negative literary book reviews, we can know clearly what readers are focused on and what aspects of literary books they like or dislike.

A large number of literary book reviews can be easily collected from social networks. However, manually annotating these reviews with sentiment labels (*i.e.*, positive, negative, and neutral) is laborious and time-consuming. In such conditions, researchers adopted several automatic labeling methods to generate the sentiment labels [6–8]. An ordinary method is mapping the star ratings to the sentiment labels, *i.e.*, low ratings indicate negative sentiment and vice versa. Although automatic labeling methods can help us derive large-scale datasets easily, they will bring a considerable number of noisy samples. In addition, unlike the common SA of product, movie, or blog reviews, most literary books have relatively high receptions. Taking Douban's ten-point rating system as an example. Among all 2,696,751 literary books, only 133 books' averaged comment scores are below the five-star level. This phenomenon indicates that literary book review data is extremely imbalanced.

Therefore, it is infeasible to directly apply existing deep learning-based SA models on literary book review data. Although deep learning models can model complex input patterns, they tend to over-fit the dataset with bias, e.g., noisy labels and class imbalance, leading to poor performance and low generalizability. To deal with the biased dataset, we consider applying meta-learning method and propose a novel meta-learning approach with *Looking Back* mechanism (MLB) to learn a robust sample weighting function. Our method consists of a main model including a BERT [9] classifier to perform sentiment classification, and a meta-model for dynamically weighting. The meta-model can guide the main model's optimization direction by controlling the weights of training samples, which is trained by a small unbiased meta dataset to ensure an appropriate optimization direction. The whole training process can be formulated as a two-level optimization problem, alternately updating the parameters of main and meta-models.

Different from [10] which uses a multi-layer perceptron (MLP) to map a sample loss to a weight simply, our method employs the historical information of each sample. First, a series of sample features are extracted by *Looking Back* mechanism and concatenated with the corresponding label encoding as the input of our meta-model. Following, our meta-model maps this time series data into the sample weight through an LSTM network (*i.e.*, the meta-model). In this way, the learned sample weighting function is more adaptable to the changes of model parameters, and the trained model is more reliable and stable for real-world data.

The main contributions of this paper can be listed as follows:

1) We propose a novel meta-learning approach with *Looking Back* mechanism, which takes advantage of historical information to learn a robust sample weighting function.
2) We demonstrate that the BERT trained with MLB can handle the data biases, including label noises and class imbalance. The results demonstrate that our method outperforms other methods on a real-world SA dataset.
3) We publish a large-scale SA dataset for Chinese literary book reviews, and manually verified parts of them as meta data. This whole dataset contains 109,286 reviews from 187 literary books.

2 Related Work

This section describes two aspects of related work. One is the SA task, and the other is meta-learning methods about alleviating the influences from noisy and imbalanced data.

Sentiment Analysis. SA is a typical natural language processing (NLP) task studied at many levels of granularity with different data sources. [11,12] regarded SA as a document level classification task, [13,14] handled this problem at the sentence level, and [15,16] focus on aspect-level SA. More related to our work, [17] proposed an approach to automatically detect sentiments on Twitter messages that by leveraging sources of noisy labels as training data and achieve a more robust effect. [18] proposed the SLCABG model, which combines the advantages of sentiment lexicon and deep learning technology and overcomes the shortcomings of the existing SA model of product reviews.

Meta-Learning Methods. In practice, training with biased data is a commonly encountered problem. For such reason, many researchers try exploring how to modeling a more robust network with these non-qualified data, which may contain corrupted labels and class imbalance [19–21]. Especially for data collected from search engines and social media, their labels are usually generated automatically. For both of these two problems, sample re-weighting is a typical approach. The core methodology explores effective re-weighting functions to map from training loss to corresponding weights and accumulate loss dynamically to update models by different weights.

There are two opposite solutions for constructing such loss-weight mapping. One is monotonically increasing weights with loss values, and another is monotonically decreasing. The solution of monotonically increasing weights focuses on records with larger loss values because these data are more likely to be uncertain samples that are hard to classify. Many classical methods follow this idea, including AdaBoost [22], focal loss [23], and hard negative mining [24]. We argue that this weight manner is more suitable for dealing with the class imbalance problem. It can force models to focus on minority categories and provide higher

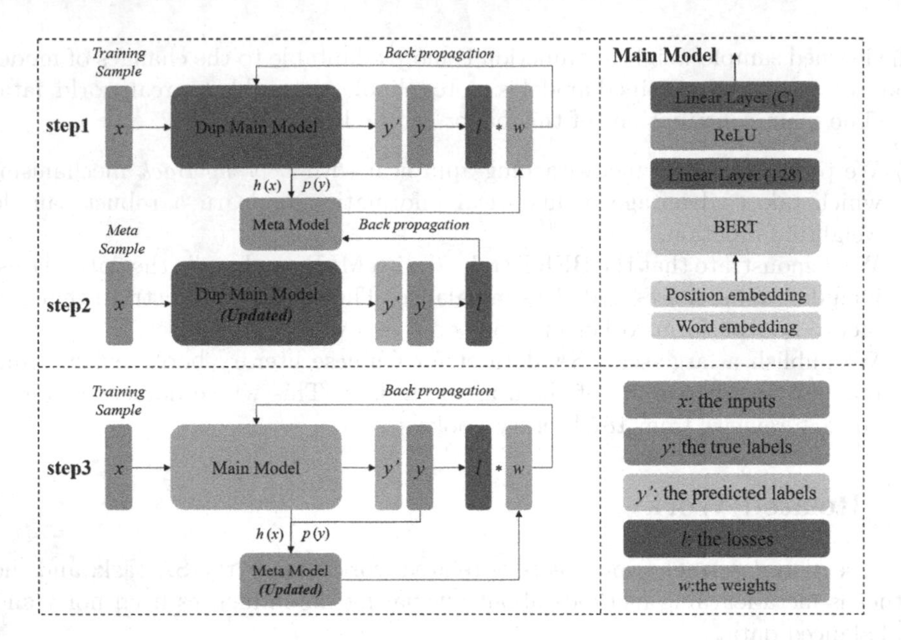

Fig. 1. A sketch of the learning process and the architecture of the main model.

training losses. Instead of paying more attention to data with more uncertainty, the monotonically decreasing solution takes samples with smaller loss values as more important ones. Self-paced learning(SPL) [25] is a typical decreasing weight solution. SPL selects the sample with a small training error and high likelihood value from all samples and then updates the model parameters. The number of samples selected in each iteration is determined by the weight parameter, introducing more samples through successive attenuation. The iterative reweighting [26] takes a similar idea, which argues that records with low loss are more likely with high-confident labels and should contribute more to a training process. These methods are usually used in noisy-label cases since it inclines to suppress the effects of extremely large loss values from incorrect labels.

However, the above two solutions have evident deficiencies when introduced to deal with our literary book review SA issue. The first one is that we have to design a specific weight mapping function. Designing an effective mapping function needs us to fully understand the data condition. However, large-scale data bring a real challenge for understanding when it comes to the deep learning field. Besides, an extra mapping function usually introduced more hyper-parameters, which further reduces performance stability and raises the complexity of neural networks. For such reason, we proposed an adaptive weighting strategy to form the mapping function automatically.

3 Method

An overview of the proposed framework is depicted in Fig. 1. The proposed framework is composed of two parts, the main model and the meta-model. To train these two models simultaneously, we conduct two data sets, *i.e.*, the training samples and the meta samples. The learning process can be concluded as three steps: (1) Updating the parameters of a duplicated main model with a batch of training samples by weighted loss, where the meta-model calculates the weights. (2) Passing a batch of unbiased meta samples through the duplicated main model and updating the parameters of the meta-model by back-propagation. (3) Updating the parameters of the main model with the same batch of training samples by weighted loss, where the weights are calculated by the updated meta-model.

3.1 BERT-Based Meta-Learning

BERT has been proven to be a powerful pre-trained encoder for downstream tasks of NLP [27,28]. To extract robust embeddings for training samples, we adopt the base BERT with an MLP classifier as the main model in our framework. The base BERT contains 12 Transformer blocks, each block has 12 heads with 768 hidden dimensions, which has 110M parameters in total. The MLP classifier consists of two hidden layers with a Relu activation. The hidden dimensions of the two layers are set to 128 and C (the number of classes).

Given a set of training samples $D = \{x, y\}^N$, where x and y indicate a sample and the corresponding label over C classes, N is the number of training samples. We denote the learnable weights of the main model as $\mathbf{W}$ and the predicted label $y' = f_{\mathbf{W}}(x)$. In order to optimize the parameters $\mathbf{W}$, we can minimize the loss of all training samples $\frac{1}{N}\sum^N L(y, f_{\mathbf{W}}(x))$. To handle the biased training data, some methods [10,23] use a sample weighting function to control the loss weight of each sample. The sample weighting function is donated as $g_{\Theta}(x)$, where Θ indicates the parameters of this function. The parameters of the main model can be optimized by minimizing the following weighted loss:

$$\mathbf{W}^* = \underset{\mathbf{W}}{argmin} \frac{1}{N} \sum^N g_{\Theta}(x) L(y, f_{\mathbf{W}}(x)) \tag{1}$$

Some works take the loss of each sample as the input of the sample weighting function $g_{\Theta}(L(y, f_{\mathbf{W}}(x)))$, where Θ can either be fixed [23] or learnable [10]. If it is learnable, it can be optimized in a meta-learning manner and regarded as the parameters of a meta-model. Unlike these works, each sample's weight is calculated based on a single loss value that relies too much on the sample's predicted result of the main model. We suppose that the original information, like the input sample x and its label y, should be considered in the meta-model. So the meta-model is re-termed as $g_{\Theta}(h(x), p(y))$, $h(x)$ indicates the learned features of sample x by the main model, $p(y)$ is the label embedding. Also, this meta-model can be regarded as a discriminator that determined whether

(x, y) is a pair and as a class balancer based on the label embedding $p(y)$. The architecture of the meta-model is shown in Fig. 2.

Given a set of unbiased meta samples $D^{meta} = \{x^{meta}, y^{meta}\}^M$, the parameters of the meta-model can be optimized by minimizing the meta loss as following:

$$\Theta^* = \underset{\Theta}{argmin} \frac{1}{M} \sum^{M} L\left(y^{meta}, f_{\mathbf{W}}(x^{meta})\right) \tag{2}$$

We adopt an iterative method to optimize two models in a single optimization loop. The optimization algorithm includes three steps:

(1) In each iteration, a batch contains n training samples $\{(x_i, y_i), i = 1, ..., n\}$ is sampled, the parameters of the main model are temporally updated as Eq. 3 via one step of gradient descent in Eq. 1.

$$\hat{\mathbf{W}}_{\Theta}^{(t)} = \mathbf{W}^{(t)} - \alpha \frac{1}{n} \sum_{i=1}^{n} g_{\Theta}^{(t)}(h^{(t)}(x_i), p(y_i)) \left. \frac{\partial L\left(y_i, f_{\mathbf{W}}^{(t)}(x_i)\right)}{\partial \mathbf{W}} \right|_{\mathbf{W}^{(t)}} \tag{3}$$

α is the learning rate of the main model.

(2) A batch of m meta samples $\{(x_j, y_j), j = 1, ..., m\}$ is also sampled for updating the parameters of the meta-model by minimizing Eq. 2. The updated function is as following:

$$\Theta^{(t+1)} = \Theta^{(t)} - \beta \frac{1}{m} \sum_{j=1}^{m} \left. \frac{\partial L\left(y_j^{meta}, f_{\hat{\mathbf{W}}_{\Theta}}^{(t)}(x_j^{meta})\right)}{\partial \hat{\mathbf{W}}_{\Theta}} \right|_{\hat{\mathbf{W}}^{(t)}} \left. \frac{\partial \hat{\mathbf{W}}_{\Theta}}{\partial \Theta} \right|_{\Theta^{(t)}} \tag{4}$$

β is the learning rate of the meta-model.

(3) Finally, the parameters of the main model are updated by one step of gradient descent of Eq. 1 with the fixed parameters $\Theta^{(t+1)}$ of the meta-model, which formulated as:

$$\mathbf{W}^{(t+1)} = \mathbf{W}^{(t)} - \alpha \frac{1}{N} \sum_{i=1}^{N} g_{\Theta}^{(t+1)}(h^{(t)}(x_i), p(y_i)) \left. \frac{\partial L\left(y_i, f_{\mathbf{W}}^{(t)}(x_i)\right)}{\partial \mathbf{W}} \right|_{\mathbf{W}^{(t)}} \tag{5}$$

Following these three steps, both the main model and the meta-model can be optimized during the training process.

3.2 Meta-Learning with *Looking Back*

In the previous section, we formulate the meta-model as $g_{\Theta}(h(x), p(y))$, where $h(x)$ is the learned features generated by the last layer of the main model in each epoch. Thus, the meta-model can only reflect the degree of matching between x and y at the current state, but will be affected by the tuning of parameters in the following model optimization process. To address this issue, we

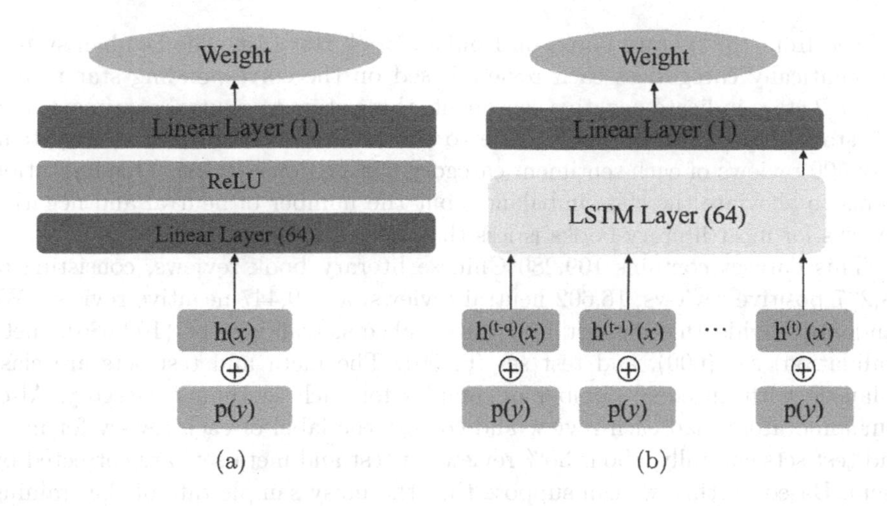

Fig. 2. The architecture of meta-models (a) without and (b) with *Looking Back*

propose a new mechanism termed as *Looking Back*, which transforms the pair matching problem into a time series classification problem. The meta-model with *Looking Back* takes a series of the learned features from last q epochs, $g_\Theta(\{h^{(t-q)}(x), ..., h^{(t)}(x)\}, p(y))$. Meanwhile, to capture the discriminative information from the sequential data, we use an LSTM with one hidden layer to replace the MLP in the meta-model, as shown in Fig. 2.

To this extent, a comprehensive meta-model with high generalization ability is constructed. It can be updated with the main model by using the algorithm introduced in the previous section.

4 Experiments

Table 1. Statistics for literary book review dataset

Reviews	Positive	Neutral	Negtive	Total
Training set	81,037	14,402	7,247	102,686
Meta (validation) set	200	200	200	600
Test set	2,000	2,000	2,000	6,000

4.1 Dataset

Considering SA for product, movie, or blog reviews, there are a lot of publicly available datasets [29–31]. As far as we know, few datasets of SA are customized for literary book reviews. To evaluate the performance of our approach, we introduce a literary book review dataset for SA. The original literary book reviews are derived from 187 literary books on Douban-Book Hot List. The Douban system automatically categorizes each review based on the corresponding star rating, *i.e.*, 0–2 stars indicate negative sentiment, three stars to neutral sentiment, and 4–5 stars to positive sentiment. Due to the limitation of the Douban system, only 500 reviews of each sentiment category can be downloaded. This limitation seems to alleviate the class imbalance, but the number of neutral and negative reviews for most literary books is less than 500 or even 100.

This dataset contains 109,286 Chinese literary book reviews, consisting of 83,237 positive reviews, 16,602 neutral reviews, and 9,447 negative reviews. We randomly divide this dataset into three subsets, training set (102,686), meta (validation) set (600), and test set (6,000). The meta and test sets are class balanced with an equal number of samples for each sentiment category. Also, four annotators read each review and correct the label of each review for meta and test sets carefully, about 30% reviews of test and meta sets are corrected by them. Based on this, we can suppose that the noisy sample rate of the training set is about 30%. In this way, two class balanced and noise-free subsets, meta set and test set are generated for training and evaluation. Note that the label of each review in the training set is derived from the Douban system without any modification, so the training set is still a dataset with the class imbalance and noisy labels. Here, the meta set can be regarded as the validation set, which is only used to selected the best parameters of each model The details of this dataset are shown in Table 1.

4.2 Settings

For comparison, we implement the following methods to perform SA on literary book reviews.

(1) **BERT**: The original BERT with modified classifier (3-class) trained on the training data only.
(2) **BERT**$_{ft}$: The BERT trained on the training data, then fine-tuned on the meta data.
(3) **BERT-MWN**: The BERT trained with the meta-learning approach proposed in [10].
(4) **BERT-MLB**$_{w/o}$: The BERT trained with our approach without *Looking Back* mechanism.

For all implemented methods, the BERT encoder is the base BERT with pre-trained weights of Chinese. We use the BERT base Chinese tokenizer to get the initial token embeddings, and the special tokens [CLS] and [SEP] are added for each sentence. The embedding of [CLS] in the last hidden layer of the BERT encoder served as the classifier's input. Also, we set 75 as the input sequence length of BERT, which means we use up to 75 tokens for each sentence.

The open-source library PyTorch 1.6 and scikit-learn 0.24.0 are adopted for implementation on a work-station with two NVIDIA 2080Ti GPUs. For the BERT encoder and classifier of all methods, the learning rates are set to 5e–5 and

1e–3, respectively. For **BERT** and **BERT**$_{ft}$, the adam optimizer is adopted for training. For meta-learning approaches, *i.e.*, **BERT-MWN**, **BERT-MLB**$_{w/o}$ and **BERT-MLB**, the stochastic gradient descent is adopted with 0.9 momentum and 5e-4 weight decay. The adam optimizer is adopted for meta-models, the learning rate is 1e-3. The training epochs of all methods are set to 50. And all learning rates are decreased to 1/10 of the original after 25 epochs. Note that **BERT-MLB** needs the historical information of the last q epochs, we set $q = 5$, and in the first q epochs, **BERT-MLB** degenerates into **BERT** to obtain the initial series of hidden features for each sample.

To measure the performance of each method, the overall accuracy, macro F1-score, precision, recall, and F1-score of each class are applied.

4.3 Result

Table 2. The comparison results, the overall accuracy (Acc), precision (P), recall (R) and F1-score (F) for SA of Chinese literary book reviews.

Model	All				Positive	Neutral	Negative
	Acc%	F%	P%	R%	F%	F%	F%
BERT	65.58	61.12	68.68	61.86	74.56	39.97	68.84
BERT$_{ft}$	67.23	63.17	70.22	63.69	75.72	43.36	70.44
BERT-MWN	68.50	66.52	66.63	66.83	77.60	49.28	72.68
BERT-MLB$_{w/o}$	69.57	66.96	71.43	66.72	78.60	52.76	69.51
BERT-MLB	**72.37**	**74.63**	**72.06**	**70.67**	**80.47**	**58.39**	**74.05**

The sentiment classification results of each method are shown in Table 2. As we expected, **BERT** shows the worst performance because it over-fits the biased training data. **BERT**$_{ft}$ achieves a little bit higher performance than **BERT** by fine-tuning on the small set of unbiased meta data. Three meta-learning approaches outperform traditional methods in almost all evaluation metrics, especially the macro F1-score. Due to the meta training strategies, *i.e.*, dynamically adjusting sample weights based on the meta data can handle the class imbalance and noisy label problem. **BERT-MLB**$_{w/o}$ outperforms **BERT-MWN** by 1% in the overall accuracy, which benefits from the detailed inputs (the hidden layer features and label embedding) of the meta-model. Moreover, our proposed approach **BERT-MLB** significantly outperforms **BERT-MLB**$_{w/o}$ by about 3% in the overall-accuracy and 8% in macro F1-score, which proves that the *Looking Back* mechanism we designed is effective for robust deep learning.

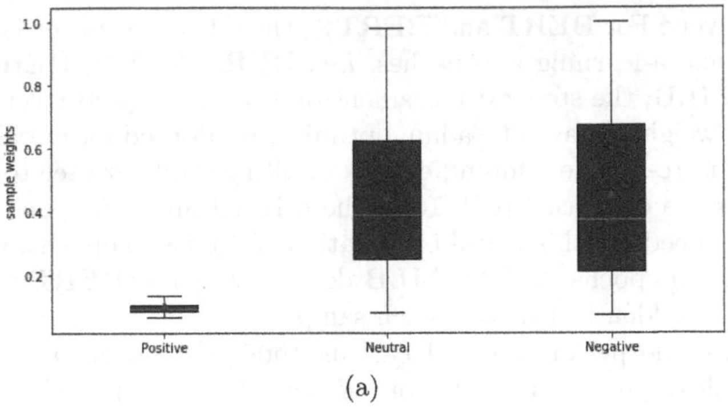

(a)

Text	Class	Weight
16 年读完，补记。	Positive	0.067
掰掰！下辈子见！	Positive	0.069
初读结构略微松散，渐入佳境后觉得张驰有度，但还是稍显乏味	Positive	0.070
不喜欢。描述了萧红小时候在地主家庭的生活...	Neutral	0.101
味道淡不说，食材也不新鲜...	Neutral	0.112
比较失望，配不上我读时这本书被宣传的名气...	Neutral	0.113
史铁生的人生经历很了不起。感情也很诚挚。	Negtive	0.109
在充满宗教性的集体想象之中出走	Negtive	0.126
很平淡地描述，又很细腻深刻的感情。	Negtive	0.130

(b)

Fig. 3. (a) The box-plot of sample weights for each class. (b) The examples of low-weight training samples.

Sample Weighting. In this paper, we emphasize that the proposed approach can learn a robust sample weighting function. So we discuss what the sample weights learned by MLB are. Firstly, we illustrate the weight distribution of each category in Fig. 3(a). From the boxes of each category, we can see that the sample weighting function of MLB gives low weights for samples from the majority class (positive) and relatively high weights for neutral and negative samples. This weight setting that automatically learned from the meta data is similar to our experience dealing with imbalanced data. Secondly, we show a few samples (due to the space limitation) with the lowest weights from different categories in Fig. 3(b). It can be seen that noisy samples would be set with low weights by the sample weighting function of MLB, which proves the effectiveness of our proposed approach for learning with noisy data.

5 Conclusion

In this paper, we proposed a novel meta-learning approach with *Looking Back* mechanism to deal with the problems of noisy labels and imbalanced categories in automatically generated SA datasets. Our method learns a robust sample weighting function from a small set of meta data, which can raise the robustness of our main network for SA task. We also introduce a Chinese literary book review dataset, with parts of manual validations as meta data, to evaluate the performance of our approach. The experimental results show that the MLB can learn an intelligible weighting function and achieve better performance than other methods. Furthermore, we hope that our approach can inspire other researchers to pay more attention to use historical information (*Looking Back*) to guide the direction of model optimization.

Acknowledgements. This work has been supported by the National Social Science Fund of China Project (19BZW024).

References

1. Boyd, R.L.: Psychological text analysis in the digital humanities. In: Hai-Jew, S. (ed.) Data Analytics in Digital Humanities. MSA, pp. 161–189. Springer, Cham (2017). https://doi.org/10.1007/978-3-319-54499-1_7
2. Clanuwat, T., Bober-Irizar, M., Kitamoto, A., Lamb, A., Yamamoto, K., Ha, D.: Deep learning for classical japanese literature. arXiv preprint arXiv:1812.01718 (2018)
3. Moreno-Ortiz, A.: Lingmotif: sentiment analysis for the digital humanities. In: Proceedings of the Software Demonstrations of the 15th Conference of the European Chapter of the Association for Computational Linguistics, pp. 73–76 (2017)
4. Cao, Y., Xu, R., Chen, T.: Combining convolutional neural network and support vector machine for sentiment classification. In: Zhang, X., Sun, M., Wang, Z., Huang, X. (eds.) CNCSMP 2015. CCIS, vol. 568, pp. 144–155. Springer, Singapore (2015). https://doi.org/10.1007/978-981-10-0080-5_13
5. Fang, X., Zhan, J.: Sentiment analysis using product review data. J. Big Data **2**(1), 1–14 (2015)
6. Lak, P., Turetken, O.: Star ratings versus sentiment analysis - a comparison of explicit and implicit measures of opinions. In: 2014 47th Hawaii International Conference on System Sciences, pp. 796–805 (2014). https://doi.org/10.1109/HICSS.2014.106
7. Haque, T.U., Saber, N.N., Shah, F.M.: Sentiment analysis on large scale amazon product reviews. In: 2018 IEEE International Conference on Innovative Research and Development (ICIRD), pp. 1–6 (2018). https://doi.org/10.1109/ICIRD.2018.8376299
8. Liu, H., Wang, J., Li, S., Li, J., Zhou, G.: Semi-supervised sentiment classification based on auxiliary task learning. In: Zhang, M., Ng, V., Zhao, D., Li, S., Zan, H. (eds.) NLPCC 2018. LNCS (LNAI), vol. 11109, pp. 372–382. Springer, Cham (2018). https://doi.org/10.1007/978-3-319-99501-4_33
9. Devlin, J., Chang, M.W., Lee, K., Toutanova, K.: Bert: Pre-training of deep bidirectional transformers for language understanding. arXiv preprint arXiv:1810.04805 (2018)
10. Shu, J., et al.: Meta-weight-net: Learning an explicit mapping for sample weighting. In: Wallach, H., Larochelle, H., Beygelzimer, A., d' Alché-Buc, F., Fox, E., Garnett, R. (eds.) Advances in Neural Information Processing Systems, vol. 32. Curran Associates, Inc. (2019)

11. Turney, P.D.: Thumbs up or thumbs down? semantic orientation applied to unsupervised classification of reviews. In: Proceedings of the Annual Meeting of the Association for Computational Linguistics (2002)
12. Pang, B., Lee, L.: A sentimental education: Sentiment analysis using subjectivity summarization based on minimum cuts. In: Proceedings of the Annual Meeting of the Association for Computational Linguistics (ACL-04), pp. 271–278 (2004)
13. Hu, M., Liu, B.: Mining and summarizing customer reviews. In: Proceedings of the Tenth ACM SIGKDD International Conference on Knowledge Discovery and Data Mining, pp. 168–177 (2004)
14. Kim, S.M., Hovy, E.: Determining the sentiment of opinions. In: COLING 2004: Proceedings of the 20th International Conference on Computational Linguistics, pp. 1367–1373 (2004)
15. Wilson, T., Wiebe, J., Hoffmann, P.: Recognizing contextual polarity in phrase-level sentiment analysis. In: Proceedings of Human Language Technology Conference and Conference on Empirical Methods in Natural Language Processing, pp. 347–354 (2005)
16. Agarwal, A., Biadsy, F., Mckeown, K.: Contextual phrase-level polarity analysis using lexical affect scoring and syntactic n-grams. In: Proceedings of the 12th Conference of the European Chapter of the ACL (EACL 2009), pp. 24–32 (2009)
17. Barbosa, L., Feng, J.: Robust sentiment detection on twitter from biased and noisy data. In: Coling 2010: Posters, pp. 36–44 (2010)
18. Yang, L., Li, Y., Wang, J., Sherratt, R.S.: Sentiment analysis for e-commerce product reviews in Chinese based on sentiment lexicon and deep learning. IEEE Access **8**, 23522–23530 (2020)
19. Goldberger, J., Ben-Reuven, E.: Training deep neural-networks using a noise adaptation layer. In: ICLR (2017)
20. Vahdat, A.: Toward robustness against label noise in training deep discriminative neural networks (2017)
21. Hendrycks, D., Mazeika, M., Wilson, D., Gimpel, K.: Using trusted data to train deep networks on labels corrupted by severe noise (2018)
22. Sun, Y., Kamel, M.S., Wong, A.K., Wang, Y.: Cost-sensitive boosting for classification of imbalanced data. Pattern Recogn. **40**(12), 3358–3378 (2007)
23. Lin, T.Y., Goyal, P., Girshick, R., He, K., Dollár, P.: Focal loss for dense object detection. In: Proceedings of the IEEE International Conference on Computer Vision, pp. 2980–2988 (2017)
24. Malisiewicz, T., Gupta, A., Efros, A.A.: Ensemble of exemplar-svms for object detection and beyond. In: 2011 International Conference on Computer Vision, pp. 89–96. IEEE (2011)
25. Kumar, M.P., Packer, B., Koller, D.: Self-paced learning for latent variable models. In: NIPS, vol. 1, p. 2 (2010)
26. De La Torre, F., Black, M.J.: A framework for robust subspace learning. Int. J. Comput. Vis. **54**(1), 117–142 (2003)
27. Qiu, X., Sun, T., Xu, Y., Shao, Y., Dai, N., Huang, X.: Pre-trained models for natural language processing: A survey. Sci. China Technol. Sci. 1–26 (2020)
28. Gao, Z., Feng, A., Song, X., Wu, X.: Target-dependent sentiment classification with bert. IEEE Access **7**, 154290–154299 (2019)
29. Socher, R., et al.: Recursive deep models for semantic compositionality over a sentiment treebank. In: Proceedings of the 2013 Conference on Empirical Methods in Natural Language Processing, pp. 1631–1642 (2013)

30. Shivaprasad, T., Shetty, J.: Sentiment analysis of product reviews: a review. In: 2017 International Conference on Inventive Communication and Computational Technologies (ICICCT), pp. 298–301. IEEE (2017)
31. Zhang, S., Wei, Z., Wang, Y., Liao, T.: Sentiment analysis of Chinese micro-blog text based on extended sentiment dictionary. Future Gener. Comput. Syst. **81**, 395–403 (2018)

ISWR: An Implicit Sentiment Words Recognition Model Based on Sentiment Propagation

Qizhi Li, Xianyong Li(✉), Yajun Du, and Xiaoliang Chen

Xihua University, Chengdu 610039, China
lixy@mail.xhu.edu.cn

Abstract. Considering that there have no efficient methods to detect implicit sentiment words, this paper proposes an implicit sentiment words recognition model based on sentiment propagation to automatically identify implicit sentiment words. In the model, we create part-of-speech trees based on Weibo texts. Based on these part-of-speech trees, we define some rules to generate word graphs. By borrowing the PolarityRank algorithm and SIS (Susceptible-Infected-Susceptible) epidemic model, we propose formulas of the sentiment propagation and non-propagation change rates for a word. Then we propagate the sentiment between the words in these word graphs to get the sentiment values of words. After normalizing the sentiment value of words, we get the implicit sentiment words. By some parameter sensitivity experiments, the optimal model parameters are chosen. Compared with SO-PMI, PolarityRank and PageRank models, the accuracy and micro-F1 of the proposed model are significantly better than the results of the three models for implicit sentiment words recognition.

Keywords: Implicit sentiment analysis · Sentiment propagation · Implicit sentiment words recognition

1 Introduction

With the development of the Internet, more and more people are accustomed to express their sentiments on social network. In their communication texts, their also carry many metaphors, Internet buzzwords, and local dialects, which

This work is partially supported by the National Natural Science Foundation (Nos. 61802316, 61872298, and 61902324), Chunhui Plan Cooperation and Research Project, Ministry of Education of China (Nos. Z2015109, Z2015100), "Young Scholars Reserve Talents" program of Xihua University, Scientific Research Fund of Sichuan Provincial Education Department (Nos. 15ZA0130, 16ZA0157), Science and Technology Department of Sichuan Province (Nos. 2021YFQ0008, 2019GFW115), Key Scientific Research Fund of Xihua University (No. z1422615).

L. Wang et al. (Eds.): NLPCC 2021, LNAI 13029, pp. 248–259, 2021.
https://doi.org/10.1007/978-3-030-88483-3_19

frequently contain implicit sentiments. According to the subjectivity and objectivity of sentiment words, Liu et al. [14] divided sentiment into explicit sentiment and implicit sentiment. In academia and industry, the explicit sentiment analysis got extensive attention [15,22]. However, implicit sentiment hidden in sentences is difficult to recognize. Since there has no implicit sentiment lexicon to querying implicit sentiment words [20], implicit sentiment analysis has attracted little attention [13,22]. Meanwhile, for the implicit sentiment analysis, researches mainly considered the implicit sentiment of entire sentences [13,19,22]. There are few results that paid attention to the implicit sentiment of words in sentences. In fact, many sentences express their sentiments by some implicit words. For example, the sentence "弗洛伊德葬礼举行 (Floyd's funeral is held)" contains explicit sentiment word and implicit word. "葬礼 (Funeral)" is a word with negative sentiment, while "弗洛伊德 (Floyd)" was a negative hot event last year. People's attitudes to this event are mostly angry and sad. The sentence "法西斯无疑了 (No doubt it is Fascism)" does not contain any explicit sentiment word, but it makes people feel angry. The reason is that the word "法西斯 (Fascism)" refers to the Axis powers and their actions during World War II more often.

Although implicit sentiment words are difficult to detect and distinguish in sentences, implicit sentiment words will co-occur with some explicit sentiment words more or less. For example, the sentence "祖国加油，一切都会好起来的相信春暖花开的那一天。(Come on, the motherland, everything will be okay, the spring will arrive someday.)" contains an implicit word "春暖花开 (spring)" with an explicit word "加油 (come on)". The original meaning of this implicit word is the arrival of spring and the blooming of flowers. When this word is used in combination with other positive words, it describes some things in a good situation. The sentence "五一不放假整这些花里胡哨，这些领导脑壳装的浆糊？(Do these bells and whistles on May 1st holiday, are these leaders' heads full of paste?)" has an implicit word "花里胡哨 (paste)" and an explicit word "浆糊 (bells and whistles)". In this sentence, the author expresses dissatisfaction with the leaders by comparing leaders' brain with the paste. So the word "paste" here is a negative word. From these sentences, it is easy to find that the sentiment source of the implicit sentiment words is other explicit sentiment word in sentences. Based on the analysis of these implicit sentiment words in sentences, the sentiment propagation is very important for recognizing the sentiment orientations of implicit sentiment words.

To the best of our knowledge, few papers adopted the sentiment propagation to recognize implicit sentiment words in sentences. In this paper, we proposed a novel implicit sentiment words recognition method (ISWR) to automatically recognize implicit sentiment words based on sentiment propagation. By setting two hyper-parameters, we measure the co-occurrence relationship between words. Some experiments show that this model is suitable for solving the recognition problem of implicit sentiment words in sentences. Hence, our model can obtain the words that carry implicit sentiment better than other models.

This paper is organized as follows. Section 2 gives a survey on implicit sentiment analysis and sentiment propagation. Section 3 demonstrates the construction method of automatic recognition of implicit sentiment word model. Section 4 presents the experimental results. Section 5 discusses the conclusion and future work.

2 Related Work

2.1 Implicit Sentiment Analysis

When people express their subjective sentiments, they will have 15%–20% of the time to express sentiments with implicit sentiment [12,13]. Liao et al. [13] constructed an implicit sentiment corpus of 20,000 labeled sentences. They found out that 3% of the implicit sentiment sentences contain metaphorical; and 10% of them have ironic. Deng et al. [3] proposed a model to identify the sentiments between the subject and object. Their model defined four rules, which can analyze implicit sentiment through the clues of explicit sentiment and related events. Chen et al. [1] constructed a double implicit corpus (implicit aspects and implicit opinions). They further proposed a method to analyze implicit sentiment orientation based on deep learning. Wei et al. [22] proposed an implicit sentiment analysis model based on BiLSTM with multi-polarity orthogonal attention to analyze implicit sentiment in sentences. Although all these models or methods mentioned above achieved great results in their tasks, they analyzed the implicit sentiment of sentences rather than words. Zhang et al. [24] referred the standard of emotion classification and orientation intensity classification of the affective lexicon ontology [23]. They convened 7 native Chinese students, changed the guidelines 4 times, and manually labeled and constructed a Chinese implicit sentiment lexicon within 9 months. This lexicon was mainly for implicit sentiment words such as metaphors and metaphors. This work constructed an implicit sentiment lexicon, but manually labeled a lexicon is a very time-consuming and laborious task.

2.2 Sentiment Propagation

Keyes [8] proposed an idea of "post-truth era". They thought that "expressing sentiments and personal beliefs are more likely to affect public opinion than stating objective truth". This revealed that the sentiment can be propagated. Hatfield et al. [7] put forward the theory of emotional contagion. They believed that the opinions and emotions expressed on social media will be imitated by other people. Under this contagion, the emotions expressed by one person will propagate to other people [4]. Zhang et al. [25] used Deepwalk algorithm and CNN-BiLSTM-Attention network to explore the coevolution of emotional contagion on Weibo. Based on the PageRank [18] and HITS [10], Cruz et al. [2] proposed the PolarityRank algorithm for the propagation of sentiments between words. They considered that in some specific situation, some nodes were affected both positively and negatively by their neighbor, so they assigned two attributes,

PR^+ and PR^-, to each node. These attributes simultaneously propagated sentiments in one network. Although this method is very good, the results of experiments show that it can only propagate explicit sentiment, not implicit sentiment. Gavilanes et al. [5] proposed a sentiment propagation model named USSPAD (Unsupervised System with Sentiment Propagation Across Dependencies). This model propagated sentiment in a syntactic graph, which generated from a syntactic tree, based on PolarityRank. By using this model, the authors have achieved good results in their subsequent work of generating an emoji sentiment lexicon [6]. SO-PMI [21] is not an algorithm to do sentiment propagation, but it uses the co-occurrences between words to analyze sentiments in words. All these methods mentioned above only consider the structures of the propagation sentences or graphs, but not consider the interference information (such as typos). Epidemic model was the earliest application in epidemiology [17]. This model used two parameters λ and μ to describe infection rate and cure rate. Some methods and fields based on epidemic model achieved great results [9,11], but it has not been applied to the sentiment propagation.

3 Implicit Sentiment Words Recognition Based on Sentiment Propagation

3.1 Construct Words Graph

Our sentiment propagation happens in word graphs, so we need to construct word graphs. The reasons why we need to construct word graphs are as follows:

1) In sentences, only nouns, verbs and adjectives carry sentiments. Adverbs only enhance or weaken the sentence's sentiments [5].
2) In the sentences that subjectively expressing sentiment, 15%–20% of the cases are expressed by implicit sentiment [12,13]. It implies that some implicit sentiment words will co-occurrence with explicit sentiment words.

There are three steps to build word graphs. We first extract nouns, verbs, and adjectives to construct word graphs by syntax analysis. Then we assign weights to the edges between words. Finally, we obtain initial sentiment value of words.

Construct Word Graphs. In order to construct a complete graph, we need to do syntax analysis to extract the part of speech of each word in a sentence. We use Stanford CoreNLP[1] to do syntax analysis [16]. Figure 1 shows a tree which constructed from a Chinese sentence: 愿逝者安息，生者奋发图强 (Wish the dead rest in peace, the living work hard).

Then we use the part-of-speeches to construct the word graph. We extract nouns, verbs, and adjectives based on the POS labels. The details of the extracted POS labels are shown in Table 1. Since the named entities of 'DATE' do not contain sentiment, we use Stanford CoreNLP to do named entity recognition to extract these named entities and remove them.

[1] https://stanfordnlp.github.io/CoreNLP/.

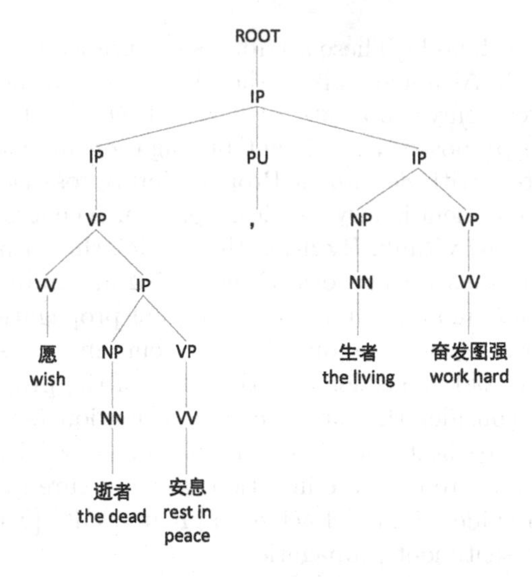

Fig. 1. Syntax tree for the running example.

Table 1. Extracted POS labels.

Nouns	Verbs	Adjectives
NN, NR, NT	MD, VV	JJ, JJR, JJS, VA

After finishing the above work, we get the words generated from the sentence. If the total number of nouns, verbs and adjectives in the sentences is less than or equal to 1, then we delete these data.

Calculate Weights. According to the idea of SO-PMI, if two words co-occur, the correlations between the two words are stronger, otherwise they are weaker. Let $F(word_i, word_j)$ be the frequency of co-occurrence of $word_i$ and $word_j$ in the corpus, and N be the number of all node pairs in the corpus. Then the weight between words i and j, denoted by w_{ij}, is defined as following.

$$w_{ij} = \frac{F(word_i, word_j)}{N}. \tag{1}$$

Obtain Initial Sentiment Value. We also need to give each node an initial sentiment value. We load the affective lexicon ontology [23] to get the initial sentiment values of words. Since the sentiment intensities of the affective lexicon ontology are "1, 3, 5, 7, 9", the sentiment intensity of explicit sentiment word i, denoted by $intensity_i$, can be normalized as follows.

$$intensity_i := \frac{intensity_i}{9}. \tag{2}$$

Based on the idea of PolarityRank, for a node v_i, we will assign the positive and negative PolarityRank values as PRP_{v_i} and PRN_{v_i}, respectively. Their initial sentiment values are assigned as follows:

1) If word does not appear in the affective lexicon ontology, or sentiment orientation of this word is 0, then $PRP_{v_i} = PRN_{v_i} = 0$.
2) If the sentiment orientation of this word is 1, then $PRP_{v_i} = intensity_i$, $PRN_{v_i} = 0$.
3) If the sentiment orientation of this word is 2, then $PRP_{v_i} = 0$, $PRN_{v_i} = intensity_i$.
4) If the sentiment orientation of this word is 3, then $PRP_{v_i} = PRN_{v_i} = intensity_i$.

Through the above three steps, we construct a word graph for each sentence. Figure 2 shows a word graph based on a running example: **大家苦口劝了这么久，还是弄这些花哨的东西，一点都不实事求是，真的恶心** (Everyone has been trying to persuade them for so long, but they still do these fancy things. They are not seeking truth from facts at all, and then they are really disgusting.)

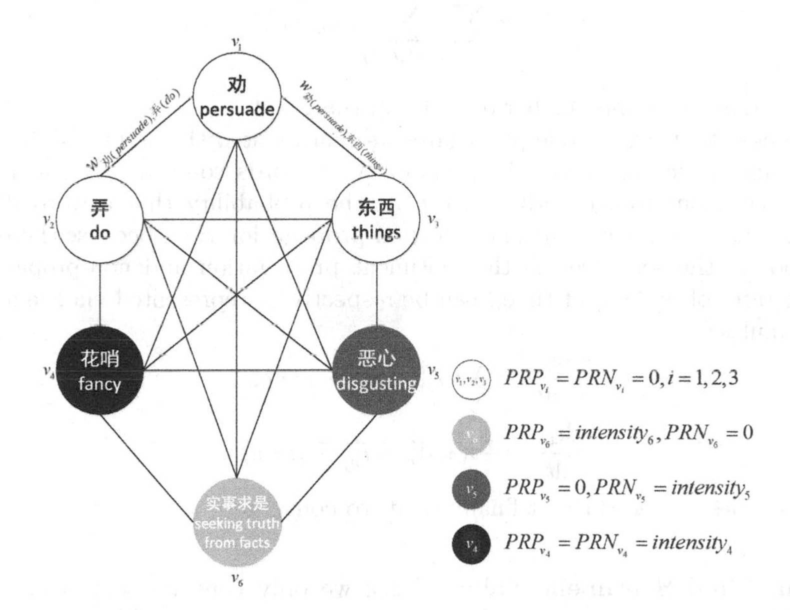

Fig. 2. Word graph for the running example.

3.2 Sentiment Propagation in Word Graph

The idea of sentiment propagation is come from SIS (Susceptible-Infected-Susceptible) model and PolarityRank algorithm. Here, the sentiments between words will propagate in one sentence. We will obtain the percentage of a word to propagate or receive sentiment value through sentiment propagation between words.

Calculate Sentiment Propagation Rate. There are n nodes (words) in a giving complete graph G. If word v_j is an explicit sentiment word, then we assume that at the initial moment the word will propagate $e^0_{v_j}$ percentage of the sentiment value; otherwise this word cannot propagate $i^0_{v_j}$ percentage of the sentiment value at the initial moment. In other words, at the initial moment, if the word carries sentiment value, it can propagate sentiment, or it cannot. And at any moment t, the sum of sentiment propagation rate and sentiment non-propagation rate by the word is 1.

$$1 = e^t_{v_j} + i^t_{v_j}. \tag{3}$$

Since each word can be affected by its neighbor nodes, we set $A^t_{v_j}$ to be the neighbor situation of node v_j at time t. Let w_{jk} be the weight (co-occurrence probability) between node j and its neighbor node k. Then we have

$$A^t_{v_j} = \sum_{j=1}^{n} \sum_{k=1,k\neq j}^{n} w_{jk} * e^t_{v_k}, \tag{4}$$

where n represents the number of nodes in the graph.

Suppose that every node propagates its sentiment with the probability λ at every time, which measures the probability of words co-occurring and propagating sentiment to each other. Let μ be the probability that a word do not propagate its sentiment value, i.e., the non-propagation rate, because there may be typos in the sentence. So the sentiment propagation and non-propagation change rates of node v_j at time t can be respectively represented via the following formulas.

$$\frac{\mathrm{d}e_{v_j}}{\mathrm{d}t} = \lambda * A^t_{v_j} * i^t_{v_j} - \mu * e^t_{v_j}, \tag{5}$$

$$\frac{\mathrm{d}i_{v_j}}{\mathrm{d}t} = -\lambda * A^t_{v_j} * i^t_{v_j} + \mu * e^t_{v_j}. \tag{6}$$

After a series of iterations, it finally tends to converge.

Obtain Final Sentiment Value. Since we only consider sentiment values that the words propagate after the value convergence in sentences, we focus on formula (5). Let $PRP^{new}_{v_j}$ and $PRN^{new}_{v_j}$ be the positive and negative sentiment values after sentiment propagation of v_j, respectively. The updating sentiment values of the words are expressed by the following.

$$PRP^{new}_{v_j} = PRP_{v_j} + \frac{1}{N-1} * \sum_{k=1,k\neq j}^{N} PRP_{v_k} * e_{v_k}, \tag{7}$$

$$PRN_{v_j}^{new} = PRN_{v_j} + \frac{1}{N-1} * \sum_{k=1,k\neq j}^{N} PRN_{v_k} * e_{v_k}, \tag{8}$$

where $1/(N-1)$ represents the probability that one word propagates sentiment to its neighbor nodes.

When the sentiment value calculations of all words in the graph are completed, we can obtain the final sentiment value SO_{v_j} of the words as following.

$$SO_{v_j} = \frac{PRP_{v_j}^{new} - PRN_{v_j}^{new}}{PRP_{v_j}^{new} + PRN_{v_j}^{new}}. \tag{9}$$

4 Experiment and Analysis

4.1 Datasets and Evaluation Index

Datasets. All the corpus data is crawled all Weibo and comments published by 13 news Weibo accounts from January 10, 2020 to June 10, 2020.

Evaluation Indices. Our main evaluation indices are accuracy and micro-F1 to evaluate the performance of model. The evaluation indices are expressed by formulas as follows.

$$acc_i = \frac{TP_i + TN_i}{TP_i + TN_i + FP_i + FN_i}, \tag{10}$$

$$acc = \frac{1}{m}\sum_{i=1}^{m} acc_i, \tag{11}$$

$$P = \sum_{i=1}^{m} \frac{TP_i}{TP_i + FP_i}, \tag{12}$$

$$R = \sum_{i=1}^{m} \frac{TP_i}{TP_i + FN_i}, \tag{13}$$

$$micro - F1 = \frac{2PR}{P+R}, \tag{14}$$

where m represents the number of categories, and acc_i represents the accuracy of the category i.

4.2 Data Preprocessing

People usually use informal texts to express their opinions, attitudes and sentiments in social media. Since social media will give the texts with many labels, we cannot process these texts directly. Hence we have to restore the informal texts into natural language before our experiments. There are five situations we need to consider.

1) There are many tags, including username links (such as "@username"), and reply contents (such as "回复 @username (reply @username)"), which are cleaned up.
2) We also cleaned up emojis, emoticons (such as ":)") and the contents that are not Chinese.
3) For the topic tags (such as "#hashtag" or "【 hashtag 】"), we cleaned these symbols, e.g. #hashtag → hashtag, 【 hashtag 】 → hashtag.
4) We converted Traditional Chinese to Simplified Chinese, e.g. 我愛中國 → 我爱中国 (I love China).
5) We split the texts into sentences according to punctuation marks ("。", ".", "！", "!", "？", and "?"). Since there may have many aspects in one paragraph, this method will improve the recognition ability of our model. For instance, "你们听见了吗？这是今日中国的声音 (Can you hear it? This is the voice of China today)" → [" 你们听见了吗？(Can you hear it?)", "这是今日中国的声音 (This is the voice of China today)"].

4.3 Model Parameters Analysis

In order to test the influence of different λ and μ on the accuracy and micro-F1 of implicit sentiment words analysis, we use grid search to do parameter sensitivity experiment. The experiment focuses on whether the judgment of implicit sentiment orientation of the word is right. Figure 3(a) demonstrates the parameter sensitivity experiment of the accuracy of analyzing implicit sentiment words; and Fig. 3(b) demonstrates the experiment of the micro-F1.

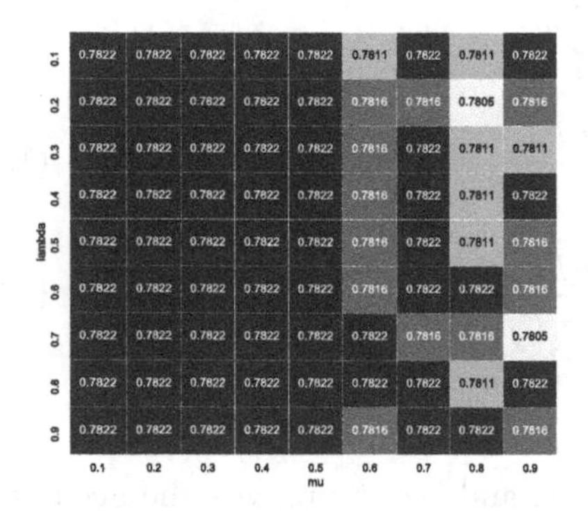

(a) The impact of different λ and μ on accuracy.

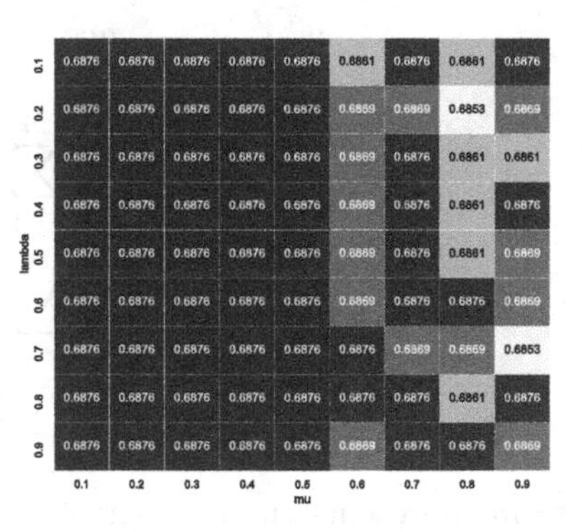

(b) The impact of different λ and μ on micro-F1.

Fig. 3. The impact of different λ and μ on sentiment propagation.

The results show that our model is not sensitive to the parameters. The reason is that we have not referred on the implicit sentiment intensity of words. As the value of μ increases, the performance imperceptibly decreases. Every weight between two words is small, and then the interaction between them is low. In the following experiment, we choose the most ideal parameters, i.e. $\lambda = 0.5$ and $\mu = 0.5$.

4.4 Compare Different Sentiment Propagation Models

We compared the ability of identifying implicit sentiment words of our model with other sentiment propagation model. Figure 4 shows the experiment result. It implies that our model achieves the best result with the accuracy of 78.22% and the micro-F1 of 68.76%. The PolarityRank is the worst. Formula (15) is the calculation method of the sentiment value SO for a word by the PolarityRank algorithm [2], where PR^+ and PR^- respectively represent the positive and negative sentiment values of the node. Our experiment shows that if the word is not an explicit sentiment word, then its $PR^+ = PR^-$ after sentiment propagation, and SO of the node is 0. The reason for a little accuracy and micro-F1 of PolarityRank is that there are some neutral sentiment words. Meanwhile, there have some neutral implicit sentiment words, causing little accuracy and micro-F1 to the result. This demonstrates that PolarityRank has no ability to identify implicit sentiment.

$$SO = \frac{PR^+ - PR^-}{PR^+ + PR^-}. \tag{15}$$

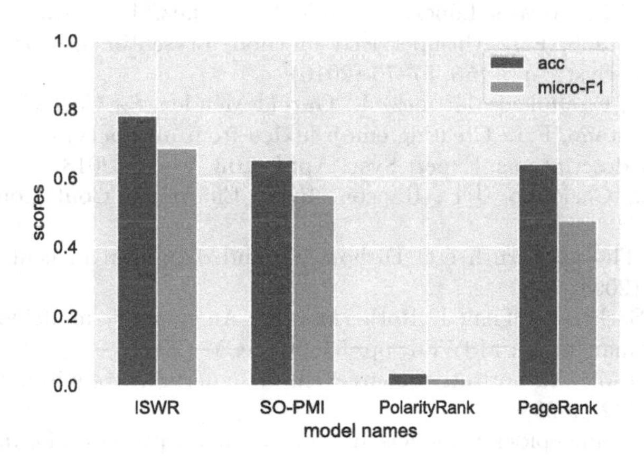

Fig. 4. The performance of different models.

5 Conclusion and Future Work

In this paper, we proposed a novel method that only use little labeled data to recognize implicit sentiment words through sentiment propagation. Since the implicit sentiment words did not contain any obvious sentiment orientation, our model focused on sentiment propagation between words to identify implicit sentiment words. Specifically, we first considered co-occurrence between words. Then we analyzed the relationship between words based on the idea of epidemic model and PolarityRank algorithm. Finally, we obtained implicit sentiment words for sentences. Our experiments analyzed the influence of different parameters on our model. They showed that our model is not sensitive to the parameters. Compared with other sentiment propagation models, our model achieved the best results on distinguishing implicit sentiment words. In the future, we will consider other methods to complete sentiment propagation, such as machine learning, deep learning, etc., to improve the accuracy of identifying implicit sentiment words.

References

1. Chen, H., Chen, H.: Implicit polarity and implicit aspect recognition in opinion mining. In: ACL (2). The Association for Computer Linguistics (2016)
2. Cruz, F.L., Vallejo, C.G., Enríquez, F., Troyano, J.A.: Polarityrank: finding an equilibrium between followers and contraries in a network. Inf. Process. Manag. **48**(2), 271–282 (2012)
3. Deng, L., Wiebe, J.: Sentiment propagation via implicature constraints. In: EACL, pp. 377–385. The Association for Computer Linguistics (2014)
4. Espinosa, K.J.P., Bernales, A.M.J.: Characterizing influence factors affecting emotion diffusion in facebook. Lecture Notes in Engineering & Computer Science, vol. 2214(1), pp. 797–802 (2014)
5. Gavilanes, M.F., Álvarez-López, T., Juncal-Martínez, J., Costa-Montenegro, E., González-Castaño, F.J.: Unsupervised method for sentiment analysis in online texts. Expert Syst. Appl. **58**, 57–75 (2016)
6. Gavilanes, M.F., Juncal-Martínez, J., García-Méndez, S., Costa-Montenegro, E., González-Castaño, F.J.: Creating emoji lexica from unsupervised sentiment analysis of their descriptions. Expert Syst. Appl. **103**, 74–91 (2018)
7. Hatfield, E., Cacioppo, J.T., Rapson, R.L.: Emotional Contagion. Wiley, Ltd (1993)
8. Keyes, R.: The post-truth era: Dishonesty and deception in contemporary life. philosophy (2004)
9. Khelil, A., Becker, C., Tian, J., Rothermel, K.: An epidemic model for information diffusion in manets. In: MSWiM, pp. 54–60. ACM (2002)
10. Kleinberg, J.M.: Authoritative sources in a hyperlinked environment. J. ACM **46**(5), 604–632 (1999)
11. Kong, Q.: Linking epidemic models and hawkes point processes for modeling information diffusion. In: WSDM, pp. 818–819. ACM (2019)
12. Liao, J., Li, Y., Wang, S.: The constitution of a fine-grained opinion annotated corpus on weibo. In: CCL. Lecture Notes in Computer Science, vol. 10035, pp. 227–240 (2016)
13. Liao, J., Wang, S., Li, D.: Identification of fact-implied implicit sentiment based on multi-level semantic fused representation. Knowl. Based Syst. **165**, 197–207 (2019)
14. Liu, B.: Sentiment analysis and opinion mining. In: Synthesis Lectures on Human Language Technologies 5.1 (2012). pp. 1–167 (2011)
15. Liu, B.: Sentiment Analysis - Mining Opinions, Sentiments, and Emotions. Cambridge University Press, Cambridge (2015)

16. Manning, C.D., Surdeanu, M., Bauer, J., Finkel, J.R., Bethard, S., McClosky, D.: The stanford corenlp natural language processing toolkit. In: ACL (System Demonstrations), pp. 55–60. The Association for Computer Linguistics (2014)
17. O., K.W., G., M.A.: A contribution to the mathematical theory of epidemics. Proc. Royal Soc. A Math. Phys. Eng. Sci. **115**(772), 700–721 (1927)
18. Page, L., Brin, S., Motwani, R., Winograd, T.: The pagerank citation ranking: Bringing order to the web. Technical report, Stanford InfoLab (11 1998)
19. Shutova, E., Sun, L., Gutiérrez, E.D., Lichtenstein, P., Narayanan, S.: Multilingual metaphor processing: experiments with semi-supervised and unsupervised learning. Comput. Linguistics **43**(1), 71–123 (2017)
20. Taboada, M., Brooke, J., Tofiloski, M., Voll, K.D., Stede, M.: Lexicon-based methods for sentiment analysis. Comput. Linguistics **37**(2), 267–307 (2011)
21. Wang, G., Araki, K.: Modifying SO-PMI for Japanese weblog opinion mining by using a balancing factor and detecting neutral expressions. In: HLT-NAACL (Short Papers), pp. 189–192. The Association for Computational Linguistics (2007)
22. Wei, J., Liao, J., Yang, Z., Wang, S., Zhao, Q.: Bilstm with multi-polarity orthogonal attention for implicit sentiment analysis. Neurocomputing **383**, 165–173 (2020)
23. Xu, L.: Constructing the affective lexicon ontology. Journal of the China Society for Scientific and Technical Information (2008)
24. Zhang, D., Lin, H., Yang, L., Zhang, S., Xu, B.: Construction of a chinese corpus for the analysis of the emotionality of metaphorical expressions. In: ACL (2), pp. 144–150. Association for Computational Linguistics (2018)
25. Zhang, Q., Zhang, Z., Yang, M., Zhu, L.: Exploring coevolution of emotional contagion and behavior for microblog sentiment analysis: A deep learning architecture. Complex. 2021, 6630811:1–6630811:10 (2021)

An Aspect-Centralized Graph Convolutional Network for Aspect-Based Sentiment Classification

Weixiang Zhao, Yanyan Zhao(✉), Xin Lu, and Bing Qin

Harbin Institute of Technology, Harbin 150001, China
{wxzhao,yyzhao,xlu,qinb}@ir.hit.edu.cn

Abstract. Recent works on aspect-based sentiment classification have manifested the great effectiveness of modeling syntactic dependency with graph neural networks (GNN). However, these works ignore the fact of sentiment information decreasing over dependency paths due to the complex syntactic structure. To tackle the above limitation, we explore a novel solution of constructing an Aspect-Centralized Graph (ACG) for each aspect. Specifically, we directly link all words in a sentence to the aspect word and create a more effective way for the interaction between aspects and opinion words. Based on it, we also incorporate syntactic information into the new graph. To achieve this, we substitute edges of ACG with weighed values, which are calculated from the syntactic relative distance between the aspect and context words on the original dependency graph. Then we propose an Aspect Centralized Graph Convolutional Network (ACGCN) to extract aspect-specific features and effectively interact them with context representations. Extensive experiments on five benchmark datasets show that our model achieves better performance over most baseline models and extensively boosts the performance with BERT.

Keywords: Aspect-based sentiment classification · Aspect-centralized graph · Syntactic relative distance

1 Introduction

Aspect-based Sentiment Classification (ASC), a fine-grained sentiment analysis task, aims at identifying the sentiment polarity (e.g. positive, negative, or neutral) of a specific aspect according to the contexts. It is obvious that just assigning a single sentiment polarity for the sentence as what sentence-level sentiment analysis does is improper now. For example, in the review *great food but the service was dreadful.*, sentiment polarities for the given two aspects *food* and *service* are positive and negative, respectively.

The challenge of ASC lies in associating aspect with its sentiment descriptor words concretely. Early studies mostly adopted machine learning methods

L. Wang et al. (Eds.): NLPCC 2021, LNAI 13029, pp. 260–271, 2021.
https://doi.org/10.1007/978-3-030-88483-3_20

to build sentiment classifier [1,2]. With the great development of deep learning techniques, previous works achieved promising performance based on neural networks [3,6,7]. To better connect aspects with their opinion words, attention mechanism was introduced into this task and boost the performance further [4,5]. However, syntactic dependency, which helps shorten the distance between aspects and opinion words and makes aspects attend to related context words more correctly, are ignored by such attention-based methods.

Recently, several appealing works adopted graph neural network (GNN) models to take dependency trees as the input and successfully incorporate syntactic information to associate the aspect term with its relevant opinion words [8–13].

However, one limitation of previous GNN based work is, when propagating to the aspect through the connections upon dependency tree, the information of associated opinion words may be weakened due to the complex syntactic structures. We use a real example to better illustrate the above problem.

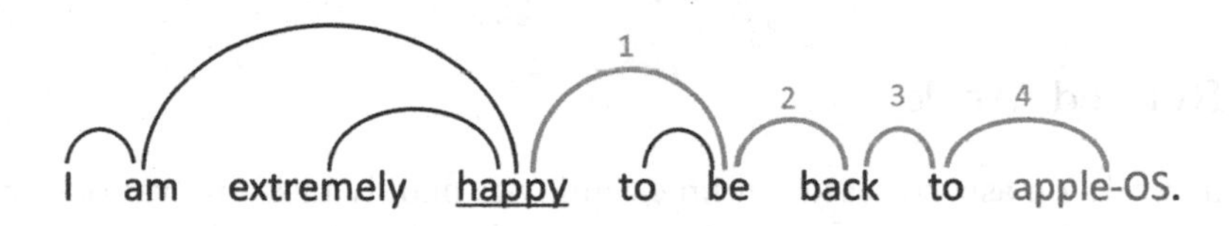

Fig. 1. Example of sentiment information decreasing over dependency paths

Given the syntactic structure of the sentence in Fig. 1 and from our human sights, for the aspect *apple-OS*, the word *happy* is the most relevant to provide sentiment clues and should be laid more consideration. But from the stand of syntactic structures, the aspect word *apple-OS* needs to get connected with its opinion words *happy* via three middle context words *to*, *back* and *be*. Due to the complicated and redundant syntactic structures, when *happy* transfer its information through the long dependency path, the information from middle words and their neighbors, which is useless for sentiment identification, would be incorporated. This results in that opinion words *happy* just take a small part of weights when passing to the aspect *apple-OS*. Thus, the complicated syntactic structure between *happy* and *apple-OS* make the propagating distance too far and weaken the interaction between them.

To better connect the aspect word with its potential opinion words, we propose a novel method to construct the dependency graph, named Aspect-Centralized Graph (ACG). To be more specific, we directly link all words in a sentence to the given aspect. With such a simplification operation, the interaction between aspect and potential opinion words become more direct and effective. In addition, we also calculated the syntactic relative distance based on the original dependency tree between each context word and aspect and pose importance weights according to it. The aim of this is that, on the one hand, context words with different distance from aspect should be assigned different weights to indicate their importance. On the other hand, different weight values can be seen as the different hop connections on the original dependency tree. Thus, to some extent, we incorporate the syntactical information into our new structure and inherit the advantages brought by it. Then we propose a graph convolutional network based model, named Aspect Centralized Graph Convolutional Network (ACGCN), to effectively capture the association between aspect and potential opinion words.

To sum up, the main contributions of our work can be summarized as follows:

- To alleviate sentiment information decreasing and better interact aspect with its opinion words, we propose a novel method to construct dependency graph, in which Aspect-Centralized Graph is proposed.
- Based on the syntactic relative distance to inherit the advantage provided by syntactic structure, we pose difference importance weights to the words in a sentence.
- Extensive experiments on five benchmark datasets show that our model beats most baseline works and extensively boosts the performance with BERT.

2 Related Work

Recent works on aspect-based sentiment analysis mainly based on various types of deep neural networks. We summarize them into two categories according to whether the syntactic structure is incorporated and briefly go through them separately.

For the works without considering syntactic structure, majority of them utilize Long Short-Term Memory (LSTM) neural networks to interact the aspect with context words in a sentence based on sequence order. [3] use two LSTMs to model the preceding and following contexts surrounding the aspect target and feature representations in both directions are used for sentiment classification. Attention mechanism, which helps model focus on the most relevant words to indicate sentiment information, is introduced to this task and achieve distinguished results combined with LSTM [4,5]. Apart from these LSTM-based methods, there also exists some other neural network attempts. [6] first adopt convolutional neural networks (CNN) for ABSA and combine it with gating structure to acquire aspect-related information from contexts. [7] incorporate aspect information into CNN with parameterized filters and parameterized gates.

More recently, syntactic information in the form of dependency tree, which helps shorten the distance between the aspect and opinion words, has been introduced into serval appealing works. We categorize them into two types: modeling upon the original dependency tree from parser and exploring new dependency structure via reshaping or pruning. As for the former one, [8–10] utilize graph convolutional network (GCN) or graph attention network (GAT) to compute specific aspect representation. Then, they interact it with the whole sentence features for classification. While the work guided by the second type of method, [12] propose a relational graph attention network to take dependency relations into consideration on a pruned dependency tree. [13] construct a heterogeneous graph to consider both aspect-focused and inter-aspect contextual dependencies.

The aforementioned GCN-based models, however, all neglect that, the information of aspect-related opinion words may diminish when propagating to the aspect. To address this, we propose a graph convolutional network based model ACGCN with a novel dependency structure ACG. Through this, not only can aspect and opinion words better interact, but also syntactic information is taken into consideration according to the syntactic relative distance.

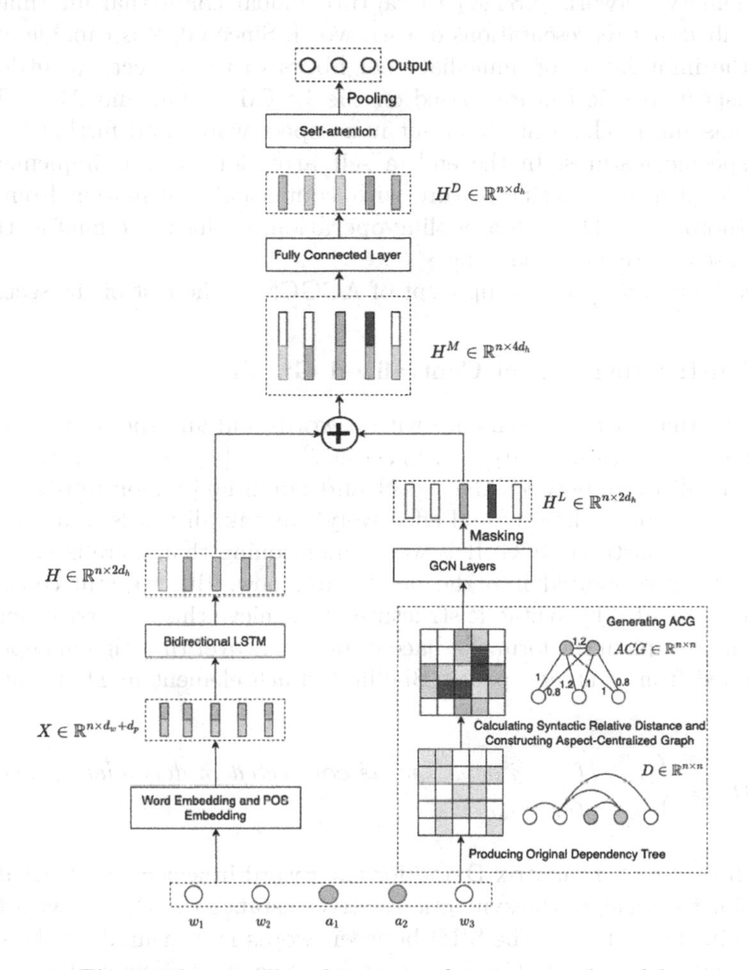

Fig. 2. The overall architecture of our proposed model

3 Methodology

Aspect-based sentiment classification task can be formulated as follows. Given the pair of a sentence consists of N words $s = \{w_1, w_2, \cdots, w_n\}$ and an aspect $a = \{a_1, a_2, \cdots, a_m\}$, the goal of this task is to predict the sentiment polarity

(Positive, Negative or Neural) of the aspect according to context information in the sentence.

Figure 2 gives the overall architecture of our proposed ACGCN model. In order to alleviate sentiment information decreasing over dependency path, we first construct Aspect-Centralized Graph (ACG) for better interaction between aspect and opinion words. And ACGCN starts with a bidirectional Long Short-Term Memory network (LSTM) to capture global contextual information on top of embedding representations of each word. Since GCN is capable of aggregating the information of immediate neighbors to the aspect, we utilize it to obtain aspect-specific features based on the LSTM output and ACG. The following masking mechanism filters out non-aspect words and further highlights aspect specific features. In the end, a self attention layer is implemented to interact local aspect-specific feature with contextual information from LSTM output thoroughly. Through a pooling operation, we have the final features for aspect-based sentiment predicting.

We will elaborate each component of ACGCN in the rest of the section.

3.1 Constructing Aspect Centralized Graph

We assume that there is a sentence with n words and an aspect consisting of m words, i.e. $s = \{w_1, w_2, \cdots, a_1, \cdots, a_m, \cdots w_n\}$, $a = \{a_1, a_2, \cdots, a_m\}$. To further shorten the distance between the aspect and potential opinion words and make the interaction more directly and effectively, we link all words in a sentence to the aspect to construct ACG. It is worth mentioning that there is no syntactic information incorporated into the new graph now. But we still consider the advantage provided by syntactic structure. To achieve this, we first generate the dependency graph in the form of adjacent matrix D over the original dependency tree derived from syntactic parser Biaffine[1]. Each element in D is represented by:

$$D_{i,j} = \begin{cases} 1, & if\ i = j\ or\ w_i,\ w_j\ is\ connected\ in\ dependency\ tree \\ 0, & otherwise \end{cases} \tag{1}$$

Based on the adjacent matrix D and for each word in aspect, we treat it as the start point to calculate the syntactic relative distance (SRD) [19] with the rest of words in the sentence. The SRD between words is measured by the shortest distance between connected nodes over adjacent matrix. Now we have $SRDs \in \mathbb{R}^{m \times n}$, derived from the original syntactic tree, to indicate different weights between context words and the aspect. In the end, for each element in the new constructed ACG:

$$ACG_{i,j} = \begin{cases} 1 - \dfrac{SRDs_{i,j} - thr}{N} & ,\quad if\ w_i\ in\ a \\ 0 & ,\quad otherwise \end{cases} \tag{2}$$

[1] The parser is implemented by https://github.com/yzhangcs/parser.

Where a is the word set of aspect, N is the sentence length and thr stands for a given threshold. That means the words within thr-hops from the aspect are reinforced while those over thr-hops are undermined. So far, a new dependency structure in the form of adjacent matrix named ACG, which is not only suitable for the interaction between the aspect and opinion words, but also inherit the superiority of syntactic information, is completely constructed.

3.2 Aspect Centralized Graph Convolutional Network

- **Embedding Module and Bidirectional LSTM**

In the left parts of Fig. 2, we convert each word in a sentence to a distributed representation, which is mapped from the embedding lookup matrix $E \in \mathbb{R}^{d_w \times |L|}$, where d_w is the embedding dimension and $|L|$ is the vocabulary size. We utilize the pre-trained word embedding GloVe [18] and BERT [14] to initialize word vectors in E and finetune BERT during the training phase.

Intuitively, different parts of speech (POS) of the same word should be treated differently. Thus, we also get POS embedding for each word by looking up a POS embedding matrix $P \in \mathbb{R}^{d_p \times |N|}$, where d_p is the POS embedding dimension and N is the number of POS types. Since most opinion words are adjective, it become one type for POS embedding. Verb can express an action which may indicate sentiment information and adverb modifies verbs and adjectives. Thus, they become the other two types. All the remaining types are categorized as others, so N is set to 4 [24]. P is initialized randomly and updated during the training process. We concatenate word embedding and POS embedding of each word to get the final representation.

According to this, for a sentence with n words, we have the embedding representation $x = \{x_1, x_2, \cdots, x_n\}$, where $x_i \in \mathbb{R}^{d_w + d_p}$. Based on word embeddings of the sentence, we feed it to the Bidirectional LSTM layer to produce context-aware hidden states vectors $H = \{h_1, h_2, \cdots, h_n\}$, where $h_t \in \mathbb{R}^{2d_h}$ represents the hidden state vector at time step t and d_h is the dimension of a LSTM cell output.

- **GCN Layers**

In GCN layers, the ACG and hidden representations from Bi-LSTM layers are taken as the input. Through this, with the information of opinion words propagating over the graph, the specific aspect will aggregate helpful sentiment features for classification. Each node in l-th GCN layer is updated as follows:

$$\tilde{h_i^l} = ACG_i \, g_i^{l-1} \, W^l \tag{3}$$

We also add normalization factor adopted by a previous GCN-based work [8]:

$$h_i^l = ReLU(\tilde{h_i^l} \, / \, (d_i + 1) + b^l) \tag{4}$$

where $g_i^{l-1} \in \mathbb{R}^{2d_h}$ is the hidden representations of the proceeding GCN layer while h_i^l is the output of current GCN layer, ACG_i is the Aspect-Centralized adjacent matrix and d_i represent the sum of a row in ACG. The weights W^l and b^l are both trainable parameters. To highlight representations of the aspect word, we mask hidden state vectors of non-aspect words. Thus, the final outcome of GCN layers is $H^L = \{0, 0, \cdots, h_s^L, \cdots, h_{s+m}^L, \cdots, 0\}$, Where s is the start index of the first word in aspect and m is the aspect length.

- **Generating Output Representation**

As shown in top parts of Fig. 2, to interact context-aware representations and aspect-specific information, we first concatenate H and H^L to attain the merged features $H^M \in \mathbb{R}^{n \times 4d_h}$. Then we utilize a self-attention layer to retrieve significant features that are semantically relevant to aspect words. Through an average pooling operation, we attain the final representations $r \in \mathbb{R}^{d_h}$ for classification. Our implementation is demonstrated as below:

$$h_i^M = h_i \oplus h_i^l \tag{5}$$

$$h_i^D = \widetilde{W} h_i^M + \widetilde{b} \tag{6}$$

$$h_i^S = SelfAttention(h_i^D) \tag{7}$$

$$r = Pooling(h_i^S) \tag{8}$$

$$y = softmax(W_o r + b_o) \tag{9}$$

Where $\widetilde{W}$, $\widetilde{b}$, W_o and b_o are all trainable parameters, $SelfAttention$ is the self-attention layer and $softmax$ is the softmax function to obtain classification output.

3.3 Model Training

The training objective is to minimize the cross entropy loss:

$$L = \sum_{i=1}^{S} \sum_{j=1}^{C} \hat{y_i}^j \cdot log(y_i^j) + \lambda ||\Theta||_2 \tag{10}$$

Where S is the size of training set and C is the number of classes. $\hat{y_i}^j$ stands for the ground-truth label of sentiments. λ is the coefficient of L_2 regularization term and Θ represents all trainable parameters.

4 Experiments

4.1 Dataset and Experiment Setting

We conduct experiments on five benchmark datasets from SemEval 2014 [20] (Restaurant14, Laptop14), SemEval 2015 [21] (Restaruarnt15), SemEval 2016 [22] (Restaurant16) and Twitter [23] dataset. Statistics of the five datasets are shown in Table 1.

In our experiments, for non-BERT models, we use GloVe vectors to initialize word embedding matrix. For POS embedding, we randomly initialize the POS embedding matrix and update it during the training process. The dimension of word embedding vector, POS embedding vector and hidden vector of LSTM cells are set to 300. When calculating syntactic relative distance, we set thr to 2. The number of GCN layers are set to 2 and attention heads in self-attention layer is set to 6. The coefficient λ of L_2 regularization item is set to 10^{-5}. Adam is utilized as the optimizer to train the model with a learning rate of 10^{-3} and 5×10^{-5} for ACGCN and ACGCN-BERT, respectively.[2]

Table 1. Dataset statistics

Dataset	Positive		Neural		Negative	
	Train	Test	Train	Test	Train	Test
REST14	2,164	728	637	196	807	196
LAP14	994	341	464	169	870	128
REST15	1,178	439	50	35	382	328
REST16	1,620	597	88	38	709	190
TWITTER	1,561	173	3,127	346	1,560	173

4.2 Comparison Model

We compared our proposed model with the following methods:

• Methods without syntactic structure

- **TD-LSTM** [3] utilizes LSTMs to model bidirectional contextual representations surrounding a given aspect.
- **ATAE-LSTM** [4] exploits aspect-specific attention mechanism into LSTM.
- **HPAN** [5] adopts hierarchical attention based on relative position order.

• Methods with syntactic structure

- **ASGCN** [8] exploits syntactical dependency structures and propose an Aspect-specific GCN model with directional and un-directional graph.
- **TD-GAT** [10] proposes a target-dependent graph attention network to leverage the syntax structure of a sentence.
- **CDT** [9] combines GCN model and Bi-LSTM model to allow information to be transferred from opinion words to aspect words.
- **Bi-GCN** [11] exploits both the syntactic and lexical graphs to capture the dependency relations in a sentence and the word co-occurrence relations in the training corpus.

[2] Our source code is available at https://github.com/circle-hit/ACGCN.

– **R-GAT** [12] reshapes ordinary dependency trees and incorporate dependency relations into a new GAT model.
– **InterGCN** [13] interactively extracting the sentiment relations within aspect words and across different aspects in the context to derive aspect-specific features.

• Methods based on BERT

– **BERT** [14] is the vanilla BERT model and take "[CLS] sentence [SEP] aspect [SEP] as the input.
– **AEN+BERT** [15] explores an attention encoder network based on pre-trained BERT.
– **SAGAT** [16] combines syntax-guide graph attention and external pre-training knowledge to attain sentiment classification features.
– **SA-GCN+BERT** [17] employs the selective attention mechanism based on GCN and BERT.
– **R-GAT+BERT** [12] is the R-GAT model integrated with pre-trained BERT.
– **InterGCN+BERT** [13] is the InterGCN model based on pre-trained BERT.
– **ACGCN** is our proposed model.
– **ACGCN+BERT** integrates our model with pre-trained BERT.

4.3 Main Results

Table 2. Main experimental results. *Acc.* represents accuracy, $F1$ represents Macro-F1 score. Best results are in bold.

Category	Model	*LAP*14		*REST*14		*REST*15		*REST*16		*TWITTER*	
		Acc.(%)	*F*1(%)	*Acc.*(%)	*F*1(%)	*Acc.*(%)	*F*1(%)	*Acc.*(%)	*F*1(%)	*Acc.*(%)	*F*1(%)
Without syntactic structure	TD-LSTM [3]	71.83	68.43	78.00	66.73	76.39	58.70	82.16	54.21	–	–
	ATAE-LSTM [4]	68.88	63.93	78.60	67.02	78.48	60.53	83.77	61.71	69.65	67.40
	HPAN [5]	77.27	–	82.23	–	–	–	–	–	–	–
With syntactic structure	ASGCN [8]	75.55	71.05	80.77	72.02	79.89	61.89	88.99	67.48	72.15	70.40
	CDT [9]	77.19	72.99	82.3	74.02	–	–	85.58	69.93	73.29	72.02
	TD-GAT [10]	73.70	–	81.10	–	–	–	–	–	72.20	70.45
	Bi-GCN [11]	74.59	71.84	81.97	73.48	81.16	64.79	88.96	70.84	74.16	73.35
	R-GAT [12]	77.42	73.76	83.3	76.08	–	–	–	–	75.57	73.82
	InterGCN [13]	77.86	74.32	82.23	74.01	81.76	65.67	89.77	73.05	–	–
	ACGCN (ours)	78.21	74.24	83.57	75.59	82.47	64.04	89.29	72.69	75.72	74.73
BERT-Based	BERT [14]	77.59	73.28	84.11	76.68	83.48	66.18	90.10	74.16	–	–
	AEN+BERT [15]	79.93	76.31	83.12	73.76	–	–	–	–	74.71	73.13
	SAGAT [16]	80.37	76.94	85.08	77.94	–	–	–	–	75.40	74.17
	SA-GCN+BERT [17]	81.70	78.80	85.80	79.70	–	–	–	–	–	–
	R-GAT+BERT [12]	78.21	74.07	86.60	81.35	–	–	–	–	76.15	74.88
	InterGCN+BERT [13]	**82.87**	**79.32**	**87.12**	81.02	85.42	71.05	91.27	78.32	–	–
	ACGCN+BERT (ours)	81.66	78.52	86.96	**82.13**	**86.16**	**74.80**	**93.51**	**79.60**	**76.45**	**75.22**

As shown in Table 2, our proposed ACGCN model achieves the best results in terms of *Acc* scores on almost all five benchmark datasets. The $F1$ scores are also among the best ones, and are only slightly worsen than SOTA models on *LAP*14, *REST*14 and *REST*16, where the difference is 0.08, 0.49 and 0.36, respectively. This demonstrates the effectiveness of ACG to alleviate sentiment information decreasing and also verifies that ACGCN is able to interact aspect words with potential opinion words more effectively.

Compared with models without considering syntactic structures, ACGCN is much better than them, indicating that syntactic dependency between words are beneficial to identify sentiment polarity. Also, better performance over other syntax-based models proves that the ACG in our ACGCN is more suitable for the propagating of sentiment information.

Moreover, when integrating with pre-trained BERT, ACGCN-BERT largely boosts the performance. It not only performs better than the vanilla BERT, but also it outperforms other comparison BERT-based models. It verifies the effectiveness of ACG further and also indicates ACGCN can be easily integrated with pre-trained BERT.

4.4 Ablation Study

To analyze the effectiveness of our proposed method and the impact of different components of ACGCN, we conduct an ablation study on five benchmark datasets. The results are shown in Table 3.

- **To verify the effectiveness of GCN layers**

We first investigate the impact of GCN's ability to extract aspect-specific features. With the removal of GCN layers and its input ACG, we just keep the remaining parts of ACGCN. Compared with the complete model, performance drops evidently, which indicates that it is of importance to capture and highlight specific aspect information for sentiment classification.

- **To verify the effectiveness of ACG**

We show the effect of ACG by replacing it with the original dependency graph derived from dependency parse as the input of GCN layers. The poorer results suggest the novel Aspect-Centralized Graph with direct connections between aspects and opinion words is more effective for the interaction among them and alleviate the problem of sentiment information decreasing.

- **To verify the effectiveness of SRD**

At last, we explore the incorporated syntactic information brought by SRD. To achieve this, we just linking other context words to the aspect, without considering their syntactic dependencies and different importance calculated by SRD. It is noticed that the results are worsen than the model with complete ACG, manifesting the necessity of introducing syntactic information to differentiate context words in a sentence. The results of both last two ablation experiments properly verify the effectiveness of our novel ACG.

Table 3. Experimental results of the ablation study

Model	*LAP*14		*REST*14		*REST*15		*REST*16		*TWITTER*	
	Acc.(%)	*F*1(%)	*Acc.*(%)	*F*1(%)	*Acc.*(%)	*F*1(%)	*Acc.*(%)	*F*1(%)	*Acc.*(%)	*F*1(%)
ACGCN w/o GCN	75.24	70.88	82.86	73.35	79.89	62.09	88.47	66.40	74.28	73.21
ACGCN w/o ACG	76.96	73.08	82.32	73.38	81.00	**65.21**	88.80	70.35	74.13	72.45
ACGCN w/o SRD	76.02	72.00	82.86	73.15	80.81	63.76	88.80	71.37	74.71	72.97
ACGCN	**78.21**	**74.24**	**83.57**	**75.59**	**82.47**	64.04	**89.29**	**72.69**	**75.72**	**74.73**

5 Conclusion

In this paper, we propose a novel method of constructing Aspect-Centralized Graph (ACG) to alleviate sentiment information decreasing over dependency path for aspect-based sentiment classification tasks. Specifically, we directly link all words within a sentence to the aspect. Further, in order to distinguish the importance of different words and inherit the advantage brought by syntactic structure, we pose different weights to each word according to the relative syntactic distance. Based on it, an Aspect Centralized Graph Convolutional Network (ACGCN) model is proposed to interact aspect-specific sentiment features with the context information. Experimental results on five benchmark datasets demonstrate the proposed ACGCN and ACGCN-BERT beat most majority of baseline models.

Acknowledgements. We thank the anonymous reviewers for their insightful comments and suggestions. This work was supported by the National Key R&D Program of China via grant 2018YFB1005103 and National Natural Science Foundation of China (NSFC) via grant 61632011 and 61772153.

References

1. Jiang, L., Yu, M., Zhou, M., Liu, X., Zhao, T.: Target-dependent twitter sentiment classification. In: ACL, pp. 151–160 (2011)
2. Mohammad, S.M., Kiritchenko, S., Zhu, X.: NRC-Canada: Building the state-of-the-art in sentiment analysis of tweets. arXiv preprint arXiv:1308.6242 (2013)
3. Tang, D., Qin, B., Feng, X., Liu, T.: Effective LSTMs for target-dependent sentiment classification. In: COLING, pp. 3298–3307 (2016)
4. Wang, Y., Huang, M., Zhu, X., Zhao, L.: Attention-based LSTM for aspect-level sentiment classification. In: EMNLP, pp. 606–615 (2016)
5. Li, L., Liu, Y., Zhou, A.: Hierarchical Attention based position-aware network for aspect-level sentiment analysis. In: CoNLL, pp. 181–189 (2018)
6. Xue, W., Li, T.: Aspect based sentiment analysis with gated convolutional networks. In: ACL, pp. 2514–2523 (2018)

7. Huang, B., Carley, K.: Parameterized convolutional neural networks for aspect level sentiment classification. In: EMNLP, pp. 1091–1096 (2018)
8. Zhang, C., Li, Q., Song, D.: Aspect-based sentiment classification with aspect-specific graph convolutional networks. In: EMNLP-IJCNLP, pp. 4568–4578 (2019)
9. Sun, K., Zhang, R., Mensah, S., Mao, Y., Liu, X.: Aspect-level sentiment analysis via convolution over dependency tree. In: EMNLP-IJCNLP, pp. 5679–5688 (2019)
10. Huang, B., Carley, K.: Syntax-aware aspect level sentiment classification with graph attention networks. In: EMNLP-IJCNLP, pp. 5469–5477 (2019)
11. Zhang, M., Qian, T.: Convolution over hierarchical syntactic and lexical graphs for aspect level sentiment analysis. In: EMNLP, pp. 3540–3549 (2020)
12. Wang, K., Shen, W., Yang, Y., Quan, X., Wang, R.: Relational graph attention network for aspect-based sentiment analysis. In: ACL, pp. 3229–3238 (2020)
13. Liang, B., Yin, R., Gui, L., Du, J., Xu, R.: Jointly learning aspect-focused and inter-aspect relations with graph convolutional networks for aspect sentiment analysis. In: COLING, pp. 150–161 (2020)
14. Devlin, J., Chang, M.W., Lee, K., Toutanova, K.: BERT: pre-training of deep bidirectional transformers for language understanding. In: NAACL, pp. 4171–4186 (2019)
15. Song, Y., Wang, J., Jiang, T., Liu, Z., Rao, Y.: Attentional encoder network for targeted sentiment classification. arXiv preprint arXiv:1902.09314 (2019)
16. Huang, L., Sun, X., Li, S., Zhang, L., Wang, H.: Syntax-aware graph attention network for aspect-level sentiment classification. In: COLING, pp. 799–810 (2020)
17. Hou, X., Huang, J., Wang, G., Huang, K., He, X., Zhou, B.: Selective attention based graph convolutional networks for aspect-level sentiment classification. arXiv preprint arXiv:1910.10857 (2019)
18. Pennington, J., Socher, R., Manning C.: Glove: global vectors for word representation. In: EMNLP, pp. 1532–1543 (2014)
19. Phan, M.H., Ogunbona, P.O.: Modelling context and syntactical features for aspect-based sentiment analysis. In: ACL, pp. 3211–3220 (2020)
20. Pontiki, M., Galanis, D., Papageorgiou, H., Manandhar, S., Androutsopoulos, I.: SemEval-2014 task 4: aspect based sentiment analysis. In: SemEval, pp. 27–35 (2014)
21. Pontiki, M., Galanis, D., Papageorgiou, H., Manandhar, S., Androutsopoulos, I.: SemEval-2015 task 12: aspect based sentiment analysis. In: SemEval, pp. 486–495 (2015)
22. Pontiki, M., Galanis, D., Papageorgiou, H., Manandhar, S., Androutsopoulos, I.: SemEval-2016 task 5: aspect based sentiment analysis. In: SemEval, pp. 19–30 (2016)
23. Dong, L., Wei, F., Tan, C., Tang, D., Zhou, M., Xu, K.: Adaptive recursive neural network for target-dependent twitter sentiment classification. In: ACL, pp. 49–54 (2014)
24. Shuang, K., Gu, M., Li, R., Loo, J., Su, S.: Interactive POS-aware network for aspect-level sentiment classification. Neurocomputing **420**(1), 181–196 (2021)

NLP Applications and Text Mining

Capturing Global Informativeness in Open Domain Keyphrase Extraction

Si Sun[1], Zhenghao Liu[2], Chenyan Xiong[3], Zhiyuan Liu[4,5,6](✉), and Jie Bao[1](✉)

[1] Department of Electronic Engineering, Tsinghua University, Beijing, China
s-sun17@mails.tsinghua.edu.cn, bao@tsinghua.edu.cn

[2] Department of Computer Science and Technology, Northeastern University, Shenyang, China
liuzhenghao@cse.neu.edu.cn

[3] Microsoft Research, New York, USA
chenyan.xiong@microsoft.com

[4] Department of Computer Science and Technology, Tsinghua University, Beijing, China

[5] Institute for Artificial Intelligence, Tsinghua University, Beijing, China
liuzy@tsinghua.edu.cn

[6] Beijing National Research Center for Information Science and Technology, Beijing, China

Abstract. Open-domain KeyPhrase Extraction (KPE) aims to extract keyphrases from documents without domain or quality restrictions, e.g., web pages with variant domains and qualities. Recently, neural methods have shown promising results in many KPE tasks due to their powerful capacity for modeling contextual semantics of the given documents. However, we empirically show that most neural KPE methods prefer to extract keyphrases with good phraseness, such as short and entity-style n-grams, instead of globally informative keyphrases from open-domain documents. This paper presents JointKPE, an open-domain KPE architecture built on pre-trained language models, which can capture both local phraseness and global informativeness when extracting keyphrases. JointKPE learns to rank keyphrases by estimating their informativeness in the entire document and is jointly trained on the keyphrase chunking task to guarantee the phraseness of keyphrase candidates. Experiments on two large KPE datasets with diverse domains, OpenKP and KP20k, demonstrate the effectiveness of JointKPE on different pre-trained variants in open-domain scenarios. Further analyses reveal the significant advantages of JointKPE in predicting long and non-entity keyphrases, which are challenging for previous neural KPE methods. Our code is publicly available at https://github.com/thunlp/BERT-KPE.

Keywords: Keyphrase extraction · Open domain · Global informativeness · Pre-trained language model

S. Sun and Z. Liu—Equal contributions.

L. Wang et al. (Eds.): NLPCC 2021, LNAI 13029, pp. 275–287, 2021.
https://doi.org/10.1007/978-3-030-88483-3_21

1 Introduction

Keyphrases that can provide a succinct summary of a document have exhibited their potential in improving many natural language processing (NLP) and information retrieval (IR) tasks [12], such as summarization [29], recommendation [28], and document retrieval [14]. High-quality keyphrases have shown two features, *phraseness* and *informativeness* [32]: *phraseness* refers to the degree to which a sequence of words can serve as a complete semantic unit in the local contexts of documents; *informativeness* indicates how well a text snippet captures the global topic or salient concept of the full document [18]. Many previous studies leverage the two features to improve the performance of KPE [10,20].

With the development of neural networks, neural KPE methods have achieved convincing performance in extracting keyphrases for scientific publications [23,25,31]. In recent years, many researchers have begun to adapt neural KPE to open-domain scenarios by considering diverse extraction domains with variant content qualities, such as the web KPE [39]. Existing neural KPE methods often formulate keyphrase extraction as the tasks of word-level sequence tagging [40], n-gram level classification [39], and span-based extraction [25]. Despite their success, these neural methods seem to focus more on modeling the localized semantic features of keyphrases, which may cause them to prioritize local *phraseness* rather than *informativeness* of the global document when extracting keyphrases. As a consequence, they are inclined to extract keyphrases with semantic integrity from open-domain documents, such as short n-grams and head-ish entities, while long-tail phrases sometimes convey more pivotal information [6].

In this paper, we propose JointKPE that servers open-domain keyphrase extraction scenarios. It can take both phraseness and informativeness into account when extracting keyphrases under a multi-task training architecture. JointKPE first resorts to powerful pre-trained language models to encode the documents and estimate the localized informativeness for all their n-grams. For the n-grams that have the same lexical strings but appear in various contexts, JointKPE further calculates their global informativeness scores in the full document. Lastly, JointKPE learns to rank these keyphrase candidates according to their global informative scores and jointly trains with the keyphrase chunking task [39] to capture both local phraseness and global informativeness.

Experiments on two large-scale KPE datasets, OpenKP [39] and KP20k [23] with web pages and scientific papers, demonstrate JointKPE's robust effectiveness with the widely used pre-trained model, BERT [7], and its two pre-training variants, SpanBERT [15] and RoBERTa [19]. Our empirical analyses further indicate JointKPE's advantages in predicting long keyphrases and non-entity keyphrases in open-domain scenarios.

2 Related Work

Automatic KeyPhrase Extraction (KPE) is concerned with automatically extracting a set of important and topical phrases from a document [26,33].

Throughout the history of KPE, the earliest annotated keyphrase extraction corpora came from scientific domains, including technical reports and scientific literature [2,3,17]. It is because scientific KPE corpora are easily curated whose keyphrases have already been provided by their authors. In addition to scientific corpora, some researchers have collected KPE corpora from the Internet and social media, e.g., news articles [34] and live chats [16], but these corpora are limited in size and lack diversity in terms of domains and topics. Recently, a large scale of open-domain KPE dataset named OpenKP [39] has been released, which includes about one hundred thousand annotated web documents with a broad distribution of domains and topics. Based on the aforementioned corpora, extensive automatic KPE technologies have been proposed and examined.

Existing KPE technologies can be categorized as unsupervised and supervised methods [26]. Unsupervised KPE methods are mainly based on statistical information [4,9,11] and graph-based ranking algorithms [21,24,30]. Supervised KPE methods usually formulate keyphrase extraction as phrase classification [5,22,36] and learning-to-rank tasks [13,20,41]. Thrived from sufficient well-annotated training data, supervised KPE methods, especially with neural networks, have significant performance advantages over state-of-the-art unsupervised methods in many KPE benchmarks [17,23,39].

The earliest neural KPE methods often treat KPE as sequence-to-sequence learning [23] and sequence labeling [40] tasks, which adopts RNN-based encoder-decoder frameworks. The performance of these methods is limited by the shallow representations of textual data. Recently, Xiong et al. [39] formulate KPE as an n-gram level keyphrase chunking task and propose BLING-KPE. It incorporates deep pre-trained representations into a convolutional transformer architecture to model n-gram representations. BLING-KPE achieves great improvement over previous methods. In addition, the recent work [1] replaces the full self-attention of Transformers with local-global attention, which significantly boosts the KPE performance for long documents. More recently, Wang, et al. [35] also show that incorporating multi-modal information in web pages, such as font, size, and DOM features, can bring further improvement for open-domain web KPE.

3 Methodology

Given a document D, JointKPE first extracts all keyphrase candidates p from the document by enumerating its n-grams and utilizes a hierarchical architecture to model n-gram representations. Based on n-gram representations, JointKPE employs the informative ranking network to integrate the localized informativeness scores of multi-occurred phrases for estimating their global informativeness scores in the full document. During training, JointKPE is trained jointly with the keyphrase chunking task [39] to better balance phraseness and informativeness.

N-gram Representation. JointKPE first leverages pre-trained language models, e.g., BERT [7], to encode the document $D = \{w_1, \ldots, w_i, \ldots, w_n\}$, and outputs a sequence of word embeddings $\boldsymbol{H} = \{\boldsymbol{h}_1, \ldots, \boldsymbol{h}_i, \ldots, \boldsymbol{h}_n\}$:

$$\boldsymbol{H} = \text{BERT}\{w_1, \ldots, w_i, \ldots, w_n\}, \tag{1}$$

where $\boldsymbol{h}_i$ is the word embedding of the i-th word w_i in the document D.

To enumerate keyphrase candidates of the document D, the word embeddings are integrated into n-gram representations using a set of Convolutional Neural Networks (CNNs) since keyphrases usually appear in the form of n-grams. The representation of the i-th k-gram $c_i^k = w_{i:i+k-1}$ is calculated as:

$$\boldsymbol{g}_i^k = \mathrm{CNN}^k\{\boldsymbol{h}_i, \dots, \boldsymbol{h}_{i+k-1}\}, \tag{2}$$

where each k-gram is composed by its corresponding CNN^k with the window size of k $(1 \leq k \leq K)$. K is the maximum length of the extracted n-grams.

Informative Ranking. To estimate the informativeness for n-gram c_i^k in local contexts, JointKPE takes a feedforward layer to project its context-specific representation $\boldsymbol{g}_i^k$ to a quantifiable score:

$$f(c_i^k, D) = \mathrm{Feedforward}(\boldsymbol{g}_i^k). \tag{3}$$

We further compute global informativeness scores for those phrases that appear multiple time in different contexts of the document. Specifically, let phrase p^k be a multi-occurred phrase of length k in the document D. This phrase occurs in different contexts $\{c_j^k, \dots, c_l^k, \dots, c_m^k\}$ of the document, which thus leads to diverse localized informativeness scores $\{f(c_j^k, D), \dots, f(c_l^k, D), \dots, f(c_m^k, D)\}$. For this multi-occurred phrase, JointKPE applies max-pooling upon its localized informativeness scores to determine the global informativeness score $f^*(p^k, D)$:

$$f^*(p^k, D) = \max\{f(c_j^k, D), \dots, f(c_l^k, D), \dots, f(c_m^k, D)\}. \tag{4}$$

After estimating the global informativeness scores for all phrases in the document D, JointKPE can learn to rank these phrases in the document level based on their global informativeness scores using the pairwise ranking loss:

$$L_{\mathrm{Rank}} = \sum_{p_+, p_- \in D} \max(0, 1 - f^*(p_+, D) + f^*(p_-, D)), \tag{5}$$

where the ranking loss L_{Rank} enforces JointKPE to rank keyphrases p_+ ahead of non-keyphrases p_- within the same document D.

Keyphrase Chunking. To enhance the phraseness measurement at the n-gram level, JointKPE combines the keyphrase chunking task [39] to directly learn to predict the keyphrase probability of n-grams by optimizing the binary classification loss L_{Chunk}:

$$L_{\mathrm{Chunk}} = \mathrm{CrossEntropy}(P(c_i^k = y_i^k)), \tag{6}$$

where y_i^k is the binary label, which denotes whether the n-gram c_i^k exactly matches the string of a keyphrase annotated in the document.

Multi-task Training. The ultimate training objective of JointKPE is to minimize the linear combination of the informative ranking loss L_{Rank} and the keyphrase chunking loss L_{Chunk}:

$$L = L_{\text{Rank}} + L_{\text{Chunk}}. \tag{7}$$

At the inference stage, the top-ranking phrases with the highest global informativeness scores are predicted as keyphrases for the given document.

Table 1. Overall accuracy of keyphrase extraction. All scores of JointKPE on OpenKP are from the official leaderboard with a blind test. The baseline evaluation results on KP20k are obtained from corresponding papers, and the asterisked baselines* are our implementations. Bold **F1@3** and **F1@5** are main evaluation metrics of OpenKP and KP20k, respectively.

Methods	OpenKP									KP20k	
	F1@1	F1@3	F1@5	P@1	P@3	P@5	R@1	R@3	R@5	F1@5	F1@10
Baselines											
TFIDF [25,39]	0.196	0.223	0.196	0.283	0.184	0.137	0.150	0.284	0.347	0.108	0.134
TextRank [25,39]	0.054	0.076	0.079	0.077	0.062	0.055	0.041	0.098	0.142	0.180	0.150
Maui [25]	n.a.	n.a.	n.a.	n.a.	n.a.	n.a.	n.a.	n.a.	n.a.	0.273	0.240
PROD [39]	0.245	0.236	0.188	0.353	0.195	0.131	0.188	0.299	0.331	n.a.	n.a.
CopyRNN [23]	0.217	0.237	0.210	0.288	0.185	0.141	0.174	0.331	0.413	0.327	0.278
CDKGEN [8]	n.a.	n.a.	n.a.	n.a.	n.a.	n.a.	n.a.	n.a.	n.a.	0.381	0.324
BLING-KPE [39]	0.267	0.292	0.209	0.397	0.249	0.149	0.215	0.391	0.391	n.a.	n.a.
SKE-Base-Cls (BERT-Base) [25]	n.a.	n.a.	n.a.	n.a.	n.a.	n.a.	n.a.	n.a.	n.a.	0.386	0.326
Span Extraction (BERT-Base)*	0.318	0.332	0.289	0.476	0.285	0.209	0.253	0.436	0.521	0.393	0.325
Sequence Tagging (BERT-Base)*	**0.321**	**0.361**	**0.314**	**0.484**	**0.312**	**0.227**	**0.255**	**0.469**	**0.563**	**0.407**	**0.335**
Our methods											
JointKPE (BERT-Base)	**0.340**	**0.376**	**0.325**	**0.521**	**0.324**	**0.235**	**0.280**	**0.491**	**0.583**	0.411	0.338
Only Informative Ranking	0.342	0.374	**0.325**	0.513	0.323	**0.235**	0.273	0.489	0.582	**0.413**	**0.340**
Only Keyphrase Chunking	0.340	0.356	0.311	0.511	0.306	0.225	0.271	0.464	0.558	0.412	0.337
JointKPE (SpanBERT-Base)	**0.359**	**0.385**	**0.336**	**0.535**	**0.331**	**0.243**	**0.288**	**0.504**	**0.603**	**0.416**	**0.340**
Only Informative Ranking	0.355	0.380	0.331	0.530	0.327	0.240	0.284	0.497	0.593	0.412	0.338
Only Keyphrase Chunking	0.348	0.372	0.324	0.523	0.321	0.235	0.278	0.486	0.581	0.411	0.338
JointKPE (RoBERTa-Base)	**0.364**	**0.391**	**0.338**	**0.543**	**0.337**	**0.245**	**0.291**	**0.511**	**0.605**	**0.419**	**0.344**
Only Informative Ranking	0.361	0.390	0.337	0.538	**0.337**	0.244	0.290	0.509	0.604	0.417	0.343
Only Keyphrase Chunking	0.355	0.373	0.324	0.533	0.322	0.235	0.283	0.486	0.581	0.408	0.337

4 Experimental Methodology

This section introduces our experimental settings, including datasets, evaluation metrics, baselines, and implementation details.

Dataset. Two large-scale KPE datasets are used in our experiments. They are OpenKP [39] and KP20k [23]. OpenKP is an open-domain keyphrase extraction dataset with various domains and topics, which contains over 150,000 real-world web documents along with the most relevant keyphrases generated by expert

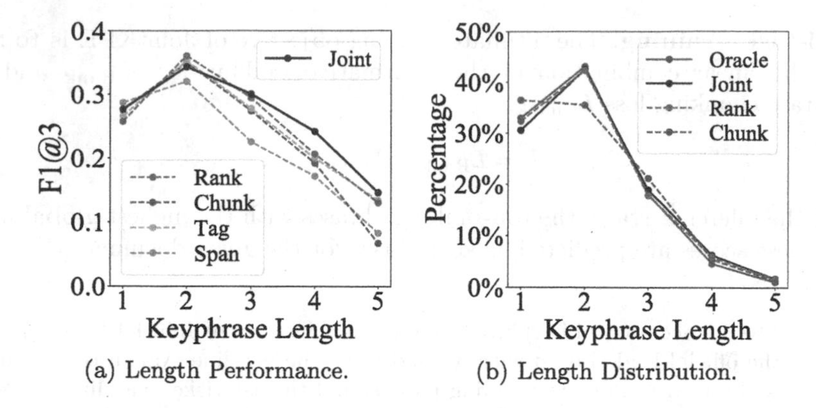

(a) Length Performance. (b) Length Distribution.

Fig. 1. Keyphrase length analyses. Figure 1(a) exhibits the F1@3 score of different neural KPE methods on OpenKP at different keyphrase lengths. Figure 1(b) illustrates the length distribution of keyphrases predicted by JointKPE (Joint) and its two ablated versions: only informative ranking (Rank) and only keyphrase chunking (Chunk). Methods of Sequence Tagging and Span Extraction are abbreviated as Tag and Span, respectively.

annotation. We consider OpenKP as our main benchmark and follow its official split of training (134k documents), development (6.6k), and testing (6.6k) sets. KP20k is a scientific KPE dataset with the computer science domain, which consists of over 550,000 articles with keyphrases assigned by their authors. We follow the original work's partition of training (528K documents), development (20K), and testing (20K) sets [23].

Evaluation Metrics. Precision (P), Recall (R), and F-measure (F1) of the top N keyphrase predictions are used for evaluating the performance of KPE methods. Following prior research [23,39], we utilize $N = \{1, 3, 5\}$ on OpenKP and $N = \{5, 10\}$ on KP20k, and consider F1@3 and F1@5 as the main metrics for OpenKP and KP20k, respectively. For KP20k, we use Porter Stemmer [27] to determine the match of two phrases, which is consistent with prior work [23].

Baselines. Two groups of baselines are compared in our experiments, including *Traditional KPE baselines* and *Neural KPE baselines.*

Traditional KPE Baselines. Keeping consistent with previous work [23,25,39], we compare JointKPE with four traditional KPE methods. They are two popular unsupervised KPE methods, TFIDF and TextRank [24], and two feature-based KPE systems that perform well on OpenKP and KP20k, named PROD [39] and Maui [22]. For these baselines, we take their reported performance.

Neural KPE Baselines. We also compare our method with six neural KPE baselines: the previous state-of-the-art method on OpenKP, BLING-KPE [39] and three advanced neural KPE methods on KP20k, including CopyRNN [23], CDKGEN [8] and SKE-Base-Cls [25], and two BERT-based KPE methods we have reconstructed, including Span Extraction and Sequence Tagging. They formulate KPE as the tasks of span extraction and sequence tagging, respectively.

Table 2. Examples of entity and non-entity keyphrases in the OpenKP dataset.

Entity keyphrase	Entity type
facebook	business.brand
the happy wanderer	music.composition
health and human services	government.government_agency
Non-Entity Keyphrase	Type
join a world	verb phrase
care tenderly	adverbial phrase
beautiful clothing	adjective phrase
opposite a monkey	prepositional phrase
i have seen the light	sentence snippet
firing wice with double click	complex phrase
landmarks and historic buildings	composition phrase

Implementation Details. The base versions of BERT [7], SpanBERT [15] and RoBERTa [19], initialized from their pre-trained weights, are used to implement JointKPE. All our methods are optimized using Adam with a 5e−5 learning rate, 10% warm-up proportion, and 64 batch size. We set the maximum sequence length to 512 and simply keep the weights of the two training losses (Eq. 7) to be the same. Following previous work [23,39], the maximum phrase length is set to 5 ($K = 5$). The training used two Tesla T4 GPUs and took about 25 h on three epochs. Our implementations are based on PyTorch-Transformers [37].

5 Results and Analyses

In this section, we present the evaluation results of JointKPE and conduct a series of analyses and case studies to study its effectiveness.

5.1 Overall Accuracy

The evaluation results of JointKPE and baselines are shown in Table 1.

Overall, JointKPE outperforms all baselines on all evaluation metrics stably on both open-domain and scientific KPE benchmarks. Compared to the best feature-based KPE systems, PROD [39] and Maui [22], JointKPE outperforms them by large margins. JointKPE also outperforms the strong neural baselines, BLING-KPE [39] and CDKGEN [8], the previous state-of-the-art on OpenKP and KP20k; their F1@3 and F1@5 are improved by 22.8% and 7.9% respectively. These results demonstrate JointKPE's effectiveness.

Moreover, even with the same pre-trained model, JointKPE still achieves better performance than those neural KPE methods mainly based on localized features, e.g., SKE-Base-Cls [25], Span Extraction, and Sequence Tagging. These

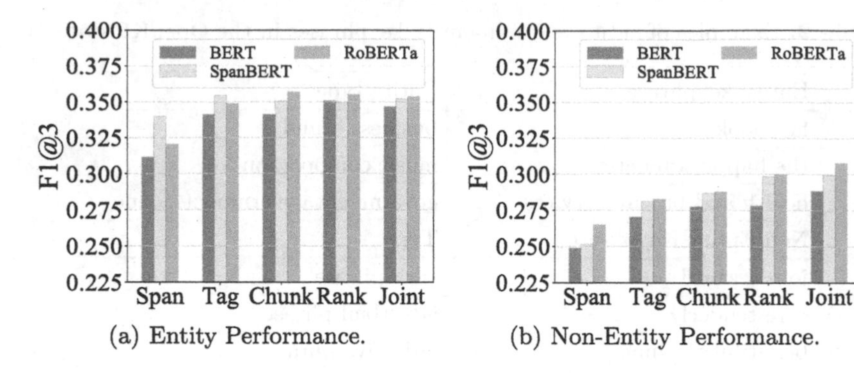

(a) Entity Performance. (b) Non-Entity Performance.

Fig. 2. Keyphrase type analyses. Figure (a) and (b) show the extraction performance for entity keyphrases and non-entity keyphrases on the OpenKP dataset, respectively. Five BERT-based KPE models, Span Extraction (Span), Sequence Tagging (Tag), the two ablated versions (Chunk and Rank), and the full model of JointKPE (Joint) are compared.

results reveal that the effectiveness of JointKPE, in addition to the benefits gained from the pre-trained model, also stems from its ability to combine local and global features when extracting keyphrases.

As shown in the ablation results of JointKPE, without either informative ranking or keyphrase chunking, its accuracy declines, especially in the absence of the informative ranking on the open-domain OpenKP dataset. The result indicates the benefit of multi-task learning and the key role of capturing global informativeness in open-domain keyphrase extraction scenarios.

Besides, JointKPE's effectiveness can be further improved with initialized from SpanBERT [15] and RoBERTa [19], the two BERT variants with updated pre-training strategies, which shows JointKPE's ability to utilize the advantages of better pre-trained models. Furthermore, we observe that our method obtains more significant improvements on RoBERTa due to the higher enhancement of informative ranking performance, and the improvements brought by SpanBERT and RoBERTa to the keyphrase chunking task are relatively close[1].

5.2 Performance w.r.t. Keyphrase Lengths

One challenge in open-domain KPE is the long-keyphrase extraction scenario [39]. In this experiment, we compare the behavior and effectiveness of JointKPE and other BERT-based KPE methods in extracting keyphrases of different lengths, using the open-domain OpenKP dataset.

As shown in Fig. 1(a), all neural KPE methods have better extraction performance for short keyphrases than longer ones. Nevertheless, compared with other methods, JointKPE has a more stable extraction ability for keyphrases

[1] Due to space limitations, more baseline results based on SpanBERT and RoBERTa can be found in https://github.com/thunlp/BERT-KPE.

of different lengths and performs best in predicting longer keyphrases (length ≥ 4). Notably, the F1@3 score of JointKPE is 17% higher than its two ablated versions, Rank (only informative ranking) and Chunk (only keyphrase chunking), at the keyphrase length of 4. These results reveal that JointKPE's joint modeling of phraseness and informativeness helps alleviate the long-keyphrase challenge in open-domain extraction scenarios.

Figure 1(b) further illustrates the length distribution of keyphrases predicted by JointKPE and its two ablated versions, Chunk and Rank. The ablated model trained only with the keyphrase chunking task (Chunk) [39] is inclined to predict more single-word phrases. By contrast, the keyphrase length distributions predicted by our full model (Joint) and its ranking-only version (Rank) are more consistent with the ground-truth length distribution. The result further shows that globally capturing informativeness can guide the neural extractor to better determine the boundaries of truth keyphrases.

5.3 Performance w.r.t. Keyphrase Types

One of the main differences between keyphrases and entities is that not all keyphrases are knowledge graph entities, especially in open-domain KPE scenarios. From our observation of OpenKP, about 40% of ground-truth keyphrases are not entities, which accounts for a significant portion. As examples exhibited in Table 2, non-entity keyphrases often have higher-variant forms and may be more complicated to identify from the contexts. This experiment studies JointKPE's effectiveness for non-entity keyphrases, where its two ablated versions and another two BERT-based KPE baselines are used as the comparison.

Figure 2(a) and 2(b) illustrate the performance of different BERT-based KPE methods on entity keyphrases and non-entity keyphrases, where entities are identified by the CMNS linker [38]. Not surprisingly, all KPE methods with pre-trained models perform well in extracting entity keyphrases. Entity-style keyphrases with distinct and independent existence are easily recognized from the contexts. Also, pre-trained models might already capture the entity information and simple layers upon them can effectively identify entity keyphrases.

Nevertheless, the accuracy of all methods drops dramatically when extracting non-entity keyphrases, which evidently shows the challenge in identifying such high-variant keyphrases. Despite the challenge, JointKPE and its informative ranking version (Rank) significantly outperform other methods in predicting non-entity keyphrases. The results reveal that capturing the global informativeness of keyphrase candidates using learning-to-rank helps overcome some difficulty of non-entity keyphrase extraction.

5.4 Case Studies

Table 3 exhibits some extraction cases of long and non-entity keyphrases from JointKPE and its two ablated versions for the OpenKP dataset.

The first two cases show the long-keyphrase extraction scenarios. In the first case, the bi-gram keyphrase "walter lagraves" has been successfully predicted by all three methods, but the longer keyphrase "mosquito control board race" has only been extracted by JointKPE. Besides, JointKPE successfully predicted the longer keyphrase "surplus lines export eligibility" in the second case, while its two ablated versions only extracted the short one "compliance corner".

The remaining cases are non-entity examples. In contrast to entity keyphrases, non-entity keyphrases have more variable forms and may appear less frequently in contexts of the given documents, e.g., entity "fire port" and non-entity "dv to laptop" in the third case, and entity "north country realty" and non-entity "the beauty of the north" in the fourth case. All three methods have successfully predicted the entity keyphrases, but the non-entity ones were only extracted by the full version of JointKPE.

Table 3. Prediction cases of long and non-entity keyphrases targeting OpenKP. Ground-truth keyphrases are **underlined** in the document. The top three predictions of JointKPE (Joint) and its two ablated versions (Rank and Chunk) are exhibited. All three methods extracted **blue keyphrases**, while red keyphrases were only predicted by JointKPE's full version.

Long Keyphrase Cases
Document: ... **walter lagraves** on **mosquito control board race** i will not vote for by **walter lagraves** dear editor ralph depalma is a candidate for the board of the florida keys mosquito control authority mc lets think about ralph depalmas record ... **@Joint: walter lagraves; mosquito control board race**; mosquito control **@Rank:** mosquito control; **walter lagraves**; ralph depalma **@Chunk: walter lagraves**; ralph depalma; mosquito control board
Document: ... you can use in the know education central **compliance corner** case notes enforcement actions ... new laws may have been enacted subsequently compliance topic of the month **surplus lines export eligibility** the department has received numerous questions from agents seeking guidance on how to comply with ... **@Joint: compliance corner**; surplus lines; **surplus lines export eligibility** **@Rank: compliance corner**; surplus lines; insurance business **@Chunk: compliance corner**; export eligibility; surplus lines
Non-Entity Keyphrase Cases
Document: how to transfer **dv to laptop** without having a **firewire port** we will be using pinnacle moviebox update it does work ... here i compare the quality of a video transferred to the laptop via **firewire port** vs pinnacle moviebox ... **@Joint: firewire port, dv to laptop**, dv **@Rank: firewire port**, laptop, dv **@Chunk: firewire port**, dv, laptop
Document: ... **north country realty** welcome we deal in **the beauty of the north** we look forward to working with you soon whether you are selling or buying north country property **north country realty** is here to be of assistance to you we will gladly offer references from past clients and customers we have served ... **@Joint: north country realty**; oscoda; **beauty of the north** **@Rank: north country realty**; oscoda; north country property **@Chunk: north country realty**; north country property; real estate

6 Conclusion

This paper presents JointKPE, a multi-task architecture built upon pre-trained language models for open-domain KPE, which can capture both local phraseness and global informativeness when extracting keyphrases. Our experiments demonstrate JointKPE's effectiveness on both open-domain and scientific scenarios and across different pre-trained models. Comprehensive empirical studies further indicate that JointKPE can alleviate the problem of preferring shorter and entity-style keyphrases in previous neural KPE methods and exhibits more balanced performance on keyphrases of different lengths and various types.

Acknowledgments. This work is partly supported by National Natural Science Foundation of China (NSFC) under grant No. 61872074 and 61772122; Beijing National Research Center for Information Science and Technology (BNR2019ZS01005).

References

1. Ainslie, J., et al.: ETC: encoding long and structured inputs in transformers. In: Proceedings of EMNLP, pp. 268–284 (2020)
2. Alzaidy, R., Caragea, C., Giles, C.L.: Bi-LSTM-CRF sequence labeling for keyphrase extraction from scholarly documents. In: Proceedings of WWW, pp. 2551–2557 (2019)
3. Boudin, F., Gallina, Y., Aizawa, A.: Keyphrase generation for scientific document retrieval. In: Proceedings of ACL, pp. 1118–1126 (2020)
4. Campos, R., Mangaravite, V., Pasquali, A., Jorge, A.M., Nunes, C., Jatowt, A.: A text feature based automatic keyword extraction method for single documents. In: Proceedings of ECIR, pp. 684–691 (2018)
5. Caragea, C., Bulgarov, F., Godea, A., Gollapalli, S.D.: Citation-enhanced keyphrase extraction from research papers: a supervised approach. In: Proceedings of EMNLP, pp. 1435–1446 (2014)
6. Dalvi, N., Machanavajjhala, A., Pang, B.: An analysis of structured data on the web. In: Proceedings of the VLDB Endowment, vol. 5, pp. 680–691 (2012)
7. Devlin, J., Chang, M.W., Lee, K., Toutanova, K.: BERT: pre-training of deep bidirectional transformers for language understanding. In: Proceedings of NAACL, pp. 4171–4186 (2019)
8. Diao, S., Song, Y., Zhang, T.: Keyphrase generation with cross-document attention. arXiv preprint arXiv:2004.09800 (2020)
9. El-Beltagy, S.R., Rafea, A.: KP-Miner: a keyphrase extraction system for English and Arabic documents. Inf. Syst. **34**(1), 132–144 (2009)
10. El-Kishky, A., Song, Y., Wang, C., Voss, C.R., Han, J.: Scalable topical phrase mining from text corpora. In: Proceedings of the VLDB Endowment, vol. 8, pp. 305–316 (2014)
11. Florescu, C., Caragea, C.: A new scheme for scoring phrases in unsupervised keyphrase extraction. In: Proceedings of ECIR, pp. 477–483 (2017)
12. Hasan, K.S., Ng, V.: Automatic keyphrase extraction: a survey of the state of the art. In: Proceedings of ACL, pp. 1262–1273 (2014)
13. Jiang, X., Hu, Y., Li, H.: A ranking approach to keyphrase extraction. In: Proceedings of SIGIR. pp. 756–757 (2009)

14. Jones, S., Staveley, M.S.: Phrasier: a system for interactive document retrieval using keyphrases. In: Proceedings of SIGIR, pp. 160–167 (1999)
15. Joshi, M., Chen, D., Liu, Y., Weld, D.S., Zettlemoyer, L., Levy, O.: SpanBERT: improving pre-training by representing and predicting spans. TACL **8**, 64–77 (2020)
16. Kim, S.N., Baldwin, T.: Extracting keywords from multi-party live chats. In: Proceedings of PACLIC, pp. 199–208 (2012)
17. Kim, S.N., Medelyan, O., Kan, M.Y., Baldwin, T.: Automatic keyphrase extraction from scientific articles. Lang. Resour. Eval. **47**(3), 723–742 (2013)
18. Liu, J., Shang, J., Han, J.: Phrase Mining from Massive Text and its Applications. Synthesis Lectures on Data Mining and Knowledge Discovery, vol. 9, no. 1, pp. 1–89 (2017)
19. Liu, Y., et al. RoBERTa: a robustly optimized BERT pretraining approach. arXiv preprint arXiv:1907.11692 (2019)
20. Liu, Z., Huang, W., Zheng, Y., Sun, M.: Automatic keyphrase extraction via topic decomposition. In: Proceedings of EMNLP, pp. 366–376 (2010)
21. Liu, Z., Li, P., Zheng, Y., Sun, M.: Clustering to find exemplar terms for keyphrase extraction. In: Proceedings of EMNLP, pp. 257–266 (2009)
22. Medelyan, O., Frank, E., Witten, I.H.: Human-competitive tagging using automatic keyphrase extraction. In: Proceedings of EMNLP, pp. 1318–1327 (2009)
23. Meng, R., Zhao, S., Han, S., He, D., Brusilovsky, P., Chi, Y.: Deep keyphrase generation. In: Proceedings of ACL, pp. 582–592 (2017)
24. Mihalcea, R., Tarau, P.: TextRank: bringing order into text. In: Proceedings of EMNLP, pp. 404–411 (2004)
25. Mu, F., et al.: Keyphrase extraction with span-based feature representations. arXiv preprint arXiv:2002.05407 (2020)
26. Papagiannopoulou, E., Tsoumakas, G.: A review of keyphrase extraction. Wiley Interdisc. Rev. Data Min. Knowl. Disc. **10**(2), e1339 (2020)
27. Porter, M.F.: An algorithm for suffix stripping. Program **40**, 211–218 (1980)
28. Pudota, N., Dattolo, A., Baruzzo, A., Ferrara, F., Tasso, C.: Automatic keyphrase extraction and ontology mining for content-based tag recommendation. Int. J. Intell. Syst. **25**, 1158–1186 (2010)
29. Qazvinian, V., Radev, D., Özgür, A.: Citation summarization through keyphrase extraction. In: Proceedings of COLING, pp. 895–903 (2010)
30. Rousseau, F., Vazirgiannis, M.: Main core retention on graph-of-words for single-document keyword extraction. In: Proceedings of ECIR, pp. 382–393 (2015)
31. Sun, Z., Tang, J., Du, P., Deng, Z.H., Nie, J.Y.: DivGraphPointer: a graph pointer network for extracting diverse keyphrases. In: Proceedings of SIGIR, pp. 755–764 (2019)
32. Tomokiyo, T., Hurst, M.: A language model approach to keyphrase extraction. In: Proceedings of ACL, pp. 33–40 (2003)
33. Turney, P.D.: Learning algorithms for keyphrase extraction. Inf. Retrieval **2**(4), 303–336 (2000)
34. Wan, X., Xiao, J.: CollabRank: towards a collaborative approach to single-document keyphrase extraction. In: Proceedings of COLING, pp. 969–976 (2008)
35. Wang, Y., Fan, Z., Rose, C.: Incorporating multimodal information in open-domain web keyphrase extraction. In: Proceedings of EMNLP, pp. 1790–1800 (2020)
36. Witten, I.H., Paynter, G.W., Frank, E., Gutwin, C., Nevill-Manning, C.G.: KEA: practical automated keyphrase extraction. In: Design and Usability of Digital Libraries: Case Studies in the Asia Pacific, pp. 129–152 (2005)
37. Wolf, T., et al.: Transformers: State-of-the-art natural language processing. In: Proceedings of EMNLP, pp. 38–45 (2020)

38. Xiong, C., Callan, J., Liu, T.Y.: Bag-of-entities representation for ranking. In: Proceedings of ICTIR, pp. 181–184 (2016)
39. Xiong, L., Hu, C., Xiong, C., Campos, D., Overwijk, A.: Open domain web keyphrase extraction beyond language modeling. In: Proceedings of EMNLP, pp. 5178–5187 (2019)
40. Zhang, Y., Li, J., Song, Y., Zhang, C.: Encoding conversation context for neural keyphrase extraction from microblog posts. In: Proceedings of NAACL, pp. 1676–1686 (2018)
41. Zhang, Y., Chang, Y., Liu, X., Gollapalli, S.D., Li, X., Xiao, C.: MIKE: keyphrase extraction by integrating multidimensional information. In: Proceedings of CIKM, pp. 1349–1358 (2017)

Background Semantic Information Improves Verbal Metaphor Identification

Wentao Qin[1,2] and Dongyan Zhao[1,2](✉)

[1] Center for Data Science, AAIS, Peking University, Beijing, China
{qinwentao,zhaody}@pku.edu.cn
[2] Wangxuan Institute of Computer Technology, Peking University, Beijing, China

Abstract. Metaphor is frequently applied in human language. In natural language processing field, metaphor identification has long been studied. In this work, we focus on the verbal metaphor identification. Based on the observation that verbal metaphor occurs on the iteration between the verb and its subject/object, we propose to leverage the abundant information of the sentences containing the verb, named as background semantic information. We devise to leverage the background knowledge to improve verbal metaphor identification, and obtain a state-of-the-art performance in two public verbal metaphor identification datasets, MOH_X and Trofi. Further experiment analyses verify the effectiveness of our proposed method.

Keywords: Metaphor identification · Knowledge enhancement · Text representation

1 Introduction

Metaphor is a common phenomenon in language. In [24], metaphor is defined as a figure of speech in which the speaker makes an implicit comparison between seemingly unrelated things which nonetheless have certain common characteristics. For example, a metaphoric sentence *he always wears a smile* contains an implicit analogy from *smile* to *something wearable*. Given its ubiquity in our everyday communication, metaphorical language poses an important problem for natural language understanding [3,27].

Over the last couple of years, there has been an increasing interest towards metaphor processing and its applications, either as a part of natural language processing (NLP) tasks such as machine translation [12], text simplification [5] and sentiment analysis [22]. The central study with metaphor lies in metaphor identification, which has long been studied [15,21,23,27]. For metaphor identification, the majority of existing approaches, as well as the available datasets pertaining to metaphor processing, focus on the metaphorical usage of verbs and adjectives. The reason is these part-of-speech(POS) types exhibit metaphoricity more frequently than others according to corpus-based analyses [27]. In this

L. Wang et al. (Eds.): NLPCC 2021, LNAI 13029, pp. 288–300, 2021.
https://doi.org/10.1007/978-3-030-88483-3_22

work, following previous studies [8, 23], we place our focus on the task of verbal metaphor identification.

In a verbal metaphoric sentence, the big gap between the current subject (or object) and the typical subject (or object) of the metaphoric verb leads to a metaphoric expression. For instance, for the metaphoric sentence *he always wears a smile*, the typical object of *wears* is something such as the cloth or pants, which is far different from the current object *smile*, thus yielding a metaphoric expression. From this observation, being aware of the common usage of the subject/verb/object is natural for verbal metaphor identification. Based on this idea, aiming at verbal metaphor identification, we devise injecting a kind of knowledge, which we name as *background semantic information*, to boost the identification performance. The background semantic information provides the usage of the verb, the subject and the object of the verb to the model. With this knowledge in mind, the model is expected to have a better judge of the metaphoricity of the verb.

Our method shares a similar idea with [15], in which they use Glove [19] embedding of the verb as the general representation of it. Our method not only consider the verb, but also consider its subject and object. [29] showed that the type of syntactic construction a verb occurs in influences its metaphoricity, and thus taking the subject and object into consideration can be beneficial to metaphor identification.

In the experiment, we test our model with two public verbal metaphor identification datasets, MOH-X [17] and Trofi [1]. Our method achieve a new state-of-the-art level in these datasets. We also make further analyses, including ablation study, so as to verify the effectiveness of the proposed method.

2 Related Work

The majority of approaches to metaphor identification model the problem as a binary classification task, i.e., classifying linguistic expressions as metaphorical or literal. Early methods used hand-engineered features and relied on manually-annotated resources to extract them, including lexical and syntactic information [11] and higher-level features such as conceptual information [18], and WortNet supersenses [30].

In order to reduce the reliance on manual annotation, other researchers experimented with sparse distributional features [26] and dense neural word embeddings [2, 25]. [21] presents the first deep learning architecture for metaphor detection. [8] shows standard BiLSTM models which operate on complete sentences work well on detecting metaphorical word use in context. They propose a model that concatenates GLove and ELMo [20] representations which are then encoded by BiLSTMs.

Some approaches leverage certain intuition behind metaphor. [15] explores two linguistic theories of selectional preferences violation [31] and metaphor identification procedure [28] on metaphor identification. These two theories analyse the relations between metaphors and their context. [23] supposes verbal multi-word expressions (VWMEs) typically convey a non-literal meanings, and bolster verbal metaphor identification by knowledge of VWMEs.

Among above studies, [2,21,25,30] are relation-level processing. They take <adjective/verb, noun> pairs as input and judge their metaphoricity. [33] also targets on relation-level processing. However, they additionally use the sentences that contain the pairs, and use the sentences as context to discern metaphoricity. Others [13,15,23] focus on sentence-level (or word-level) processing, i.e., given a sentence, judging the metaphoricity of it (or certain words in it).

3 Method

3.1 Problem Formulation

Our work aims at sentence-level metaphor identification. Given an example $x = \{x_1, x_2, \ldots, x_n\}$ in a metaphor identification dataset, where n is the length of the example, and a target verb index i, the aim is to predict a binary label l to indicate the metaphoricity of the target verb x_i.

3.2 Pre-processing

Given an example, we first retrieve external sentences that contain the target verb. Then we seek out the subject and object of the target verb in the example and the retrieved sentences. We detail them in following two subsections.

Context Retrieval. We use Wikipedia corpus[1] as the source of the external sentences. To this end, we first split the Wikipedia corpus into sentences. In order to collect sentences that contain the target verb, we can naively apply exact matching, i.e., choosing those sentences with existence of the target verb. However, there are two defects with this approach. First, some verbs have extra POS. Second, a verb often occurs with certain tense in sentences, causing that the exact matching may miss some sentences. To conquer these problems, we stem the target verb and each word in the external sentences. Then we use the stemmed target verb to conduct exact matching. Meanwhile, we obtain POS of words in the sentences, and only keep the sentences that actually contain the target verb as a verb. The stemmer and POS tager tools are from NLTK library[2]. By now, for each target verb in the dataset, we retrieve up to K sentences that contain it.

[1] http://nlp.stanford.edu/data/WestburyLab.wikicorp.201004.txt.bz2.
[2] Natural Language Toolkit (NLTK, http://www.nltk.org/).

Subject and Object Extraction. The subject and object of a verb is helpful to identify its metaphoricity [29,33]. [34] employs Stanford dependency parser [4] to identify grammar relations, and extracts the verb-direct object (i.e., *dobj*) relations. However, this practice fails on many sentences, especially those with complex structure. In [34], the authors just discard those sentences without a recognizable *dobj* relation. To get a more complete extraction of subjects and objects, we turn to a more specialized technology, i.e., semantic role labeling, which is more suitable to automatically locate the subjects and objects.

In detail, we adopt practNLPTools[3] for semantic role labeling. This tool can recognize each verb given a sentence, and for each verb, it can locate its subject and object (if exist). In experiment, we discover that this tool produces empty results for a small part of sentences, such as when mistaking the target verb as a wrong POS. For those sentences that practNLPTools cannot handle, we propose a heuristic method. The heuristic method is based on a sketchy assumption, i.e., the subject is probably used before the verb, and the object appears probably after the verb. We thus design a program that takes the closest noun-phrase before the verb as the subject, and similarly, takes the closest noun-phrase after the verb as the object.

For examples in the dataset and retrieved sentences, we use the above approach to extract the subject and object for each target verb. Out of the complexity of language, the result provided by the extraction method is not perfectly precise, and thus may inject noise to our model. However, the correctness of the majority of extraction is able to weaken the effect of the noise [6].

3.3 Model

Input. We first describe the input to our model. After the pre-processing step, for each example in the dataset, we obtain up to K sentences from the Wikipedia corpus which all contain the target verb. Besides, for each target verb, we locate its subject and object in these sentences. For the convenience of narration, we define several variables as follow. For an example x^k in the dataset, it has a target verb v^k. The position of v^k in x^k is denoted as pos_v^k. In the processing step, we obtain the subject (s^k) and object (o^k) of v^k, whose positions are denoted as pos_s^k and pos_o^k. It should be noted that pos_s^k and pos_o^k are both lists, because they can be noun phrase containing more than one word. In addition, pos_s^k and pos_o^k might be empty, because in some sentences, there exists no subject or object for the target verb. We use a three-dimensional 0-1 vector w^k to indicate the existence of s^k, v^k and o^k, whose indices in w^k are 1, 2, 3 respectively. For instance, $w^k = [0, 1, 1]$ represents the example has no subject for the target verb.

As claimed before, for each example x^k in the dataset, we collect up to K sentences containing its target verb. This set of sentences is denoted as $\mathcal{C}^k$. Here, we aim to obtain vector representations for the target verb, its subjects and objects in $\mathcal{C}^k$. Compared with traditional word-specific embedding (e.g., word2vec [16]) in which each token is assigned a global representation, recent work, such as

[3] https://github.com/biplab-iitb/practNLPTools.

BERT [7], is able to produce contextualized representation. In this work, we choose the pre-trained BERT to get contextualized vector representation for each sentence. We group the vector representation of subjects, target verbs and objects in $\mathcal{C}^k$ into three groups, which are denoted as S^k, V^k and O^k. Above all, we will use x^k, pos_v^k, pos_o^k, pos_s^k, w^k, S^k, V^k, O^k as input to our model, and predict the metaphoricity of the target verb in x^k.

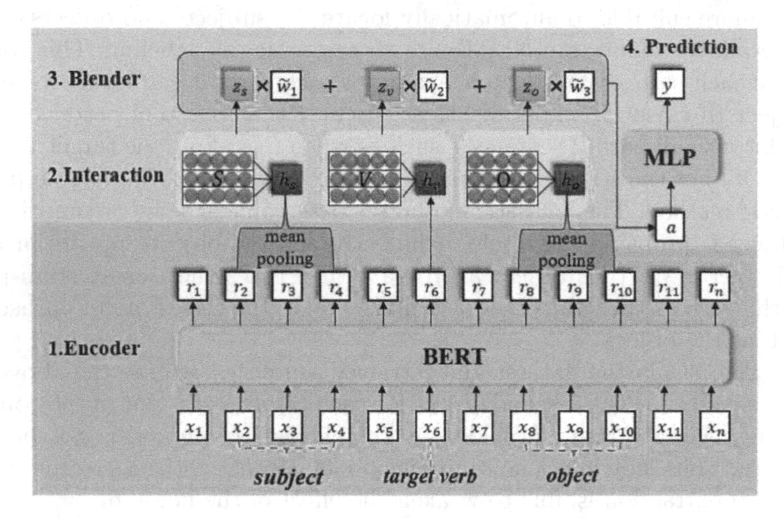

Fig. 1. The overview of our model. Our model can be roughly divided into four modules: 1. Encoder; 2. Interaction; 3. Blender; 4. Prediction. The detail of Interaction module is shown in Fig. 2. Note that in the this example, the subject is word sequences 'x_2 x_3 x_4', the target verb is x_6, and the object is word sequences 'x_8 x_9 x_{10}'. Note that their positions are different among different examples. For clear presentation, we omit the superscript k of variables in this figure. $\widetilde{w}_i$ is the normalized value of w_i.

Model. The model overview is shown in Fig. 1. Our model has four modules. We give detailed descriptions of each module in following subsections.

1. Encoder. We adopt BERT to encode the sentence x_k into vectors. The input to BERT is a sequence of words, corresponding to $x_1, x_2, \ldots, x_n$ in Fig. 1 (we omit the superscript k of variables for simplicity henceforth). For each word x_i, BERT will produce a vector r_i. For the target verb, we have its index pos_v, and thus the vector representation of the target verb, denoted as h_v, is equal to r_{pos_v}. For the subject and the object, their positions pos_s and pos_o are both lists. We then use mean pooling strategy to obtain their vector representation h_s and h_o.

2. Interaction. In this module, external vectors S, V and O are injected into our model. The interaction module takes S and h_s (or V and h_v, O and h_o) as input, and output a vector z_s (or z_v, z_o). Through the interaction operation, external information will be infused into the model. The Interaction module is

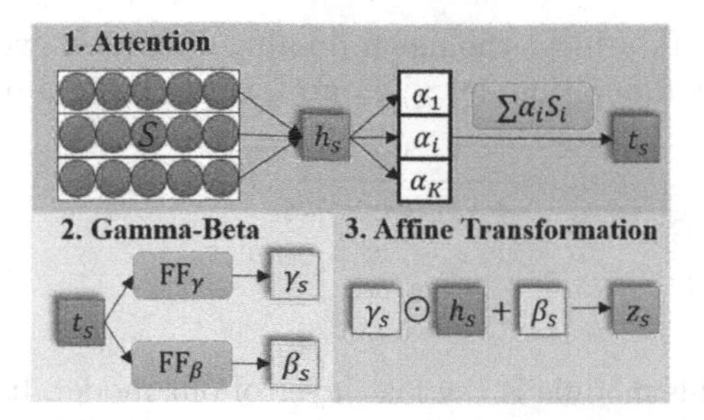

Fig. 2. The Interaction module. Here we take the S and h_s as input. Note that it's the same for V and h_v, O and h_o.

illustrated in Fig. 2. This module consists of three parts: 1. Attention; 2. Gamma-Beta; 3. Affine Transformation. The input is $S = \{S_1, \ldots, S_K\}$ and h_s. In the Attention phase, we compute the inner product between each S_i and h_s, and apply a softmax function to obtain a series of weights α_i:

$$p_i = Inner\_Product(S_i, h_s), 1 \leq i \leq K,$$
$$\alpha_i = \frac{exp(p_i)}{\sum_{k=1}^{K} exp(p_k)},$$

Afterwards, we obtain a vector t_s by weighted summation of S_i, i.e.,

$$t_s = \sum_{k=1}^{K} \alpha_i S_i.$$

Then, we input t_s into two different feed forward networks FF_γ and FF_β, whose output dimension is the same as t_s, to produce two vector γ_s and β_s. Afterwards, in the Affine Transformation part, we use γ_s and β_s to obtain z_s by:

$$z_s = \gamma_s \odot h_s + \beta_s,$$

where $\odot$ is an element-wise multiplication between two vectors. For the target verb and object part, we take the same occupation with two different Interaction modules, and then obtain z_v and z_o. Through Interaction module, we inject external information into the representations of subject, the target verb and object, thus obtaining representations of richer information.

3. Blender. Now that we have the representations for the subject, target verb and the object, denoted as z_s, z_v and z_o. We need to blend these vectors for the following label prediction. Here, we take mean pooling strategy to blend them. Noted that for some examples in the dataset, they don't have a subject or object. For this situation, the value of z_s or z_o is meaningless. Therefore, we leverage the three-dimensional indicator vector w (refer previous narration) to avoid the problem. To achieve the mean pooling operation, we first normalize w, and apply a weighted addition on z_s, z_v and z_o to obtain a output vector a:

$$\widetilde{w}_i = \frac{w_i}{w_1 + w_2 + w_3}, i = 1 \sim 3,$$
$$a = \widetilde{w}_1 z_s + \widetilde{w}_2 z_v + \widetilde{w}_3 z_o,$$

4. Prediction. This module is the last layer of our model. It takes the vector a as input, and goes through a Multi-Layer Perceptron (MLP) layer and a softmax layer to produce a 2-dimension vector y. The MLP in our model is a two-layer feed forward layers. In the training phase, we adopt cross entropy loss function on y. The loss $\mathcal{L}$ is:

$$\mathcal{L} = Cross\_Entropy(y, y'),$$

where y' indicates the true label. For the test phase, we use argmax operation on y to get the predicted label.

3.4 Training

Before training our model with loss $\mathcal{L}$, we initialize the parameters of the Encoder module with the pre-trained BERT module. Then we update the whole parameters until convergence of loss.

4 Experiments

4.1 Data Preparation

We apply our model on two different metaphor datasets: MOH-X [17], and TroFi [1]. Both datasets contain sentences in which a single verb token is labelled as metaphorical or not. There is also an index indicating the location of the target verb in the sentence. The statistics of the datasets are shown in Table 1.

Table 1. Statistics of MOH-X and Trofi datasets. # Expl., % Metaphor, # Uniq. Verb and Avg # Sent. Len refer to number of examples, sentence-level percentage, number of unique verbs and average sentence length respectively.

Datasets	# Expl.	% Metaphor	# Uniq. Verb	Avg # Sent. Len
MOH-X	647	49%	214	8.0
TroFi	3,737	43%	50	28.3

4.2 Baselines

We adopt several representative methods, including the state-of-the-art one as baselines to compare. The baselines can be divided into two categories: non-BERT (traditional approaches) [8,15] and BERT based [23].

RNN-ELMO [8] is a model that uses GloVe and ELMo as features for sequential metaphor identification. GloVe and ELMo are concatenated and encoded by BiLSTM, classified by a softmax classifier.

RNN-HG [15] adopt a linguistic theory named Metaphor Identification Procedure that a metaphor is identified if the literal meaning of a word contrasts with the meaning that word takes in this context [9,28]. They use Bi-LSTM hidden states as the contextual meaning representations, and pre-trained Glove as literal meaning representation.

RNN-MHCA [15], similar to RNN-HG, inspired by another linguistic theory that a metaphor is identified by noticing a semantic contrast between a target word and its context, which is the basis of Selectional Preference Violation [31]. They compare a target word representation and its attentive context representation to predict the metaphoricity of the target word.

BERT [7] Like in [23], BERT baseline is a vanilla pre-trained BERT with a classification layer added on top. The classifier uses the contextualized representation of the target verb for prediction.

BERT-GCN [23] This method use GCN [10] that uses dependency parse information on the top of BERT. Then like BERT baseline, it use the GCN representation of the target verb for prediction.

BERT-MWE [23], which acts as the state-of-the-art method in verbal metaphor identification. This method is inspired by an assumption that a metaphor classification model can be bolstered by knowledge of verbal multi-word expressions (MWE). They adopt another GCN to encode MWE knowledge on BERT-GCN, then use the concatenation of outputs from two GCNs for prediction.

4.3 Setup

In the pre-processing part, we retrieve up to 100 sentences for each example. During training, we truncate each example to length 96. The BERT in our model is the basic version of Google BERT, which is the same as in [23]. For all experiments, we use the pre-trained BERT of bert-base-uncased version from the *transformers* library [32]. As for optimizer, we use AdamW [14]. The batch size is 25. The learning rate grows linearly from 0 to 2e−5 in the first one-tenth steps, then decreases linearly to 0 in the remaining steps.

4.4 Result

Following common practice, we report the results in terms of accuracy, precision, recall and F1-score, macro averaged over the measures obtained from 10 fold cross-validation. The evaluation result is shown in Table 2. The results of other

Table 2. Evaluation on MOH-X and TroFi.

Model	MOH-X				TroFi			
	Acc	P	R	F1	Acc	P	R	F1
RNN-ELMO [8]	78.5	75.3	**84.3**	79.1	73.7	68.7	74.6	72.0
RNN-HG [15]	79.7	79.7	79.8	79.8	74.9	67.4	**77.8**	72.2
RNN-MHCA [15]	79.8	77.5	83.1	80.0	75.2	68.6	76.8	72.4
BERT [7]	78.04	78.38	77.87	77.82	70.38	70.54	68.89	68.84
BERT-GCN [23]	79.44	79.79	79.36	79.31	72.01	72.32	70.45	70.65
BERT-MWE [23]	80.47	79.98	80.40	80.19	73.45	73.78	71.81	72.78
Ours	**81.91**	**82.75**	81.78	**81.64**	**75.03**	**74.69**	74.15	**74.26**

models are copied from the corresponding papers. The result of BERT is copied from [23]. It can be seen that our proposed model outperforms the baselines and also surpasses the state-of-the-art in terms of every metric in both datasets. Compared to the state-of-the-art model, our model gains a 1.45 F-1 points growth in MOH-X and 1.48 F-1 points growth in TroFi, achieving a new state-of-the-art level. Compared with BERT baseline, our model has a big improvement, 3.77 F-1 points growth in MOH-X and 5.42 F-1 points growth in TroFi. Our model is built on the BERT baseline, with two additions. One is leveraging the position information of the subject and object. Another is injecting the background information. The results strongly demonstrate the effectiveness of our proposed model.

4.5 Analyses

Ablation. We do an ablation of different parts of our model to validate their usefulness. We mainly focus on two parts: whether to use the background information, and whether to use the position information of the subject and object. The result is presented in Table 4. The first block shows the result without using background information, including four settings: BERT, with Sub., with Obj. and with Sub.+Obj.. BERT baseline doesn't consider background information (Know.), position information of subject (Sub.) and object (Obj.). Compared with BERT, the scores of other settings are higher, which indicates the position information of the subject and object is helpful for metaphor identification. The growth brought by Obj. is larger than Sub.. The reason probably lies in that the object is easier to recognize in the pre-processing phase, as Table 3 shows. When combined with both subject and object information, the model gains a better result. The second block shows models using background information. The injection of background information brings large improvements compared with the models without it. The last row, which contains not only the background information but also the subject and object information, i.e., our whole model, gains the best result.

Table 3. The number of examples having Subject (Sub.), Object (Obj.) and both.

Datasets	MOH-X	TroFi
# Sub.	409	2,420
# Obj.	578	3,232
# Both	340	1,931

Table 4. Ablation Results. The first Column `Use Know.` represents whether to use background information. `w Sub.`, i.e., `with Subject`, refers to whether to use subject representation. Similarly, `w Obj.` refers to whether to use object representation. `w Sub.+Obj.` considers both.

Use Know.	Model	MOH-X				TroFi			
		Acc	P	R	F1	Acc	P	R	F1
No	BERT	78.04	78.38	77.87	77.82	70.38	70.54	68.89	68.84
	w Sub.	78.82	79.48	78.59	78.58	70.92	70.87	69.32	69.45
	w Obj.	79.28	79.83	79.34	79.15	71.12	71.04	69.76	69.78
	w Sub.+Obj.	79.75	80.18	79.63	79.57	73.29	70.09	71.53	70.37
Yes	BERT	80.53	81.27	80.43	80.31	73.80	73.75	72.39	72.62
	w Sub.	80.84	81.21	80.81	80.72	74.15	73.81	73.22	73.31
	w Obj.	81.00	81.72	81.01	80.83	76.63	74.44	73.48	73.67
	Ours (w Sub.+Obj.)	**81.91**	**82.75**	**81.78**	**81.64**	**75.03**	**74.69**	**74.15**	**74.26**

The Effect of K. During the pre-processing phase, we collect up to K sentences from Wiki corpus for each example. In this subsection, we aim to verify the effect of the size of external sentences. We set different K values. In each setting, we train our model and test it on the test dataset. Figure 3 shows the scores on MOH-X and Trofi with different K. The maximum value of K stops at 100, because for some verbs, it's hard to obtain more than 100 appropriate sentences containing them. It can be seen that the scores are positively correlated with K. In fact, bigger K means more background information, thus bringing richer semantic information into the model and leading better performance. In the same time, bigger K needs more external knowledge, and consume more computation resource. Therefore, choosing the size of K is a trade-off.

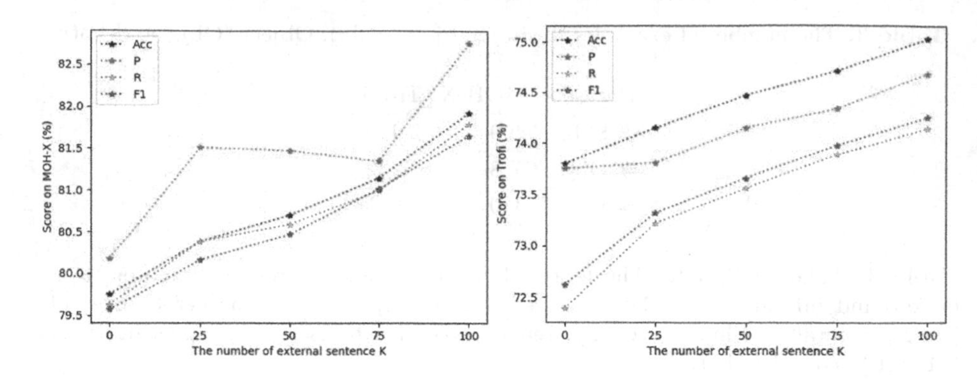

Fig. 3. The effect of K.

5 Conclusion

In this work, we explore leveraging external semantic information to improve the performance of metaphor identification. Based on the observation that verbal metaphor occurs on the iteration between the verb and its subject/object, we propose to leverage the abundant information of the sentences containing the verb, in order to boost metaphor identification. We obtain a state-of-the-art performance in two verbal metaphor identification datasets. Further analyses verify the effectiveness of our proposed method.

Acknowledgments. We thank the reviewers for their valuable comments. This work was supported by the National Key Research and Development Program of China (No. 2020AAA0106602).

References

1. Birke, J., Sarkar, A.: A clustering approach for nearly unsupervised recognition of nonliteral language. In: 11th Conference of the European Chapter of the Association for Computational Linguistics (2006)
2. Bulat, L., Clark, S., Shutova, E.: Modelling metaphor with attribute-based semantics (2017)
3. Cameron, L.: Metaphor in educational discourse. A&C Black (2003)
4. Chen, D., Manning, C.: A fast and accurate dependency parser using neural networks. In: Proceedings of the 2014 Conference on Empirical Methods in Natural Language Processing (EMNLP) (2014)
5. Clausen, Y., Nastase, V.: Metaphors in text simplification: to change or not to change, that is the question. In: Proceedings of the 14th Workshop on Innovative Use of NLP for Building Educational Applications, pp. 423–434 (2019)
6. Dai, H., Song, Y.: Neural aspect and opinion term extraction with mined rules as weak supervision (2019)
7. Devlin, J., Chang, M.W., Lee, K., Toutanova, K.: BERT: pre-training of deep bidirectional transformers for language understanding. arXiv preprint arXiv:1810.04805 (2018)

8. Gao, G., Choi, E., Choi, Y., Zettlemoyer, L.: Neural metaphor detection in context. arXiv preprint arXiv:1808.09653 (2018)
9. Pragglejaz Group : MIP: a method for identifying metaphorically used words in discourse. Metaphor & Symbol (2007)
10. Kipf, T.N., Welling, M.: Semi-supervised classification with graph convolutional networks (2016)
11. Klebanov, B.B., Leong, C.W., Gutierrez, E.D., Shutova, E., Flor, M.: Semantic classifications for detection of verb metaphors. In: Proceedings of the 54th Annual Meeting of the Association for Computational Linguistics (Volume 2: Short Papers), pp. 101–106 (2016)
12. Koglin, A., Cunha, R.: Investigating the post-editing effort associated with machine-translated metaphors: a process-driven analysis. J. Specialised Transl. **31**, 38–59 (2019)
13. Leong, C.W., Klebanov, B.B., Shutova, E.: A report on the 2018 vua metaphor detection shared task. In: Proceedings of the Workshop on Figurative Language Processing, pp. 56–66 (2018)
14. Loshchilov, I., Hutter, F.: Fixing weight decay regularization in ADAM (2018)
15. Mao, R., Lin, C., Guerin, F.: End-to-end sequential metaphor identification inspired by linguistic theories. In: Proceedings of the 57th Annual Meeting of the Association for Computational Linguistics, pp. 3888–3898 (2019)
16. Mikolov, T., Chen, K., Corrado, G.S., Dean, J.: Efficient estimation of word representations in vector space (2013)
17. Mohammad, S., Shutova, E., Turney, P.: Metaphor as a medium for emotion: an empirical study. In: Proceedings of the 5th Joint Conference on Lexical and Computational Semantics, pp. 23–33 (2016)
18. Mohler, M., Bracewell, D., Tomlinson, M., Hinote, D.: Semantic signatures for example-based linguistic metaphor detection. In: Proceedings of the 1st Workshop on Metaphor in NLP, pp. 27–35 (2013)
19. Pennington, J., Socher, R., Manning, C.D.: GloVe: global vectors for word representation. In: Proceedings of the 2014 Conference on Empirical Methods in Natural Language Processing (EMNLP), pp. 1532–1543 (2014)
20. Peters, M.E., et al.: Deep contextualized word representations. arXiv preprint arXiv:1802.05365 (2018)
21. Rei, M., Bulat, L., Kiela, D., Shutova, E.: Grasping the finer point: a supervised similarity network for metaphor detection. arXiv preprint arXiv:1709.00575 (2017)
22. Rentoumi, V., Vouros, G.A., Karkaletsis, V., Moser, A.: Investigating metaphorical language in sentiment analysis: a sense-to-sentiment perspective. ACM Trans. Speech Lang. Process. (TSLP) **9**(3), 1–31 (2012)
23. Rohanian, O., Rei, M., Taslimipoor, S., et al.: Verbal multiword expressions for identification of metaphor. In: Proceedings of the 58th Annual Meeting of the Association for Computational Linguistics, pp. 2890–2895 (2020)
24. Shutova, E.: Models of metaphor in NLP. In: Proceedings of the 48th Annual Meeting of the Association for Computational Linguistics, pp. 688–697 (2010)
25. Shutova, E., Kiela, D., Maillard, J.: Black holes and white rabbits: metaphor identification with visual features. In: Proceedings of the 2016 Conference of the North American Chapter of the Association for Computational Linguistics: Human Language Technologies, pp. 160–170 (2016)
26. Shutova, E., Sun, L., Korhonen, A.: Metaphor identification using verb and noun clustering. In: Proceedings of the 23rd International Conference on Computational Linguistics, Coling 2010, pp. 1002–1010 (2010)

27. Shutova, E., Teufel, S.: Metaphor corpus annotated for source-target domain mappings. In: LREC, vol. 2, p. 2. Citeseer (2010)
28. Steen, G.: A Method for Linguistic Metaphor Identification: From MIP to MIPVU, vol. 14. John Benjamins Publishing (2010)
29. Stowe, K., Palmer, M.: Leveraging syntactic constructions for metaphor identification. In: Proceedings of the Workshop on Figurative Language Processing (2018)
30. Tsvetkov, Y., Boytsov, L., Gershman, A., Nyberg, E., Dyer, C.: Metaphor detection with cross-lingual model transfer. In: Proceedings of the 52nd Annual Meeting of the Association for Computational Linguistics (Volume 1: Long Papers), pp. 248–258 (2014)
31. Wilks, Y.: A preferential, pattern-seeking, semantics for natural language inference. Artif. Intell. **6**(1), 53–74 (1975)
32. Wolf, T., et al.: Huggingface's transformers: state-of-the-art natural language processing (2019)
33. Zayed, O., McCrae, J.P., Buitelaar, P.: Contextual modulation for relation-level metaphor identification. arXiv preprint arXiv:2010.05633 (2020)
34. Zayed, O., Mccrae, J.P., Buitelaar, P.: Adaptation of word-level benchmark datasets for relation-level metaphor identification. In: The 2nd Workshop on Figurative Language Processing (2020)

Multimodality and Explainability

Towards Unifying the Explainability Evaluation Methods for NLP

Diana Lucaci(✉) and Diana Inkpen

University of Ottawa, Ottawa, ON, Canada
{dluca058,diana.inkpen}@uottawa.ca

Abstract. Neural models make decisions based on the features learned from the training data, each prediction being mostly influenced by one or more similar training data points. While extractive approaches of explainability for Deep Learning text classifiers focus on highlighting the most important parts of the input text, they do not take into consideration the training set. Furthermore, the interpretability methods proposed by the research community vary not only in methodology, but also in adopting different metrics for evaluating the explanations, which makes it difficult for users to compare and choose an explainable method appropriate for their use case. We propose a novel deep learning interpretability approach and a method-agnostic evaluation metric for measuring the added value of a feature-based explanation for natural language processing applications. The model architecture selects the phrase that influences a given prediction the most out of a set of key phrases extracted in an unsupervised manner from the training set. The evaluation approach enables one to compare natural language explanations based on the quantification of the faithfulness to the model, metric that we refer to as the contribution metric.

Keywords: Interpretability · Explainability · Deep learning · Natural language processing · Text classification

1 Introduction

The ubiquitous presence of automated decision making systems that have a performance comparable to humans brought the attention towards the need of interpretability of the generated decisions.

Although the terms explainability and interpretability are often used interchangeably in the literature, there is a subtle difference between them. While the explainability task focuses on explaining the inner working of a Machine Learning model, interpretability tries to draw insights about it by analyzing the impact of different input or parameter alterations such as feature subset selection, adversarial instances, or attention weights manipulation.

The present work applies the proposed evaluation metric on a novel explanation generation method that falls in the interpretability category of approaches,

L. Wang et al. (Eds.): NLPCC 2021, LNAI 13029, pp. 303–314, 2021.
https://doi.org/10.1007/978-3-030-88483-3_23

as it draws conclusions about the classifier by assessing the influence of a new phrase (previously extracted from the training set) on the prediction.

The architecture presented in this paper uses neural predictions to identify the phrase that has the highest contribution for a given input text, phrase that is later used for drawing insights about what the model considers relevant for making a prediction. When the generated explanation not only has a high contribution value, but it is also relevant to the context of the classified text, the model makes an informed decision and correctly captures the semantics of the text. In contrast, when a high-contribution explanation is unrelated to the context, the model is highly influenced by features that are irrelevant for making the correct prediction or even having opposite meaning relative to the task that the model is evaluated on. Such conclusions allow one to make decisions for semantic model comparison, to determine the most influential topics and keywords of the training set identified by the trained model, and to predict a model's behavior.

This novel research direction detaches from the previous extractive approaches that explain or interpret the model's behavior through the instance under classification or feature subsets of it, by exploiting the training dataset and selecting the most influential phrases for a given prediction.

2 Related Work

Previous approaches of models that output an explanation for their classification prediction are either supervised approaches for the explanation generation task, having human annotators providing gold-truth explanations [1] or use granular information to justify the prediction, approach that is common in multi-aspect sentiment analysis tasks [6], using datasets that contain both granular and overall scores. While such use cases benefit from having gold truth explanations, gathering labels for tasks such as document classification is often infeasible when new datasets are acquired.

Our proposed method differs from previous unsupervised explanation generation approaches [5] by generating a probability distribution over the most important features extracted from the training set, that are independent from the given classified instance. Our goal (new task) is not to detect what parts of the instance leads to a correct prediction, but rather what features of the training set are representative and help improving the classification confidence of a given classifier. Thus, this method reveals insights into both the most important features for a prediction and the patterns of the training dataset that the model relies on when making predictions such as true or false positives and negatives.

3 Explanation Generation

3.1 Classification Task and Dataset

The explained text classifier is trained for the binary sentiment analysis task, since this allows one to automatically determine the polarity of the phrases and measure its consistency with the prediction and with the gold truth label, using Sentiment Analyzers such as Valence Aware Dictionary sEntiment Reasoner – VADER [2].

The employed dataset is the IMDB [7] collection of movie reviews, that contains 25000 training and validation, and 25000 testing instances. This is a balanced dataset with reviews having an average length of 268 tokens. The number of positive labeled instances is approximately equal to the negative labeled instances, but the training set also contains lengthy reviews that contain both positive appreciations and criticism about the movies.

3.2 Corpus Exploitation

The training data is often overlooked by the current XAI approaches. The present research work addresses this by building a dictionary of explanation candidates extracted in an unsupervised manner from the collection of training instances, prior to the next stages of training the explanation generation model. For our experiments, we have used Rapid Automatic Keyword Extraction – RAKE [10] algorithm to build a dictionary of 590 4-word phrases – explanations candidates from which the generator chooses the most influential one for a given prediction of the classifier. The length of each explanation needs to be compelling and coherent. For the current dataset we have found that the length of 4 is appropriate, since the 2 or 3-word phrases do not contain enough information – leading to small contribution values that do not significantly influence the classifier – and longer phrases lose coherence as the algorithm combines nouns that are extracted from the same neighborhood, after removing linking and stop words. A large dimension of the dictionary leads to a higher semantic variety that can lead to more interpretable explanations for a large variety of new classified instanced, while a small dictionary reduces the complexity of the trained models, but, according to our experiments, do not cover the semantic variety of the training set, resulting in numerous negative contribution values (the maximum value contribution among the possible candidates is not faithful to the model's prediction).

3.3 Architectural Design

Bidirectional Recurrent Neural Networks [11] capture the context preceding and following the current token. For our experiments we derive insights for a bidirectional LSTM (biLSTM) text classifier – f, trained on the sentiment analysis task. With its weights frozen, the pretrained classifier is further used to obtain the prediction probability denoted by $f(x)$ for a classified instance $x = (x_1, x_2, ..., x_n)$. The two directions of the biLSTM lead to forward hidden sequence $\overrightarrow{h_t} = (\overrightarrow{h_1}, \overrightarrow{h_2}, ..., \overrightarrow{h_n})$ and a backward hidden sequence $h_t = (h_1, h_2, ..., h_n)$. The biLSTM representation (also represented in Fig. 1), which consists of the concatenation of the final forward and backward outputs $z_t = [\overrightarrow{h_t}, \overleftarrow{h_t}]$, is the input of the

MLP generator (G) that generates the probability distribution over the explanation candidates from the extracted dictionary. The probability distribution is a one-hot vector obtained by Categorical Reparametrization using Gumbel-Softmax [3], resulting in one explanation generated. We have also performed experiments with generating a soft distribution over the entire dictionary, but this approach leads to noisy representations of the entire dictionary that are ultimately ignored by the classifier at prediction time.

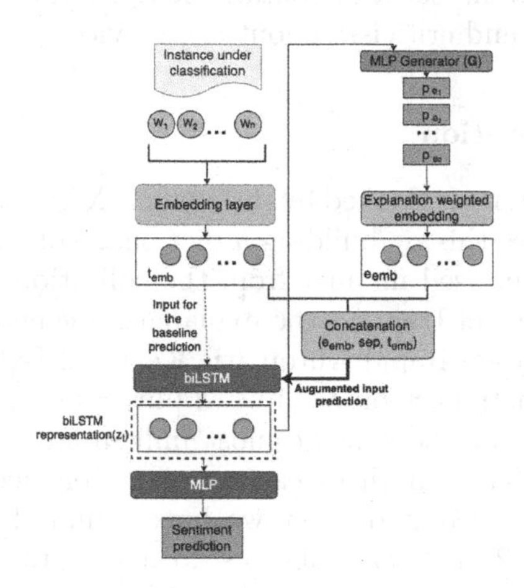

Fig. 1. Explanation generation architecture diagram. $w_1, w_2, ..., w_n$ – the words of a classified instance with n words, embedded into the text embedding t_{emb}. biLSTM – the explained text classifier. G - the MLP generator outputting the probability distribution $p_{e_1}, p_{e_2}, ..., p_{e_d}$, where d is the size of the dictionary of explanation candidates. e_{emb} – the weighted embedding of the explanation (according to the generated probability distribution). *sep* is the 0-vector separator between the original text and the newly added phrase used in order to minimize the impact of the contextual disruption. The biLSTM representation is the concatenation between the last hidden states: of the forward and backward LSTMs.

The MLP generator is trained in an unsupervised manner for the interpretability task, using as training guidelines (enforced through the loss function) the following desiderata:

1. to preserve the overall performance of the classifier when adding the explanations to the text instance;
2. to improve the confidence of the classifier for a given instance on the text classification task.

To achieve these desiderata, we propose a loss function that has two parts. The first one is the binary cross entropy applied between the target labels (y) and the predicted label after the explanation is added for classification: $\mathcal{L}_1(x) = BCE(f(e_x \cdot x), y)$ – where $e_x \cdot x$ denotes the concatenation between the explanation and the original input text (Fig. 1), and the second one is a custom loss function $\mathcal{L}_2 = 1 - C(e_x)$ that aims to maximize the contribution of the explanation e_x.

The final loss function is parametrized by a hyperparameter, α, that weights the two desiderata: $\mathcal{L} = \alpha\mathcal{L}_1 + (1 - \alpha)\mathcal{L}_2$.

The contribution of an explanation that has been added to the instance under classification x represents the added value in the correct prediction, a similar approach to the Shapley values from game theory [9]. For the binary sentiment analysis task, we propose the formula in Eq. 1, where f is the baseline classifier's prediction, e_x is the generated explanation for the instance x using the generator's transformation function, g, and $\cdot$ represents the concatenation operation. The higher the contribution coefficient, the more impact the explanation has on improving the confidence of the classifier towards the correct predicted label.

$$C(e_x) = \begin{cases} f(x \cdot e_x) - f(x), & \text{label} = \text{positive} \\ f(x) - f(x \cdot e_x), & \text{label} = \text{negative} \end{cases} \quad (1)$$

The proposed method compares the prediction confidence for a given test instance with the prediction for the instance concatenated with the corresponding generated explanation. Note that the presented formalization leads to a positive contribution value irrespective of the label, when the classifier is influenced towards the correct class (either positive or negative). For example, if the classifier predicts with a confidence of 0.6 that a given test instance is positive ($f(x) = 0.6$) and the instance together with the generated explanation is classified as 0.8 positive ($f(x \cdot e_x) = 0.8$), then the explanation's contribution is 0.2. While if the second prediction is 0.3, the contribution is -0.3, as the classifier makes a wrong prediction when the phrase is added. Similarly, for a negative instance, when $f(x) = 0.3$ and $f(x \cdot e_x) = 0.2$, then $C(e_x) = 0.1$, but when $f(x \cdot e_x) = 0.4$, $C(e_x) = -0.1$, since the explanation contributes towards the positive class.

4 Experimental Setting

For evaluating the proposed approach for the task of semantic model comparison, we choose two biLSTM models with comparable accuracy results, but with different hyperparameter settings.

The baseline classifiers: VLSTM and VLSTM+ are biLSTM models with 2 hidden layers of 64 neurons and 356, respectively, trained using a dropout rate of 0.3, and 0.5, respectively. The training of both models has been optimized using Adam Optimizer [4]. We trained embeddings initialized with the uniform distribution (using the seed 0 across all experiments).

eVLSTM and eVLSTM+ refer to the models enhanced with the explanation generator models and use the same MLP architecture: with 30 hidden layers with 512 neurons and 30 hidden layers with 256 neurons, trained using the dropout rate 0.7, Adam Optimization with decoupled weight regularization of 0.001 and 0.01, respectively, and learning rate 0.01, and 0.001 respectively. The hyperparameter α has been determined empirically by performing grid search, the optimal value being 0.7. During the generator's training, both eVLSTM and eVLSTM+ use the output of their corresponding explained pretrained classifier (VLSTM and VLSTM+) that have their weights frozen.

5 Evaluation

For evaluating the proposed approach, we design both quantitative and qualitative experiments, employing automatic and human evaluation methods for the model interpretability task, also assessing the capacity of the model to generalize and obtain positive contribution values on unseen data.

We further focus our analysis on the explanations with the highest contribution values in order to derive insights about the two considered models, since a higher contribution value translates into a higher degree of faithfulness to the model, thus the explanations carry more information relevant for interpretability. Irrespective of the relatedness to the context, or of the conveyed sentiment, the explanations that have the highest contribution values influence the classifier the most towards the correct prediction. Thus, when the model highly relies on an unrelated explanation or an opposite polarity explanation to make a prediction, we can conclude that the model is not able to capture the semantic meaning of the classified instance.

5.1 Quantitative Analysis

The two desiderata for the generator are being evaluated through automatic methods.

Classification Performance. Firstly, maintaining the performance for the sentiment analysis task is validated by the accuracy rate the models obtain on the test set for each model (Table 2). Both generators manage to preserve the prediction rate of the classifiers when the augmented reviews are used. Although the improvement is not significant compared to the baseline, both the eVLSTM and eVLSTM+ models manage not only to preserve the original correct classifications, but also to obtain a higher number of correctly changed predictions compared to the predictions that were influenced to make the wrong prediction. In addition to the changed predictions reported in Table 2, out of the 3036 instances misclassified by VLSTM+, 52.23% are converted to the correct prediction. While both of the classifiers have a rate of more than 54% correctly produced changes (out of all the changes), the eVLSTM+ benefits from a higher number of correctly changed predictions even though the missclassified instances are fewer.

Individual Contribution Values. Secondly, the individual contribution values obtained for the 25 000 test instances (Table 3), reflect the ability of the models to generate positive contribution values for unseen data points and prove their generalization power. Since the range of the bigger model (eVLSTM+) contains higher positive contribution values, the model seems to be able to capture more granular semantic information, being more influenced by the explanation candidates compared to the eVLSTM model. It is also worth noting that the review average length is of 268 tokens, while the added explanation has only 4 tokens, thus low values for the contribution metric are expected.

Table 1. Consistency of the sentiment polarity of the positive contribution explanations with the 1. label and 2. prediction of the enhanced models (eVLSTM, eVLSTM+). The values represent percentages of the positive contribution values for the correctly, incorrectly, and all the classified test instances that have the same sentiment polarity with the prediction and label, respectively.

Consistency		eVLSTM	eVLSTM+
With the prediction	All	56.65%	53.07%
	Correct	61.75%	53.93%
	Incorrect	23.42%	44.03%
With the label	All	63.73%	54.10%
	Incorrect	76.58%	55.97%

Table 2. Classification performance using the IMDB test set.

Model	Acc. (%)	Changed pred.	% Correct changes
VLSTM	85.69	–	–
eVLSTM	85.81	333	54.04%
VLSTM+	86.08	–	–
eVLSTM+	87.78	2751	57.65%

Table 3. Contribution values for the test set instances: range of values and the percentage of positive contributions out of all the test cases.

Model	Contribution range	Positive cont. (%)
eVLSTM	[−0.5004, 0.2810]	55.04
eVLSTM+	[−0.5682, 0.5549]	61.55

Semantic Model Selection Through Sentiment Polarity Analysis. For comparing the two models semantically, we analyze the percentages of the correctly and incorrectly classified instances with the prediction and label, reporting the results using the VADER Sentiment Analyzer for the sentiment evaluation of the explanation phrases in Table 1. For the incorrectly classified instances, the majority of the explanations are consistent with the label, rather than the prediction, which means that the added phrases have an added value insufficient for correcting the prediction. This phenomenon is more pronounced in the eVLSTM model's case, which translates into the model being influenced by phrases with the same sentiment as the expected label, while for the larger model, eVLSMT+, for almost half of the incorrect prediction, the model is very confident in making the wrong prediction. This fact is also supported by the higher consistency rate of the eVLSTM model for the correctly classified test instances. Overall, eVLSTM is inclined, in over 63% of the cases, to rely on expressions that have the same polarity with the gold truth, while the larger model that has a slightly better performance on the test set, in terms of accuracy, seems to be unsure and influenced by phrases with both polarities.

This analysis helps us make an informed decision when comparing the two models from a semantic point of view. eVLSTM is more confident in distinguishing the correct sentiment even for reviews that are more difficult to classify such as the ones that contain both positive and negative sentiment phrases, while eVLSTM+ is influenced by phrases that convey the opposite sentiment compared to the sentiment of the classified review.

5.2 Qualitative Analysis

For the qualitative analysis, we further explore the generated explanations and the predictive power they give to humans for interpretability. The human evaluation experiments described below involve two annotators (one of the authors and one MCS student). To remove any potential bias, we report below the metrics obtained by the external annotator, along with the agreement between the two annotators.

Model Interpretability
Experimental Setting. We select a sample of 100 highest contribution test instances. To also facilitate the model comparison experiment, we ensure that 50 of those instances have different predictions between the two classifiers and we use this smaller sample for the model comparison evaluation task as well.

Results. For the eVLSTM model, 86% human predictions matched the true label, while 50% matched the model's output. The low confidence of the classifier, together with an insufficient added value brought by the generated explanation lead to a lower predictive power for the annotators. On the other hand, for the eVLSTM+ model, 81% matched the true label, and 74% the model's output. The confidence of the larger model and the higher contribution values of the explanation lead to an increase in the changed predictions, which is also reflected by the higher percentage of cases where the annotators were able to predict the classifier's output.

The **agreement between the two annotators** obtained for the eVLSTM output's predictions is 82%, with a Cohen's kappa score [8] of 0.3539 (fair agreement) and for the eVLSTM+ model, the agreement is 89%, with a Cohen's kappa score of 0.6746 (substantial agreement).

Model Comparison
Experimental Setting. We use the sample of 50 instances for which the two models produced different outputs and we ask the annotators to choose, based on the generated explanations and their contribution values, the model that correctly classifies a given review.

Results. The challenge of this experiment was to choose examples that have high contribution values for both models, yet different predictions. The annotators found that 13 of those examples had less informative explanations, making it difficult to make an informed decisions. From the remaining, in 75.68% of the cases, the better performing model (the one correctly classifying the current instance) has been chosen, the agreement between the two annotators being of 74%.

This approach of model comparison allows one to choose the model based on the semantic interpretation of the generated explanations, rather than merely the accuracy rate obtained on a held-out dataset. Humans can employ this method

Table 4. Most influential explanations according to the prediction and true label for eVLSTM (most frequent explanations from the top 100 phrases with the highest contribution values). The emphasized phrases convey the same sentiment as the true label.

		True label	
		Positive	**Negative**
Prediction	Pos.	respectable affluent rich america wide white ready grin **leslie still look great** **king give impressive performances** **definitely worth seeing despite** raging alcoholic matt sorenson **great comedy writing team** **great curtis mayfield wrote** chance like john gulager	respectable affluent rich america wide white ready grin **rape scenes might seem** **depression following almost ruined** **subsequent cuban missile crisis** **less interesting stuff worthwhile** **whole affair tremendously disappointing** **eventually several people die** kung fu u scenes
	Neg.	wide white ready grin respectable affluent rich america **leslie still look great** **king give impressive performances** **physically impressive man capable** kung fu u scenes **indeed amazing filmic imagery**	wide white ready grin respectable affluent rich america leslie still look great beloved younger sister nettie **sending apparent death threats** **subsequent post traumatic stress** **average predictable romantic comedy**

for comparing the performance of competing models on a subset of data points and drawing insights about whether or not the model will be able to generalize on new unseen data.

For our task, we analyze the most frequently generated explanations from the annotated sample, according to the prediction and gold-truth label for each of the two models: eVLSTM (Table 4) and eVLSTM+ (Table 5). Note that these explanations add a certain weight (contribution) to the final prediction which may or may not change it. While for the VLSTM+ model, the most frequent explanations are very granular, we can continue to see the phenomenon of it relying on phrases that are in agreement with the prediction (even for the false positive and false negative cases). The model seems to be confident in what it had learned, but it does not always generalize well on more difficult instances. In contrast, the eVLSTM model relies on phrases that have the same sentiment as the gold truth for both the false positive and false negative predictions.

Table 5. Most influential explanations according to the prediction and true label for eVLSTM+ (most frequent explanations from the top 100 phrases with the highest contribution values)

		True label	
		Positive	**Negative**
Prediction	**Pos.**	really bad buddy movie one real bad movie post traumatic stress syndrome **anyone could easily enjoy** **big budget blockbuster faire** deliberately shoot people dead really stupid ending tacked **really great indie films** pretty painless ninety minutes **really well done honestly** events seem incredibly forced	glitzy hollywood plastic fantastic showcase magnificent special effects talented visual directors working **incredibly bad process work** **two uncomfortably prolonged assault** really well done honestly **horrifying shock therapy session** talented visual directors working **sending apparent death threats**
	Neg.	two uncomfortably prolonged assaults deliberately shoot people dead one particularly terrifying scene **gorgeous female actresses got** old man getting sick **greatest movies ever made**	**horrifying shock therapy session** heroine get happily married **events seem incredibly forced** former award winning actresses **deliberately shoot people dead** **fake death ending coming**

6 Conclusions

We proposed a novel approach of unsupervised neural explanation predictions for individual instances classified by biLSTM text classifiers, by generating the phrases of the training corpus that influence the prediction the most. Similarly to how humans justify and make decisions based on past experiences, the presented approach attempts to offer a better understanding of what features a pre-trained model uses from the training set to make a prediction.

Our methodology employs unsupervised key phrase extractions algorithms for the set of possible explanations, allowing one to obtain rich explanations that capture contextual semantics and also to adapt the method to other text classification tasks. We evaluated our approach through both quantitative analysis with the goal of preserving the initial classification rate of the text classifier that is being explained and through qualitative analysis for the generated explanations. Firstly, the newly-introduced contribution metric assesses the impact of the added explanation to the text that is being classified also measuring its faithfulness to the original model. Secondly, the coherence rate measures the agreement between the generated explanation's polarity and the input text's label. Furthermore, the human evaluation, together with the examples shown in Table 6, allow the user to get a better understanding of how the models perform and analyze samples of test instances for which the classifier does not provide the correct prediction.

Table 6. Examples of explanations from the model comparison task, correctly classified by the annotators. The raw predictions refers to the output of the original classifier. The bold text corresponds to the context similar to the explanation of the model with the correct prediction.

Review ID	VLSTM explanation and confidence	VLSTM prediction & Raw prediction	VLSTM+ explanation and confidence	VLSTM+ prediction & Raw prediction	Gold label
1	Though often considered Peter Sellers' worst film, it is in fact an excellent send-up of medical corporate corruption and abuses of power. [...] This film had **excellent performances** by Jo Ann Pflug and Pat Morita [...], but was marked by its ribaldary, irreverence, and total madcap demolition of the medical industry. [...] Peter Sellers movies are in demand by fans, but this effort, Where Does It Hurt?, has by its nature become almost impossible to find				
	Surprisingly strong performance playing (0.1726)	Positive (0.6404)	Talented actor producer director (0.0623)	Negative (0.3638)	Positive
2	Paul Naschy as a ghostly security guard in this is scarier than most of his fur-and-shoe-polish werewolf guises. [...] The thing is, that one of **these kid's fathers** did the same thing years ago but he's now deceased [...]. This is fairly well done for films of this type, and there's an air of mystery [...]. This moves along at a fairly good clip and doesn't let you lose interest like a lot of films do, and the oddball story is compelling enough to keep you interested too [...]. The ending is rather abrupt and I suppose is left mostly to your imagination, but then again it doesn't out-stay its welcome either. 7 out of 10, check it out				
	Wonderful character old wise (0.1174)	Positive (0.6676)	True romantic comedy classics (0.1702)	Negative (0.096)	Positive
3	Grand is the cost of a new car. A new car that Jake West now needs to escape the hordes of angry villagers desperate for his blood. Some may say this film could attract" So bad it's good" status. In my Opinion it is the proud owner of the "**So bad it's Bad**" label				
	Terribly boring new age (0.1472)	Positive (0.8527)	Truly stunned moved inspired (0.0499)	Negative (0.1180)	Negative
4	Clyde Bruckman borrows the premise of this short from Buster Keaton's' "Seven Chances," recently tepidly remade as "The Bachelor." [...] In this version, **musical teacher** Prof. Shemp has only 7 h (After all, it is a short!). This is one of the better Stooges shorts due to the storyline and wonderful routines [...]. I'm not a huge Stooges fan, but this one should be noted by any student of comedy as one of their very best since the early 30s shorts				
	Sympathetic grumpy shop owner (0.0552)	Negative (0.4383)	**Talented singer song writer (0.554)**	Positive (0.5985)	Positive

The presented results can be further extended and used to answer research questions about the model behavior for other text classification tasks such as multi-aspect sentiment analysis by analyzing subsets of explanations for correctly and incorrectly classified instances, contribution values and statistics about the generated explanations.

References

1. Camburu, O., Rocktäschel, T., Lukasiewicz, T., Blunsom, P.: e-SNLI: natural language inference with natural language explanations. CoRR abs/1812.01193 (2018)
2. Hutto, C., Gilbert, E.: VADER: a parsimonious rule-based model for sentiment analysis of social media text, January 2015

3. Jang, E., Gu, S., Poole, B.: Categorical reparameterization with Gumbel-Softmax. CoRR abs/1611.01144 (2016)
4. Kingma, D., Ba, J.: Adam: a method for stochastic optimization. In: International Conference on Learning Representations, December 2014
5. Lei, T., Barzilay, R., Jaakkola, T.S.: Rationalizing neural predictions. CoRR abs/1606.04155 (2016)
6. Liu, H., Yin, Q., Wang, W.Y.: Towards explainable NLP: a generative explanation framework for text classification. CoRR abs/1811.00196 (2018)
7. Maas, A.L., Daly, R.E., Pham, P.T., Huang, D., Ng, A.Y., Potts, C.: Learning word vectors for sentiment analysis. In: Proceedings of the 49th Annual Meeting of the Association for Computational Linguistics: Human Language Technologies, June 2011
8. McHugh, M.: Interrater reliability: the kappa statistic. Biochemia medica **22**, 276–282 (2012). https://doi.org/10.11613/BM.2012.031
9. Molnar, C.: Interpretable Machine Learning (2019). https://christophm.github.io/interpretable-ml-book/
10. Rose, S., Engel, D., Cramer, N., Cowley, W.: Automatic keyword extraction from individual documents, pp. 1–20 (03 2010). https://doi.org/10.1002/9780470689646.ch1
11. Schuster, M., Paliwal, K.: Bidirectional recurrent neural networks. IEEE Trans. Sig. Process. **45**(11), 2673–2681 (1997)

Explainable AI Workshop

Detecting Covariate Drift with Explanations

Steffen Castle[1(✉)], Robert Schwarzenberg[1], and Mohsen Pourvali[2]

[1] German Research Center for Artificial Intelligence (DFKI), Berlin, Germany
{steffen.castle,robert.schwarzenberg}@dfki.de

[2] Lenovo Research, Beijing, China
mpourvali@lenovo.com

Abstract. Detecting when there is a domain drift between training and inference data is important for any model evaluated on data collected in real time. Many current data drift detection methods only utilize input features to detect domain drift. While effective, these methods disregard the model's evaluation of the data, which may be a significant source of information about the data domain. We propose to use information from the model in the form of explanations, specifically *gradient times input*, in order to utilize this information. Following the framework of Rabanser et al. [11], we combine these explanations with two-sample tests in order to detect a shift in distribution between training and evaluation data. Promising initial experiments show that explanations provide useful information for detecting shift, which potentially improves upon the current state-of-the-art.

Keywords: Data drift · XAI · Two-sample tests

1 Introduction

Many AI models are trained on data that is gathered during an initial collection period and evaluated on data that is collected in real time. Real-time data is often subject to drift due to changes in data collection methodology, sampling differences, or a drift in underlying variables over time. Such drift can be problematic and may result in a degradation in the performance of the model. Therefore, detecting drift is important to ensure that a model performs optimally.

Covariate drift, which is a drift in the input domain, is commonly detected by comparing inputs from the training domain to the test domain using statistical tests. However, using only information from the inputs disregards another significant source of information about the data: The model itself, which is trained in the input domain.

One way to represent this information from the model is with explanations. Explanations are generated for a model to provide insight into a model's evaluation of an input. One type of explanation, attribution, denotes the importance

L. Wang et al. (Eds.): NLPCC 2021, LNAI 13029, pp. 317–322, 2021.
https://doi.org/10.1007/978-3-030-88483-3_24

of each input feature to the model's output on a data point. Many attribution techniques use model gradients as feature importance values, or more specifically, the gradient of the class output neuron with respect to the input feature [14]. Gradient times input [13] simply multiplies the attribution provided by the gradient by the input itself.

In order to use information from the model to detect drift, we introduce a model-based drift detection method in the framework of Rabanser et al. [11] that employs attributions in the form of gradient times input to detect data drift.

1.1 Related Work

Much work has already been done on detecting data drift. The main source of inspiration for our work, Rabanser et al. [11] employs two-sample tests, which compare datasets as samples from probability distributions in order to identify any divergence. Distribution-free tests are used to compare the distance between distributions and, based on this distance, determine the probability that the samples come from the same distribution. These tests are applied to representations of the input consisting of various dimensionality reductions.

Our method is not the first to use the model's representation of the data in detecting drift. Other methods, such as Black Box Shift Detection (BBSD)[9], make use of model output. BBSD was originally defined to detect prior-probability drift, but has also been applied to detect covariate drift [11]. Other methods use additional information from the model such as Elsahar et al. [2], where model confidence and reverse classification accuracy are used to detect drift conditioned on the model.

2 Methods

Our goal is to detect whether newly collected data has drifted compared to the initial training dataset. Formally, we compare samples from two distributions $X = \{x_1, x_2, ..., x_n\} \sim P_{train}(x)$ and $X' = \{x'_1, x'_2, ..., x'_n\} \sim P_{test}(x)$ to test the null hypothesis H_0 that the two samples come from the same distribution.

To accomplish this, we employ the MMD test [4] as employed by Rabanser et al. [11]. The MMD statistic represents the squared distance between the embedding means of distributions, that is:

$$\mathrm{MMD}(\mathcal{F}, p, q) = \|\mu_{\mathbf{P_{train}}} - \mu_{\mathbf{P_{test}}}\|_{\mathcal{F}}^2 \tag{1}$$

where $\mu_{p_{train}}$ and $\mu_{p_{test}}$ are the mean embeddings of the distributions P_{train} and P_{test} in a reproducing kernel Hilbert space $\mathcal{F}$. The MMD test is distribution-free, meaning that it does not require any prior knowledge of the distribution types of P_{train} and P_{test}. The MMD statistic on X and X' should be large when P_{train} and P_{test} are different; the MMD kernel matrix can also be used to calculate a p-value for H_0 using a permutation test. A threshold α is chosen such that if the p-value $\leq \alpha$, H_0 can be rejected. We choose the standard $\alpha = 0.05$.

Our contribution to this framework is to introduce attributions to the two-sample drift detection procedure. We perform two-sample tests on representations of the input as in Rabanser et al. [11]. However, instead of dimensionality reduction, we substitute the original inputs with attribution maps consisting of the gradient times input. The attribution map $\phi(x)$ for a data point x and model output $f(x)$ is defined for gradient times input as

$$\phi(x) = \frac{\partial f(x)}{\partial x} \cdot x \tag{2}$$

3 Experiments

Following the framework of Rabanser et al. [11], shifts are artificially induced in the inputs of the test set and representations are produced, which are then compared to the original validation set with two-sample tests[1]. In the image domain, we test the two main harmful types of drift identified by Rabanser et al. [11]: An image shift consisting of a random image translation of 5% or less, rotation of 10 degrees or less, and scaling of 10% or less (denoted small image shift in [11]), along with an adversarial shift, which replaces the test set with adversarial samples from FGSM [3]. For gradient times input, the model used to generate the explanations is a ResNet-50 model [5] trained on the train set of MNIST [8] with early stopping.

Input representations compared are those outlined by Rabanser et al. [11] plus the gradient times input methods. These representations consist of:

- No reduction (**NoRed**): A simple baseline consisting of original, unmodified inputs.
- Principal Components Analysis (**PCA**), Sparse Random Projection (**SRP**): Standard dimensionality reduction techniques detailed in [11]. As in the original paper, the input dimensionality is reduced to a size of 32.
- Autoencoder, Trained (**TAE**) and Untrained (**UAE**): The latent space of an autoencoder which has either been trained on the input domain, or has not received any training.
- Black Box Shift Detection, Softmax (**BBSDs**): The softmax-layer output of a model trained for classification on the training set.
- Gradient times input (**GradxInput**): The attribution produced by the gradient of the output neuron with highest activation with respect to the input multiplied by the input as in Eq. 2.

We aim to detect drift with high sensitivity, that is, with as low a number of samples as possible. Thus we compare results from different methods with varying random sample sizes $s \in \{10, 20, 50, 100, 200, 500, 1000\}$. Each test is performed 15 times and the mean p-value is then determined.

We also evaluate on a simple sentiment classification task consisting of a DistilBERT [12] model pretrained on the Large Movie Review Dataset (IMDB)

[1] Code available at https://github.com/DFKI-NLP/xai-shift-detection.

[10] of movie reviews labeled as either positive or negative. In this case, the shift consists of adversarial reviews provided by TextFooler [6]. Representations are compared at the embedding level, with the gradient times input calculated with respect to the embedding. For this experiment, the autoencoder representations were not evaluated. Additionally, since only 1000 adversarial examples were available, $s = 1000$ was not tested for this task.

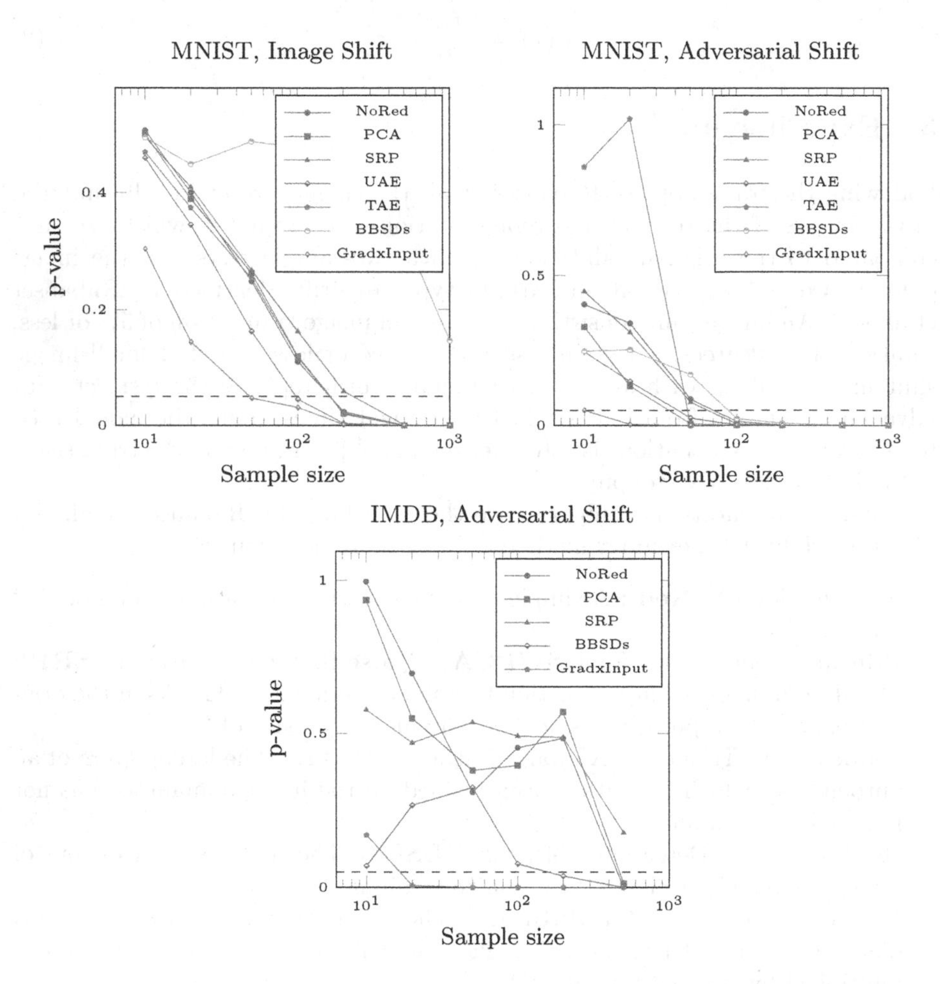

Fig. 1. Top: Mean p-values over 15 random runs of the MMD test for small image shift (left) and adversarial shift (right) with the MNIST dataset. In all experiments, the null hypothesis is rejected with a lower number of samples when input times gradient is used as the representation of the input. H_0 is rejected when the p-value ≤ 0.05. **Bottom:** Mean p-values for the IMDB adversarial experiment. A significant shift is detected with gradient times input at a lower sample size than any of the other representations.

4 Results

Results are shown in Fig. 1. The gradient times input methods generally outperform other methods tested in sensitivity: They are able to reject H_0 with an order of magnitude fewer samples. This method therefore improves over the baseline in sensitivity of detecting drift. The gradient times input seems especially useful for detecting adversarial samples with MNIST, as it detects these with very high sensitivity compared to other methods. It also greatly outperforms other representations when used to detect adversarial shift in the text task.

5 Discussion and Future Work

While initial results show promise, additional testing is needed to fully compare results to Rabanser et al. and establish the sensitivity of the method. Our experiments differ from Rabanser et al. in a few ways: First, we only use the multivariate MMD test, as the authors found it to have similar shift detection performance to the univariate KS test which they also evaluated. Future findings should also provide results on the KS test. In addition, we perturb all test samples, rather than fractions of the test set. Thus we do not evaluate the performance when only a subset of samples has shifted. We also do not evaluate on all types of shift identified by Rabanser et al. A more definitive comparison can only be established after all tests from the original paper are assessed.

Testing on a wider variety of model architectures and datasets would also provide more opportunity to demonstrate the strong performance of this method. In further testing, models such as BERT [1] should be evaluated in the text domain for other tasks such as question answering. In addition, other types of explanations such as PatternAttribution [7] or SmoothGrad [15] can be evaluated in place of gradient times input, as they have been found to reduce noise in explanations. Comparing against other baselines [2][9] is also important to show that this method is state-of-the-art.

6 Conclusion

We demonstrate initial promising results which show improvement upon the framework employed by Rabanser et al. Representing data using explanations from the model in the form of gradient times input provides additional information about the data domain for two-sample tests, and helps improves shift detection sensitivity beyond the performance seen by the baseline.

References

1. Devlin, J., Chang, M.W., Lee, K., Toutanova, K.: Bert: pre-training of deep bidirectional transformers for language understanding. arXiv preprint arXiv:1810.04805 (2018)

2. Elsahar, H., Gallé, M.: To annotate or not? predicting performance drop under domain shift. In: Proceedings of the 2019 Conference on Empirical Methods in Natural Language Processing and the 9th International Joint Conference on Natural Language Processing (EMNLP-IJCNLP), pp. 2163–2173 (2019)
3. Goodfellow, I.J., Shlens, J., Szegedy, C.: Explaining and harnessing adversarial examples. arXiv preprint arXiv:1412.6572 (2014)
4. Gretton, A., Borgwardt, K.M., Rasch, M.J., Schölkopf, B., Smola, A.: A kernel two-sample test. J. Mach. Learn. Res. **13**(1), 723–773 (2012)
5. He, K., Zhang, X., Ren, S., Sun, J.: Deep residual learning for image recognition. In: Proceedings of the IEEE Conference on Computer Vision and Pattern Recognition, pp. 770–778 (2016)
6. Jin, D., Jin, Z., Zhou, J.T., Szolovits, P.: Is bert really robust? natural language attack on text classification and entailment. arXiv preprint arXiv:1907.11932 (2019)
7. Kindermans, P.J., et al.: Learning how to explain neural networks: patternnet and patternattribution. arXiv preprint arXiv:1705.05598 (2017)
8. LeCun, Y., Cortes, C.: MNIST handwritten digit database (2010). http://yann.lecun.com/exdb/mnist/
9. Lipton, Z., Wang, Y.X., Smola, A.: Detecting and correcting for label shift with black box predictors. In: International Conference on Machine Learning, pp. 3122–3130. PMLR (2018)
10. Maas, A.L., Daly, R.E., Pham, P.T., Huang, D., Ng, A.Y., Potts, C.: Learning word vectors for sentiment analysis. In: Proceedings of the 49th Annual Meeting of the Association for Computational Linguistics: Human Language Technologies, pp. 142–150. Association for Computational Linguistics, Portland (2011). http://www.aclweb.org/anthology/P11-1015
11. Rabanser, S., Günnemann, S., Lipton, Z.C.: Failing loudly: an empirical study of methods for detecting dataset shift. arXiv preprint arXiv:1810.11953 (2018)
12. Sanh, V., Debut, L., Chaumond, J., Wolf, T.: Distilbert, a distilled version of BERT: smaller, faster, cheaper and lighter. CoRR abs/1910.01108 (2019). arXiv:1910.01108
13. Shrikumar, A., Greenside, P., Kundaje, A.: Learning important features through propagating activation differences. In: Proceedings of the 34th International Conference on Machine Learning, vol. 70, pp. 3145–3153. JMLR. org (2017)
14. Simonyan, K., Vedaldi, A., Zisserman, A.: Deep inside convolutional networks: visualising image classification models and saliency maps. CoRR abs/1312.6034 (2014)
15. Smilkov, D., Thorat, N., Kim, B., Viégas, F., Wattenberg, M.: Smoothgrad: removing noise by adding noise. arXiv preprint arXiv:1706.03825 (2017)

A Data-Centric Approach Towards Deducing Bias in Artificial Intelligence Systems for Textual Contexts

Shipra Jain(✉), Sreya Dey, Prateek Bajaj, and A. S. Rohit Kumar

SAP Labs India Pvt. Ltd., Bangalore, India
{shipra.jain01,sreya.dey01,prateek.bajaj,rohit.kumar.a.s}@sap.com

Abstract. With the ever-growing usage of artificial intelligence in business decisions, organizations, and the day-to-day lives of the society at large, it is imperative that different forms of biases that human beings are privy to do not get trickled to the machine learning and artificial intelligence systems being developed and used in abundance.

This paper proposes the use of explainable AI features to quantify data bias, which can assist in better decision-making. First, this paper provides references to the kinds of biases that exist in the world today, and how they impact how decisions are made in the real-world. Next, the paper maps the underlying causes of such biases that get trickled to artificial intelligence systems through the forms of data and how they impact decision making of the end-users. Once this context has been defined, the paper proposes a list of approaches to understand the causes and effects of such biases in data. Explainable AI and natural language processing techniques are applied to determine and quantify bias in the form of bias index in textual data.

With the help of said bias indices, a data scientist using such data can use such indices to determine whether there are biases existing in the datasets being using to train the machine learning models. This step allows the data scientists to make informed decisions on how to process such data to avoid several biases even before the actual training of the machine learning models begin.

Keywords: Explainable AI · Gender bias · Data bias

1 Introduction

1.1 Problem Statement

Artificial Intelligence and Machine Learning model generation relies heavily on the data that is actually being used to train such models. One of the most important characteristics of such techniques of prediction is the quality (as well as the quantity) of the data that is used to train the models. As rightly pointed out in [1], the importance of the quality of data is beyond just the core pre-processing. There are a lot of factors that need to be kept in mind while handling

L. Wang et al. (Eds.): NLPCC 2021, LNAI 13029, pp. 323–333, 2021.
https://doi.org/10.1007/978-3-030-88483-3_25

such data to make sure that the predictions being made by the machine learning models trained by such data are as expected by the data scientists.

What is also important while looking at the quality of data is the actual 'sanity' of data, as discussed by Talman et al., 2021 in [2]. From a machine's perspective, there are very few ways of defining whether the data that are being fed into a model are 'sane', i.e., the data does not contain any malice, any bias, or any other form of discrimination for or against a particular characteristic or a group of characteristics in the dataset. However, human beings should be responsible for making sure that the sanity of such data is kept before actually training a machine to predict outcomes from the same. This is a debate that has been far discussed for years, with abundant research on the subject, as explored in [3–5]

Although there is various research available on how biases can be detected, mitigated, and in turn resolved from artificial intelligence systems, this paper looks at a quantifiable approach that takes into account various characteristics of what datasets provide along with explainable artificial intelligence techniques and indexing of various kinds of biases that exist in the real-world to formulate an algorithm that suggests whether the data that is being used to train a particular model are biased.

1.2 Research Proposal

In this research, the authors propose to first understand the core principle and definition of the term 'bias' with respect to data and algorithms used to build machine learning models, then understand the causes for such bias. Next, the paper understands different types of biases and how they impact the outputs and predictions of such models differently.

The topic of Explainable AI (henceforth referred to as XAI) has also been investigated in-depth to formulate multiple ways to understand the behaviour of machine learning models and their training data, as comprehensively consolidated in [6] and [7]. There are multiple ways to explain model and data behaviour, and this paper considers a few common practices and open-source technologies that can be leveraged for the same.

Explainable AI can be a combination of both technical and manual ways of understanding what a particular model would predict, and why the model is predicting such outputs. This paper leverages multiple such available and calculated metrics for different data sets. Once such metrics are calculated, the aim of the paper is to define a quantifiable metric (say, a bias score) for each data set. This technique is used for textual data by applying explainable AI techniques with the addition of natural language processing techniques to determine if bias exists in textual data. Further, the results are validated by detecting bias indices for tabular data as well.

2 Existing Literature

2.1 Defining Bias

The following is the definition of bias determined as per existing and relevant research:

The concept of bias in artificial intelligence and machine learning originates from the common term 'bias' that is used in human terminology as anything that leads to any form of discrimination originating for or against a person, group, or characteristic. More so, in terms of decision making, analyses that are different in ethics based on certain sensitive attributes can also qualify as biased. The different types of biases can be broadly categorized as follows:

1. **Interaction Bias:** This bias talks about the basic concept of bias added to a machine learning model by way the human interacts with it. For instance, if a person is asked to draw a shoe, each person would have their own way of drawing what a shoe looks like for them. However, there may be more ways of defining a shoe than some particular styles. This kind of data skews the model into only a specific set of types.

2. **Latent Bias:** This bias can be defined by way of explaining the fact that some algorithms/data may intrinsically associate some terms with certain races, genders, etc. As highlighted in [8], a hiring algorithm only recommended male employees because it was trained on historical patterns in which men were primarily hired. This is a rampant bias widespread in datasets due to historical reasons and prejudice, mainly with respect to gender and race.

3. **Selection Bias:** This bias talks about training the algorithm with an over-represented set of data. Such inherently biased datasets contribute to skewed and inaccurate results as explored in [9], wherein the bias in the dataset was unintentionally propagated to the trained model as well.

The perception and conceptualization of bias tend to vary from author-to-author and can substantially influence bias detection techniques as discussed in [10].

2.2 Understanding Explainable AI

Explainable AI is another important topic that needs to be looked at while discussing bias deducing techniques. This is important because this research relies heavily on using explainable AI metrics to deduce how data behaves.

Explainable AI is aimed at providing human-interpretable insights into the performance and working of complex systems and algorithms. It generally employs the use of associating fundamental concepts of interpretability, feature importance, completeness etc., to the trained model by attempting to create a simplified and high-level representation of the system. From a broad overview, some of the most relevant Explainable AI features applicable to this paper are:

1) **Interpretability:** A core tenet of Explainable AI is to produce explanations and outputs in a simple manner that humans can understand. Not only does this help in easy creation of white-box models but also assists in enhancing the fairness of the model, as elaborated in [11].
2) **Feature Importance:** Assigning scores to the input features that indicate the relative importance in predicting the target feature.
3) **Human evaluation:** The results and inferences must be judged by humans for efficacy and validation.

3 Our Proposed Work

In this paper, the authors have considered gender bias to demonstrate our methodology for latent biases in textual data. The authors propose that similar methods can be applied for other forms of biases and dataset types as well.

3.1 Quantifying Bias in Textual Data

Models trained on biased text can display implicit biases in many forms such as gender, race, emotion, propaganda etc. Most work done on textual bias consists of detection and mitigation in the intermediate representation of the learning models. In this paper, we approach the problem at the grassroots by attempting to identify bias at the sentence level i.e., at the dataset level. At sentence-level detection, the approaches for each bias become quite different and non-generic. Hence, we mainly look at gender bias throughout the course of the paper.

Widespread NLP applications such as question-answering, analogy identification etc., that are trained on textual data, are prey to innate gender biasing. As evident in [12], models trained on biased text displayed sexist behaviour, few of which are re-narrated below.

E.g.: The words **babysit**, **sewing** have higher affinity towards the female gender and analogy queries such as **fighting: man :: babysit:__** would be answered with the female word from a pair of correlated male-female words i.e., **grandmother** would be selected as the better choice among [**grandmother, grandfather**] pair. Similarly, **king : man :: doctor :__**, was answered with the male gendered words.

Such innate biases are learnt by models from sexist and gender-biased sentences present in the learning data. A few examples are:

- A **doctor** must always be alert and humble, no matter the amount of **his** past success.
- A **nurse**'s duty towards **her** patients is one of compassion and reassurance.

When such biased sentences are converted to intermediate representation such as vector-space models or word embeddings, the vectors for such neutral nouns and pronouns associate closer with the biased gender and start accumulating gender bias. We propose the following approach to identify and quantify such gender generalizations and explicit sexist sentences.

3.1.1 Proposed Methodology

3.1.1.1 ***Pre-processing:*** Standard text pre-processing techniques were applied to improve performance. We remove stop-words and punctuations and expand contractions of auxiliary verbs. Lemmatization and Stemming were avoided to preserve the structure of the words.

3.1.1.2 ***Identify Key Sentences:*** Since it's infeasible to check for gender bias in every sentence, we hypothesize that it is sufficient to check only in the most important and relevant sentences. To identify the key sentences, we first identify the keywords/key n-grams using an unsupervised keyword extraction method based on various statistical features of the text. For this purpose, we use **YAKE** [13], but other extraction methods based on feature importance will also be sufficient. This brings in the concept of explainable AI into our research. This technique of detecting the most relevant keywords/phrases from the text input is referred to as deriving feature importance from the data.

The concept of feature importance is used to derive the most important keywords from the text input by assigning weights to the vectors being generated for each of the word inputs. Next, we pick the highest ranking keywords from the dataset to find out the most relevant features, and in turn the most relevant sentences to which these features belong. We identify this list of sentences from the text that contain at least one keyword/key n-gram. Figure 1 shows the list of keywords and their weights for the sample text in Fig. 2

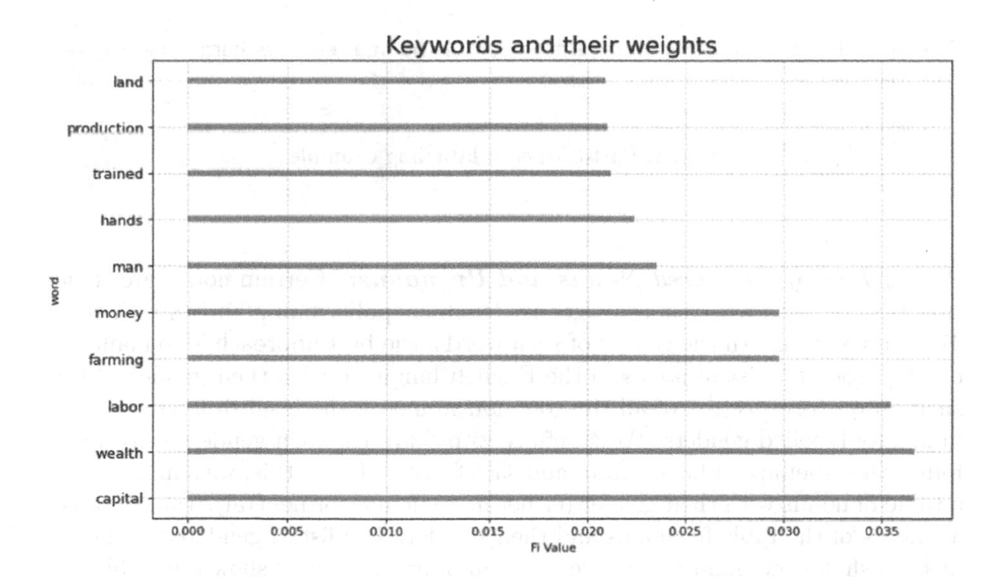

Fig. 1. Keywords and their weights

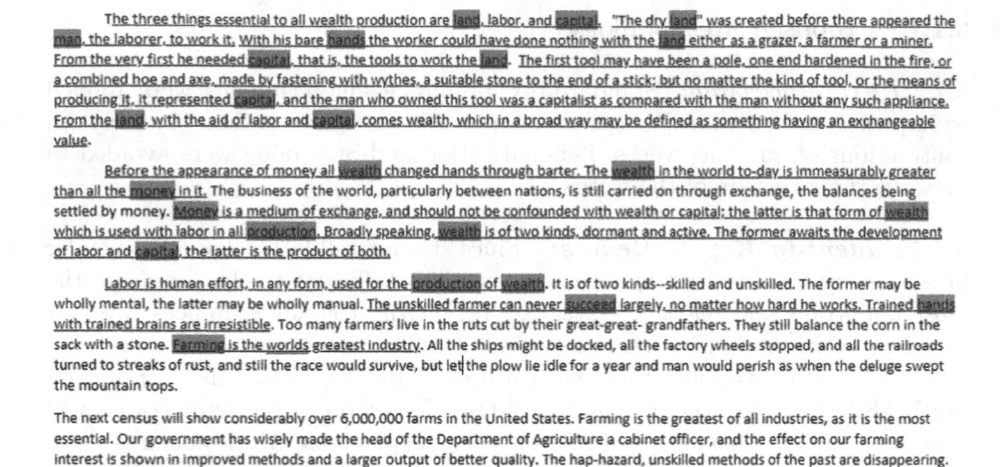
The three things essential to all wealth production are land, labor, and capital. "The dry land" was created before there appeared the man, the laborer, to work it. With his bare hands the worker could have done nothing with the land either as a grazer, a farmer or a miner. From the very first he needed capital, that is, the tools to work the land. The first tool may have been a pole, one end hardened in the fire, or a combined hoe and axe, made by fastening with wythes, a suitable stone to the end of a stick; but no matter the kind of tool, or the means of producing it, it represented capital, and the man who owned this tool was a capitalist as compared with the man without any such appliance. From the land, with the aid of labor and capital, comes wealth, which in a broad way may be defined as something having an exchangeable value.

Before the appearance of money all wealth changed hands through barter. The wealth in the world to-day is immeasurably greater than all the money in it. The business of the world, particularly between nations, is still carried on through exchange, the balances being settled by money. Money is a medium of exchange, and should not be confounded with wealth or capital; the latter is that form of wealth which is used with labor in all production. Broadly speaking, wealth is of two kinds, dormant and active. The former awaits the development of labor and capital, the latter is the product of both.

Labor is human effort, in any form, used for the production of wealth. It is of two kinds--skilled and unskilled. The former may be wholly mental, the latter may be wholly manual. The unskilled farmer can never succeed largely, no matter how hard he works. Trained hands with trained brains are irresistible. Too many farmers live in the ruts cut by their great-great- grandfathers. They still balance the corn in the sack with a stone. Farming is the worlds greatest industry. All the ships might be docked, all the factory wheels stopped, and all the railroads turned to streaks of rust, and still the race would survive, but let the plow lie idle for a year and man would perish as when the deluge swept the mountain tops.

The next census will show considerably over 6,000,000 farms in the United States. Farming is the greatest of all industries, as it is the most essential. Our government has wisely made the head of the Department of Agriculture a cabinet officer, and the effect on our farming interest is shown in improved methods and a larger output of better quality. The hap-hazard, unskilled methods of the past are disappearing. Science is lending her aid to the tiller of the soil, and the wise ones are reaching out their hands in welcome.

Fig. 2. Sample text to demonstrate the algorithm's working. Keywords are highlighted in yellow. Sentences containing keywords are underlined

3.1.1.3 ***POS-Tagging:*** On each of the key sentences, perform Part-Of-Speech (POS) tagging to extract the nouns and pronouns. We use the open-source module spacy to accomplish this but any standard library will suffice. Figure 3 depicts part-of-speech labelling on an example sentence.

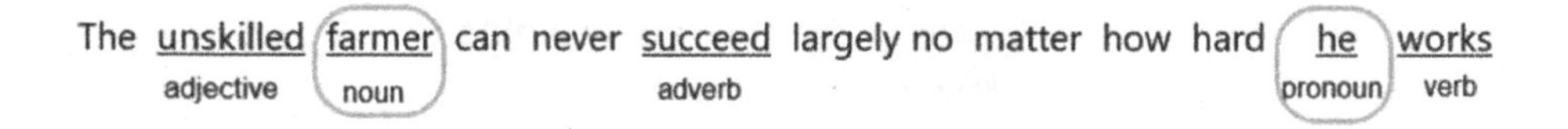

Fig. 3. Part-of-speech labelling example

3.1.1.4 ***List of Gendered Nouns and Pronouns:*** Certain nouns are structurally embedded with a specific gender (such as 'policeman', 'air hostess', etc.). To be able to discern the gender of such words, the best approach is to manually create/procure a list of nouns in the English language with their gender. [14] is such an open-source list of all the common nouns in the English language with manually labelled genders. We carefully went through each gendered word and found the labelling to be accurate and satisfactory. Using this, we simply create a table of nouns with their gender (either male, female or neutral). Table 1 shows a sample of the table for nouns and their gender. The list of gendered pronouns in English is very small and hence was manually created as shown in Table 2.

3.1.1.5 ***Cluster Identification:*** We then perform coreference resolution using the open-source coreference-resolution technique based on [15], on the sentence

Table 1. Example of the list of gendered nouns

astronaut	neutral
chariman	male
headmistress	female

Table 2. List of gender-specific pronouns

Male	['He', 'His', 'Him', 'Himself']
Female	['She', 'Her', 'Hers', 'Herself']

to identify the pronouns in the sentence and the nouns to which they map to. The output at this stage would look like the following example:

Sentence: The unskilled farmer can never succeed largely no matter how hard he works.
Cluster(s): [he, farmer]

3.1.1.6 ***Identify Gender Mismatches:*** In the clusters generated for every sentence, if at least one pronoun exists, perform the following two checks:

(i) **Male-pronoun and non-male noun**

E.g.: [he, laborer], [him, housewife]. In both the cases, the pronoun is a male word. But in the former's case, it refers to the neutral word 'farmer', while in the latter's case, it refers to the female word 'housewife'.

(ii) **Female pronoun and non-female noun**

E.g.: [she, chairman], [her, nurse]. The former has the female pronoun referring to a male noun while the latter has it referring to a gender-neutral noun.

In case of no pronouns existing in the sentence, we assume it to be unbiased with respect to gender. Figure 4 shows the detected biases for the text in Fig. 2. Finally, the authors calculate the percentage of sentences in which at least one gender bias was identified to quantify the overall "gender-bias" metric of the text. The algorithm, as demonstrated in Algorithm 1 explains how the bias index is calculated.

```
detected bias : male
{pronoun : noun}, {' his ' :' worker ' }
sentence : With his bare hands the worker could have done nothing with the land either as a grazer a farmer or a miner
-----------

detected bias : male
{pronoun : noun}, {' he ' :' farmer ' }
sentence : The unskilled farmer can never succeed largely no matter how hard he works
-----------

detected bias : female
{pronoun : noun}, {' her ' :' Science ' }
sentence : Science is lending her aid to the tiller of the soil and the wise ones are reaching out their hands in welcome
```

Fig. 4. Biases detected in the text in Fig. 2

The authors make use of Explainable AI to extract the most important keywords from the text input with the help of feature importance. With this, an array of keywords is generated along with feature importance (weights) that signifies how important/relevant a keyword is for the sentence. Next, the algorithm determines all the sentences in which these aforementioned keywords are present, and assign weights to these sentences based on the weights associated with their keywords. This is derived by finding out the maximum weight from all the weights in the list of keywords of the sentence. This step is important because there can be more than one keyword in a sentence that are important/relevant. Once this association is calculated, each sentence is picked and natural language processing techniques are applied on it. The authors use the concept of part-of-speech (POS) tagging to determine noun-pronoun combination(s) in each of the sentences. Once all such relevant, feature-important noun-pronoun associations are derived, the algorithm determines all gender-biased associations from this list, and adds the corresponding sentence weight (weight assigned to that sentence containing the keyword) to deduce the sum of bias. To calculate the final bias index, we divide this sum by the total sum of the sentence weights for all the noun-pronoun associations, which gives a value between 0 and 1 (0 suggesting no bias, 1 suggesting high bias).

This **Bias Index**, a number between 0 and 1, suggests that higher the score, more the chances that the text input is gender biased.

3.2 Results on Sample Text Extracts

Table 3 shows the working of the algorithm on the sample text in Fig. 2. Furthermore, the algorithm was run on ten extracts taken from varied sources such as articles, instruction manuals, travelogues etc. Figure 5a depicts the Bias Indices deduced for the ten extracts. Figure 5b illustrates the various number of bias instances identified in the extracts.

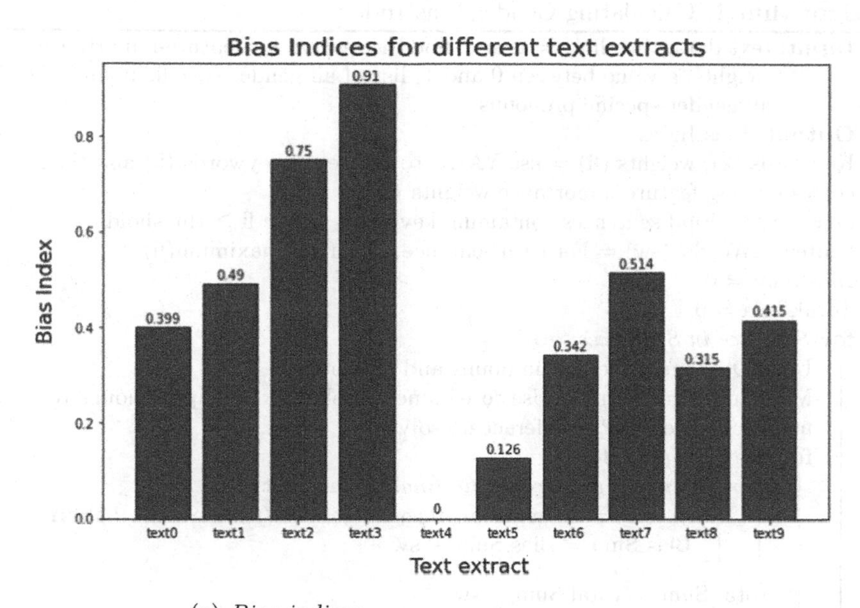

(a) *Bias indices*

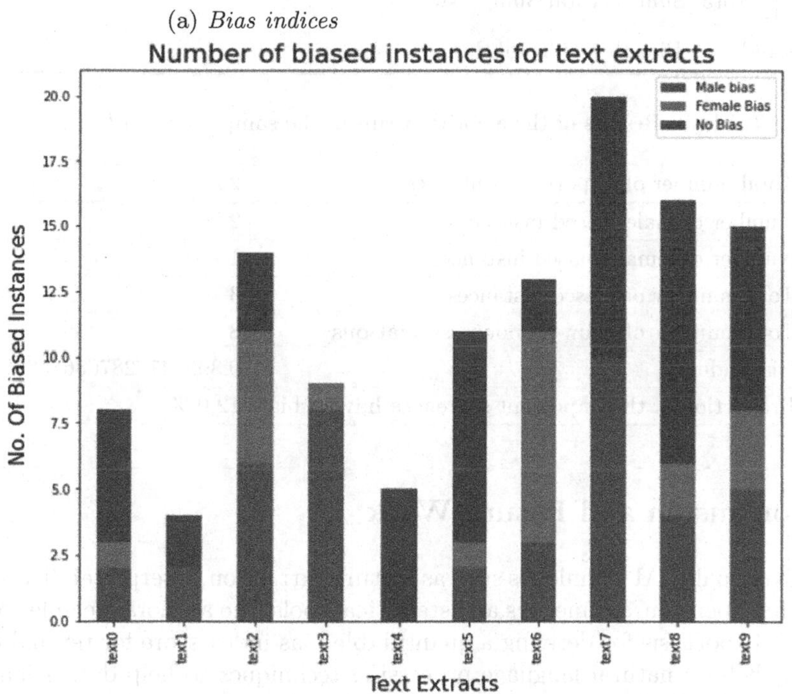

(b) *Number of Bias instances*

Fig. 5. Results on ten text extracts

Algorithm 1: Calculating Gender Bias Index

```
   Input: text data, threshold = user-defined threshold for feature-importance
          weights: a value between 0 and 1, list of all gender-specific nouns, list of
          all gender-specific pronouns
   Output: Bias_Index
 1 Keywords (k), weights (fi) = use YAKE to get set of keywords (k) and their
   corresponding feature importance weights (fi)
 2 Sentences = Find sentences containing keywords where fi ≥ threshold
 3 Sentence_Weight (sw) = For each sentence, calculate maximum(fi)
 4 Bias_Sum = 0
 5 Total_Sum = 0
 6 for Sentence in Sentences: do
 7     Use POS-Tagging to obtain nouns and pronouns
 8     Map(Pronoun, Noun) = Use coreference resolution to map pronouns to
       noun (eg. AllenNLP CoreferenceResolver)
 9     for Entry in Map do
10         if pronoun == gender-specific (male/female): then
11             if If noun == neutral or opposite-gender (female/male): then
12                 Bias_Sum = Bias_Sum + sw
13         Total_Sum = Total_Sum + sw
14 Bias_Index = Bias_Sum / Total_Sum
```

Table 3. Results of the algorithm run in the sample text in Fig. 2

Total number of important sentences	25
Number of male biased instances	2
Number of female biased instances	1
Total number of biased instances	3
Total number of noun-pronoun associations	8
Bias index	0.3991477287686358
Proportion of the important sentences having bias	12.0 %

4 Conclusion and Future Work

With Explainable AI techniques such as feature extraction, interpretability along with other open-sourced metrics and statistical tools, the authors are able to formulate a hypothesis for deriving a quantifiable bias index score for textual data with the help of natural language processing techniques to help data scientists understand the level of bias from a gender-bias perspective in their training data. The authors have also used a similar technique from a tabular data perspective to validate whether such Explainable AI metrics work for other forms of biases as well.

Similar methodologies can be applied to the quantification of other latent biases in text pertaining to race, propoganda or any other glaring demographic disparities. Furthermore, promising research in the domain of entity-level sentiment detection can be incorporated to extract entity-level biases.

References

1. "The Effects of Data Quality on Machine Learning Algorithms.", Conference: Proceedings of the 11th International Conference on Information Quality, MIT, Cambridge, MA, USA, 10–12 November, 2006
2. Talman, A., Apidianaki, M., Chatzikyriakidis, S., Tiedemann, J.: NLI Data Sanity Check: Assessing the Effect of Data Corruption on Model Performance. arXiv preprint arXiv:2104.04751 (2021)
3. Shah, D., Schwartz, H.A., Hovy, D.: Predictive biases in natural language processing models: A conceptual framework and overview. arXiv preprint arXiv:1912.11078 (2019)
4. Sun, T., et al.: Mitigating gender bias in natural language processing: Literature review. arXiv preprint arXiv:1906.08976 (2019)
5. Manzini, T., Lim, Y.C., Tsvetkov, Y., Black, A.W.: Black is to criminal as caucasian is to police: detecting and removing multiclass bias in word embeddings. arXiv preprint arXiv:1904.04047 (2019)
6. Gilpin, L.H., Bau, D., Yuan, B.Z., Bajwa, A., Specter, M., Kagal, L.: Explaining explanations: an overview of interpretability of machine learning. In: 2018 IEEE 5th International Conference on data science and advanced analytics (DSAA), pp. 80–89. IEEE, October 2018
7. Danilevsky, M., Qian, K., Aharonov, R., Katsis, Y., Kawas, B., Sen, P.: A survey of the state of explainable AI for natural language processing. arXiv preprint arXiv:2010.00711 (2020)
8. https://www.reuters.com/article/us-amazon-com-jobs-automation-insight-idUSKCN1MK08G
9. Zhang, G., et al.: Selection bias explorations and debias methods for natural language sentence matching datasets. arXiv preprint arXiv:1905.06221 (2019)
10. Blodgett, S.L., Barocas, S., Daumé III, H., Wallach, H.: Language (technology) is power: a critical survey of "bias" in nlp. arXiv preprint arXiv:2005.14050 (2020)
11. Linardatos, P., Papastefanopoulos, V., Kotsiantis, S.: Explainable ai: a review of machine learning interpretability methods. Entropy **23**(1), 18 (2021)
12. Bolukbasi, T., Chang, K.W., Zou, J.Y., Saligrama, V., Kalai, A.T.: Man is to computer programmer as woman is to homemaker? debiasing word embeddings. Adv. Neural. Inf. Process. Syst. **29**, 4349–4357 (2016)
13. Campos, R., et al.: Yet Another Keyword Extractor (Yake). https://github.com/LIAAD/yake. Accessed 15 July 2021
14. Gendered Words Dataset. https://github.com/ecmonsen/gendered_words. Accessed 15 July 2021
15. Lee, K., He, L., Lewis, M., Zettlemoyer, L.: End-to-end neural coreference resolution. arXiv preprint arXiv:1707.07045 (2017)

References

Student Workshop

Enhancing Model Robustness via Lexical Distilling

Wentao Qin[1,2] and Dongyan Zhao[1,2(✉)]

[1] Center for Data Science, AAIS, Peking University, Beijing, China
{qinwentao,zhaody}@pku.edu.cn
[2] Wangxuan Institute of Computer Technology, Peking University, Beijing, China

Abstract. Humans are prone to making typos in writing, which, though, doesn't affect understanding the whole sentence. However, neural models in natural language processing(NLP) would collapse when confronted with such tiny mistakes. This problem results from that neural models incline to entangle information, i.e., replacing a single aspect of the input text leads to significant changes in all components of the representation. Therefore, a trivial noise in a sentence can bring about a dramatic performance drop of the model. In this paper, we propose a novel and general framework to enhance the robustness of a model. The whole framework is trained in an adversarial style, which enables the model to encode the original sentence and the sentence refined by a lexical distiller to a similar sentence representation. We verify the effectiveness of the proposed framework in auto-encoder task. Experimental results show that our framework enhances the robustness of the model in different aspects.

Keywords: Model architecture · Robustness · RNN

1 Introduction

Deep learning model has achieved remarkable effect on various tasks, especially on supervised ones, such as text classification [8], machine translation [11], machine comprehension [10]. For instance, in the field of machine translation, [1,9] exploit a seq2seq (sequence-to-sequence) model with the attention mechanism, which brings neural machine translation to the state-of-the-art level. Among these supervised learning tasks, the paradigm of the deep model can be roughly divided into two procedures: first, mapping the input layer to a hidden layer(or several hidden layers); second, mapping the hidden layer to the output layer. By designing an appropriate loss function and training via back propagation, the deep model is expected to learn a good probability distribution which will assigns a high probability to correct outputs given a source input.

This elegant and powerful deep model makes deep learning extraordinarily popular these years. However, the merit that deep learning can learn to fit any

L. Wang et al. (Eds.): NLPCC 2021, LNAI 13029, pp. 337–344, 2021.
https://doi.org/10.1007/978-3-030-88483-3_26

complex distribution throws up a severe problem. The above-mentioned mapping procedure from the input layer, which comes from a huge vector space(e.g., text space), to the hidden layer, which comes from a relatively much dense vector space, is equal to "densely compressed representation". Whereas densely compressed representation tends to entangle information, i.e., changing a single aspect of the input yields significant changes in all components of the representation [14], and this may lead to unexpected or even unwanted outcomes in the output layer. [4,12,13] likewise illustrate that the deep model is not robust.

Some researches have been done to equip the deep model with robustness. [2] create perturbed translation pairs to fulfill a more robust NMT(Neural Machine Translation) model. Similarly, [7] and [15] construct adversarial input by adding distractors to the input, to build a robust machine comprehension model.

Manually creating adversarial input provides an angle to train a robust model. However, there exists a fatal weakness in them. Because their models' training is built on adversarial input which are finite and specific(e.g., replace certain words, add a distractor to the input). Their models will perform well when confronted with similar adversarial input, but it's hard for them to handle those unseen types of adversarial input.

Now that it is intractable to add all possible adversarial input to enlarging the training set, a new approach is in need. In this paper, we propose a novel and general architecture of robust model which needn't extra manual adversarial input. The core idea of the proposed model is to force the model to distinguish the less important part of the input text and concentrate on the mapping from the main part of the input text to the intermediary hidden layer. Therefore, after the model is well trained when given a noisy input, the model will only take its main part into consideration, with the noise being ignored, and thus reducing the effect of the noise.

Section 3 walks the reader along more details of the proposed framework. Before that, we introduce some related works in Sect. 2. In Sect. 4, we take an experiment on an auto-encoder model built on our framework, and validate its stronger robustness than the trivial auto-encoder. We conclude our work in Sect. 5.

2 Related Work

Early researches on building a robust model arose in computer vision. In the image classification task, [4] builds a more robust and error-resistant model by training with adversarial examples. Following this idea, preexisting works in NLP to enhance the robustness of a deep model mainly concentrate on constructing manual perturbed input. In [2], the authors construct perturbed input by replacing some words of input or perturbing the words' embeddings. In [7] and [15], they construct a distractor and add it to the input to produce an adversarial example.

In our framework, the idea of forcing the encoder to produce similar intermediary representations for the input and perturbed counterpart derives from

GAN [3], and is similar to [2]. The novel part of our frame is that we leverage a lexical distiller. The distiller plays an important role in refining the input by decreasing the effect of the noisy words. To our knowledge, we are the first one to propose a robust framework which doesn't need extra manual perturbed input.

3 The Model

Our framework comprises three parts: an RNN-based model, a lexical distiller, and a discriminator. To facilitate discussion, we employ the auto-encoder(AE) as the model to discuss. Figure 1 illustrates the model of our approach. Details are discussed below.

3.1 RNN-based Model: Auto-Encoder

The **Embedding**, **Encoder** and **Decoder** in Fig. 1 make up an usual AE model. The **Embedding** maps the input text $\mathbf{x} = (x_1, x_2, \ldots, x_n)$ to $\mathbf{emb} = (e_1, e_2, \ldots, e_n)$. The **Encoder** is a bi-directional Long Short Term Memory (LSTM) [5]. For each e_i, **Encoder** produces an output vector out_i and a hidden vector h_i. In Fig. 1, $\mathbf{out} = (out_1, out_2, \ldots, out_n)$, and $\mathbf{H} = h_n$. **Decoder** contains an unidirectional LSTM. Decoding starts with <SOS> representing the start of a sentence, and **H**. At each decoding step, **Decoder** outputs a probability distribution on the vocabulary and a hidden vector. The word with the highest probability and the hidden vector will act as next input to **Decoder**. The goal of AE is to reconstruct the input text as accurately as possible. Loss_{ori} is the loss function of AE which reconstructs output from **emb**. We take the negative log likelihood to compute Loss_{ori}:

$$\text{Loss}_{\text{ori}} = -\sum_{i=1}^{n} \log p(x_i)$$

where $p(x_i)$ is the probability of word x_i in the i-th decoding step.

3.2 Lexical Distiller

The lexical distiller is corresponding to **Distiller** in Fig. 1. **Distiller** accepts **emb** and **out** as input, and produces $\mathbf{emb'} = (e'_1, e'_2, \ldots, e'_n)$ as output. $\mathbf{emb'}$ has the same shape as **emb**. **Distiller** ensures that for some indexes i, $e'_i = e_i$; for other, $e'_i = \mathbf{0}$. The detail of **Distiller** is illustrated in Fig. 2.

Firstly, **out0** is produced in the same way as **out** in Fig. 1. **out0** is then fed to **W**, which is a fully-connected layer with the output dimension being 2. Therefore, **out1** has a shape of $n \times 2$. To decide whether to set e'_i to e_i or $\mathbf{0}$, we can naively compare $out1_{i0}$ with $out1_{i1}$. If $out1_{i0}$ is larger, set e'_i to e_i; otherwise, set e'_i to $\mathbf{0}$. However, this operation is not differentiable. To conquer this problem, we employ Gumbel Softmax operation [6] on **out1** to produce **out2**. Gumbel Softmax operation is differentiable, and is able to output an

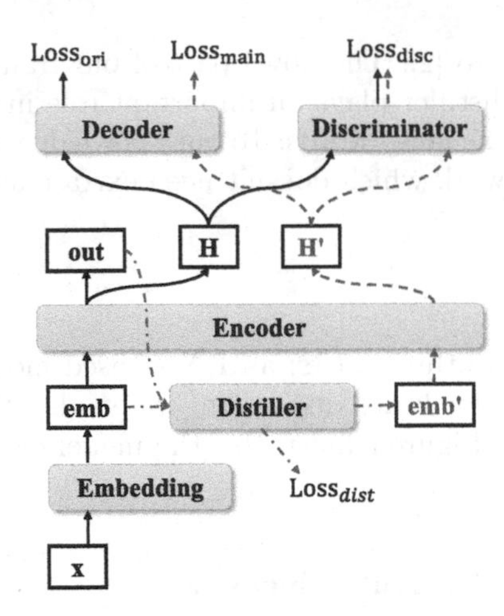

Fig. 1. Data flow of our model. The black lines represent data flow of **x**, the red dashed lines represent data flow of **emb′** which comes from **emb** refined by **Distiller**, and the blue dashed lines represent data flow related with **Distiller**. The four Loss_{xxx}s indicate four loss functions.

one-hot vector, therefore each $out2_i$ is a two-dimensional one-hot vector. We refer $(out2_{10}, out2_{20}, \ldots, out2_{n0})$ as **out2[:, 0]** in Fig. 2. **emb′** is computed by multiplying **out2[:, 0]** and **emb**. This multiplication operation is defined as:

$$e'_i = out2_{i0} \times e_i,\ for\ i = 1, \ldots, n$$

It is worth noting that the way we distill a word x_i out from **x** is setting e_i to an all-zero vector, instead of removing it directly. There are two reasons for our choice: (1) compared with removing x_i, setting e_i to an all-zero vector is comparable on eliminating the effect of x_i; (2) setting e_i to an all-zero vector is more practical in back propagation training.

Our framework encourages **Encoder** to produce similar representations for **emb** and **emb′** (cf. Sect. 3.3). This will lead **Distiller** to learning an identical transformation. To prevent this from happening, we employ a loss function called $\text{Loss}_{\text{dist}}$, to force **Distiller** to distill a certain ratio of words. $\text{Loss}_{\text{dist}}$ is computed as:

$$\text{Loss}_{\text{dist}} = [\text{sum}(out2[:, 0])/n - kw\_ratio]^2$$

where $\text{sum}(out2[:, 0])$ is the amount of 1 in $out2[:, 0]$, and kw_ratio is a hyper parameter representing the keep ratio of words.

After **emb′** is obtained, another loss $\text{Loss}_{\text{main}}$ can be computed in the same way as Loss_{ori}:

$$\text{Loss}_{\text{main}} = -\sum_{i=1}^{n} \log p'(x_i)$$

where $p'(x_i)$ is the probability of word x_i in the i-th decoding step related with the red dashed data flow in Fig. 1.

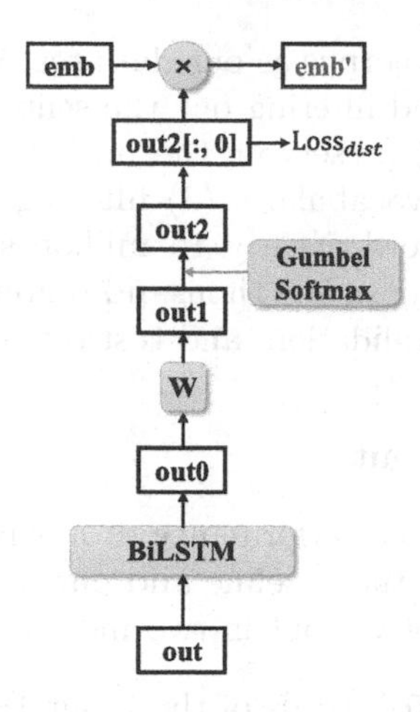

Fig. 2. Details of our lexical distiller. It accepts **out** and **emb** as input, and produce **emb′** as output.

3.3 Discriminator

The discriminator is corresponding to **Discriminator** in Fig. 1. **Discriminator** encourages **Encoder** to encode **emb** and **emb′** to representations that are similar to each other. This idea derives from [3]. In our framework, **Encoder** acts as the Generator, who is trying to produce similar hidden representations, **H** and **H′**, and **Discriminator** tries to correctly distinguish these two representations. **Discriminator** produces a scalar ranging from 0 to 1 for **H** and **H′**, denoted by $D(\mathbf{H})$ and $D(\mathbf{H}')$. The loss $\text{Loss}_{\text{disc}}$ is:

$$\text{Loss}_{\text{disc}} = \log(1 - D(\mathbf{H})) + \log D(\mathbf{H}')$$

3.4 Training

In the training procedure, the objective function is a linear combination of the four loss functions as mentioned, i.e.,

$$\begin{aligned}\text{Loss} = &\alpha_1 \text{Loss}_{\text{ori}} + \alpha_2 \text{Loss}_{\text{main}} + \\ &\alpha_3 \text{Loss}_{\text{dist}} - \alpha_4 \text{Loss}_{\text{disc}}\end{aligned}$$

where $\alpha_1 \sim \alpha_4$ are hyper parameters to adjust the weight of four loss values. We train $\text{Loss}_{\text{disc}}$ and the discriminator alternatively in a two-step adversarial fashion until $\text{Loss}_{\text{disc}}$ is converged.

4 Experiments

4.1 Data Set

To validate the effectiveness of our framework to enhance a neural model's robustness, we choose the classical auto-encoder(AE) as the baseline model.

We employ the English wiki corpus as our data set. We clean this corpus by (1) splitting it into sentences and filtering out the sentences whose length is under than 5 or more than 20; (2) preserving the most frequent 50,000 words in the remaining sentences as the vocabulary; (3) filtering out the sentences with the OOV (out-of-vocabulary) word. Nearly 15 million sentences remain after pre-processing. We randomly choose 500 thousand sentences as the whole data set. and divide it into training, validation, and test set by the proportion of 8: 1: 1.

4.2 Construct Noisy Input

The deep model will encounter performance drop when the input has noise. We evaluate the performance of the baseline and our model on noisy input. In our experiment, we construct noisy input in two methods:

- **Redundancy**: Repeat 15% words of the input 1$\sim$3 times.
- **Typos**: Replace 15% words of the input with <pad> token. (In practice, deep model regards the OOV word as <pad> or <unk>.)

We denote the input without manual noise as **Clean** input.

4.3 Evaluation on Robustness

We build an AE model in a conventional way (a BiLSTM as Encoder, a LSTM as Decoder). And we establish our model by equipping the baseline with **Distiller** and **Discriminator**. In the training procedure of our model, the hyper parameters *kw_ratio*, α_1, α_2, α_3, α_4 are set as 0.75, 1, 1, 20, 1, and Loss only contains Loss_{ori} in the first training epoch. The training epoch times for the baseline and our model are both 10.

In the test procedure, the forward data flow of our model is from **Embedding** to **Distiller**, then **Encoder** and lastly **Decoder**.

We define a metric named reconstruction accuracy for quantitative analysis. The definition is:

Given a golden sentence $\mathbf{s} = (s_1, s_2, \ldots, s_n)$, and a reconstructed sentence $\mathbf{p} = (p_1, p_2, \ldots, p_m)$, denote the total of i that satisfies $p_i = s_i$ as *hits*, and the reconstruction accuracy is defined as $hits/n$.

The result of reconstruction accuracy is showed in the column Acc of Table 1. Compared to row **Clean**, each model on noisy input(row **Redundancy** and **Typos**) encounters a performance drop, which proves that noisy input will decrease the performance of deep model. Our model gains a performance improvement on both clean and noisy input compared with the baseline.

We also analyze the change rate of hidden intermediary representation of the baseline and our model. Denote $\boldsymbol{h1}$ and $\boldsymbol{h2}$ as the intermediary hidden representations of input $\mathbf{x}$ and $\mathbf{x}$ with noise in an AE model. A robust AE model

Table 1. Column Acc illustrates the reconstruction accuracy results of normal AE(Baseline) and our model(Robust) on clean input(Clean), and test set with two kinds of noisy input (Redundancy, Typos). Column CR illustrates the change rate from noisy input to clean input

		Acc	CR
Clean	**Baseline**	0459	0
	Robust	0.616	0
Redundancy	**Baseline**	0.305	0.101
	Robust	0.349	0.010
Typos	**Baseline**	0.347	0.053
	Robust	0.492	0.044

should make $\boldsymbol{h2}$ closer to $\boldsymbol{h1}$. We define a changing rate(CR) from $\boldsymbol{h1}$ to $\boldsymbol{h2}$ as following:

$$CR = \frac{\sum_{i=1}^{n}(h1_i - h2_i)^2}{\sum_{i=1}^{n} h1_i^2}$$

The result of change rate from noisy input to clean input is shown in the column CR of Table 1. The CR values of our model are lower than the baseline, which indicates that our model can render an AE model more stable when facing with noisy input.

5 Conclusion

We propose a novel framework to enhance the robustness of deep model in NLP, and make a preliminary exploration on auto-encoder task. Experiment results demonstrate its effectiveness to build a robust auto-encoder. Our framework is general and not limited to auto-encoder. It is applicative to many tasks that handle natural languages, such as machine translation, dialogue system, machine comprehension, etc. In the future, we will conduct a deeper analysis of our model, and validate its effectiveness on more NLP tasks.

Acknowledgments. We thank the reviewers for their valuable comments. This work was supported by the National Key Research and Development Program of China (No. 2020AAA0106602).

References

1. Bahdanau, D., Cho, K., Bengio, Y.: Neural machine translation by jointly learning to align and translate. arXiv preprint arXiv:1409.0473 (2014)
2. Cheng, Y., Tu, Z., Meng, F., Zhai, J., Liu, Y.: Towards robust neural machine translation. arXiv preprint arXiv:1805.06130 (2018)
3. Goodfellow, I.J., et al.: Generative adversarial nets, pp. 2672–2680 (2014)

4. Goodfellow, I.J., Shlens, J., Szegedy, C.: Explaining and harnessing adversarial examples. arXiv preprint arXiv:1412.6572 (2014)
5. Hochreiter, S., Schmidhuber, J.: Long short-term memory. Neural Comput. **9**(8), 1735–1780 (1997)
6. Jang, E., Gu, S., Poole, B.: Categorical reparameterization with gumbel-softmax. In: International Conference on Learning Representations (2017)
7. Jia, R., Liang, P.: Adversarial examples for evaluating reading comprehension systems. arXiv preprint arXiv:1707.07328 (2017)
8. Lai, S., Xu, L., Liu, K., Zhao, J.: Recurrent convolutional neural networks for text classification. In: AAAI, vol. 333, pp. 2267–2273 (2015)
9. Luong, M.T., Pham, H., Manning, C.D.: Effective approaches to attention-based neural machine translation. arXiv preprint arXiv:1508.04025 (2015)
10. Rajpurkar, P., Zhang, J., Lopyrev, K., Liang, P.: Squad: 100,000+ questions for machine comprehension of text. arXiv preprint arXiv:1606.05250 (2016)
11. Sutskever, I., Vinyals, O., Le, Q.V.: Sequence to sequence learning with neural networks. In: Advances in neural information processing systems, pp. 3104–3112 (2014)
12. Szegedy, C., et al.: Intriguing properties of neural networks. arXiv preprint arXiv:1312.6199 (2013)
13. Tang, Y., Eliasmith, C.: Deep networks for robust visual recognition. In: Proceedings of the 27th International Conference on Machine Learning (ICML-10), pp. 1055–1062. Citeseer (2010)
14. Vincent, P., Larochelle, H., Lajoie, I., Bengio, Y., Manzagol, P.A.: Stacked denoising autoencoders: Learning useful representations in a deep network with a local denoising criterion. J. Mach. Learn. Res. **11**, 3371–3408 (2010)
15. Wang, Y., Bansal, M.: Robust machine comprehension models via adversarial training. arXiv preprint arXiv:1804.06473 (2018)

Multi-stage Multi-modal Pre-training for Video Representation

Chunquan Chen[1(✉)], Lujia Bao[1], Weikang Li[2], Xiaoshuai Chen[2], Xinghai Sun[2], and Chao Qi[2]

[1] Beijing University of Post and Telecommunications, Beijing, China
{ccq1996,baolj}@bupt.edu.cn
[2] Tencent Inc, Beijing, China
{wavewkli,sheltonchen,xinghaisun,chaoqi}@tencent.com

Abstract. Multi-modal networks are usually challenging to train because of their complexity. On the one hand, multi-modal networks are often prone to underfitting due to their heterogeneous data formats of different modalities. On the other hand, data from different domains have different distributions and domain differences could be difficult to eliminate in joint training. This paper presents a Multi-Stage Multi-Modal pre-training strategy (MSMM) to train multi-modal joint representation effectively. To eliminate the difficulty of multi-modal end-to-end training, MSMM trains different Uni-modal network separately and then jointly trains multi-modal. After multi-stage pre-training, we can get a better multi-modal joint representation and better uni-modal representations. Meanwhile, we design a multi-modal network and multi-task loss to train the whole network in an end-to-end style. Extensive empirical results show that MSMM can significantly improve the multi-modal model's performance on the video classification task.

Keywords: Multi-modal pre-training · Video representation

1 Introduction

With the recent advances in multi-modal learning, multi-modal pre-training techniques play a vital role in learning multi-modal representation. The paradigm is to pre-train the model on large scale general data and fine-tune on the downstream tasks using task-specific data.

Inspired by the BERT [2] model's success for NLP tasks, numerous multimodal image-language pre-training models have been proposed. Different from previous text pre-training or image-language pre-training, videoBERT [15] is the first pioneer to investigate high-level visual-linguistic pre-training on instructional videos. Unlike using videos as training data, VisualBERT [7], ImageBert [12] take images and captions as input and then learn image and caption joint representation. Nevertheless, the BERT-style objective methods of learning modal transformation have some constraints. They need to collect large-scale datasets and guarantee the sequence of multi-modal sample pair is aligned.

L. Wang et al. (Eds.): NLPCC 2021, LNAI 13029, pp. 345–353, 2021.
https://doi.org/10.1007/978-3-030-88483-3_27

Besides, the visual samples are larger than the text samples, causing the multi-modal pre-training data to consume more storage and hinder the further expansion of the dataset.

To solve the problem, we need to make full use of the large scale domain-related uni-modal dataset and uni-modal pre-trained parameters to initialize multi-modal networks. Specifically, we propose MSMM, a Multi-Stage Multi-Modal pre-training strategy. In the first stage, we first pre-train each modality on general large-scale datasets. In the second stage, we further pre-train each modality on domain-specific datasets. In the third stage, the aforementioned two-stage pre-trained uni-modals are then fine-tuned jointly on task datasets to learn a final multi-modal representation. We found that the final multi-modal joint training's stability and speed are improved through the previous two-stage training. Not only the entire model has a better result, but also each modal has a better uni-modal feature representation. In summary, our contributions are as follows:

- We propose a general pre-train strategy (MSMM) to train a multi-modal network, which includes three stages: (1) firstly pre-train each modality's model on large-scale general datasets; (2) further pre-train each uni-modal network on domain-specific datasets; (3) jointly train the multi-modal representation on task specific datasets.
- We also explore different multi-modal fusion strategies and end-to-end jointly training strategies, which demonstrates the importance of multi-modal joint representation and multi-task training.
- Experiments show that each stage of pre-training can improve the performance of multi-modal representation, and meanwhile the performance of each modal and training speed are both greatly improved.

2 Proposed Method

2.1 Model Architecture

In this paper, we denote general pre-training datasets as $D_G = D_{g_1} \bigcup D_{g_2} \bigcup \cdots \bigcup D_{g_k}$, where D_{g_i} refers to modality i in general datasets. The domain-specific pre-training datasets are denoted as $D_S = D_{s_1} \bigcup D_{s_2} \bigcup \cdots \bigcup D_{s_k}$, where D_{s_i} refers to modality i in domain-specific datasets. Given a target multi-modal task $D_T = \{(X_{m_1}, X_{m_2}, \cdots, X_{m_k}), Y\}$, where X_{m_i} is as modality i and Y is the target label. We first pre-train each uni-modal network separately on general datasets D_G, and then further pre-train each uni-modal network separately on domain-specific datasets D_S to obtain better uni-modal networks. Finally, We initialize the multi-modal model with each pre-trained uni-modal parameters, and fine-tune the multi-modal model on the target datasets D_T.

Figure 1 presents the multi-modal network architecture. The multi-modal model independently extracts coordinated representations of video, text and audio using various uni-modal encoders. Specifically, we adopt InseptionResnetV2 [10], ALBERT [6] and Vggish [4] respectively as video, text and audio

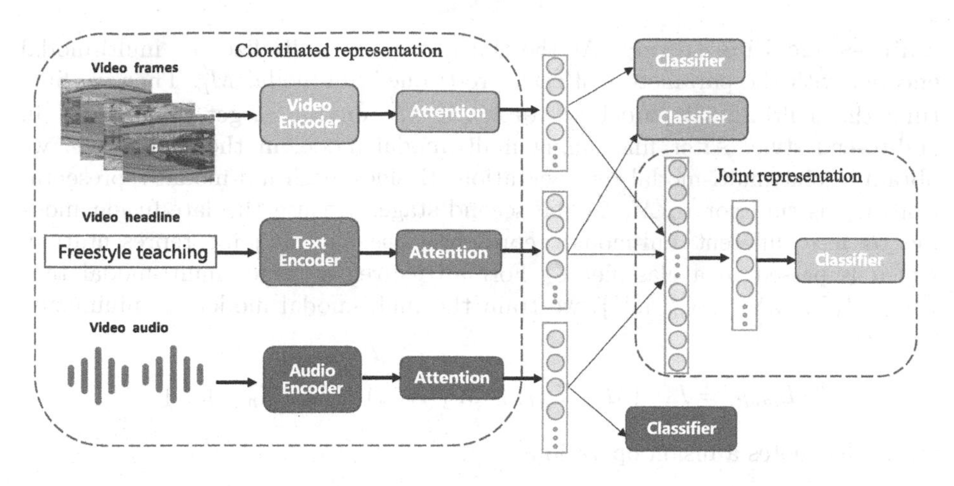

Fig. 1. Model architecture of multi-modal network

encoders. Then a transformer encoder is used to temporal attention to the video and audio vectors to obtain the multi-modal feature. Meanwhile, we explore the coordinated uni-modal representation and joint multi-modal representation, both representations are end-to-end framework. The difference is the coordinated representation has separate representations which means each modality encoder is parallelism and they only share the final loss function. The late-fusion module merges and projects uni-modal representations together into a multi-modal space and generates a joint feature.

2.2 MSMM Pre-training and Fine-Tuning

First-stage Pre-training. In the text-only pre-training, we perform the standard mask-completion and sequence-order prediction tasks on general text datasets D_{g_1}, which obtain a language model M_1. In the audio-only pre-training, we conduct the multi-label classification task on general soundtracks dataset D_{s_2}, which obtains an audio model M_2. In the image-only pre-training, we carry out the single-label classification task on general image dataset D_{s_3}, which obtains a video model M_3. Specifically, given a uni-modal dataset $D_{s_i} = \{X_j, Y_j\}$, we train the uni-modal model by minimizing the empirical loss:

$$L_{uni} = L(C(M(X)), Y) \tag{1}$$

where M is a uni-model neural network and C is a fully connected layer, L is the cross entropy loss in classification task. Minimizing Eq. 1 gives a solution M and C.

Second-Stage Pre-training. At the second stage, our method further pre-trains each uni-modal model $M_i^{'}$ separately on domain-specific dataset D_S after first-stage pre-training. The target is to learn a transform $F : M_{X_i}^{'} \rightarrow Y_i$, which is a video-level classifiction in our case.

Third-stage Fine-tuning. At the third stage, we initialize the multi-modal encoder with the parameters of each pre-trained uni-modal $M_i^{'}$. Then we fine-tune the multi-modal model $M_1^{''} \oplus M_2^{''} \oplus M_3^{''}$ on the target dataset D_T in end-to-end style. After fine-tuning multi-modal model in the final stage, we obtain a joint multi-modal representation. Besides, each uni-modal representation $M_i^{''}$ is superior to $M_i^{'}$ in the second-stage. We use the late-fusion module to fuse different uni-modal representations as the joint representation, which is passed to a classifier C. Formally, given a target multi-modal task $D_T = \{(X_{m_1}, X_{m_2}, X_{m_3}), Y\}$, we train the multi-modal model by minimizing the loss:

$$L_{multi} = L(C(M_1^{''}(X_{m_1}) \oplus M_2^{''}(X_{m_2}) \oplus M_3^{''}(X_{m_3})), Y) \quad (2)$$

where $\oplus$ denotes a fusion operation.

3 Experiment

3.1 Dataset

First-stage uni-modal general datasets come from multiple large scale general datasets. Text source domain dataset contains nearly one billion training sentences, which are collected from the Wikipedia. Video source domain dataset is the ImageNet [13], which contains nearly 1.28 million training images. The audio source domain dataset is the AudioSet [3], which contains nearly 2 million human-labeled 10-second sound clips drawn from YouTube videos. Second-stage domain-specific datasets are sampled from the massive standardized video data of Tencent Video platform. The text side uses 200 million video titles as pre-training data. Due to the large storage of structured video data, only 3 million videos are sampled on the audio and video sides. The second-stage pre-training data distribution comes from the same domain as the target domain dataset, and their label systems are the same. Third-stage task dataset is collected from the Tencent Video platform which contains nearly one million untrimmed videos. Each video contains multi-modal information, which consists of video title, voice signal and video frames. The label system is a hierarchical multi-class classification with two-level, of which the first-level contains 28 labels and the second-level contains 175 labels. We randomly sample nearly ten thousand videos from the dataset as validation set and test set respectively.

3.2 Implementation Details

We consider three modalities: video, text, and audio. For video, we uniformly sampled video frames with a fixed temporal length as the inputs. Let's denote the spatio-temporal shape of video clips as (L × 3 × H × W), where L is the number of continuous frames, 3 refers to the number of RGB channels, H and W are the spatial height and width of the video frames. In our experiments, L = 30, and H = W = 224 were set. For audio, we used log-Mel spectrograms with

30 temporal frames by 40 Mel filters. Backbone architecture: we used Albert as our text backbone model, InceptionResnetV2 as our visual backbone for RGB, and Vggish as our audio model. We used a fully connected layer network on concatenated features from visual, text, and audio backbones, followed by one prediction layer for fusion. Our loss function is softmax activations with cross-entropy loss for multi-class target datasets, and we report Top-1 accuracy.

3.3 Analysis

In order to express it more clearly, in the following experiments, we will use accuracy as an evaluation indicator to compare and measure the performance of different models.

Performance Improvement via MSMM. We mainly studied the effect of MSMM pre-training strategy. Table 1 shows the effect of training the best multi-modal joint representation model on the three-stage datasets. S_1 and S_2 refer to first-stage and second-stage pre-training respectively. Moreover, *first* and *second* refer to the the two level of labels' tasks. We also took out all the uni-modal models from a trained multi-modal model, and tested them on the same dataset with the corresponding single modal input.

Table 1. Ablation study of different pre-train stage in MSMM.

	Text		Video		Audio		Fusion	
	First	Second	First	Second	First	Second	First	Second
uni-modal	70.9	58.5	75.5	62.8	41.2	32.4	–	–
MSMM *w/o* S_1 *and* S_2	65.2	53.2	57.2	46.1	28.7	19.6	71.3	59.5
MSMM *w/o* S_2	69.5	56.9	60.7	48.9	35.1	26.9	80.4	69.5
MSMM	73.2	61.8	77.8	66.2	46.7	36.7	**85.2**	**75.7**

MSMM without *S1* and *S2* means the uni-modal model's weight is randomly initialized. In first-stage pre-training, we trained the uni-modal models on a large scale general dataset. In second-stage, we further pre-trained the uni-modal models in the Tencent video dataset. After that, we finetuned the multi-modal model on mulit-modal dataset based on the pre-trained or randomly initialized uni-modal models. The first row of the Table 1 presents the performance of uni-modal models trained throughout two pre-training stages and finetuned on uni-modal dataset. After first-stage pre-training, our framework improves first level top-1 accuracy by 9.1% in multi-modal fusion predictions. After two-stage pre-training, our framework also improves the Top-1 accuracy from 80.4% to 85.2%. Meanwhile, each uni-modal also gets a better performance in both first and second level label prediction. It is also worth noting that, after multi-modal training, every uni-modal model gains a considerable accuracy growth, compare to their counterparts trained on uni-modal datasets. The aforementioned phenomenon promises a way to fuse with information from multi-modalities into a uni-modal model, which we left for further study. Hence, we conclude that more diverse multi-modal representation can enrich the feature space and acquire better recognition results.

Coordinated Representation vs Joint Representation. As depicted in Fig. 1, we conduct experiments on two multi-modal representation. In the first case, multiple uni-modal encoders are trained in end-to-end style separately, and the coordinated representation is based on every modal's classifier's logits. Moreover, the joint representation is a later fusion of features from all uni-modal encoders in the second case. The results on the test set of the Tencent Video dataset are shown in Table 2.

Table 2. Comparison of multi-modal representation.

Representation	First	Second
Ensemble	76.6	65.5
Coordinated	79.6	68.5
Joint	85.2	75.7

The ensemble is separate training three best uni-modal networks, and weighting the logits of uni-modals. The coordinated representation and joint representation are training multi-modal end-to-end, but the latter adds a multi-modal fusion module and multi-modal loss based on the former.

All experiments were performed after the second stage pre-training. This result shows the importance of end-to-end training of multi-modal on task-specific datasets, which is better than ensemble learning in both coordinated representation and joint representation, and multi-modal joint representation is better than coordinated representation. It is worth noting that to learn the joint representation, we also calculated losses from each uni-modal classifiers, and weighted them with multi-modal loss to form the final loss.

Does Every Modality Make a Contribution in Third-stage? We mainly explored the contribution of different source types: Wikipedia, ImageNet, the AudioSet. Results in Table 3 reveal that every general uni-modal dataset can improve the accuracy on the task dataset. When these sources are combined, the performance is further improved. To evaluate each modal's role in multi-modal learning, we used ensemble learning as the baseline and design two network structures: joint representation and coordinated representation. The coordinated representation only has three uni-modal losses and uni-modal logits, joint representation further concatenate the embedding feature of multi-modal, and passed to a classifier. In Table 3, TAV refers to Text, Audio and Video encoders, AV refers to Audio and Video feature early fusion before Text embedding feature.

4 Related Work

Multi-modal Representation. The most fundamental challenge is to represent the multi-modal data in a way that leverage the complementarity of multiple modalities. The multi-modal representation could be summarized into two categories: (1) Joint representations that project uni-modal representations together into a multi-modal space. Antol et al. [1] proposed pre-training on data from related domains. The model proposed by Ngiam et al. [8] tackling denoising autoencoders represent each modality individually and fused them into a multi-modal representation using another autoencoder layer. Similarly, Silberer et al. [14] proposed a multi-modal autoencoder for task of semantic concept grounding. (2) Coordinated Representations consist of separate representations learned for each modality and coordinate through a constraint. Pan et al. [9] constructed a coordinated space between videos and sentences using language model and a deep video model, minimizing the distance of the video semantic embedding and text semantic embedding.

Table 3. Ablation study of different modality fusion and loss.

Modality	Loss	First	Second
T	L_T	71.3	59.8
V	L_V	73.5	60.5
A	L_A	42.4	32.8
TAV	$L_{coordinated}$	80.5	69.1
TA+V	L_{joint}	80.7	69.4
TV+A	L_{joint}	81.4	70.2
AV+T	L_{joint}	83.4	73.2
TAV	L_{joint}	**85.2**	**75.7**

Multi-modal Pre-training. Our work is related to previous research on multi-modal pre-train [5], these works generally adopt a joint multi-modal representation and demonstrate the effectiveness of joint training multi-modal network.
Multi-modal Domain Adaption. Most of the previous domain adaptation works is designed for uni-modal data. Qi et al. [11] solves multi-modal domain adaption by designing attention block and loss function. However, it is hard to design a general model to adapt all the datasets with different tasks. Sun et al. [16] found that multi-phase adaptive pre-training offers considerable gains in task performance.

5 Conclusion

This paper proposes a novel Multi-Stage Multi-Modal pre-training framework (MSMM), an efficient yet effective multi-modal pre-train strategy. First, we pre-train each uni-modal network separately on general uni-modal datasets to enhance the generalization ability of the model. Second, we further pre-train uni-modal network on domain-specific data to bridge the domain gap. Finally, we fine-tune multi-modal on specific task datasets. Extensive experiments demonstrate that our proposed method can indeed improve over a variety of baselines and yields comparable results. Except for the video classification task, our proposed method also obtains good performances on several video tagging tasks.

References

1. Antol, S., et al.: Vqa: visual question answering. In: Proceedings of the IEEE International Conference on Computer Vision, pp. 2425–2433 (2015)
2. Devlin, J., Chang, M.-W., Lee, K., Toutanova, K.: Bert: Pre-training of deep bidirectional transformers for language understanding. arXiv preprint arXiv:1810.04805 (2018)
3. Gemmeke, J.F., et al.: Audio set: an ontology and human-labeled dataset for audio events. In: 2017 IEEE International Conference on Acoustics, Speech and Signal Processing (ICASSP), pp. 776–780. IEEE (2017)
4. Hershey, S., et al.: Cnn architectures for large-scale audio classification. In 2017 IEEE International Conference on Acoustics, Speech and Signal Processing (ICASSP), pp. 131–135. IEEE (2017)
5. Huang, G., Pang, B., Zhu, Z., Rivera, C., Soricut, R.: Multimodal pre-training for dense video captioning. arXiv preprint arXiv:2011.11760 (2020a)
6. Lan, Z., et al.: Albert: a lite bert for self-supervised learning of language representations. arXiv preprint arXiv:1909.11942 (2019)
7. Li, L.H., Yatskar, M., Yin, D., Hsieh, C.-J., Chang, K.-W.: Visualbert: a simple and performant baseline for vision and language. arXiv preprint arXiv:1908.03557 (2019)
8. Ngiam, J., et al.: Multimodal deep learning. In ICML (2011)
9. Pan, Y., Mei, T., Yao, T., Li, H., Rui, Y.: Jointly modeling embedding and translation to bridge video and language. In: Proceedings of the IEEE Conference on Computer Vision and Pattern Recognition, pp. 4594–4602 (2016)
10. Piergiovanni, A.J., Angelova, A., Toshev, A., Ryoo, M.S.: Evolving space-time neural architectures for videos. In: Proceedings of the IEEE International Conference on Computer Vision, pp. 1793–1802 (2019)
11. Qi, F., Yang, X., Xu, C.: A unified framework for multimodal domain adaptation. In: Proceedings of the 26th ACM International Conference on Multimedia, pp. 429–437 (2018)
12. Qi, D., Su, L., Song, J., Cui, E., Bharti, T., Sacheti, A.: Imagebert: Cross-modal pre-training with large-scale weak-supervised image-text data. arXiv preprint arXiv:2001.07966 (2020)
13. Russakovsky, O., et al.: Imagenet large scale visual recognition challenge. Int. J. Comput. Vision **115**(3), 211–252 (2015)
14. Silberer, C., Lapata, M.: Learning grounded meaning representations with autoencoders. In: Proceedings of the 52nd Annual Meeting of the Association for Computational Linguistics (Volume 1: Long Papers), pp. 721–732 (2014)
15. Sun, C., Myers, A., Vondrick, C., Murphy, K., Schmid, C.: Videobert: a joint model for video and language representation learning. In Proceedings of the IEEE International Conference on Computer Vision, pp. 7464–7473 (2019a)

16. Sun, C., Qiu, X., Xu, Y., Huang, X.: How to Fine-Tune BERT for text classification? In: Sun, M., Huang, X., Ji, H., Liu, Z., Liu, Y. (eds.) CCL 2019. LNCS (LNAI), vol. 11856, pp. 194–206. Springer, Cham (2019). https://doi.org/10.1007/978-3-030-32381-3_16
17. Wang, W., Tran, D., Feiszli, M.: What makes training multi-modal classification networks hard? In: Proceedings of the IEEE/CVF Conference on Computer Vision and Pattern Recognition, pp. 12695–12705 (2020)

Nested Causality Extraction on Traffic Accident Texts as Question Answering

Gongxue Zhou(✉), Weifeng Ma, Yifei Gong, Liudi Wang, Yaru Li, and Yulai Zhang

School of Information and Electronic Engineering, Zhejiang Unviersity of Science and Technology, Hangzhou, China
{221901852064,mawf,zhangyulai}@zust.edu.cn

Abstract. As an important type of relationship, causality plays a vital role in relation reasoning. Therefore, causality extraction from natural language texts is a crucial task, especially in the field of traffic. For example, we can quickly discover the causes of traffic accidents and the correlation between the cause events. Existing methods utilize machine learning models to extract simple causality, however, there is nested causality in the traffic accident sentences, which is important for us to reason the cause and effect. In order to extract the causality successfully, we simplify the complex nested causality structure to the pairwise causality structure. On this basis, we propose a method that contains two steps. First, we extract the cause events from the input sentence, second, combine the extracted cause with the incomplete question template to obtain a complete question sentence, then we adopt the way of question answering tasks to extract the effect events. Experiments on the traffic accident dataset show the effectiveness of our model. However, we observe that due to the small training set, there is still room for improvement in the extraction accuracy of nested causality.

Keywords: Causality extraction · Nested causality · Question answering

1 Introduction

Texts in daily life often contain a lot of relational information. Extracting relations from texts can help us better understand the nature of events. Extracting causal relations from texts is more conducive to discovering the development and evolution of events. Therefore, this article focuses on the traffic accident text, and our aim is to dig out the causal relationship between traffic incidents in the traffic accident text.

Figure 1(a) shows an example of a causal sentence in the text of a traffic accident, the sentence contains a cause event "严重超员(seriously overcrowding)" and an effect event "伤亡扩大(increased casualties)", and forms a causal triplet: { "严重超员(seriously overcrowding)", cause-effect, "伤亡扩

L. Wang et al. (Eds.): NLPCC 2021, LNAI 13029, pp. 354–362, 2021.
https://doi.org/10.1007/978-3-030-88483-3_28

大(increased casualties)" }. However, due to the complexity and diversity of the traffic accident texts, the causality relations in the sentences are also complex and diverse. As shown in Fig. 1(b), the sentence contains two cause events, forms two triplets. In the above examples, the effect events are all triggered by cause events, however, the complexity of natural language texts makes the effect events may be triggered by triplets. As the sentence shown in Fig. 1(c), a causal triplet({ "天气炎热 (hot weather)" , cause-effect, "爆胎(flat tire)" }) leads to "车辆侧翻(vehicle rollover)" , the difference from the example in Fig. 1(b) is that the effect event is caused by different causes, one is an event and another is a triplet. We call this kind of causality nested causal relations, that a pair of cause-effect can be the cause of another higher-level causality. The goal of nested causality mining is to extract the nested causality structure from each sentence. In view of the particularity of traffic accident texts, accidents are caused by the evolution of many factors. Therefore, the text contains a large number of nested causal relations, and we cannot know the depth of the nested causality structure. [1,2] finds that the nested causality structure can be transformed into a graph of pairwise causality between sentence segments, and the cause-effect pair in a causal relation can be represented by its effect. Therefore, in order to dig out the nested causal relations in the text, we simplify the complex nested causality structure and use the pairwise causality structure to express the causal relationship in the sentence. As shown in Fig. 1(d), when e2 is caused by e1, it is no longer uses the entire triplet to trigger e3, and we use the effect of the triplet to represent the triplet.

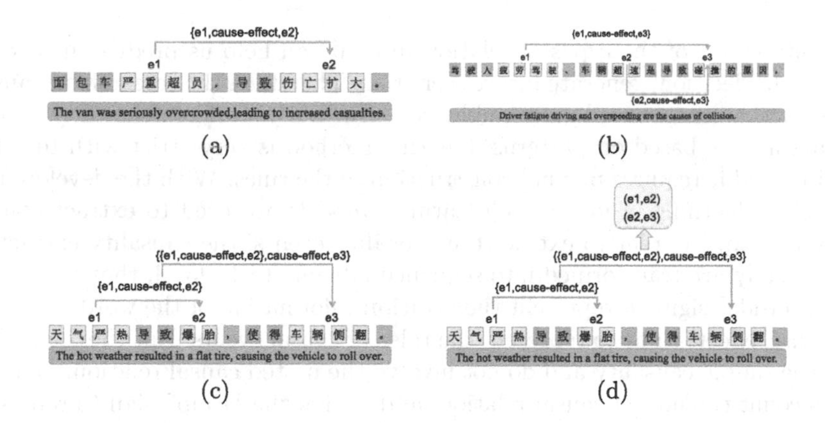

Fig. 1. Several causal relationships in traffic accident texts. (a): An example of a causal sentence contains a causal triplet. (b): An example of a causal sentence contains multiple cause events and an effect event. (c): An example of a causal sentence contains nested causal relastion. (d): An example of nested causality structure to pairwise causality structure.

Recently, the extraction of causality is mainly divided into two types: the way based on pipeline and joint extraction. The pipeline method is divided into

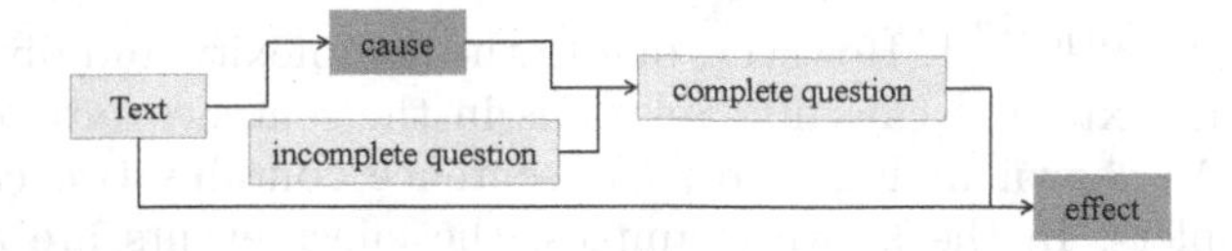

Fig. 2. The specific process of extracting the causal relastions in the sentences.

two separate task: named entity recognition and relation classification, the joint method is to directly extract the triples. The pipeline method first recognizes all entities in the sentence and then judges whether the entities belong to the causality relationship, but it suffers one main problem, if the causal events are extracted incorrectly, it will lead to an error in relation classification. To avoid this problem, we choose to extract the causality of traffic accidents based on joint extraction, which unifies the two models into one model, and uses the potential correlation of the two tasks to alleviate the problem of error propagation. We first extract the cause and then transform the extraction task into a question answering task, using the cause information to identify effects spans from the context. The specific process is shown in Fig. 2, after extracting the cause events, we combine the cause events with the incomplete question (What is the effect of xx?) to constitute the complete question (What is the effect of cause?), and then we use the question and the input sentence to extract the effect events.

2 Related Works

Causality is one of the types of relationships, it can help us predict the events, process the decision, generate future scenarios. Therefore, the research on causality extraction has aroused great public concern. Early works in causality extraction are always based on patterns, but this method is very strict with text formatting, and it requires manual construction of the rules. With the development of machine learning, a lot of deep learning models are used to extract causality extraction. In order to extract the causality events, the causality extraction tasks usually are transformed into sequence labeling tasks [3,4], they use "B, I" (begin, inside) signs to represent the position information of the words, and "C, E" (cause, effect) signs to represent the roles of causal events. But they only consider the simple causality and do not involve the nested causal relation. [5] takes into account the nested causal relation, and design the "Emb" sign to represent the embedded causality. However, this method can only find one level of nested causal relation, if there are multiple nested causal relationships in the sentence, the signs which represent the embedded causalities are prone to be messed up, and unable to complete the extraction tasks.

In relational extraction tasks, for efficiently extracting the relation triples in which a sentence contains multiple relation triples that overlap with each other, many methods are proposed. [6] proposed a novel cascade binary tagging framework, instead of treating relations as discrete labels as in previous works, their framework models relations as functions that map subjects to objects in a sentence. [7] cast the task as a multi-turn question answering problem, the extraction of entities and relations is transformed to the task of identifying answer spans from the context. Inspired by their novel ideas, we use the strategy from machine reading comprehension (MRC) models to find the cause events and effect events which is always be used to extract text span in passages given queries.

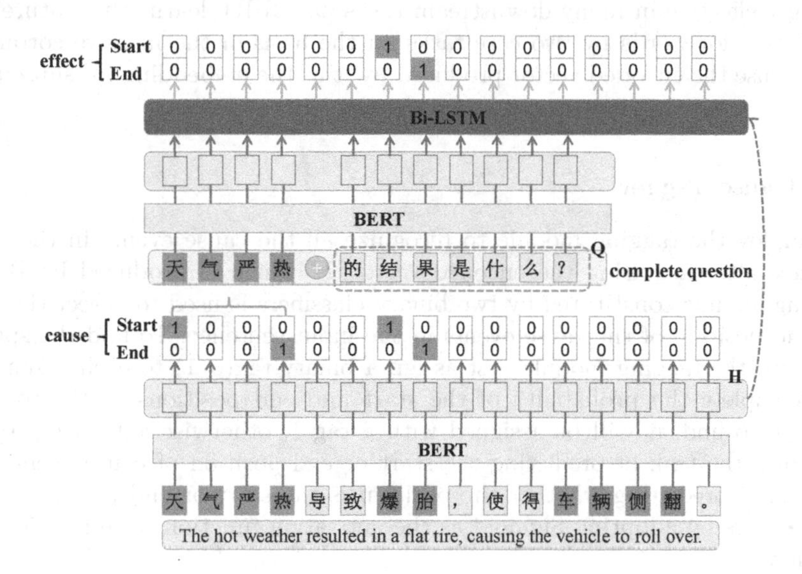

Fig. 3. An overview of our model.

3 Method

In our work, the way to extract causality relation is divided into two steps. First, we extract the cause events from the input sentence, and then combine the extracted cause events with the incomplete question sentence template to transform the causality relationship extraction task into a question answering task. The overall structure of our model is shown in Fig. 3. To successfully extract the causal events and their relationships, after inputting the sentence about the description of the traffic accident, we hope to get the position of the causal events in the sentence, so that we refer to the method of extracting relational triple in [6], by designing a subject tagger to find the cause events' start and end positions from the sentences. Through the way, we finally get 4 vectors to respectively denote the start position of the cause events, the end position of the cause events, the start position of effect events and the end position of effect events.

3.1 BERT

We apply the pre-trained BERT model [8] to encode the input sentence, which is stacking by n-layers Transformer blocks. It has caused a sensation since be proposed and has been widely applied in various natural language processing tasks [11,12].

BERT is designed to learn deep representations by jointly conditioning on both left and right context of each word, and it has recently been proven surprisingly effective in many downstream tasks [9]. BERT learns the feature representation for words by two core tasks on the basis of the massive corpus, so that we use the encoded vector produced by it as the embedding for subsequent tasks.

3.2 Cause Tagger

We employ the tagging module to recognize all the cause events in the input sentence after acquiring the encoded vector $\boldsymbol{H}$ which is produced by BERT. Tagging module constituted by two binary classifiers is used to detect the start and end position of the cause events in the input sentence. To find the specific positions, the tagging module first assign a binary tag(0/1) to each token, and then complete the probability of the start and end positions, if the token is we hope to find, it will be assigned with a tag 1, otherwise with a tag 0. We transform the task of predicting the start or end position of cause events into the task of predicting whether the position is the start or end position of the cause events by adopting Sigmoid as the activation function. The formulas are as follows:

$$p_i^{cause\_start} = \sigma(W_{cause\_start}x_i + b_{cause\_start}) \tag{1}$$

$$p_i^{cause\_end} = \sigma(W_{cause\_end}x_i + b_{cause\_end}) \tag{2}$$

Where σ is the Sigmoid activation function, $p_i^{cause\_start}$ and $p_i^{cause\_end}$ respectively represent the probability of the i-th token as the start and end position of a cause event in the input sentence. If the probability exceeds a certain threshold(the threshold will be discussed in the experiment), this token will be considered to be the start or end position, and it will be assigned with a tag 1. x_i represents the encoded vector which produced by the pro-trained BERT model, W_{cause_start} and W_{cause_end} represent trainable weight, b_{cause_start} and b_{cause_end} represent the bias.

3.3 Effect Extraction

It is easy to observed that the text of our task is similar to the text of the question answering task, the answers what we want to find are text spans. So we apply the method of question answering task to extract the effect events in the sentences.

Since there are multiple cause events in one sentence, we randomly select an effect event and the incomplete question template to generate a question sentence during the training processing. Because the length of the question sentence is different from the input sentence, we fill the lacking characters with 0 to the same length as the input sentence. And we employ the pre-trained BERT model to encode the sentences, so the encoded vector $\boldsymbol{Q}$ be produced. Then we concatenate the vector $\boldsymbol{Q}$ and the vector $\boldsymbol{H}$ and feed them into a bidirectional long and short-term memory networks (Bi-LSTM) layer, Bi-LSTM helps us to obtain contextual information at a longer distance, and it uses a forward LSTM [13] and a backward LSTM for each sequence to obtain two separate hidden states: $\overrightarrow{l_i}$, $\overleftarrow{l_i}$, the output is formed by concatenating two separate hidden states:

$$l_i = [\overrightarrow{l_i}, \overleftarrow{l_i}] \tag{3}$$

Then we use the tagging module to recognize all the effect events similarly:

$$p_i^{effect\_start} = \sigma(W_{effect\_start} l_i + b_{effext\_start}) \tag{4}$$

$$p_i^{effect\_end} = \sigma(W_{effect\_end} l_i + b_{effect\_end}) \tag{5}$$

where $p_i^{effect\_start}$ and $p_i^{effect\_end}$ respectively represent the probability of the i-th token as the start and end position of the effect events in the input sentence, W_{effect_start} and W_{effect_end} represent trainable weight, b_{effect_start} and b_{effect_end} represent the bias.

As shown in Fig. 3, after the input sentence being encoded by BERT, we first use the cause tagger to extract the cause events, and then select the event "“天气炎热(hot weather)" to combine with the incomplete question sentence, so we get the complete question "天气炎热的结果是什么? (What is the effect of hot weather?)", then we concatenate the encoded question vector and encoded input sentence vector, and feed them into Bi-LSTM to obtain the contextual information, finally by employing the effect tagger we get the effect event "爆胎(flat tire)".

4 Experiment

4.1 Dataset

We collected thousands of traffic accident reports from the Internet as the corpus, then we cleaned and processed the reports to obtain a traffic accident dataset. When we label the tags, since the judgment of causality is strongly subjective, in order to ensure the objectivity of the dataset, we choosed four annotators to label the causality relations in the sentence at the same time, and only the consistent results will be selected as the final tag. Our dataset obtains 1256 causality sentences, the dataset is divided into the training set and validation set respectively obtain 1098 sentences and 158 sentences. Due to the small number of the training set, we adopted Easy Data Augmentation (EDA) [10] to enhance the training data. After data enhancement, a total of 2155 training data were obtained. The specific quantity description of the dataset is shown in Table 1.

Table 1. The specific quantity description of the dataset.

Dataset	Causal sentences	Causal triplets	The Sentence consist nested casal relation
Training Set	2155	3261	904
Validation Set	158	480	63

4.2 Results

Comparison. [6] is proposed to solve the overlapping triple problem where multiple relational triples in the same sentence share the same entities, they also employ the strategy of point net to extract the subject and object. Their work gives us much help, and their goals are similar to ours, so we took this as a baseline. The experiment for comparison between baseline and ours is shown as results in Table 2. Compared with the baseline, our model performs better on classification for cause events and effect events, F1-score is improved by 8.71% on the cause events and 9.32% on the effect events on the dataset which contains all sentences. It shows our method is stronger in extracting events. In addition, F1-score and precision respectively improved by 19.22% and 12.98% on the same dataset when extracting the all causal-effect pairs in the sentences, it reveals that our model successfully extracts more cause-effect pairs, this reflects the effectiveness of our method. However, we see the poor extraction power on the nested dataset. It may be caused by follow reasons: (i) Our dataset is too small, leads to our model has low performance. (ii) Due to the particularity of the traffic accident text, a sentence even contains 14 cause-effect pairs, its complexity makes it is difficult for us to extract all relations accurately.

Table 2. Results(%) of evaluation for comparison

Method	Dataset	Cause Events			Effect events			Complete[a]		
		P	R	F	P	R	F	P	R	F
Baseline	Nested[b]	75.00	25.00	37.50	64.58	15.74	25.31	25.00	8.89	13.11
	All[c]	71.92	46.88	56.76	79.41	35.76	49.35	51.47	32.99	40.20
Ours	Nested	88.64	30.71	45.61	83.08	27.14	40.91	60.00	20.96	31.06
	All	85.92	52.89	65.47	84.05	45.07	58.67	**70.69**	**42.62**	**53.18**

[a] Completely extract all the causal-effect pairs in the sentences.
[b] Only contains sentences with nested causal relations.
[c] Contains all sentences in our dataset.

Discussion on the Thresholds. By two binary classifiers we can get the probability of the tokens as the start and end position of the cause events and effect events in input sentences, when the probability exceeds a certain threshold, the tokens will be considered to be the start or end positions. Therefore, the value of threshold is very important. If it is too large, numerous events that should

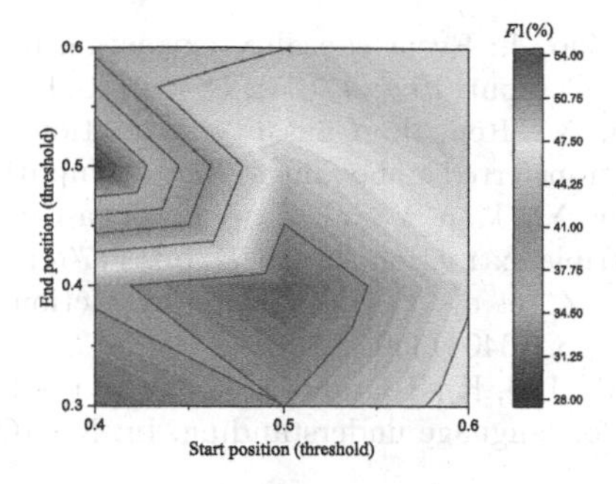

Fig. 4. F1-score of extracting cause-effect pairs from sentences with different threshlod. The redder the color, the better the extraction effect.

have been successfully extracted will be missed. If it is too small, many wrong events will be extracted. So we need to discuss the value of the threshold. We controlled the threshold of start position and end position range from 0.3 to 0.6, and compare the value of F1-score, the specific results are shown in Fig. 4. We can see when the threshold of start position is 0.5, and the threshold of end position is 0.4, the extraction result is best.

5 Conclusion

In this work, we proposed a model to solve the problem of extracting nested causal relations in the text. We divide the extraction method into two steps. First, extract all the cause events in the sentence, and then combine the cause events with the incomplete question template, so that the task of extracting the result events is completed by the question answering method. Experiments have proved that our proposed model can effectively finish the extraction of nested causality. Since our dataset is small, the accuracy of extraction needs to be improved. In future work, we will continue to enrich our dataset, and combine the characteristics of the field of traffic accidents to better improve the accuracy.

Acknowledgement. This work was supported by the NSFC (No. 61803337).

References

1. Chen, D., Cao, Y., Luo, P.: Pairwise causality structure: towards nested causality mining on financial statements. In: Zhu, X., Zhang, M., Hong, Yu., He, R. (eds.) NLPCC 2020. LNCS (LNAI), vol. 12430, pp. 725–737. Springer, Cham (2020). https://doi.org/10.1007/978-3-030-60450-9_57
2. Cao, Y., Chen, D., Li, H., Luo, P.: Nested relation extraction with iterative neural network. In: CIKM, pp. 1001–1010. ACM (2019)
3. Xu, J., Zuo, W., Liang, S., Wang, Y.: Causal relation extraction based on graph attention networks. J. Comput. Res. Dev. **57**(1), 159–174 (2020)

4. Zheng, Q., Wu, Z., Zuo, J.: Event causality extraction based on two-layer CNN-BiGRU-CRF model. Comput. Eng. **47**(05), 58–64 (2021)
5. Li, Z., Li, Q., Zou, X., Ren, J.: Causality extraction based on self-attentive BiLSTM-CRF with transferred embeddings. Neurocomputing **423**, 207–219 (2021)
6. Wei, Z., Su, J., Wang, Y., Tian, Y., Yi, C.: A novel cascade binary tagging framework for relational triple extraction. In: ACL, pp. 1476–1488 (2020)
7. Li, X., Yin, F., Sun, Z., et al.: Entity-relation extraction as multi-turn question answering. In: ACL, pp. 1340–1350 (2019)
8. Devlin, J., Chang, M., Lee, K., Toutanova, K.: Bert: pre-training of deep bidirectional transformers for language understanding. In: NAACL-HLT, pp. 4171–4186 (2019)
9. Zhong, P., Wang, D, Miao, C.: Knowledge-enriched transformer for emotion detection in textual conversations. In: EMNLP-IJCNLP, pp. 165–176 (2019)
10. Wei, J., Zou, K.: EDA: easy data augmentation techniques for boosting performance on text classification tasks. In: EMNLP-IJCNLP, pp. 6382–6388 (2019)
11. Li, X., Feng, J., Meng, Y., et al.: A unified MRC framework for named entity recognition. In: ACL, pp. 5849–5859 (2020)
12. Gu, Y., Yang, M., Lin, P.: Lightweight multiple perspective fusion with information enrich for BERT-based answer selection. In: NLPCC 2020, pp. 543–554(2020)
13. Hochreiter, S., Schmidhuber, J.: Long short-term memory. Neural Comput. **9**(8), 1735–1780 (1997)

Evaluation Workshop

MSDF: A General Open-Domain Multi-skill Dialog Framework

Yu Zhao, Xinshuo Hu, Yunxin Li, Baotian Hu(✉), Dongfang Li, Sichao Chen, and Xiaolong Wang

Harbin Institute of Technology, Shenzhen, China
hubaotian@hit.edu.cn, xlwangsz@hit.edu.cn

Abstract. Dialog systems have achieved significant progress and have been widely used in various scenarios. The previous researches mainly focused on designing dialog generation models in a single scenario, while comprehensive abilities are required to handle tasks under various scenarios in the real world. In this paper, we propose a general **M**ulti-**S**kill **D**ialog **F**ramework, namely **MSDF**, which can be applied in different dialog tasks (e.g. knowledge grounded dialog and persona based dialog). Specifically, we propose a transferable *response generator* pre-trained on diverse large-scale dialog corpora as the backbone of MSDF, consisting of BERT-based encoders and a GPT-based decoder. To select the response consistent with dialog history, we propose a *consistency selector* trained through negative sampling. Moreover, the flexible copy mechanism of external knowledge is also employed to enhance the utilization of multiform knowledge in various scenarios. We conduct experiments on knowledge grounded dialog, recommendation dialog, and persona based dialog tasks. The experimental results indicate that our MSDF outperforms the baseline models with a large margin. In the Multi-skill Dialog of 2021 Language and Intelligence Challenge, our general MSDF won the 3rd prize, which proves our MSDF is effective and competitive.

Keywords: Multi-skill dialog · Knowledge grounded dialog · Conversational recommendation · Persona based dialog

1 Introduction

Propelled by the acquisition of large-scale dialog corpora and the advance of pre-training technology, dialog generation models have made great progress, and dialog systems have been applied in various scenarios, such as chit-chat, knowledge grounded dialog, and conversational recommendation. However, most of the previous works only focused on modeling the dialog system within a single scenario, which is difficult to be applied in other scenarios directly. It does not

Y. Zhao and X. Hu—Contribute equally to this work.

L. Wang et al. (Eds.): NLPCC 2021, LNAI 13029, pp. 365–376, 2021.
https://doi.org/10.1007/978-3-030-88483-3_29

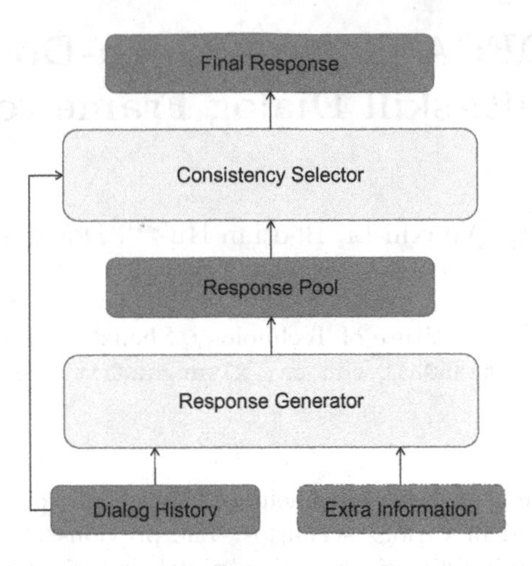

Fig. 1. Overall Multi-Skill Dialog Framework (MSDF). The response generator first generates diverse responses through a sampling-based decoding algorithm conditioned on the dialog history and optional extra information (e.g. knowledge, user profile and/or machine persona). Then, the consistency selector selects the most contextually consistent response as the final response.

satisfy the requirements of practical application in the real world, where the dialog model needs to generate responses in various scenarios. To handle different tasks in various scenarios, multiple dialog skills are requested, such as knowledge utilization, commodities recommendation, and persona understanding skills. It is of great necessity to model a general multi-skill dialog framework that can be flexibly applied in various scenarios.

How to model the general multi-skill dialog systems that can effectively use information from diverse sources still remains challenging. On the one hand, the model needs to use various information (e.g. structured and unstructured knowledge, persona, conversation topics), and the previous works usually design complex models [6,11] to utilize specific information in a single scenario, which results in lacking universality. On the other hand, the previous works used complicated data processes and training processes [9] to optimize models on specific dialog corpus, thus, the model is difficultly transferred to other scenarios.

In this work, we propose the general multi-skill dialog framework **MSDF** to address the above problems, which consists of a pre-trained dialog *response generator* and a *consistency selector*. As depicted in Fig. 1, MSDF generates responses in two stages: 1) generating diverse responses as the candidate *response pool*, via *dialog history* (and *extra information*); 2) selecting the *final response* by consistency selector. Specifically, we first pre-train a universal and transferable encoder-decoder based model on various diverse dialog corpora, including chit-chat, knowledge dialog, and recommendation dialog, to obtain a general model with multiple coarse-grained skills enhanced. Then, we apply multiple identical encoders to encode different source information and equip the decoder with the multi-source information fusion module, which can be flexibly transferred to different application scenarios. Moreover, a dialog history can be mapped into multiple acceptable responses, which is also known as the one-to-many mapping [13], especially in the open domain. Thus, we introduce a BERT-based consistency selector to choose the most contextually consistent response with dialog history from the response pool, which is trained via negative sampling to distinguish consistent and inconsistent responses with dialog history.

Experiments are conducted to evaluate the performance of our MSDF in knowledge grounded dialog, recommendation dialog, and persona dialog tasks. The experimental results indicate that our MSDF outperforms the baseline models with a large margin in terms of F1 and BLEU scores. In the Multi-skill Dialog of 2021 Language and Intelligence Challenge, our MSDF won the third prize, with 6th rank in the human evaluation of the finals. Both automatic and human evaluation results indicate that our MSDF is effective and competitive.

Our contributions are as bellows.

- We propose a general multi-skill dialog framework that solves various tasks in different scenarios, namely MSDF, consisting of a dialog *response generator* with multi-source information encoders and a *consistency selector*.
- We pre-train an encoder-decoder based dialog generation model on various types of large-scale open-domain dialog corpora, which can be effectively transferred into our MSDF to solve different tasks.
- The experimental results indicate that our MSDF outperforms the baseline models with a large margin in knowledge dialog, recommendation dialog, and persona dialog tasks, demonstrating the effectiveness of MSDF.

2 Related Work

The Multi-Skill Dialog of 2021 Language and Intelligence Challenge focuses on three kinds of dialog generation tasks, including the knowledge grounded dialog, recommendation dialog, and persona dialog. The previous works always modeled dialog systems to solve a single task in a scenario. For knowledge grounded dialogs, the model generates responses conditioned on dialog history and the given external knowledge. [8] employed posterior probability distribution of knowledge to guide the learning of prior distribution during training. [6] proposed the sequential decision making method to select knowledge used in response. And [12] modeled the knowledge selection as walking over knowledge graphs. For the persona dialogs, the model is required to understand and utilize the given persona to generate personalized responses. [9] improved the performance of persona understanding through persona negative sampling. [22] trained a VAE (Variational Auto-encoder) model to produce persona-related topic words jointly generating responses. For recommendation dialogs, the model needs to make recommendations through conversations, which is usually separated to two tasks: recommendation and dialog generation. [4,27] incorporated the common sense knowledge graph to improve the user profile understanding and recommendation dialog generation. [28] incorporated topic planning to enhance the recommendation dialog. Following the work of [3], we also propose a general multi-skill dialog framework that can handle various tasks.

3 Method

3.1 Multi-skill Dialog

The multi-skill dialog task aims to construct a general dialog generation model with multiple skills to solve various tasks in various scenarios. In our work, we focus on the dialog modeling task with multiple skills, including three sub-tasks: 1) knowledge dialog, 2) recommendation dialog, and 3) persona dialog. The descriptions of the three sub-tasks are as follows[1] (Fig. 2).

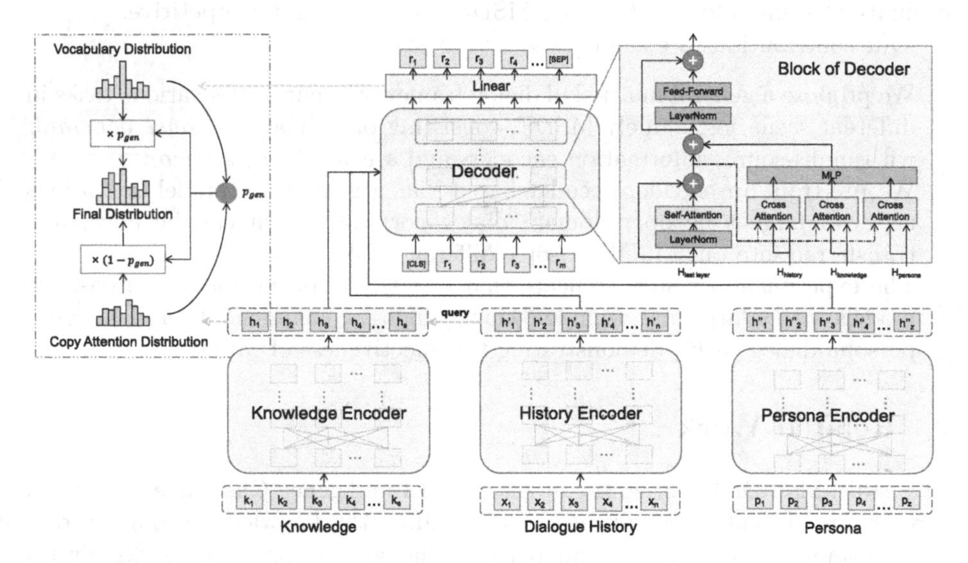

Fig. 2. Responses generation process in MSDF. The top-right corner is the detail of decoder block, and the top-left corner shows the process of copy mechanism.

Knowledge Dialog. The inputs of the knowledge dialog include dialog goals $\mathbf{G} = [\mathbf{g}_1, \mathbf{g}_2, \ldots, \mathbf{g}_r]$, knowledge information $\mathbf{K} = [\mathbf{k}_1, \mathbf{k}_2, \ldots, \mathbf{k}_s]$, and dialog history $\mathbf{H} = [\mathbf{x}_1, \mathbf{x}_2, \ldots, \mathbf{x}_n]$, where $\mathbf{g}_i$ is a description of dialog target or a conversation topic, $\mathbf{k}_i$ is the form of a triple or an unstructured sentence, and $\mathbf{x}_i$ consists of a sequence of words. The knowledge grounded dialog generation task requires the model to use appropriate knowledge to generate the response $\mathbf{R} = [r_1, r_2, \ldots, r_m]$, where r_i denotes the i-th word in response, m denotes the length of response.

[1] Details in: https://aistudio.baidu.com/aistudio/competition/detail/67.

Recommendation Dialog. The inputs of the recommendation dialog include user profiles $\mathbf{U} = [\mathbf{u}_1, \mathbf{u}_2, \ldots, \mathbf{u}_t]$, conversation scenario $\mathbf{S}$, dialog targets $\mathbf{G}$, knowledge information $\mathbf{K}$, and dialog history $\mathbf{H}$, where $\mathbf{u}_i$ is an aspect of the user profile in a key-value form. The recommendation dialog generation task requires the model to make recommendations based on the user profile and the scenario through conversation.

Persona Dialog. The inputs of the persona dialog include persona information of machine $\mathbf{P} = [\mathbf{p}_1, \mathbf{p}_2, \ldots, \mathbf{p}_z]$, and dialog history $\mathbf{H}$, where $\mathbf{p}_i$ is a sentence describing the persona. The persona dialog generation task requires the model to generate natural, fluent and informative machine response in line with dialogue history and persona information.

3.2 Overall Framework

We propose a general multi-skill dialog framework MSDF, mainly consisting of a response generator and a history-response consistency selector, as shown in Fig. 1. MSDF generates the response in two steps: response generator is applied to generate diverse responses as the response pool, and then consistency selector chooses the most consistent response from the response pool as the final output.

Response Generator. The previously proposed generation models, such as GPT2 [14], PLATO2 [1], and BART [7], have shown their great generation performance obtaining from large-scale corpora. To obtain a universal pre-trained dialog model that can be flexibly transferred to the three sub-tasks, we pre-train an encoder-decoder-based model, which consists of a BERT-based encoder and a GPT-based decoder, and the optimization objective is to generate a target response conditioned on dialog history. The six large-scale dialog datasets, including LCCC [19], Weibo [17], Douban [21], DuConv [20], KdConv [26] and DuRecDial [11], are used to pre-train the model. Specifically, we process all the data to a general dialog form: for the multi-turn conversations, we split them into history-response pairs; for the conversations with extra knowledge or user profiles, we ignore them during pre-training. Moreover, to accelerate the pre-training process, we use the parameters of DialoGPT [24] to initialize both the encoder and decoder. We find that the model pre-trained with all data tends to generate short responses, thus, we additionally select the half of data with longer responses to pre-train it in the latter part of pre-training phase. In each sub-task, we duplicate the encoder of pre-trained model to encode specific information, and equip the decoder with an information fusion module.

Denote $\mathbf{H}_{l-1}^{D} \in \mathbb{R}^{L_d \times d}$ as the output of $(l-1)$-th decoder block, where L_D is the length of decoding sequence and d is the hidden size of decoder blocks, the calculating process in each decoder block is described as follows:

$$\mathbf{H}_l^{SA} = SelfAttention\left(LayerNorm(\mathbf{H}_{l-1}^{D})\right) + \mathbf{H}_{l-1}^{D}, \quad (1)$$

$$\mathbf{H}_l^{FA} = Fusion\left(\mathbf{H}_l^{SA}, \mathbf{H}_{history}, \mathbf{H}_{knowledge}, \mathbf{H}_{persona}\right), \quad (2)$$

$$\mathbf{H}_l^D = FFN\left(LayerNorm(\mathbf{H}_l^{FA})\right) + \mathbf{H}_l^{FA}, \tag{3}$$

where $\mathbf{H}_{history} \in \mathbb{R}^{n \times d}$, $\mathbf{H}_{knowledge} \in \mathbb{R}^{s \times d}$ and $\mathbf{H}_{persona} \in \mathbb{R}^{z \times d}$ denotes the encoded information, $\mathbf{H}_l^{SA} \in \mathbb{R}^{L_d \times d}$ denotes the result of self-attention, and $\mathbf{H}_l^{FA} \in \mathbb{R}^{L_d \times d}$ denotes the fused multi-source information. The $Fusion$ function is calculated in the following way:

$$\mathbf{A}_l(\mathbf{H}^E) = softmax\left(\frac{\left(LayerNorm(\mathbf{H}_l^{SA})\mathbf{W}_l^Q\right)(\mathbf{H}^E\mathbf{W}_l^K)^T}{\sqrt{d}}\right)(\mathbf{H}^E\mathbf{W}_l^V), \tag{4}$$

$$\mathbf{H}_l^{FA} = [\mathbf{A}_l(\mathbf{H}_{history}); \mathbf{A}_l(\mathbf{H}_{knowledge}); \mathbf{A}_l(\mathbf{H}_{persona})]\mathbf{W}_l^P + \mathbf{H}_l^{SA}, \tag{5}$$

where [;] indicates concatenation on the last dimension, $\mathbf{W}_l^Q$, $\mathbf{W}_l^K$ and $\mathbf{W}_l^V$ are learnable projection matrices and $\mathbf{W}_l^P \in \mathbb{R}^{3d \times d}$ is a learnable attention fusing matrix (we get $\mathbf{W}_l^P \in \mathbb{R}^{2d \times d}$ when there are only two cross-attention to be fused).

Finally, the hidden state $\mathbf{H}_l^D \in \mathbb{R}^{L_D \times d}$ of the l-th decoder layer is output by the FFN layer and residual layer:

$$\mathbf{H}_l^D = FFN(LayerNorm(\mathbf{H}_l^{FA})) + \mathbf{H}_l^{FA}. \tag{6}$$

With the great performance of hybrid pointer-generator network [15] in summarization, we utilize the attention-based copy mechanism to generate knowledge-enhanced response. Decoder hidden states are put into linear language model to get original vocabulary distribution $\mathbf{P}_{vocab} \in \mathbb{R}^{L_D \times L_V}$,

$$\mathbf{P}_{vocab} = softmax(\mathbf{H}^D\mathbf{W}^{LM}), \tag{7}$$

where $W^{LM} \in \mathbb{R}^{d \times L_V}$ is the learnable language model head and L_V denotes the vocabulary length. And then attention-based copy mechanism is utilized to generate extra knowledge enhanced response. We obtain knowledge copy attention $\mathbf{A}_{copy} \in \mathbb{R}^{L_D \times L_K}$, via cross-attention of decoded hidden states and encoded knowledge (or persona) hidden state:

$$\mathbf{A}_{copy} = softmax((\mathbf{H}^D\mathbf{W}^Q)(\mathbf{H}_{knowledge}\mathbf{W}^K)^T), \tag{8}$$

where $\mathbf{W}^Q$ and $\mathbf{W}^K$ are learnable projection matrices, and L_K denotes the length of the knowledge sequence. Then generation probability $\mathbf{p}_{gen} \in [0, 1]$ can be calculated:

$$\mathbf{p}_{gen} = sigmoid([\mathbf{A}_{copy}\mathbf{H}_{knowledge}; \mathbf{H}^{\mathbf{D}}]\mathbf{W}^{mlp}), \tag{9}$$

where $W^{mlp} \in R^{2d \times 1}$ is learnable matrices. We obtain the following probability distribution over the merged vocabulary to predict word w:

$$\mathbf{P}(w) = \log(\mathbf{p}_{gen}\mathbf{P}_{vocab}(w) + (1 - \mathbf{p}_{gen})\mathbb{I}(w_i = w)A_{copy}(w)), \tag{10}$$

where $\mathbb{I}(\cdot)$ is an indicator function, and $A^{copy}(w)$ is the element at w-th column in the copy attention matrix.

Consistency Selector. Inspired by the BERT next sentence prediction [5], we propose a history-response consistency selector, since the response should be the next sentence by dialog history. The consistency selector is a binary classifier to distinguish the consistent and inconsistent response with dialog history. We construct the consistency selector by a pre-trained model RoBBERTa [10], plus a linear head layer. We first concatenate the dialog history and response as the pre-trained model input to get context representation, and then put the first token (usually known as [CLS] token) to the linear head to get a binary classification score. The consistency selector is trained on positive and negative sampling examples from training data, where inconsistent responses are randomly sampled. During inference, we get the positive score from the consistency selector output as the consistency score.

3.3 Data Processing

We separate different input resources into four categories: dialog history, knowledge information, persona information, and current reply (or previously generated response during inference). The data preprocessing and reprocessing are demonstrated as follows.

The knowledge dialog generation model consists of a dialog history encoder, a knowledge encoder and a decoder. During data preprocessing of DuConv, we reformat the knowledge graph to a pseudo unstructured knowledge sentences, by concatenating the subject, predicate and object, and join all knowledge sentence with special token "[SEP]". Since the dialog topics are limited in movies and film stars and there are at most two topics during a dialog, we introduce special tokens to format the subject in the knowledge graph, including "[movie1]", "[movie2]", "[star1]", and "[star2]", which will be restored in data reprocessing after response generated. And we also add speaker tokens "[speaker1]" and "[speaker2]" to distinguish speakers in the dialog history. We also reprocess personal pronouns and figures due to the knowledge, such as outcome date of movies, and birthday of stars.

The recommendation dialog generation model consists of a dialog history encoder, a knowledge encoder, a persona encoder and a decoder. To facilitate optimization, we make the knowledge encoder and persona encoder share parameters. The knowledge graph in DuRecDial is also reformatted in the same way as DuConv, with all the candidate recommendation goal planning concatenated at the end. Since the recommendation goal is labeled with each response, we introduce a new special token "[goal]" as a separator, and join the golden goal with response e.g. "[goal]问User性别[goal]我该称呼您是先生还是女士" ("[goal] Asking about the user's gender [goal] Should I call you Mr. or MS"). This golden goal prefix in the response performs as the semantic guidance, which is typically like conditional generation. The dialog situation is viewed as extra knowledge concatenated before knowledge information. We also replace the user name with a new special token "[uname]" and add speaker tokens. Personal pronouns and figures will be corrected and goal information will be removed during data reprocess.

The persona dialog generation model consists of a dialog history encoder, a persona encoder and a decoder. We trained a simple Word2Vec by open-source gensim implementation[2], on CPC sentences to estimate the similarity between the response and persona by cosine similarity of average word vector, and randomly drop training examples with persona information similarity less than 0.7. It is observed that the generator prefers to generate longer sentences and there are too many responses whose topic is about the user job, so we drop some examples with too long responses and "工作" ("work") in dialog utterances. Considering that conversations in CPC are more like chit-chat than task-oriented dialog, we abandon the copy mechanism in this generator.

4 Experiment

4.1 Experimental Settings

Dataset. For pre-training, we use the Weibo dataset [17], Douban multi-round conversation [21], LCCC dataset [19], Emotional Conversational Dataset [25], retrieval-assisted conversational dataset [2], and Kdconv dataset [26]. For fine-tuning, we use DuConv [20] for knowledge grounded dialog, DuRecDial [11] for the conversational recommendation, and Chinese persona chat (CPC) for persona dialog. Statistics of all dataset are summarized in Table 1.

Table 1. Statistics of all datasets. † denotes the datasets for fine-tuning, and the others are for pre-training.

	Train	Dev	Test
DuConv† [20]	19858	2000	5000
DuRecDial† [11]	6618	946	4645
Chinese Persona Chat†	23000	1500	3000
Weibo dataset [17]	3103764	443394	886790
Douban multi-round conversation [21]	5000000	25001	1186
LCCC dataset [19]	11987759	20000	10000
Emotional Conversational Dataset [25]	899207	110000	110000
retrieval-assisted conversational dataset [2]	5498480	107332	156706
Kdconv dataset [26]	3000	300	2751

Implementation Details. For the response generator, the encoders and decoder settings follow the DialoGPT [24], where the hidden size is 768 for 12 layers, the maximum input length is 512, and there are 12 heads in multi-head attention. For all dropout layers, the dropout rate is set to 0.1. We adopt

[2] https://radimrehurek.com/gensim/intro.html.

cross-entropy loss as our loss function, and the parameters are saved in term of the minimum cross-entropy loss on the development datasets. The consistency selector is implemented alike a NLI model as described in Sect. 3.2, which is also optimized through cross-entropy loss function.

4.2 Evaluation

The competitive baseline models are compared with the proposed MSDF on different quantitative metrics. We use BLEU-1 and BLEU-2 to evaluate the n-gram lexical similarity, F1 to evaluate the Chinese character level similarity, and DISTINCT1 and DISTINCT-2 to evaluate the diversity of generated responses. The total SCORE for the automatic evaluation is calculated by averaging all the F1/BLEU1/BLEU2 scores for the subtasks. To exhibit the improvement of our MSDF, we also implement and test four baseline models: Seq2Seq [18], HRED [16], DialoGPT [24] and BERT-GPT [23]. All these baseline models only take dialog history as inputs, without aquiring extra information. Since test datasets are not released, we evaluate all the models on LUGE platform[3]. The performance of baseline models and our MSDF is presented in Table 2, including ablation experiments.

With respect to the human evaluation, we refer readers to the competition leaderboard[4] for the single-turn or multi-turn dialog evaluation results.

Table 2. Automatic evaluation of test set B on baseline models and our MSDF (ranked by total score). All the results of F1, BLEU, and DISTINCT are average scores from three sub-tasks.

	SCORE	F1	BLEU1/2	DISTINCT1/2
Seq2Seq [18]	0.522	22.33	0.202/0.096	0.038/0.100
HRED [16]	0.565	23.66	0.220/0.108	0.038/0.105
DialoGPT [24]	0.373	17.05	0.141/0.061	**0.079/0.313**
BERT-GPT [23]	0.573	24.51	0.215/0.113	0.061/0.214
MSDF	**0.934**	**38.62**	**0.333/0.215**	0.057/0.183
-consistency selector	0.872	36.00	0.314/0.198	0.051/0.173
-extra knowledge	0.705	28.98	0.271/0.145	0.049/0.162

4.3 Discussion

Our MSDF outperforms the baseline models with a large margin, even though without extra information, which strongly presents the effectiveness. The consistency selector preferred to select the most common response. It reduces the

[3] LUGE: https://aistudio.baidu.com/aistudio/competition/detail/48.

[4] Leaderboard: https://aistudio.baidu.com/aistudio/competition/detail/67.

variance of generating performance and significantly improves automatic evaluation results, without considering the limited diversity of responses in human evaluation. Besides, the attention-based copy mechanism is of great importance for generating knowledge-enhanced response. According to our observation, DialoGPT and BERT-GPT still talk rubbish after fine-tuning, despite resulting in higher DISTINCT scores. In the competition, our MSDF got 9-th rank in automatic evaluation and 6-th rank in human evaluation. It increased by 3 ranks in human evaluation compared to the automatic evaluation. We attribute this to the multi-skill pre-training, from which our model could generate more human-like responses, in spite of the mismatch with the golden references in automatic evaluation.

5 Conclusion

This paper describes our general multi-skill dialog framework MSDF, consisting of a response generator and a dialog history consistency selector. We first pre-train the basic encoder-decoder on diverse datasets and then fine-tune it on the specific dataset to construct the response generator with a strong specific skill. We won the third prize in Multi-task Dialog of 2021 Language and Intelligence Challenge. Experiments are also conducted on several baseline models. The vast improvement over the baseline models indicates that our framework is effective and competitive. In future work, we will experiment with our MSDF on more tasks to evaluate the comprehensive skills of our framework and further improve its performance.

Acknowledgement. We appreciate the beneficial and insightful feedback from the anonymous reviewers and Baidu Inc. This work is jointly supported by grants: Natural Science Foundation of China (No. 62006061 and 61872113), Strategic Emerging Industry Development Special Funds of Shenzhen (JCYJ20200109113403826 and JCYJ20200109113441941) and Stable Support Program for Higher Education Institutions of Shenzhen (No. GXWD20201230155427003-20200824155011001).

References

1. Bao, S., et al.: Plato-2: towards building an open-domain chatbot via curriculum learning. arXiv preprint arXiv:2006.16779 (2020)
2. Cai, D., Wang, Y., Bi, W., Tu, Z., Liu, X., Shi, S.: Retrieval-guided dialogue response generation via a matching-to-generation framework. In: Proceedings of the 2019 Conference on Empirical Methods in Natural Language Processing and the 9th International Joint Conference on Natural Language Processing (EMNLP-IJCNLP), pp. 1866–1875 (2019)
3. Cao, Y., Bi, W., Fang, M., Tao, D.: Pretrained language models for dialogue generation with multiple input sources. In: Proceedings of the 2020 Conference on Empirical Methods in Natural Language Processing: Findings, pp. 909–917 (2020)

4. Chen, Q., et al.: Towards knowledge-based recommender dialog system. In: Proceedings of the 2019 Conference on Empirical Methods in Natural Language Processing and the 9th International Joint Conference on Natural Language Processing (EMNLP-IJCNLP), pp. 1803–1813 (2019)
5. Devlin, J., Chang, M.W., Lee, K., Toutanova, K.: Bert: pre-training of deep bidirectional transformers for language understanding. In: Proceedings of the 2019 Conference of the North American Chapter of the Association for Computational Linguistics: Human Language Technologies, Volume 1 (Long and Short Papers), pp. 4171–4186 (2019)
6. Kim, B., Ahn, J., Kim, G.: Sequential latent knowledge selection for knowledge-grounded dialogue. In: International Conference on Learning Representations (2019)
7. Lewis, M., et al.: Bart: Denoising sequence-to-sequence pre-training for natural language generation, translation, and comprehension. In: Proceedings of the 58th Annual Meeting of the Association for Computational Linguistics, pp. 7871–7880 (2020)
8. Lian, R., Xie, M., Wang, F., Peng, J., Wu, H.: Learning to select knowledge for response generation in dialog systems. In: IJCAI International Joint Conference on Artificial Intelligence, p. 5081 (2019)
9. Liu, Q., et al.: You impress me: dialogue generation via mutual persona perception. In: Proceedings of the 58th Annual Meeting of the Association for Computational Linguistics, pp. 1417–1427 (2020)
10. Liu, Y., et al.: Roberta: a robustly optimized bert pretraining approach. arXiv preprint arXiv:1907.11692 (2019)
11. Liu, Z., Wang, H., Niu, Z.Y., Wu, H., Che, W., Liu, T.: Towards conversational recommendation over multi-type dialogs. In: Proceedings of the 58th Annual Meeting of the Association for Computational Linguistics, pp. 1036–1049 (2020)
12. Moon, S., Shah, P., Kumar, A., Subba, R.: Opendialkg: explainable conversational reasoning with attention-based walks over knowledge graphs. In: Proceedings of the 57th Annual Meeting of the Association for Computational Linguistics, pp. 845–854 (2019)
13. Ni, J., Young, T., Pandelea, V., Xue, F., Adiga, V., Cambria, E.: Recent advances in deep learning-based dialogue systems. arXiv preprint arXiv:2105.04387 (2021)
14. Radford, A., Wu, J., Child, R., Luan, D., Amodei, D., Sutskever, I.: Language models are unsupervised multitask learners. OpenAI blog **1**(8), 9 (2019)
15. See, A., Liu, P.J., Manning, C.D.: Get to the point: summarization with pointer-generator networks. In: Proceedings of the 55th Annual Meeting of the Association for Computational Linguistics (Volume 1: Long Papers), pp. 1073–1083 (2017)
16. Serban, I.V., Sordoni, A., Bengio, Y., Courville, A., Pineau, J.: Building end-to-end dialogue systems using generative hierarchical neural network models. In: Thirtieth AAAI Conference on Artificial Intelligence (2016)
17. Shang, L., Lu, Z., Li, H.: Neural responding machine for short-text conversation. In: Proceedings of the 53rd Annual Meeting of the Association for Computational Linguistics and the 7th International Joint Conference on Natural Language Processing (Volume 1: Long Papers), pp. 1577–1586 (2015)
18. Sutskever, I., Vinyals, O., Le, Q.V.: Sequence to sequence learning with neural networks. Adv. Neural. Inf. Process. Syst. **27**, 3104–3112 (2014)
19. Wang, Y., Ke, P., Zheng, Y., Huang, K., Jiang, Y., Zhu, X., Huang, M.: A large-scale Chinese short-text conversation dataset. In: CCF International Conference on Natural Language Processing and Chinese Computing, pp. 91–103. Springer (2020)

20. Wu, W., Guo, Z., Zhou, X., Wu, H., Zhang, X., Lian, R., Wang, H.: Proactive human-machine conversation with explicit conversation goal. In: Proceedings of the 57th Annual Meeting of the Association for Computational Linguistics, pp. 3794–3804 (2019)
21. Wu, Y., Wu, W., Xing, C., Zhou, M., Li, Z.: Sequential matching network: a new architecture for multi-turn response selection in retrieval-based chatbots. In: Proceedings of the 55th Annual Meeting of the Association for Computational Linguistics (Volume 1: Long Papers), pp. 496–505 (2017)
22. Xu, M., et al.: A neural topical expansion framework for unstructured persona-oriented dialogue generation. arXiv preprint arXiv:2002.02153 (2020)
23. Zeng, G., et al.: Meddialog: a large-scale medical dialogue dataset. In: Proceedings of the 2020 Conference on Empirical Methods in Natural Language Processing (EMNLP), pp. 9241–9250 (2020)
24. Zhang, Y., et al.: Dialogpt: large-scale generative pre-training for conversational response generation. In: Proceedings of the 58th Annual Meeting of the Association for Computational Linguistics: System Demonstrations, pp. 270–278 (2020)
25. Zhou, H., Huang, M., Zhang, T., Zhu, X., Liu, B.: Emotional chatting machine: Emotional conversation generation with internal and external memory. In: Thirty-Second AAAI Conference on Artificial Intelligence (2018)
26. Zhou, H., Zheng, C., Huang, K., Huang, M., Zhu, X.: Kdconv: A chinese multi-domain dialogue dataset towards multi-turn knowledge-driven conversation. In: Proceedings of the 58th Annual Meeting of the Association for Computational Linguistics. pp. 7098–7108 (2020)
27. Zhou, K., Zhao, W.X., Bian, S., Zhou, Y., Wen, J.R., Yu, J.: Improving conversational recommender systems via knowledge graph based semantic fusion. In: Proceedings of the 26th ACM SIGKDD International Conference on Knowledge Discovery & Data Mining. pp. 1006–1014 (2020)
28. Zhou, K., Zhou, Y., Zhao, W.X., Wang, X., Wen, J.R.: Towards topic-guided conversational recommender system. In: Proceedings of the 28th International Conference on Computational Linguistics. pp. 4128–4139 (2020)

RoKGDS: A Robust Knowledge Grounded Dialog System

Jun Zhang[1], Yuxiang Sun[1], Yushi Zhang[1], Weijie Xu[1], Jiahao Ying[1], Yan Yang[1,2](✉), Man Lan[1,2], Meirong Ma[3], Hao Yuan[3], and Jianchao Zhu[3]

[1] School of Computer Science and Technology, East China Normal University, Shanghai, China
{51194506048,51194506082,51205901031,51194506040}@stu.ecnu.edu.cn
{yanyang,mlan}@cs.ecnu.edu.cn

[2] Shanghai Key Laboratory of Multidimensional Information Processing, East China Normal University, Shanghai 200241, China

[3] Shanghai Transsion Co. Ltd., Shanghai, China
{MEIRONG.MA,HAO.YUAN,JIANCHAO.ZHU}@TRANSSION.COM

Abstract. In this paper, we propose a pre-training based **R**obust **K**now-ledge **G**rounded **D**ialog **S**ystem (**RoKGDS**) to enhance the performance of the model in unknown scenarios, which is easily generalized to various knowledge grounded dialog tasks, such as persona dialog, knowledge dialog, recommendation dialog. We use a bucket encoder to efficiently extract all kinds of knowledge information (e.g. profile, knowledge graph, and dialog goal). To improve the robustness of the model, we develop a hybrid decoder with a hybrid attention and a copy mechanism. The hybrid attention is an adaptation scheme to apply the pre-trained language model to our model and the copy mechanism is a gate mechanism to control generating a word from generic vocabulary or the input knowledge. Experiments show that our model is more robust than the other baseline models. Furthermore, we use visualization to explain the effectiveness of the hybrid attention compared to other two adaptation schemes. In the 2021 Language and Intelligence Challenge: Multi-Skill Dialog task, our best model ranked 3rd in the automatic evaluation stage and 5th in the human evaluation stage.

Keywords: Knowledge grounded dialog system · Transfer learning

1 Introduction

Open-domain dialog system has recently attracted much attention [6], thanks to the success of end-to-end deep models trained with large-scale dialog data. However, these data-driven models face the problem of generating meaningless or repeated responses [8,9]. To improve the informativeness of generated responses, previous works grounded the conversation with topic [18], persona [19], knowledge graph [4,17] or dialog goal [10]. These knowledge grounded dialog tasks aim

L. Wang et al. (Eds.): NLPCC 2021, LNAI 13029, pp. 377–387, 2021.
https://doi.org/10.1007/978-3-030-88483-3_30

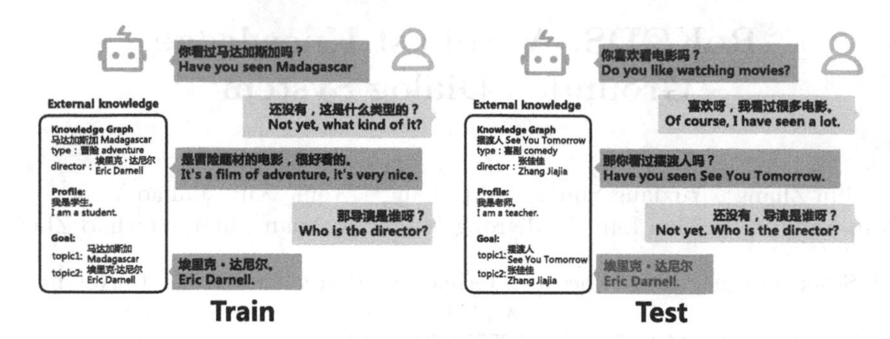

Fig. 1. A sample of knowledge grounded dialog. The external knowledge includes user profile, knowledge graph, and dialog goal. In the test scenario, the model cannot effectively use the knowledge that has never been exposed before, which leads to the failure of the dialog.

to use structured or unstructured knowledge information to accomplish persona consistent dialog, knowledge-accurate dialog, or purposeful dialog.

Although there are many dialog datasets with additional information to address knowledge grounded dialog tasks, such as *PersonaChat* [19], *DuConv* [4] and *DuRecDial* [10], these datasets are relatively small compared to open-domain dialog datasets without external knowledge, which leads the model to overfit the training data and not generalize well to the test data. Due to the small-scale training data, the model will mistakenly learn the mapping of the context and the knowledge that often appears in the responses, rather than the mapping of the context and the corresponding knowledge in the knowledge graph. For example in Fig. 1, the model tends to generate the response with the greatest likelihood in training data rather than the correct knowledge in the test scenario.

Recently, the pre-trained language models have achieved great success in the dialog generation task, which has significantly improved the model's robustness. However, most Chinese pre-trained language models (e.g. ERNIE [14], GPT-2 [12], BERT [3], RoBERTa [2], etc.) are based on either decoder-only or encoder-only Transformer [15]. Budzianowski and Vulić [1] use a common encoding method that concatenates all the additional information together (called *naive encoding*) as the input of the Transformer to apply these models to dialog task with additional information. However, it is time-consuming and space-consuming. It is difficult to apply these pre-trained language models to the dialog generation tasks with a large amount of knowledge.

To address the problems above, we propose a general robust knowledge grounded dialog system (RoKGDS) based on RoBERTa. Since there is a lot of input knowledge, such as user profile, knowledge graph, or dialog goal, RoKGDS first uses a bucket encoder to efficiently encoding the knowledge information. Then a hybrid decoder is developed to extract useful knowledge information conditioned on the dialog context through a hybrid attention. Finally, RoKGDS use a copy mechanism to copy correct knowledge from the input knowledge

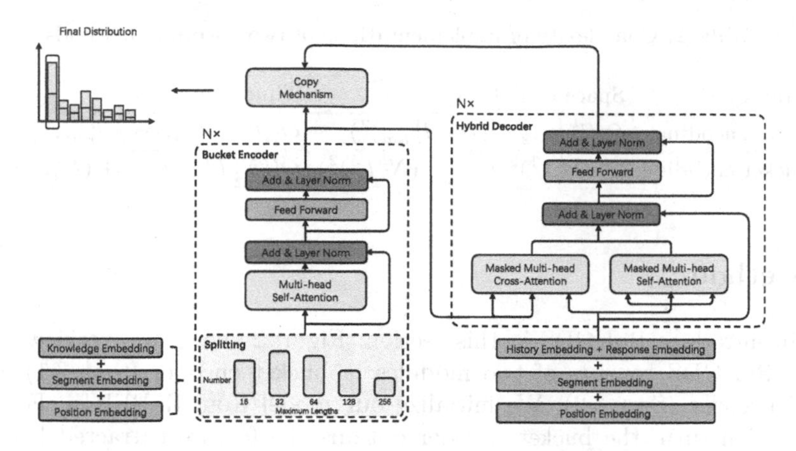

Fig. 2. The architecture of the RoKGDS

and avoid generating unrelated knowledge. Both automatic and manual evaluations on *DuConv*, *DuRecDial*, *CPC* all show that RoKGDS is more robust than other baseline models. Moreover, we visualize the attention score between context and knowledge to study the behavior of the hybrid attention compared to the other two baseline adaptation schemes in response generation, indicating that RoKGDS can effectively incorporate the given knowledge and context in the test scenario.

2 Related Work

Recently, the pre-trained language models such as BERT, GPT-2, ERNIE, etc. have led to state-of-the-art results in many tasks. In the dialog generation task, the pre-trained language models improve the model's ability to understanding and generating natural language. Wolf et al. [16] proposed to combine transfer learning with multi-task learning to improve discriminative language understanding. Budzianowski and Vulić [1] joined the belief state, context, and database state together as the GPT-2's input to accomplish task-oriented dialog. Golovanov et al. [5] studied using different finetune strategies to apply GPT-2 to improve the quality of the generated responses on persona dialog task.

Compared to the previous works, we propose a general model to improve the model's robustness in knowledge grounded dialog, which is easy to be generalized to other similar tasks. Additionally, our work evaluates the adaptation schemes of the pre-trained model on three tasks with additional knowledge – knowledge dialog, recommendation dialog, and persona dialog. Furthermore, in addition to automatic evaluation and manual evaluation, we study in detail the model's concerns during the generation process through visualization.

Table 1. Complexity of implementations of two encoding methods

Strategy	Space complexity	Time complexity
Naive encoding	$O((l_1^k + l_2^k + \cdots + l_{num}^k)^2)$	$O((l_1^k + l_2^k + \cdots + l_{num}^k)^2)$
Bucket encoding	$O((N_1^b L^1)^2 + \cdots + (N_5^b L^5)^2)$	$O((N_1^b L^1)^2 + \cdots + (N_5^b L^5)^2)$

3 Method

We will introduce RoKGDS in this section. Figure 2 shows an overview of our model. RoKGDS consists of two modules: a bucket encoder (Sect. 3.1) and a hybrid decoder (Sect. 3.2). We initialize our model from RoBERTa. For each conversation turn, the bucket encoder obtains the feature extracted from the external knowledge, such as user profile, knowledge graph, and dialog goal. Then the hybrid decoder incorporates the context and knowledge features to generate the final dialog system response.

3.1 Bucket Encoder

In the knowledge grounded dialog task, the knowledge involved is always massive and diverse. To efficiently encode knowledge, we use a *bucket encoding* method. We first concatenate structured knowledge into the form of natural language (e.g. knowledge triple (h, r, t) to $[h; r; t]$). Then we put knowledge texts of similar length into the same bucket. We set five buckets with maximum lengths L of 16, 32, 64, 128, and 256, respectively. We use a Transformer to process knowledge in the same bucket in parallel. Specifically, the process is as follows:

$$K_i = \{k_j \big| L_{i-1} < |k_j| \leq L_i\} \quad (1)$$

$$B_i = \text{Transformer}(K_i) \quad (2)$$

where k_j is j-th knowledge text. Finally, the bucket encoder outputs E_K (the concatenation of all the encoded knowledge feature $e_{z,i}^k$ contained in B_i) for next stage processing. Given the knowledge texts $[k_1, k_2, ..., k_{num}]$ with the lengths $[l_1^k, l_2^k, ..., l_{num}^k]$ and the number of knowledge texts contained in each bucket after splitting operation $[N_1^b, N_2^b, ..., N_5^b]$. The complexities of naive encoding and bucket encoding are summarized in Table 1. Table 1 shows that bucket encoder is more efficient to encoding knowledge than naive encoding.

3.2 Hybrid Decoder

To enhance the robustness of the model, we develop a hybrid attention in the hybrid decoder to adapt the Chinese pre-trained language model RoBERTa to a knowledge grounded dialog system. Moreover, we use a copy mechanism [13] to copy the given knowledge. The hybrid attention and copy mechanism help to generate appropriate responses with correct knowledge for scenarios that never

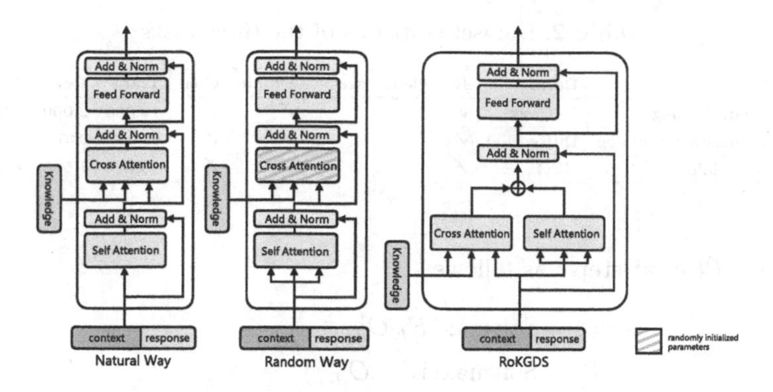

Fig. 3. Adaptation schemes

appeared during the training process. The input to the hybrid decoder is the knowledge feature E_K and the concatenation of a context and response $s = (x_1, ..., x_N, [sep], y_1, ..., y_M)$.

Adaptation Schemes. Considering that there is a gap between the RoBERTa pre-training task and the knowledge grounded dialog task, we propose the hybrid attention to combine our model with the pre-trained language model. As shown in the right of Fig. 3, the hybrid attention processes self attention and cross attention in the same layer, and the averaged output is input in the feed-forward network. Additionally, we study the other two adaptation schemes for comparison. First, a natural way to adapt RoBERTa to the Transformer based encoder-decoder framework is successively processing context and knowledge information by the multi-head attention with the pre-trained parameters. Although this natural way seems effective, it is difficult to fit the training data. So we study the second way (called random way) that we randomly initialize the parameters of the cross-attention. The random way performs better than the natural way shown in experiments.

Copy Mechanism. Due to the small amount of training data, the model tends to overfit the training data, leading to the model not be able to make good use of the knowledge it has never seen before. Although we input knowledge related to the current scenario, the model will produce responses with unrelated knowledge. To address this problem, we use the copy mechanism to copy the input knowledge appropriately. Specifically, we compute the word generation

Table 2. Dataset statistics of the three tasks

Task	Dataset	Knowledge graph	Profile	Goal	Train	Dev	Test
Knowledge dialog	DuConv	✓	✗	✓	19,858	2,000	5,000
Recommendation dialog	DuRecDial	✓	✓	✓	6,618	946	4,645
Persona dialog	CPC	✗	✓	✗	23,000	1,500	1,500

probability $P(w)$ at step t as follows:

$$score = \text{Softmax}(E_K O_{dec}^t) \tag{3}$$

$$P_g = \text{Softmax}(W_{voc} O_{dec}^t) \tag{4}$$

$$\alpha = \text{Sigmoid}(W_k O_{dec}^t) \tag{5}$$

$$P(w) = \alpha * P_g(w) + (1 - \alpha) \sum_{i:w=w_i} score_i \tag{6}$$

where W_{voc} denotes the word embedding matrix and W_k is a trainable parameter. O_{dec}^t is the outputs of the Transformer in hybrid decoder at step t. $\alpha \in [0, 1]$ controls the weight of generating a general word.

Training Objective. We optimize the following cross-entropy loss between the predicted word distribution P and ground-truth distribution o:

$$L = -\sum_{j=1}^{M} o_j \log\left(P(y_j)\right) \tag{7}$$

4 Experiments

4.1 Dataset

2021 Language and Intelligence Challenge: Multi-skills Dialog Task contains the tasks of the three major aspects of the dialog: knowledge dialog [17], recommendation dialog [10], and persona dialog. The knowledge dialog task requires the model to generate fluent, appropriate responses containing correct knowledge. The recommendation dialog task requires the model to appropriately respond to user requests, and proactively make recommendations based on user interests. The persona dialog task requires the model to generate fluent, appropriate, and persona-consistent responses. We show the details of the datasets used in the above three tasks in Table 2.

4.2 Implement Details

We implement RoKGDS in PyTorch[1]. Both the bucket encoder and hybrid decoder contain 12 Transformer blocks. Each Transformer block uses 12 attention

[1] Code will be available at https://github.com/z562/RoKGDS.

Table 3. Training details

Task	Learning rate	Batch size	Epoch	GPU	Training time
Knowledge dialog	4.0e−5	4	8	16 G Tesla P100 GPU	33 h
Recommendation dialog	3.5e−5	4	6		30 h
Persona dialog	2.5e−5	12	8		80 h

Table 4. Automatic evaluation

Method	DuConv	DuRecDial	Persona
	F1/BLEU-1/BLEU-2	F1/BLEU-1/BLEU-2	F1/BLEU-1/BLEU-2
Transformer	36.99%/0.302/0.176	56.40%/0.429/0.358	27.58%/0.270/0.145
GPT-2	43.87%/0.344/0.202	60.02%/0.494/0.419	34.11%/0.319/0.185
RoKGDS	**46.74%/0.376/0.252**	**63.43%/0.545/0.468**	**35.00%/0.329/0.201**
RoKGDS+Nat	0.57%/0.000/0.000	1.43%/0.000/0.000	1.91%/0.000/0.000
RoKGDS+Ran	44.59%/0.350/0.223	61.13%/0.513/0.441	33.62%/0.313/0.186
RoKGDS-Copy	46.04%/0.372/0.249	62.37%/0.527/0.451	34.64%/0.320/0.192

heads. The word embedding size is set to 768. The vocabulary size is 21,128. The maximum context length is 512. We use Adam [7] algorithm with the mini-batch method for optimization. We show the settings on different tasks in Table 3.

4.3 Baseline Models

We compare RoKGDS against several baselines:

- **Transformer:** We implement the Transformer model as introduced by Radford et al. [15].
- **GPT-2:** We initialize RoKGDS without the copy mechanism from the GPT-2 [12].

To further study the impact of various adaptation schemes and components on the model, we conduct the following ablation experiments: (1) **RoKGDS+Nat**: with the adaptation scheme *natural way*; (2) **RoKGDS+Ran:** with the adaptation scheme *random way*; (3) **RoKGDS-Copy**: without the copy mechanism.

4.4 Automatic Evaluation

Metrics. We use F1 [10] and BLEU [11] to evaluate the models. F1 measures the token-level similarity between the predicted response and the ground truth, while BLEU evaluates the word-level similarity.

Results. We present the results of automatic evaluation in Table 4. Our model achieves a significant improvement over other baseline models. This indicates that with the bucket encoder and the hybrid decoder, our proposed model effectively extracts the knowledge and incorporates the knowledge into the answer generation, lead to high robustness on the test datasets. We can also find that

Table 5. Human evaluation

Method	DuConv			DuRecDial			Persona		
	Info.	Cohe.	Know.	Info.	Cohe	Know.	Info.	Cohe.	Pers.
Transformer	1.325	1.105	1.065	1.570	1.510	1.360	1.190	1.025	1.005
GPT-2	1.590	1.545	1.460	1.625	1.590	1.685	1.505	1.430	1.105
RoKGDS	**1.660**	**1.736**	**1.520**	**1.845**	**1.895**	**1.740**	**1.550**	**1.580**	**1.315**

the score of the model with randomly initialized parameters has a big gap compared to the other models, indicating the pre-trained language model (GPT-2 or RoBERTa) can improve the model's performance in all kinds of knowledge grounded tasks. We show the results of the ablation study in 4. Notably, the performance of RoKGDS+Nat has extremely poor results. After our careful inspection, the RoKGDS+Nat is difficult to fit the training data. This is because the model cannot extract context information effectively. We use visualization to conduct detailed analysis in Sect. 4.6. We also find without the copy mechanism, the performance of our model has declined on all tasks, especially on the tasks that require rich knowledge to generate responses, indicating that the copy mechanism improves the utilization of input knowledge.

4.5 Human Evaluation

Metrics. We conduct human evaluations to compare three models, Transformer, GPT-2, and RoKGDS. We randomly sample 200 examples from the test set of the three tasks. Given the context and related knowledge, we generate the responses use all three models. Then we employ three human evaluators to assess the responses. For knowledge dialog and recommendation dialog, the human evaluators measure the responses in terms of informativeness, coherence, knowledge consistency using a 3-point Likert scale. For persona dialog, the human evaluators measure the responses in terms of informativeness, coherence, persona consistency. The informativeness measures if the generated responses are contentful. The coherence measures if the generated responses are appropriateness and coherence of the whole dialog flow. The knowledge consistency measures if the knowledge used in responses is correct and proper. The persona consistency measures if the responses match the given profile.

Results. The results of human evaluation are shown in Table 5. Our model outperforms all the baselines, indicating that our model can generate more appropriate and informative responses. We also observe that our model has a better performance in knowledge-rich dialog tasks (knowledge dialog and recommendation dialog). This is because our model can build a good connection between knowledge and context so that it has better performance in test scenarios.

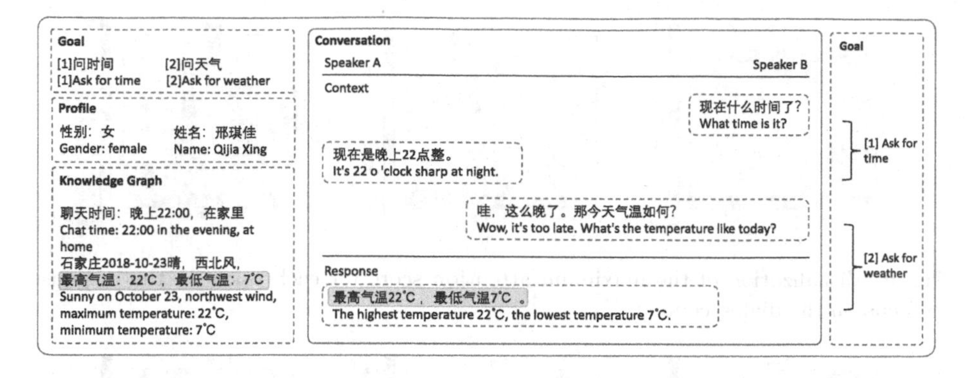

Fig. 4. A recommendation dialog sample for visualization.

4.6 Visualization

To show the behavior of the three adaptation schemes in response generation, we provide a recommendation dialog sample. As shown in Fig. 4, the model needs to use the knowledge graph information to generate the correct response *"the maximum temperature 22 ℃, the minimum temperature 7 ℃"*. We visualize the maximum response-context attention score in Fig. 5 and the maximum response-knowledge attention score in Fig. 6. As shown in the left group of Fig. 5, RoKGDS+Nat has very low maximum attention scores for the context in the second to seventh layers of Transformer compared to the first layer, indicating that the model cannot effectively use context information in these layers. This is because in the pre-training stage the RoBERTa only contains one multi-head attention layer, while in the fine-tuning stage our model uses two multi-head attention layers to process context and knowledge in turn. This leads to the continuous dilution of the dialog history information after the knowledge information is input in the first layer. In the early stage of model training, it will have a destructive effect on model fitting, resulting in very poor performance. The middle groups in Fig. 5 shows that RoKGDS+Ran can extract contextual information normally. This is because randomly initialized cross-attention weights alleviate the influence of the input of knowledge information and reduce the gap between the pre-training stage and fine-tuning stage. However, RoKGDS+Ran pays less attention to the knowledge information and tends to use the dialog context information to generate responses, which reduces the robustness of the model. It can also be reflected in the results of automatic and human evaluations. Notably, RoKGDS focuses on the dialog context and the input knowledge information simultaneously, which suggests that the model has higher robustness than the other models, and can use the knowledge that the model has never been seen before to generate correct and appropriate responses based on the context.

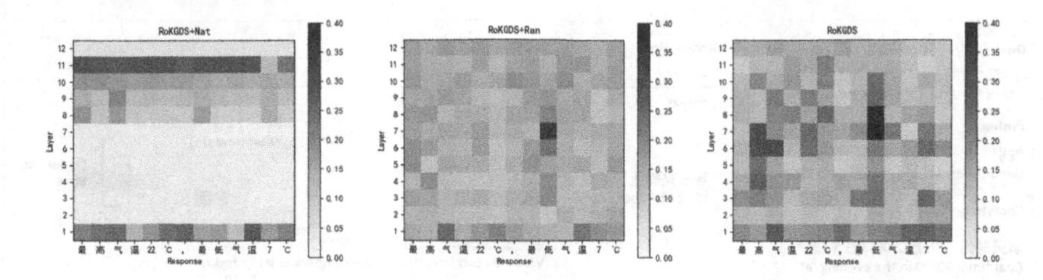

Fig. 5. Visualization of the maximum attention score for each token in the response to focus on the dialog context

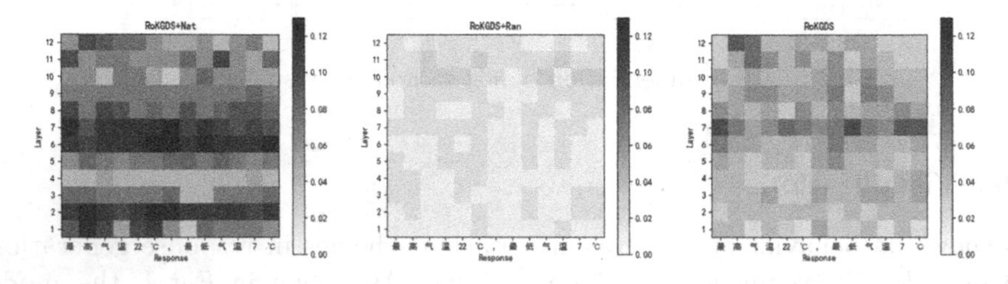

Fig. 6. Visualization of the maximum attention score for each token in the response to focus on the related knowledge

4.7 Conclusions

In this work, we propose the RoKGDS to enhance the model's robustness in knowledge grounded dialog task. We use a bucket encoder to efficiently extract knowledge features, such as profile, knowledge graph, and dialog goal. Then we propose a hybrid decoder with a hybrid attention and a copy mechanism to generate informative and proper responses in the unseen scenarios. Both automatic and human evaluation verifies the effectiveness of the proposed model. To further study the three adaptation schemes, we use the visualization and explain how the model performs in automatic evaluation.

Acknowledgements. This research is funded by the Science and Technology Commission of Shanghai Municipality (20511101205), Shanghai Key Laboratory of Multidimensional Information Processing, East China Normal University, No. 2020KEY001.

References

1. Budzianowski, P., Vulić, I.: Hello, it's GPT-2-how can I help you? Towards the use of pretrained language models for task-oriented dialogue systems. In: Proceedings of the 3rd Workshop on Neural Generation and Translation, pp. 15–22 (2019)
2. Cui, Y., Che, W., Liu, T., Qin, B., Wang, S., Hu, G.: Revisiting pre-trained models for Chinese natural language processing. In: Proceedings of the 2020 Conference on Empirical Methods in Natural Language Processing: Findings, pp. 657–668 (2020)

3. Devlin, J., Chang, M.W., Lee, K., Toutanova, K.: BERT: pre-training of deep bidirectional transformers for language understanding. arXiv preprint arXiv:1810.04805 (2018)
4. Ghazvininejad, M., et al.: A knowledge-grounded neural conversation model. arXiv preprint arXiv:1702.01932 (2017)
5. Golovanov, S., Kurbanov, R., Nikolenko, S., Truskovskyi, K., Tselousov, A., Wolf, T.: Large-scale transfer learning for natural language generation. In: Proceedings of the 57th Annual Meeting of the Association for Computational Linguistics, pp. 6053–6058 (2019)
6. Huang, M., Zhu, X., Gao, J.: Challenges in building intelligent open-domain dialog systems. ACM Trans. Inf. Syst. (TOIS) **38**(3), 1–32 (2020)
7. Kingma, D.P., Ba, J.: Adam: a method for stochastic optimization. arXiv preprint arXiv:1412.6980 (2014)
8. Li, J., Galley, M., Brockett, C., Gao, J., Dolan, W.B.: A diversity-promoting objective function for neural conversation models. In: Proceedings of the 2016 Conference of the North American Chapter of the Association for Computational Linguistics: Human Language Technologies, pp. 110–119 (2016)
9. Li, J., Galley, M., Brockett, C., Spithourakis, G., Gao, J., Dolan, W.B.: A persona-based neural conversation model. In: Proceedings of the 54th Annual Meeting of the Association for Computational Linguistics (Volume 1: Long Papers), pp. 994–1003 (2016)
10. Liu, Z., Wang, H., Niu, Z.Y., Wu, H., Che, W., Liu, T.: Towards conversational recommendation over multi-type dialogs. In: Proceedings of the 58th Annual Meeting of the Association for Computational Linguistics, pp. 1036–1049 (2020)
11. Papineni, K., Roukos, S., Ward, T., Zhu, W.J.: BLEU: a method for automatic evaluation of machine translation. In: Proceedings of the 40th annual meeting of the Association for Computational Linguistics, pp. 311–318 (2002)
12. Radford, A., Wu, J., Child, R., Luan, D., Amodei, D., Sutskever, I.: Language models are unsupervised multitask learners. OpenAI blog **1**(8), 9 (2019)
13. See, A., Liu, P.J., Manning, C.D.: Get to the point: summarization with pointer-generator networks. In: Proceedings of the 55th Annual Meeting of the Association for Computational Linguistics (Volume 1: Long Papers), pp. 1073–1083 (2017)
14. Sun, Y., et al.: ERNIE: enhanced representation through knowledge integration. arXiv preprint arXiv:1904.09223 (2019)
15. Vaswani, A., et al.: Attention is all you need. In: Advances in Neural Information Processing Systems, pp. 5998–6008 (2017)
16. Wolf, T., Sanh, V., Chaumond, J., Delangue, C.: Transfer Transfo: a transfer learning approach for neural network based conversational agents. In: NIPS2018 CAI Workshop (2018)
17. Wu, W., et al.: Proactive human-machine conversation with explicit conversation goal. In: Proceedings of the 57th Annual Meeting of the Association for Computational Linguistics, pp. 3794–3804 (2019)
18. Xing, C., et al.: Topic aware neural response generation. In: Proceedings of the AAAI Conference on Artificial Intelligence, vol. 31 (2017)
19. Zhang, S., Dinan, E., Urbanek, J., Szlam, A., Kiela, D., Weston, J.: Personalizing dialogue agents: I have a dog, do you have pets too? In: ACL, pp. 2204–2213 (2018)

Enhanced Few-Shot Learning with Multiple-Pattern-Exploiting Training

Jiali Zeng(✉), Yufan Jiang, Shuangzhi Wu, and Mu Li

Tencent Cloud Xiaowei, Beijing, China
{lemonzeng,garyyfjiang,frostwu,ethanlli}@tencent.com

Abstract. The NLPCC 2021 Few-shot Learning for Chinese Language Understanding Evaluation (FewCLUE) shared task seeks for the best solution to few-shot learning tasks with pre-trained language models. This paper presents Tencent Cloud Xiaowei's approach to this challenge, which won the 2st place in the contest. We propose a Multiple-Pattern-Exploiting Training method (MPET) for the challenge. Different from the original PET, MPET constructs multiple patterns to enhance the model's generalization capability. We take the MPET as an auxiliary task, and jointly optimize classification and MPET. Empirical results show that our MPET is effective to few-shot learning tasks.

Keywords: Few-shot learning · Pre-trained language model · Pattern-exploiting training.

1 Introduction

Pre-trained language models (PLMs) [2,6–8,16] have been proved effective in many natural language processing tasks such as text classification [12], machine reading comprehension [5,9] and inference [1]. The commonly used paradigm is pretraining and finetuning, where a language model is first pretrained on massive text corpora, followed by finetuning on downstream tasks [2]. Usually, the performance of downstream tasks rely on the number of available training examples.

However, large amount of annotating data is hardly available in many NLP fields. In practice, it is common for NLP tasks to utilize only a small number of labeled examples, making few-shot learning a highly important research area. To push the development of this area, the NLPCC 2021 few-shot learning for Chinese Language Understanding Evaluation (FewCLUE) [15] provides researchers with both platforms and data to investigate the problem more thoroughly in Chinese language. The goal of FewCLUE task is to explore the capability of PLMs conditioned on a few training data for the downstream task. In this paper,

J. Zeng and Y. Jiang—contributed equally to this work.

L. Wang et al. (Eds.): NLPCC 2021, LNAI 13029, pp. 388–399, 2021.
https://doi.org/10.1007/978-3-030-88483-3_31

Table 1. Statistics of datasets.

Task	EPRSTMT	CSLDCP	TNEWS	IFLYTEK	OCNLI
Train size	32	536	240	928	32
Dev size	32	536	240	690	32
Test size	610	1784	2010	1749	2520
# Labels	2	67	15	119	3
Task	BUSTM	ChID	CSL	WSC	–
Train size	32	42	32	32	–
Dev size	32	42	32	32	–
Test size	1772	2002	2838	976	–
# Labels	2	-	2	2	–

we present the Tencent Cloud Xiao's approach to the FewCLUE challenge. We propose a novel **M**ultiple-**P**attern-**E**xploiting **T**raining (MPET) method. Different from the original PET [10], MPET constructs multiple patterns, aiming to enhance the model's generalization capability. The different patterns are treated as individual tasks, which are optimized as auxiliary tasks together with the original sequence classification objectives. To balance and emphasize effect of different patterns, we introduce a confidence factor, which is used as a weight to select more effective patterns. Specifically, our final approach to FewCLUE can be divided into three steps. First, we post-train PLM with pattern-augmented task-related data. Then, we finetune the post-trained PLM with our MPET method. Finally, we utilize a multi-stage training method to exploit unlabeled data.

This paper is organized as follows. Section 2 describes the task, as well as the corresponding data statistics. Section 3 introduces our frameworks. Section 4 presents the experimental results. We summarize in Sect. 5.

2 Background

2.1 Task Definition

FewCLUE has a total of 9 tasks, each task containing 5 datasets and 1 public test set. The statistics of each datasets is the same, a few-shot training set and a development set. We show the statistics of one of the dataset and the public test set of each task in Table 1. In the following, we simply describe the 9 tasks.

EPRSTMT [15]. E-commerce product review sentiment analysis aims to recognize the emotional polarity of the review and labeling them as positive or negative.

CSLDCP [15]. Chinese scientific literature discipline classification is a long text classification task, which classifies which professional literature the text belongs to.

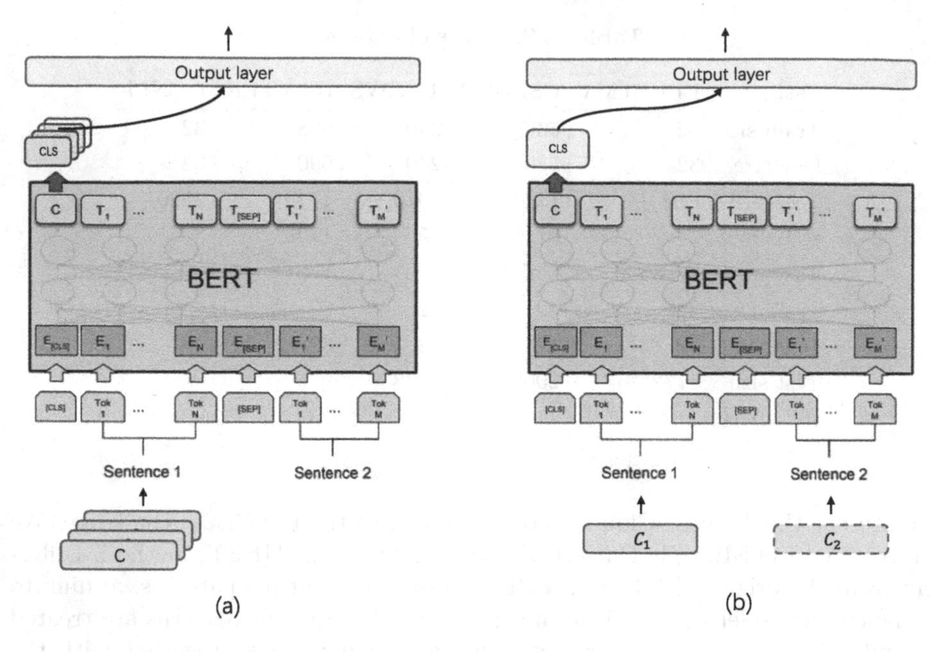

Fig. 1. The architecture of (a) a multi-choice MRC task and (b) a sequence classification task.

TNEWS [14]**.** Toutiao short text classification for news is a short text classification task, which classifies the topic of the text.

IFLYTEK[1]**.** Similar to CSLDCP, IFLYTEK is a long text classification task, that recognizes which topic of the text belongs to.

OCNLI [4]**.** Original Chinese natural language inference focuses on the problem of deciding whether two statements are connected via an entailment, a contradiction, or a natural.

BUSTM[2]**.** XiaoBu dialogue short text matching aims to decide whether two sentences express the same meaning.

ChID [17]**.** Chinese idiom for cloze text is a cloze-style reading comprehension, which studies the comprehension of idiom, a unique language phenomenon in Chinese.

CSL [14]**.** CSL is a keyword recognition task, that decides whether the keywords are all real keywords of the article.

CLUEWSC [14]**.** Winograd scheme challenge is a pronoun disambiguation task, which aims to determine which noun the pronoun in a sentence refer to.

---

[1] http://challenge.xfyun.cn/2019/gamelist..
[2] https://github.com/xiaobu-coai/BUSTM..

2.2 Formulation

We treat ChID as a multi-choice MRC task and other tasks as sequence classification tasks.

Formally, ChID can be presented as a pair $< c, A >$, where $c = \{w_0, w_1, ..., w_m\}$ is a passage with a token "*#idom*", which is replaced by a set of candidate idioms $A = \{a_0, a_i, ..., a_n\}$ to construct a set of candidate passage $C = \{c_0, c_i, ..., c_n\}$, where m denotes the length of the passage and n denotes the number of the idioms in candidate set. Only one idiom in A is correct. As illustrate in Fig. 1(a), the PLM encodes each candidate passage and obtains a set of contextual representations $H = \{h_{CLS_0}, h_{CLS_i}, ..., h_{CLS_n}\}$. Then, we stack a single full connection layer on the contextual representations to conduct classification:

$$p(y|H) = W \cdot H + b, \quad (1)$$

where y is the index of correct answer in candidates, W and b are parameters. The loss function is defined as:

$$\mathcal{L}_{cla} = \text{CE}(p(y|H), y). \quad (2)$$

As shown in Fig. 1(b), other tasks can be represented as a pair $< C, y >$, where $C = \{c_0, c_1, ..., c_J\}$, $J \in \{0, 1\}$, is a set of multiple sentences, and y is the label. We use "[SEP]" to concatenate the sentences. Next, the concatenated sequence is fed into the PLM to obtain its contextual representation. Similarly, we conduct classification by applying a full connection layer on the hidden state h_{CLS} (Eq. (1)) and define the loss function for sequence classification as:

$$\mathcal{L}_{cla} = \text{CE}(p(y|h_{CLS}), y). \quad (3)$$

3 Methodology

3.1 MPET: Multiple-Pattern-Exploiting Training

Our framework is a multi-task training framework, which jointly optimizes classification and PET. Furthermore, we introduce multiple patterns, where different patterns are viewed as different tasks, and propose a confidence factor, which weights the importance of each pattern.

Here, we first simply introduce PET [10]. PET converts an example into a cloze-style question, similar to the input format used during pre-training. The query-form in PET is defined by a Pattern-Verbalizer Pair (PVP). Particularly, each PVP consists of a **pattern**, which describes the way to convert the inputs into a cloze-style question with "[MASK]" token; and a **verbalizer** which describes how to convert the classes into the output space of tokens. We illustrate a PVP for BUSTM task in Fig. 2(a). Here, we convert sentences s_0="你能嘴甜一点吗" and s_1="说点好听的" into a cloze-style question with the pattern: ["s_0", [MASK][MASK], "s_1"]. The verbalizer maps "True or False" to

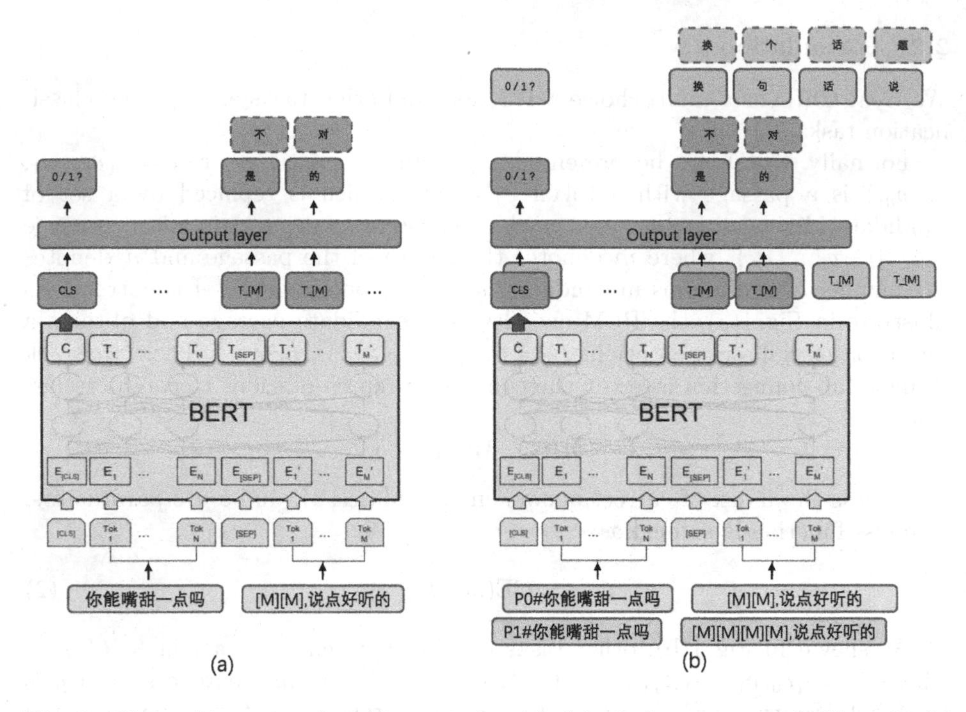

Fig. 2. The architecture of our framework with (a) single pattern and (b) multiple patterns.

" 是的 or 不对 ". After hand-designing a PVP for a given task, PET obtains logits from the PLM. Given the space of output tokens $\mathcal{Y}$[3], a softmax layer is applied to the logits and the loss is computed as following:

$$q(y|x) = \text{softmax}(\text{PLM}(x)), \tag{4}$$

$$\mathcal{L}_{pet} = \text{CE}(q(y|x), y). \tag{5}$$

A key challenge for PET is that when the number of labels is large (such as CSLDCP, IFLYTEK and ChID), it is more difficult to learn an optimal decision boundary due to the large search space. Therefore, in this paper, we address this by using PET as an auxiliary task, and jointly optimize classification and PET. Formally, with the traditional classification loss $\mathcal{L}_{cla}$ and the PET loss $\mathcal{L}_{pet}$, we compute the loss as

$$\mathcal{L} = \mathcal{L}_{cla} + \alpha \cdot \mathcal{L}_{pet}. \tag{6}$$

Here, we set α as 1.0.

Furthermore, we introduce to train the model with multiple patterns. We formulate this as a multi-task training problem, where different patterns are

[3] PET computes class probabilities using the logits that correspond to the labels for a specific task. In contrast, inspired by ADAPET [11], we computes the probability of each token in the vocabulary tokens in this paper.

viewed as different tasks. As shown in Fig. 2(b), we design $N = 2$ PVPs for the BUSTM task. For the first PVP, we convert sentences s_0="你能嘴甜一点吗" and s_1="说点好听的" into a cloze-style question with the pattern: [P0#"s_0", [MASK][MASK], "s_1"] and maps "True or False" to "是的 or 不对". For the second PVP, we convert sentences into a cloze-style question with the pattern: [P1#"s_0", [MASK][MASK][MASK][MASK], "s_1"] and maps "True or False" to "换句话说 or 换个话题". We fed the set of inputs into the PLM to conduct classification, and obtain a set of classification loss $\{\mathcal{L}^j_{cla}\}$ and PET loss $\{\mathcal{L}^j_{pet}\}$ by Eq. (3) and Eq. (5).

As mentioned in [11], the multi-task multi-pattern training somehow hurts the performance. We hypothesize that the PLM is sensitive to the language style of the pattern, and multiple hand-designed patterns may not always have a positive effect for downstream tasks. Therefore, we introduce a confidence factor c to weight the importance of each pattern. The confidence factor is a learnable parameter. In order to give model "hints" during training, the softmax prediction probabilities are adjusted by interpolating between the original predictions $p^j(y|h^j_{CLS})$ and the target probability distribution y. Formally, for j-th pattern, given the confidence c^j, we construct the probability $p^j(y|h^j_{CLS})$ as

$$\hat{p}^j(y|h^j_{CLS}) = c^j \cdot p^j(y|h^j_{CLS}) + (1 - c^j) \cdot y. \tag{7}$$

To prevent the model from minimizing the task loss by always choosing c^j=0 and receiving the entire ground truth, we add a log penalty to the loss function:

$$\hat{\mathcal{L}}^j_{cla} = \mathrm{CE}(\hat{p}^j(y|h^j_{CLS}), y) - \beta \cdot \log(c^j), \tag{8}$$

where β is a hyper-parameter, and we set it as 1.0 in this paper.

During training, the final training objective over an instance becomes

$$\mathcal{L} = \sum_{j=1}^{N} (\hat{\mathcal{L}}^j_{cla} + \alpha \cdot \mathcal{L}^j_{pet}). \tag{9}$$

During testing, the classification probability is computed as the weighted sum of the probabilities of multiple patterns:

$$p(y|h^0_{CLS}, .., h^j_{CLS}) = \frac{1}{N} \sum_{j=1}^{N} c^j \cdot p^j(y|h^j_{CLS}). \tag{10}$$

3.2 Post-training

BERT based pre-trained language models achieve strong performance on many downstream tasks by fine-tuning the model on task-specific corpus. However, the commonly used pre-trained language models are only trained on general corpus like Wikipedia and Book corpus, which lacks enough supervision of task-specific words and phrases discovered by [13]. In addition, [3] has also found that task-adaptive pretraining on task-relevant corpus provides a large performance boost for PLMs. Inspired by their works, we post-train MacBERT on the FewCLUE corpus to help model better understand the task. During post-training, the whole word masked language objective is used. For task TNEWS, IFLYTEK, OCNLI and CLUEWSC, we also collect these tasks' corpus included in CLUE and add them into the post-training corpus.

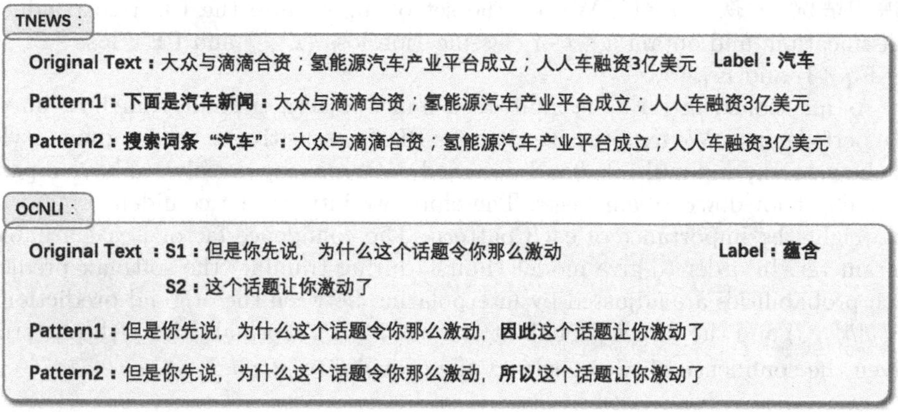

TNEWS：

Original Text：大众与滴滴合资；氢能源汽车产业平台成立；人人车融资3亿美元 Label：汽车

Pattern1：下面是汽车新闻：大众与滴滴合资；氢能源汽车产业平台成立；人人车融资3亿美元

Pattern2：搜索词条 "汽车"：大众与滴滴合资；氢能源汽车产业平台成立；人人车融资3亿美元

OCNLI：

Original Text：S1：但是你先说，为什么这个话题令你那么激动 Label：蕴含

S2：这个话题让你激动了

Pattern1：但是你先说，为什么这个话题令你那么激动，因此这个话题让你激动了

Pattern2：但是你先说，为什么这个话题令你那么激动，所以这个话题让你激动了

Fig. 3. The example of the handcrafted patterns of TNEWS and OCNLI for post-training.

To ensure consistency and better exploit the representation of pattern words, we further construct a pattern-augmented corpus by adding pattern words to each plain text according to the labels and mix this corpus with original post-training corpus (as shown in Fig. 3). From post-training, the MacBERT can learn contextual representations from the corpus more effectively.

3.3 Multi-stage Training

Finally, we propose multi-stage training to train the model with unlabeled data. Specifically, first, we use the model trained in Sect. 3.1 to generate the label of unlabeled data to construct a synthetic data. Second, we freeze the encoder parameters and the language model parameters of the PLM, and randomly initialize the parameters of classifier, and fine-tune the model. During fine-tuning, we mix synthetic data into the original training data. Third, we use the new model to generate new label of unlabeled data, and fine-tune the new model with new mixed data consists of the new synthetic data and the original training data. We iteratively conduct above process until the model convergences.

4 Experiment

4.1 Settings

Our code is implementated in Pytorch using HuggingFace[4]. We use the MacBERT-Large as the PLM. All the models are trained on NVIDIA V100 GPUs. We use different learning rate for the parameters of the PLM and the classifier. For PLM parameters, we set the learning rate as 1e−5. For EPRSTMT, OCNLI, BUSTM, CSL, and WSC, we set the training steps as 1000, the warmup

Table 2. The PVP of each task.

Task	PVPs
EPRSTMT	$P_0(\cdot)$=[P0# “sentence”用户对商品_满意。] V_0(Positive)=很, V_0(Negative)=不 $P_1(\cdot)$=[P1# “sentence”这个商品_。] V_1(Positive)=非常好, V_1(Negative)=质量差
CSLDCP	$P_0(\cdot)$=[P0# “content”为_专业文献。] V_0(label)=label $P_1(\cdot)$=[P1# 搜索词条:_，内容:“content”。] V_1(label)=label
TNEWS	$P_0(\cdot)$=[P0# 下面是_新闻:“sentence”该文章的关键词: “keywords”] V_0(label)=description of label $P_1(\cdot)$=[P1# 搜索词条:_。内容:“sentence”该文章的关键词: “keywords”] V_1(label)=description of label
IFLYTEK	$P_0(\cdot)$=[P0# “sentence”是关于_的文章。] V_0(label)=description of label $P_1(\cdot)$=[P1# 搜索词条:_。内容:“sentence”。] V_1(label)=description of label
OCNLI	$P_0(\cdot)$=[P0# “sentence1” _, “sentence2”。] V_0(contradiction)=不对, V_0(entailment)=所以, V_0(neutral)=嗯嗯 $P_1(\cdot)$=[P1# “sentence1” _, “sentence2”。] V_1(contradiction)=不是, V_1(entailment)=因此, V_1(neutral)=哦哦
BUSTM	$P_0(\cdot)$=[P0# “sentence1” _, “sentence2”。] V_0(0)=不是呢, V_0(1)=对的呀 $P_1(\cdot)$=[P1# “ s ”。这个商品[MASK]。] V_1(0)=换个话题, V_1(1)=换句话说
CSL	$P_0(\cdot)$=[P0#文章“abst”_:keywords] V_0(0)=随机选项, V_0(1)=关键单词 $P_1(\cdot)$=[P1#文章“abst”_:keywords] V_1(0)=干扰项, V_1(1)=相关词
WSC	$P_0(\cdot)$=[text,P0#pronoun_query。] V_0(true)=指的是, V_0(false)=不代表 $P_1(\cdot)$=[text,P1#pronoun_query。] V_1(true)=可以改为, V_1(false)=不能替换

[4] https://github.com/huggingface/transformers.

Table 3. Experimental results on public test set.

Model	EPRSTMT	CSLDCP	TNEWS	IFLYTEK	OCNLI
FT(B)	75.24	57.57	59.65	47.60	36.15
SPET(B)	80.98	59.41	63.23	49.45	48.05
SPET(F)	87.54	**61.15**	63.13	52.20	44.05
MPET(F)	**89.18**	60.14	63.88	**53.77**	45.95
MPET(P)	88.36	59.42	64.53	53.04	65.12
MPET+DA(P)	**89.18**	59.02	**64.73**	53.04	**66.82**
Model	BUSTM	ChID	CSL	WSC	–
FT(B)	57.62	38.36	59.51	50.51	–
SPET(B)	63.09	34.76	**63.03**	53.89	–
SPET(F)	62.13	60.14	55.57	53.38	–
MPET(F)	62.97	60.14	59.02	56.45	–
MPET(P)	**68.34**	56.49	57.65	**62.39**	–
MPET+DA(P)	67.27	**61.34**	57.65	–	–

as 200, the batch size as 8, and the learning rate of classifier parameters as 1e−6. For CSLDCP, TNEWS, ChID and IFLYTEK, the training steps is 1500, the warmup is 300, the batch size is 16, and the learning rate of classifier parameters is 1e−5. The number of PVP of each task is 2 except ChID[5]. We show PVPs used for all tasks in Table 2.

During post-training, we use the plain text and pattern-augmented text of FewCLUE and CLUE corpus, the max length is set to 512. We post-train MacBERT for 10 epochs and we set the learning rate as 1e−4 and batch size as 2k. The mask rules are the same with BERT, except we adopt whole word masking.

4.2 Results

Table 3 shows the results of different models fine-tuned on the first few-shot training set of each task and test on the public test set. All of our models are based on the original huggingface MacBERT-Large model. We compare the performance of several models listed as follows:

FT(B). We directly fine-tuned MacBERT-Large on downstream tasks.

SPET(B). We directly fine-tuned MacBERT-Large by our framework with single pattern.

SPET(F). We first post-trained MacBERT-Large on plain text of post-training corpus. Then, we fine-tuned the post-trained model by our framework with single pattern.

[5] We replace the token "#idom" in *content* with "[MASK]", and make it as a regular MLM objective.

MPET(F). We first post-trained MacBERT-Large on plain text of post-training corpus. Then, we fine-tuned the post-trained model by our framework with multiple patterns.

MPET(P). We first post-trained MacBERT-Large on pattern-augmented text of post-training corpus. Then, we fine-tuned the post-trained model by our framework with multiple patterns.

MPET + DA(P). We further trained **MPET(P)** by multi-stage training method with unlabeled data.

From Table 3, we arrive at the following conclusions: First, **SPET(B)** performs better than **FT(B)** on all tasks, demonstrating that taking PET as an auxiliary task benefits the few-shot learning. Second, compared with **SPET(B)**, **SPET(F)** obtains significantly improvement in EPRSTMT, CSLDCP, IFLYTEK, ChID, and achieves competitive performance on TNEWS, BUSTM, and WSC. This observation demonstrates the effectiveness of task-adaptive post-training of LM. Third, **MPET(F)** performs better than **SPET(F)** on most of tasks, showing that enhancing the models with multiple patterns improves the model's generalization capability. Fourth, **MPET(P)** outperforms **MPET(F)**. This observation indicates that our pattern-augmented post-training of LM truly benefits the pattern-exploiting fine-tuning. Finally, from the results of **MPET(P)** and **MPET + DA(P)**, except for OCNLI and ChID tasks, the unlabeled data of other tasks does not bring much improvements. We leave how to better utilize unlabeled data as future work.

Table 4. Experimental results on public test set.

Data	EPRSTMT	CSLDCP	TNEWS	IFLYTEK	OCNLI
0	89.18	59.02	64.73	53.04	66.82
1	88.20	60.31	62.39	51.88	67.34
2	88.68	58.74	63.78	52.61	59.76
3	89.84	57.84	63.43	52.17	66.23
4	90.65	60.03	73.73	53.62	69.29
All	88.03	64.57	65.02	54.04	71.39
Data	BUSTM	ChID	CSL	WSC	–
0	67.27	61.34	57.65	62.39	–
1	71.05	53.75	51.62	64.75	–
2	67.04	54.90	56.98	64.03	–
3	70.6	60.00	54.02	68.64	–
4	71.73	59.44	55.74	59.84	–
All	75.68	69.93	68.71	77.97	–

In total, we take **MPET + DA(P)** as our final solution for FewCLUE challenge. We illustrate the results of our method trained on each training set of each task and evaluated on public test set in Table 4.

5 Conclusion

In this paper, we propose a multi-pattern-exploiting training (MPET) method for FewCLUE. MPET constructs multiple patterns, which are treated as individual tasks, and optimized as auxiliary tasks together with the original sequence classification objectives. Furthermore, we introduce a confidence factor to balance and emphasize different effect of different patterns. Experimental results demonstrate that our method achieves significantly improvements.

There is sill plenty of room for improvement and investigation. We will explore continuous patterns in our framework and exploit strong semi-supervised approaches to utilize unlabeled data. Also, future work could look into better masking strategies for labeled conditioned MLM during post-training, such as masking important tokens or semantic units based on the gradients of the logits for the instance.

References

1. Bowman, S.R., Angeli, G., Potts, C., Manning, C.D.: A large annotated corpus for learning natural language inference. In: Màrquez, L., Callison-Burch, C., Su, J., Pighin, D., Marton, Y. (eds.) Proceedings of the 2015 Conference on Empirical Methods in Natural Language Processing, EMNLP 2015, 17–21 September 2015, Lisbon, Portugal, pp. 632–642. The Association for Computational Linguistics (2015). https://doi.org/10.18653/v1/d15-1075
2. Devlin, J., Chang, M.W., Lee, K., Toutanova, K.: BERT: pre-training of deep bidirectional transformers for language understanding. arXiv preprint arXiv:1810.04805 (2018)
3. Gururangan, S., et al.: Don't stop pretraining: adapt language models to domains and tasks. arXiv preprint arXiv:2004.10964 (2020)
4. Hu, H., Richardson, K., Xu, L., Li, L., Kübler, S., Moss, L.S.: OCNLI: original Chinese natural language inference. In: Cohn, T., He, Y., Liu, Y. (eds.) Proceedings of the 2020 Conference on Empirical Methods in Natural Language Processing: Findings, EMNLP 2020, Online Event, 16–20 November 2020. Findings of ACL, vol. EMNLP 2020, pp. 3512–3526. Association for Computational Linguistics (2020). https://doi.org/10.18653/v1/2020.findings-emnlp.314
5. Lai, G., Xie, Q., Liu, H., Yang, Y., Hovy, E.H.: RACE: large-scale reading comprehension dataset from examinations. In: Palmer, M., Hwa, R., Riedel, S. (eds.) Proceedings of the 2017 Conference on Empirical Methods in Natural Language Processing, EMNLP 2017, 9–11 September 2017, Copenhagen, Denmark, pp. 785–794. Association for Computational Linguistics (2017). https://doi.org/10.18653/v1/d17-1082
6. Lan, Z., Chen, M., Goodman, S., Gimpel, K., Sharma, P., Soricut, R.: ALBERT: a lite BERT for self-supervised learning of language representations. arXiv preprint arXiv:1909.11942 (2019)

7. Lewis, M., et al.: BART: denoising sequence-to-sequence pre-training for natural language generation, translation, and comprehension. arXiv preprint arXiv:1910.13461 (2019)
8. Liu, Y., et al.: RoBERTa: A robustly optimized BERT pretraining approach. arXiv preprint arXiv:1907.11692 (2019)
9. Rajpurkar, P., Zhang, J., Lopyrev, K., Liang, P.: SQuAD: 100, 000+ Questions for machine comprehension of text. In: Su, J., Carreras, X., Duh, K. (eds.) Proceedings of the 2016 Conference on Empirical Methods in Natural Language Processing, EMNLP 2016, 1–4 November 2016, Austin, Texas, USA, pp. 2383–2392. The Association for Computational Linguistics (2016). https://doi.org/10.18653/v1/d16-1264
10. Schick, T., Schütze, H.: Exploiting cloze-questions for few-shot text classification and natural language inference. In: Merlo, P., Tiedemann, J., Tsarfaty, R. (eds.) Proceedings of the 16th Conference of the European Chapter of the Association for Computational Linguistics: Main Volume, EACL 2021, Online, 19–23 April 2021, pp. 255–269. Association for Computational Linguistics (2021). https://www.aclweb.org/anthology/2021.eacl-main.20/
11. Tam, D., Menon, R.R., Bansal, M., Srivastava, S., Raffel, C.: Improving and simplifying pattern exploiting training. CoRR abs/2103.11955 (2021). https://arxiv.org/abs/2103.11955
12. Wang, A., Singh, A., Michael, J., Hill, F., Levy, O., Bowman, S.R.: GLUE: a multi-task benchmark and analysis platform for natural language understanding. In: 7th International Conference on Learning Representations, ICLR 2019, New Orleans, 6–9 May 2019, LA, USA. OpenReview.net (2019). https://openreview.net/forum?id=rJ4km2R5t7
13. Whang, T., Lee, D., Lee, C., Yang, K., Oh, D., Lim, H.: Domain adaptive training BERT for response selection. arXiv preprint arXiv:1908.04812 (2019)
14. Xu, L., et al.: CLUE: a Chinese language understanding evaluation benchmark. In: Scott, D., Bel, N., Zong, C. (eds.) Proceedings of the 28th International Conference on Computational Linguistics, COLING 2020, 8–13 December 2020, Barcelona, Spain (Online), pp. 4762–4772. International Committee on Computational Linguistics (2020). https://doi.org/10.18653/v1/2020.coling-main.419
15. Xu, L., et al.: FewCLUE: a Chinese few-shot learning evaluation benchmark. CoRR abs/2107.07498 (2021). arXiv: 2107.07498
16. Yang, Z., Dai, Z., Yang, Y., Carbonell, J., Salakhutdinov, R.R., Le, Q.V.: XLNet: generalized autoregressive pretraining for language understanding. In: Advances in Neural Information Processing Systems, vol. 32 (2019)
17. Zheng, C., Huang, M., Sun, A.: ChiD: a large-scale Chinese idiom dataset for cloze test. In: Korhonen, A., Traum, D.R., Màrquez, L. (eds.) Proceedings of the 57th Conference of the Association for Computational Linguistics, ACL 2019, 28 July–2 August 2019, Florence, Italy, vol. 1, Long Papers, pp. 778–787. Association for Computational Linguistics (2019). https://doi.org/10.18653/v1/p19-1075

BIT-Event at NLPCC-2021 Task 3: Subevent Identification via Adversarial Training

Xiao Liu[1,3], Ge Shi[2], Bo Wang[1,3], Changsen Yuan[1,3], Heyan Huang[1,3](✉), Chong Feng[1,3], and Lifang Wu[2]

[1] School of Computer Science and Technology, Beijing Institute of Technology, Beijing, China
{xiaoliu,yuanchangsen,hhy63,fengchong}@bit.edu.cn, bo_wang0@yeah.net

[2] Faculty of Information Technology, Beijing University of Technology, Beijing, China
{shige,lfwu}@bjut.edu.cn

[3] Southeast Academy of Information Technology, Beijing Institute of Technology, Beijing, China

Abstract. This paper describes the system proposed by the BIT-Event team for NLPCC 2021 shared task on Subevent Identification. The task includes two settings, and these settings face less reliable labeled data and the dilemma about selecting the most valid data to annotate, respectively. Without the luxury of training data, we propose a hybrid system based on semi-supervised algorithms to enhance the performance by effectively learning from a large amount of unlabeled corpus. In this hybrid model, we first fine-tune the pre-trained model to adapt it to the training data scenario. Besides, Adversarial Training and Virtual Adversarial Training are combined to enhance the effect of a single model with unlabeled in-domain data. The additional information is further captured via retraining using pseudo-labels. On the other hand, we apply Active Learning as an iterative process that starts from a small number of labeled seeding instances. The experimental results suggest that the semi-supervised methods fit the low-resource subevent identification problem well. Our best results were obtained by an ensemble of these methods. According to the official results, our approach proved the best for all the settings in this task.

Keywords: Subevent identification · Active learning · Adversarial training · Semi-supervised

1 Introduction

A subevent is the event that happens as a part of the other event (i.e., parent event) spatio-temporally [1]. Subevents commonly elaborate and expand an event, and widely exist in event descriptions and understanding. Extracting typical subevents plays an important role in understanding the internal structure of

L. Wang et al. (Eds.): NLPCC 2021, LNAI 13029, pp. 400–411, 2021.
https://doi.org/10.1007/978-3-030-88483-3_32

the event and its properties. Therefore, Subevent Identification has great potential to benefit natural language processing tasks such as information extraction, question answering [2], and contradiction detection [3].

While being in NLPCC 2021 shared task, three subevent named COVID-19, Trade War, and Tokyo Olympic Games are defined. With a large number of event-related corpus and a few labeled seed data, the task aims to build an IE system that may identify the target sub-events. The final evaluation is the average accuracy of two settings, few sample problem sets, and data selection problem setting. The former prohibits the additional annotation of data, while the latter allows only a limited number of data to annotate.

Few Sample Problem Setting: Pre-trained language models have received great interest in leveraging the learned knowledge for a down-stream task [4,5], which also developed multilingual applications [6–8]. Typically, the language model is trained by maximizing the log-likelihood of a sentence. Hence, The original model would focus more on general language features and lack understanding of the corresponding domain knowledge. Therefore, we fine-tune these models by adopting Chinese corpus crawled from Weibo[1]. The following is to take advantage of unlabeled data provided via semi-supervised techniques and Adversarial Training (AT).

Data Selection Problem Setting: Due to the limited number of extra manual annotations, highly qualified and strongly representative data need to sift through for each subevent. Active learning is a technique that can help to reduce the amount of annotation required to train a good model [9–12]. From another point, the technique could help auxiliary screen the worth annotating data in this setting. Accordingly, we apply Active Learning (AL) as an iterative process that starts from a small number of labeled seeding instances.

The main contribution of this study is that the experimental results demonstrate that semi-supervised and AT are effective under the low-resource setting of Subevent Identification task. Additionally, the results of integrating multiple and pre-trained language models show that more complete text features are extracted from different perspectives according to these models.

2 BIT-Event's System

The framework of our hybrid system is shown as Fig. 1, which consists of several subsystems. At the very beginning of the system, we crawled the data on Weibo. The data contains not only the text corresponding to three sub-events but also the data outside the event, which is used to fine-tune the several pre-trained language models (BERT [13], Roberta [14], and MacBERT [15]).

After obtaining the fine-tuned model, we acquire the additional annotation data for **Data Selection Problem Setting** using AL. Then several single models, which use different pre-trained models, initialize weight or training data, are further trained and ensembled for **Few Sample Problem Setting**. We use a

[1] https://weibo.com.

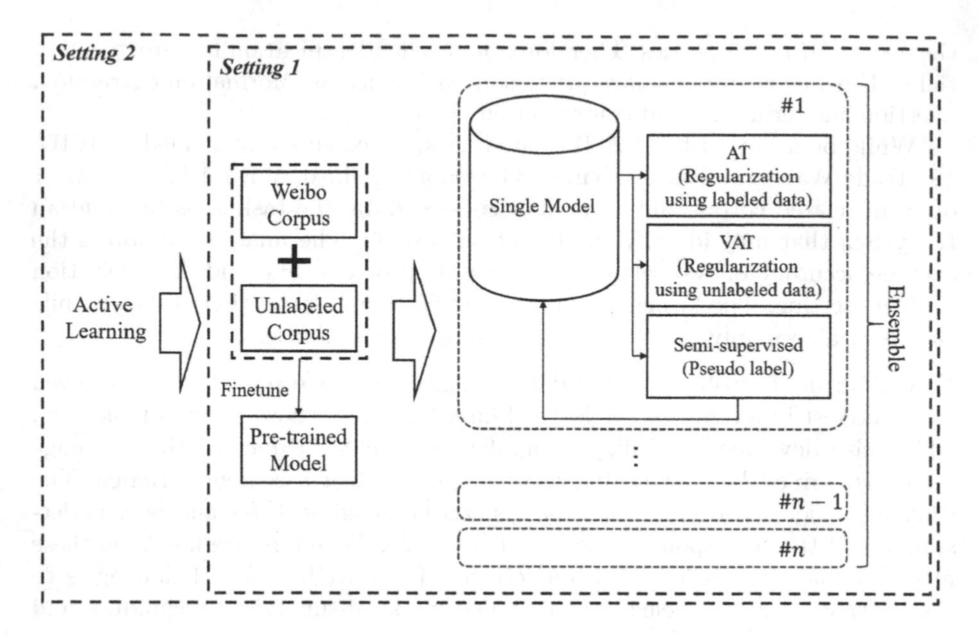

Fig. 1. The framework of BIT-Event's system.

variety of techniques to enhance the effect of a single model, including Adversarial Training (AT), Virtual Adversarial Training (VAT), and Semi-Supervised Learning. The specific practices of each technique are shown in detail below.

2.1 Active Learning

The performance of deep neural networks always improves with more annotated data. We follow the best practices to develop our active learning system, which attached a small parametric module named "loss prediction module" to a target network and learning to predict target losses of unlabeled inputs. Then, this module can suggest data that the target model is likely to produce a wrong prediction [16].

As shown in Fig. 2, the loss prediction part is connected to several layers of the target model to take the multi-level knowledge into consideration for loss prediction. While the module obtains the loss prediction and target prediction, the target prediction and the target annotation are used to compute a target loss to learn the target model. Then, the target loss is regarded as a ground-truth loss for the loss prediction module and used to calculate the loss-prediction loss. The structure is used to assist in screening the worthy labeling data, and we annotate data iteratively until the amount of data meets the requirement.

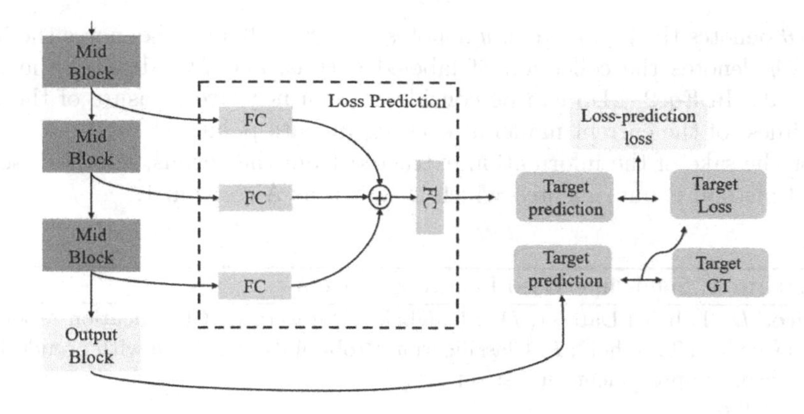

Fig. 2. The architecture of the active learning module.

2.2 Adversarial Training

Due to limited training resources, the model may face the problem of overfitting. Therefore, AT is introduced to mitigating the impact, which is a crucial way to enhance the robustness of neural networks. In the course of confrontation training, the sample is mixed with some minor perturbations, which is small but likely to cause misclassification, and then the neural network is trained to adapt to this change. Thereby, the model is robust to the confrontation sample. The loss function of AT is summarized as the following maximal and minimal formula Eq. (1):

$$\min_{\theta} \mathbb{E}_{(Z,y)\sim\mathcal{D}} \left[\max_{\|\delta\|\leq\epsilon} L\left(f_{\theta}(\boldsymbol{X}+\boldsymbol{\delta}), y\right) \right], \tag{1}$$

where X denotes the embedding extracted according to the pre-trained model, δ denotes the perturbations superimposed on the X, $f_\theta(*)$ means the network with parameter θ and y denotes the groundtruth. We use Free Large Batch Adversarial Training (FreeLB) [17] to acquire the perturbations that maximize the loss function, which performs multiple Projected Gradient Descent iterations to craft adversarial examples, and simultaneously accumulates the "free" parameter gradients $\nabla_{\boldsymbol{\theta}} L$ in each iteration. After that, it updates the model parameter θ all at once with the accumulated gradients.

2.3 Semi-supervised Learning

To make use of unlabeled corpus, two semi-supervised learning method is applied. We start this technology with VAT, "virtual" labels that are probabilistically generated from $p(yx, \theta)$ in place of unknown labels to the user, and compute adversarial direction based on the virtual tags. The final loss is defined as

$$L(\theta) = \sum_{x,y\in D_l} L(y, x, \theta) + \frac{1}{N} \sum_{x\in(D_l, D_{\mathrm{ul}})} \mathrm{LDS}(x, \theta), \tag{2}$$

where θ denotes the parameters, y denotes the groundtruth, x denotes the input data, D_l denotes the collection of labeled corpus, and D_ul denotes the unlabeled part. In Eq. 2, LDS can be considered as a negative measure of the local smoothness of the current model at each input data point.

For the sake of the information extracted from the corpus, another pseudo-labeled method is implemented which is shown as Algorithm 1.

Algorithm 1. Semi-supervised Learning Process

Require: D_l: Labeled Dataset; D_{ul}: Unlabeled Dataset; M: Classification Model; K: Per-defined Threshold; P: Classification Probability; D_s: Data with pseudo-label; S: F_1 Score Improvement on test set;
for $S > 0$ **do**
  Use D_l to retrain the M;
  Use M to obtain P for D_{ul};
  Retain the data $(P > K) \rightarrow D_s$;
  Concatenate D_s and $D_l \rightarrow D_l$;
  Delete D_s from D_{ul}
  Update S
end for

2.4 Ensemble

Three sub-events and negative samples are finally formed into a four-classification task, which got a label set $\{0, 1, 2, 3\}$. For a single piece of test data, the weighted voting ensemble method is applied to extract final result from our pool, whose formula is as

$$L = argmax(\sum_{\lambda_0 \in D_0} \lambda_0, \sum_{\lambda_1 \in D_1} \lambda_1, \sum_{\lambda_2 \in D_2} \lambda_2, \sum_{\lambda_3 \in D_3} \lambda_3), \quad (3)$$

where D_i means the collection of a weight set (the weight is corresponding to the model that predicted the i-th label), λ_i denotes the value of weight and L is the final label of our ensemble model.

3 Experiments

3.1 Dataset

In this restricted track, we used the corpora listed in 1 for two settings. The in-domain or in-genre Chinese corpus **Weibo** crawled is used to finetune existing considerable pretraining weight in the task of MaskLM. According to the result of Active Learning, **ALDATA** corpus was selected from unlabeled corpus, which is manually labeled by our group. **SUBEVENT** is the seed dataset and unlabeled dataset provided by the sponsor.

Table 1. Corpus on the two settings of subevent identification shared task.

Settings	Corpus
Data selection problem setting	**Weibo & SUBEVENT**
Few sample problem setting	**Weibo & ALDATA & SUBEVENT**

3.2 Experiment Setting

We implemented the pre-trained model based on Transformer[2] and UER[3]. The hyperparameters used in hybrid system are listed in Tables 2, 3, 4, and 5, which shows the corresponding ones used in pre-trained model, AT, VAT and common hyperparameters in several subsystems, respectively. The weight of ensemble models were initialized with different values.

Table 2. Hyperparameters of per-trained model.

Parameter	Value
Layer size (base)	12
Layer size (large)	24
Max token size	512
Learning rate	0.0001
Epoch	4
Optimizer	Adam

Table 3. Hyperparameters of AT.

Parameter	Value
Adv_init_mag	1e−1
Adv_steps	3
Adv_lr	1e−1
Adv_max_norm	0
Epoch	6
Optimizer	Adam

Table 4. Hyperparameters of VAT.

Parameter	Value
Alpha	1.0
Xi	1e−6
Eps	8
Ip	2
Epoch	6
Optimizer	Adam

Table 5. Shared hyperparameters.

Parameter	Value
Word embedding size	500
Multi-head number	8
Max token size	512
Ensemble size	24
Batch size	32
Dropout	0.3

The system submits to the shared task were evaluated using the average accuracy of two settings, Few Sample Problem Setting and Data Selection Problem Setting. Three sub-events will be tested together with negative samples. To

[2] https://github.com/huggingface/transformers.
[3] https://github.com/dbiir/UER-py.

be specific, each piece in the test dataset is labeled by 3, 2, 1, and 0, which represent Trade War, Tokyo Olympic Games, COVID-19, and negative samples, respectively.

3.3 Results

Table 6 presents the results of our system (BIT-Event) and others on two settings and the overall effects on test data for the NLPCC 2021 Subevent Identification shared task. Our system ranked 1 out of 10 teams on three metrics. The noise may be further introduced in data annotation, which causes a 0.03% decrease of F_1 on **Few Sample Problem Setting** compared with another one.

Table 6. Results of systems with the highest F_1(%) on two settings and overall one vs ours' on shared task 3 on official test data. **Setting1** means **Data Selection Problem Setting**, **Setting2** means **Few Sample Problem Setting**, and **Overall** means the final results.

Team name	Setting1	Setting2	Overall
Rank1 (ours)	**79.13**	**79.10**	**79.12**
Rank2	77.98	76.50	77.24
Rank3	76.22	76.05	76.14
Rank4	74.95	76.08	75.52
Rank5	77.40	73.38	75.39

Table 7. Ablation study of hybrid system, the settings are continued from Table 6.

	Setting1	Setting2	Overall
Hybrid system	**79.13**	**79.10**	**79.12**
-W/O AT	78.58	78.51	78.55
-W/O VAT	77.42	77.67	77.55
-W/O Pseudo-label	76.78	76.10	76.44

3.4 Discussions

To further analyze our system, we also conduct ablation analysis to illustrate the effectiveness of different modules and mechanisms in hybrid system. We show the results of the study in Table 7. In the table, the hybrid system with all technologies had a more excellent performance on test set. This is because that each module in our system can make up for the deficiency in different aspect. These ablation results show that AT, VAT, and pseudo-label are effective. The 0.57% reduction in results is due to the removal of AT. The decrease shows AT improved robustness of our model On the basis of using two semi-supervised

Table 8. Ablation study of hyperparameters, the settings are continued from Table 6.

	Setting1	Setting2	Overall
Run 3	**79.13**	**79.10**	**79.12**
Run 2	79.06	79.08	79.07
Run 1	78.96	78.88	78.92

learning methods. Besides, the system achieved merely 77.55% of the F_1 score in the absence of VAT, which is 1.57% below our best result. The major reason is that regularization information for unlabeled data is dropped without VAT, and the information enhance the model effect by using unlabeled data in the case of small samples. To assess the effectiveness, pseudo-label is removed finally. The **Overall** F_1 score are reduced by 2.68%, which gives evidence of that the pseudo-label technology learns information hidden in unlabeled data and makes the decision boundary more accurate.

During the integration process, we randomly changed the seeds and epoch number and add the output file into our ensemble pool. The results called **Run 1**, **Run 2**, and **Run 3** are shown in Table 8, which demonstrate multiple rounds of iteration training further enhance the ensemble effect of our system.

4 Related Work

A few studies have focused on identifying subevent relations in text. [18] built a logistic regression model to classify the relation between two events. [1] trained a logistic regression classifier using a range of lexical and syntactic features and then used Integer Linear Programming (ILP) to enforce document-level coherence for constructing coherent event hierarchies from news. Recently, [19] outperformed previous models for subevent relation prediction using a linear SVM classifier, by introducing several new discourse features and narrative features. Considering the generalizability issue of supervised contextual classifiers trained on small annotated data, our pilot research on subevent knowledge acquisition [20] relies on heuristics The recent work [21,22] uses generative language models to generate subevent knowledge among many other types of commonsense knowledge.

Active Learning(AL) is a technique that can help to reduce the amount of annotation required to train a good model by multiple times [9,10]. To focus on low-resource task, several techniques developed strong in natural language processing. Pre-training has always been an effective strategy to learn the parameters of deep neural networks, which are then fine-tuned on downstream tasks. Recently, the very deep PTMs have shown their powerful ability in learning universal language representations: e.g., BERT(Bidirectional Encoder Representation from Transformer) [13], Ernie(Continual pre-training framework) [23], roberta(Robustly optimized bert) [14], macbert [15] and Xlnet(Generalized autoregressive pretraining) [24]. Adversarial examples have been explored primarily in the image domain, and received many attention in text domain. Previous work on text adversaries focused on some specific task [25–28], then AT is observed to robust language model [29–31]. Semi-supervised learning [32] seeks to largely alleviate the need for labeled data by allowing a model to leverage unlabeled data. There is a wide literature on Semi-supervised learning techniques, such as (e.g., graph-based methods [33], generative modeling [34,35], etc.). [36] proposed VAT to make AT applicability to semi-supervised learning task.

5 Conclusion

In this paper, we described our hybrid system, which is based on AT, AL, Semi-supervised, and pre-trained language model. We evaluated our approach on the NLPCC 2021 Subevent Identification shared task. The experimental results demonstrated that mixing a variety of low-resource enhancement algorithms can improve the classification quality.

In this work, we only considered the fundamental pseudo-label approach. We will try to integrate the current mainstream of semi-supervised learning methods into a unified framework for better performance.

Acknowledgement. This work was supported by the Funds of the Integrated Application Software Project. We would like to thank the anonymous reviewers for their thoughtful and constructive comments. And We would like to thank Zewen Chi for his insightful comments to the improvement in technical contents and paper presentation.

References

1. Glavas, G., Snajder, J.: Constructing coherent event hierarchies from news stories. In: Proceedings of the 9th Workshop on Graph-based Methods for Natural Language Processing, pp. 34–38 (2014). https://doi.org/10.3115/v1/w14-3705
2. Narayanan, S., Harabagiu, S.M.: Question answering based on semantic structures. In: Proceedings of the 20th International Conference on Computational Linguistics (2004)
3. de Marneffe, M., Rafferty, A.N., Manning, C.D.: Finding contradictions in text. In: Proceedings of the 46th Annual Meeting of the Association for Computational Linguistics, pp. 1039–1047 (2008)
4. Reimers, N., Schiller, B., Beck, T., Daxenberger, J., Stab, C., Gurevych, I.: Classification and clustering of arguments with contextualized word embeddings. In: Proceedings of the 57th Conference of the Association for Computational Linguistics, pp. 567–578 (2019). https://doi.org/10.18653/v1/p19-1054
5. Sun, C., Qiu, X., Xu, Y., Huang, X.: How to fine-tune BERT for text classification? In: Sun, M., Huang, X., Ji, H., Liu, Z., Liu, Y. (eds.) Proceedings of the Chinese Computational Linguistics - 18th China National Conference. Lecture Notes in Computer Science, vol. 11856, pp. 194–206 (2019). https://doi.org/10.1007/978-3-030-32381-3_16

6. Chi, Z., et al.: InfoXLM: an information-theoretic framework for cross-lingual language model pre-training. In: Proceedings of the 2021 Conference of the North American Chapter of the Association for Computational Linguistics: Human Language Technologies, pp. 3576–3588 (2021)
7. Chi, Z., Dong, L., Zheng, B., Huang, S., Mao, X., Huang, H., Wei, F.: Improving pretrained cross-lingual language models via self-labeled word alignment. CoRR abs/2106.06381 (2021)
8. Chi, Z., Dong, L., Wei, F., Wang, W., Mao, X., Huang, H.: Cross-lingual natural language generation via pre-training. In: The Thirty-Fourth AAAI Conference on Artificial Intelligence. pp. 7570–7577 (2020)
9. Settles, B., Craven, M.: An analysis of active learning strategies for sequence labeling tasks. In: Proceedings of the 2008 Conference on Empirical Methods in Natural Language Processing, pp. 1070–1079 (2008)
10. Marcheggiani, D., Artières, T.: An experimental comparison of active learning strategies for partially labeled sequences. In: Proceedings of the 2014 Conference on Empirical Methods in Natural Language Processing, pp. 898–906 (2014). https://doi.org/10.3115/v1/d14-1097
11. Shen, Y., Yun, H., Lipton, Z.C., Kronrod, Y., Anandkumar, A.: Deep active learning for named entity recognition. In: Proceedings of the 6th International Conference on Learning Representations (2018)
12. Shelmanov, A., et al.: Active learning for sequence tagging with deep pre-trained models and Bayesian uncertainty estimates. In: Proceedings of the 16th Conference of the European Chapter of the Association for Computational Linguistics, pp. 1698–1712. Association for Computational Linguistics (2021)
13. Devlin, J., Chang, M., Lee, K., Toutanova, K.: BERT: pre-training of deep bidirectional transformers for language understanding. In: Proceedings of the 2019 Conference of the North American Chapter of the Association for Computational Linguistics: Human Language Technologies, pp. 4171–4186 (2019). https://doi.org/10.18653/v1/n19-1423
14. Liu, Y., et al.: RoBERTa: a robustly optimized BERT pretraining approach. CoRR abs/1907.11692 (2019)
15. Cui, Y., Che, W., Liu, T., Qin, B., Wang, S., Hu, G.: Revisiting pre-trained models for Chinese natural language processing. In: Proceedings of the 2020 Conference on Empirical Methods in Natural Language Processing: Findings, pp. 657–668 (2020). https://doi.org/10.18653/v1/2020.findings-emnlp.58
16. Yoo, D., Kweon, I.S.: Learning loss for active learning. In: Proceedings of the IEEE Conference on Computer Vision and Pattern Recognition, pp. 93–102 (2019). https://doi.org/10.1109/CVPR.2019.00018
17. Zhu, C., Cheng, Y., Gan, Z., Sun, S., Goldstein, T., Liu, J.: FreeLB: enhanced adversarial training for natural language understanding. In: Proceedings of the 8th International Conference on Learning Representations, ICLR 2020, Addis Ababa, Ethiopia, 26–30 April 2020 (2020)
18. Araki, J., Liu, Z., Hovy, E.H., Mitamura, T.: Detecting subevent structure for event coreference resolution. In: Proceedings of LREC, pp. 4553–4558 (2014)
19. Aldawsari, M., Finlayson, M.A.: Detecting subevents using discourse and narrative features. In: Proceedings of the 57th Conference of the Association for Computational Linguistics, pp. 4780–4790 (2019). https://doi.org/10.18653/v1/p19-1471
20. Badgett, A., Huang, R.: Extracting subevents via an effective two-phase approach. In: Proceedings of the 2016 Conference on Empirical Methods in Natural Language Processing, pp. 906–911 (2016). https://doi.org/10.18653/v1/d16-1088

21. Bosselut, A., Rashkin, H., Sap, M., Malaviya, C., Celikyilmaz, A., Choi, Y.: COMET: commonsense transformers for automatic knowledge graph construction. In: Proceedings of the 57th Conference of the Association for Computational Linguistics, pp. 4762–4779 (2019). https://doi.org/10.18653/v1/p19-1470
22. Sap, M., et al.: ATOMIC: an atlas of machine commonsense for if-then reasoning. In: Proceedings of the 33rd AAAI Conference on Artificial Intelligence, pp. 3027–3035 (2019). https://doi.org/10.1609/aaai.v33i01.33013027
23. Sun, Y., et al.: ERNIE 2.0: a continual pre-training framework for language understanding. In: Proceedings of the 34th AAAI Conference on Artificial Intelligence, pp. 8968–8975 (2020)
24. Yang, Z., Dai, Z., Yang, Y., Carbonell, J.G., Salakhutdinov, R., Le, Q.V.: XLNet: generalized autoregressive pretraining for language understanding. In: Advances in Neural Information Processing Systems 32, pp. 5754–5764 (2019)
25. Jia, R., Liang, P.: Adversarial examples for evaluating reading comprehension systems. In: Palmer, M., Hwa, R., Riedel, S. (eds.) Proceedings of the 2017 Conference on Empirical Methods in Natural Language Processing, pp. 2021–2031 (2017). https://doi.org/10.18653/v1/d17-1215
26. Zhao, Z., Dua, D., Singh, S.: Generating natural adversarial examples. In: Proceedings of the 6th International Conference on Learning Representations (2018)
27. Belinkov, Y., Bisk, Y.: Synthetic and natural noise both break neural machine translation. In: Proceedings of the 6th International Conference on Learning Representations (2018)
28. Iyyer, M., Wieting, J., Gimpel, K., Zettlemoyer, L.: Adversarial example generation with syntactically controlled paraphrase networks. In: Proceedings of the 2018 Conference of the North American Chapter of the Association for Computational Linguistics, pp. 1875–1885 (2018). https://doi.org/10.18653/v1/n18-1170
29. Ebrahimi, J., Rao, A., Lowd, D., Dou, D.: HotFlip: white-box adversarial examples for text classification. In: Proceedings of the 56th Annual Meeting of the Association for Computational Linguistics, pp. 31–36 (2018). https://doi.org/10.18653/v1/P18-2006
30. Ribeiro, M.T., Singh, S., Guestrin, C.: Semantically equivalent adversarial rules for debugging NLP models. In: Proceedings of the 56th Annual Meeting of the Association for Computational Linguistics, pp. 856–865 (2018). https://doi.org/10.18653/v1/P18-1079
31. Cheng, Y., Jiang, L., Macherey, W.: Robust neural machine translation with doubly adversarial inputs. In: Proceedings of the 57th Conference of the Association for Computational Linguistics, pp. 4324–4333 (2019). https://doi.org/10.18653/v1/p19-1425
32. Chapelle, O., Schölkopf, B., Zien, A. (eds.): Semi-Supervised Learning. The MIT Press, Cambridge (2006). https://doi.org/10.7551/mitpress/9780262033589.001.0001
33. Liu, B., Wu, Z., Hu, H., Lin, S.: Deep metric transfer for label propagation with limited annotated data. In: Proceedings of the 2019 IEEE/CVF International Conference on Computer Vision Workshops, pp. 1317–1326 (2019). https://doi.org/10.1109/ICCVW.2019.00167
34. Kingma, D.P., Mohamed, S., Rezende, D.J., Welling, M.: Semi-supervised learning with deep generative models. In: Ghahramani, Z., Welling, M., Cortes, C., Lawrence, N.D., Weinberger, K.Q. (eds.) Advances in Neural Information Processing Systems 27, pp. 3581–3589 (2014)

35. Pu, Y., et al.: Variational autoencoder for deep learning of images, labels and captions. In: Advances in Neural Information Processing Systems 29, pp. 2352–2360 (2016)
36. Miyato, T., Maeda, S., Koyama, M., Ishii, S.: Virtual adversarial training: a regularization method for supervised and semi-supervised learning. IEEE Trans. Pattern Anal. Mach. Intell. **41**(8), 1979–1993 (2019). https://doi.org/10.1109/TPAMI.2018.2858821

Few-Shot Learning for Chinese NLP Tasks

Liang Xu(✉), Xiaojing Lu, Chenyang Yuan, Xuanwei Zhang, Hu Yuan, Huilin Xu, Guoao Wei, Xiang Pan, Junyi Li, Jianlin Su, Zhenyu Yang, Renfen Hu, and Hai Hu

CLUE Team, Beijing, China
https://www.CLUEbenchmarks.com

Abstract. In the paper, we report the results for the NLPCC2021 shared-task of Few-shot Learning for Chinese NLP. This shared task is proposed in the context of pre-trained language models, where models only have access to limited human-labeled data. The goal of the task is to compare different learning schemes. The task includes nine sub-tasks and three task forms: single sentence classification, sentence pair classification, and machine reading comprehension. In order to accommodate the properties of few-shot learning, we sampled the examples using various sampling methods, some with 32 examples in total for one dataset, while others with 4 to 16 examples per class. Ninety teams registered for the shared task, employing a wide range of learning schemes, including data augmentation, utilizing multiple templates rather than a single template, using unlabeled data for pre-training or semi-supervised training. The best model achieved 65.3 in the mean accuracy, compared with the human score of 83.9. This result is 8 points higher than our baseline model (using the PET scheme). We believe our few-shot learning tasks and results demonstrate the potential of the recently introduced few-shot learning methods and provide guidance and important empirical evidence for future research.

Keywords: Chinese language processing · Pre-trained language models · Few-shot learning · Prompt-based method

1 Introduction

Recent work has shown that pretraining on a large amount of data and then fine-tuning on comparatively fewer, annotated in-domain data can achieve impressive results on many natural language processing (NLP) tasks [3,9,11]. However, for fine-tuning to be effective, thousands or even tens of thousands of labeled data are still needed, far more than the few examples humans rely on to perform the same NLP task. One of the most pressing issues in NLP research is how

L. Xu, X. Lu, C. Yuan, X. Zhang, H. Yuan and H. Xu—Contributed equally.

L. Wang et al. (Eds.): NLPCC 2021, LNAI 13029, pp. 412–421, 2021.
https://doi.org/10.1007/978-3-030-88483-3_33

to achieve good performance while only utilizing a few hand-annotated labeled data.

Despite the tremendous progress in Artificial Intelligence (AI), an AI agent is far behind learning from limited training examples and generalizing unseen data than humans. Unlike a human learner, for whom the mastery of addition will make multiplication very easy to learn. For example, several photos of a cup are enough to learn to recognize other images of different cups.

Few-shot learning is a research area in machine learning where only limited (usually a few dozen) human-annotated training examples are available for supervised learning. Thus, few-shot learning, as a way to reduce the high cost of human annotation and intensive computation, has received much attention in NLP research recently.

For instance, the GPT3 model [1] provides a method that does not require any gradient update or fine-tuning, using a few or even zero human-labeled data. PET, as a semi-supervised learning method, converts an input text into a template-based cloze test format to solve different NLP tasks [13,14]. LM-BFF [5] builds on PET and uses a generative model to construct the template for each task automatically. Methods such as Ptuning [10], ADAPET [15] and EFL [16] explored other ways for few-shot learning.

To facilitate the research on few-shot learning in Chinese NLP, we organized the FewCLUE shared-task, which includes nine natural language understanding (NLU) tasks, ranging from single sentence classification, sentence pair classification, and machine reading tasks. Apart from the training/dev/test sets, we also provide 20,000 unlabelled data for each task (except for CLUEWSC). Furthermore, to deal with the instability inherent in few-shot learning, we provide multiple training/dev sets. In addition, we provide competitive baselines using state-of-the-art few-shot learning methods and a comprehensive metric to calculate each team's final score. Additional tools and a benchmark are available at our website: https://github.com/CLUEbenchmark/FewCLUE.

We find that while different methods show various levels of performance, template-based methods, in general, outperform the classic pre-train-and-finetune paradigm. However, all models are still far behind human performance, suggesting that there is much room for improvement. In addition, zero-shot learning has achieved good performance on our benchmark.

2 Task Description

2.1 Task Overview

This task aims to explore the capability of a pre-trained language model conditioned on a few training data for the downstream task. CLUE designed this task based on the CLUE benchmark to promote more research and applications in Chinese NLP few-shot learning. This task covers nine different downstream sub-tasks. Submitted models need to be able to predict labels for each sub-task. The sub-tasks will be described in detail in Sect. 3.

Table 1. Metrics for the report score

Item	Criterion	Full score
Creativity	The innovation points of the algorithm in the design and implementation process, such as the performance improvement brought by model construction and algorithm transformation	100
Practice, interpretable, reproducible	Extraction of important features and interpretation in the scene, combined with theoretical or practical quantitative indicators to give insights on model training; The model has high reproducibility	100
Presentation	The organization, systematicness, completeness of the speech, and the performance of answering questions	100

We have released a tutorial for the task at https://github.com/CLUEbenchmark/FewCLUE/tree/main/baselines.

2.2 Evaluation Criteria

There is two part of the evaluation. One is the online score of the model, and the other is the result of defense. The results of the online score are evaluated using accuracy for each subtask. The final online score of this task is calculated by taking the average of scores over all subtasks.

$$\text{model\_score} = \frac{\sum \text{acc\_on\_each\_subtask}}{9}$$

For the result of the defense, each team will report their solution, and five reviewers will give scores according to specific criteria, including the performance of creativity, practice & reproducibility, and presentation. Each team will have 10 min for the oral report and 5 min for QA.

In Table 1, we present the metrics for the report score.

The final score is a weighted sum of the model score and report score.

$$\text{final\_score} = \text{model\_score} \times 0.65 + \text{report\_score} \times 0.35$$

3 Dataset Description

This section introduces the datasets used in this shared task. As shown in Table 2, there are a total of nine Chinese tasks, including four text classification tasks, two sentence pair tasks, and three reading comprehension tasks. From the perspective of text-domain, these datasets include daily language, news text, literary works, and academic literature. Apart from the CLUEWSC dataset, the other eight datasets all include unlabeled data for semi-supervised learning or other methods. In order to verify the stability of few-shot learning, we provide five randomly sampled training/verification sets of equal size for each dataset.

Table 2. Description and statistics of the nine datasets in the shared-task. EPRSTMT, CSLDCP and BUSTM are new tasks; other tasks are take from the CLUE benchmark [18].

Dataset	Train	Dev	Test_pub	Test_priv	#Labels	Unlabeled	Task	Metric	Source
Single sentence tasks									
EPRSTMT	32	32	610	753	2	19565	SntmntAnalysis	Acc	E-CommrceReview
CSLDCP	536	536	1784	2999	67	18111	LongTextClassify	Acc	AcademicCNKI
TNEWS	240	240	2010	1500	15	20000	ShortTextClassify	Acc	NewsTitle
IFLYTEK	928	690	1749	2279	119	7558	LongTextClassify	Acc	AppDesc
Sentence pair tasks									
OCNLI	32	32	2520	3000	3	20000	NLI	Acc	5Genres
BUSTM	32	32	1772	2000	2	4251	SemanticSmlarty	Acc	AIVirtualAssistant
Reading comprehension									
CHID	42	42	2002	2000	7	7585	MultipleChoice, idiom	Acc	Novel, EssayNews
CSL	32	32	2828	3000	2	19841	KeywordRecogntn	Acc	AcademicCNKI
CLUEWSC	32	32	976	290	2	0	CorefResolution	Acc	ChineseFictionBooks

EPRSTMT is a sentiment analysis dataset of e-commerce product reviews, with positive and negative labels. The training set contains 32 samples, 32 verification samples, 610 public test samples, and 753 non-public test samples. In addition, the data set also includes 19565 unlabeled samples.

CSLDCP is a long text classification dataset of Chinese scientific literature, which includes 67 labels of literature. These labels come from 13 categories, ranging from social sciences to natural sciences. The text of CSLDCP is the Chinese abstract of literature. There are 536 samples in the training set and verification set, 1784 samples in the public test set, 2999 samples in the private test set, and 18111 samples without a label.

TNEWS is a dataset containing titles of toutiao news. It has a total of 15 categories of news, including tourism, education, finance, military, etc., The training set and verification set have 240 samples; the public test set has 2010 samples; the private test set contains 1500 samples, and unlabeled sample is 20000.

IFLYTEK includes long text of application description and the corresponding category of the application, with various application topics related to daily-life [7]. It contains 119 categories of application, such as taxi/map navigation/women/business/etc. The dataset includes 928 training samples, 690 verification samples, 1749 public test samples, and 2279 in the private test set. In addition, the data set also includes 7558 unlabeled samples.

OCNLI, the original Chinese natural language inference data set, is the first large-scale original Chinese natural language inference data set using original, non-translated Chinese [6]. There are 32 samples in the training set and verification set, 2520 samples in the public test set, 3000 samples in the private test set, and 20000 samples without labels.

BUSTM is a semantic matching dataset of dialogue text, which is originated from XiaoBu assistant[1] [12]. There are 32 samples in the training set and verification set, 1772 samples in the public test set, 2000 samples in the private test set, and 4251 samples without labels.

ChID is a data set of idiom reading comprehension, which is implemented in the form of idiom cloze [19]. Many idioms in this paper are masked, and the candidates contain seven idioms with similar meanings. There are 42 samples in the training set and verification set, 2002 samples in the public test set, 2000 samples in theprivate test set and 7585 samples without a label.

CSL is a dataset of Chinese scientific and technological literature, which is extracted from abstracts and keywords of Chinese papers [18]. Papers are selected from some core journals of Chinese Social Sciences and natural sciences. The task is to judge whether all keywords are real keywords according to abstracts. There are 32 samples in the training set and verification set, 2828 samples in the public test set, 3000 samples in the private test set, and 19841 samples without a label.

Winograd scheme challenge (WSC) is a pronoun disambiguation task, which is to determine which noun a pronoun in a sentence refers to. The data comes from the literary works of contemporary Chinese writers. Nouns and pronouns are extracted from the texts, and then manually selected and marked by language experts. We use the CLUEWSC2020 dataset [18]. The training set and verification set are 32 samples, and the public test set 976 samples, the private test set 290 samples, not including unlabeled samples.

4 Baselines

We adopt the few-shot learning methods PET as our baseline. Baseline is implemented using RoBERTa and GPT, specifically RoBERTa-wwm-ext [2] and NEZHA-Gen [17]. Our baseline achieve 57.36. Below we briefly introduce our baselines.

PET reformulates the downstream task into cloze-syle question similar to MLM, to achieve good results using a limited number of samples. In detail, each sample is constructed by a pattern-verbalizer pair. A pattern is used to reformulate the input into cloze-style, which usually includes one or more MASK tokens. The

[1] It is a voice assistant developed by OPPO for mobile phones and IOT devices.

verbalizer maps each label to a word in the vocabulary. The model is trained to predict the label word on the MASK position. Most of the verbalizers in our task map each label to one character according to the meaning. Since the labels of CSLDCP and IFLYTEK are mostly more than three characters and have various lengths, they are mapped to two characters by the verbalizer.

5 Submissions

Our tasks are diverse in data sources, task types, and forms. Therefore, it attracted contestants from different areas and institutions. They have explored and tried various aspects such as text preprocessing, Pre-Training methods, template-based optimization, data augmentation, and the innovation of training methods. Their thinking and experiment tried to understand the task and explore the task from different aspects and optimize the solution, which can be a great treasure for the practical feature implementation and research direction. We give some brief descriptions as below.

5.1 Alibaba DAMO Academy and Computing Platform PAI (阿里巴巴达摩院)

They adopted the traditional Pre-train-Finetune mechanics and used a self-developed knowledge-enhanced general pre-training model, KEBERT as backbone. In the pre-training phase, they train the EFL task using the CMNLI data; and train the MLM task of fusion extraction MRC with the unlabeled data and additional news data: after segmentation of unlabeled text, a part of the keyword glossary is randomly selected; Several groups of equal length words are randomly selected from the keyword glossary, combined the original text to form the input of the extractive MRC paradigm, which enhances the sensitivity of the pre-training model to MRC task prompts. Furthermore, at the same time, the ChID task's effect is to increase the ratio of four-character words during MLM masking and to learn the expression of four-character words better.

In the fine-tuning stage, they improved the existing PET algorithm to the multi-label PET method: During training, multiple candidate words are copied multiple times as input and choose the summed probability candidate as the final result. For the CLUEWSC task, the word referring to the task position is masked for prediction, and the optimization task is closer to the task form of MLM, which improves the model performance. The CSL task designed the sentencing phase scheme to solve large data loss when a longer text is truncated, enabling the model to process more informative sentences more efficiently in the pre-training and fine-tuning phases. You can find more detail here [20].

5.2 Tencent Cloud Dingdang Education (腾讯云小微教育)

They used the two-stage learning strategy: in the first stage, all tasks are defined as classification tasks (ChID is multi-choice classification, and the other tasks are single sentence-level classification), and proposed PET-based fine-tuning strategy, which uses the full-text information and language model context information contained in the [CLS] characters in the pre-trained model. Since different PET templates have a significant impact on the model effect, multiple patterns are used for model training simultaneously, and a learnable confidence parameter is set for each pattern. The weighted sum of the multi-pattern classification probabilities and the corresponding confidence parameters is used to get the final prediction result during the prediction phase. In the second stage, they use the first-stage training model to predict the unlabeled data to generate pseudo labeled training examples, using the backbone of the first-stage model except for the classifier layer and retrain the classifier. Finally, bootstrap the model until it is converged. You can find more detail here [8].

5.3 Changhong AI Laboratory (长虹 AI 实验室)

They applied the mask-based data augmentation(DA) for few-shot samples: using [MASK] or token in the model dictionary to replace any token in the original sentence and replace the strategy n times to get an n-times augmented dataset. The template-based method used the PET method with adversarial training, which adds perturbation into the embedding layer. Such adversarial training improves the robustness and generalization performance of the model. In the meantime, they used the Sparse-CrossEntropy to prevent the overfitting under a few-shot setting. They considered and tried contrastive learning and semi-supervised learning. However, these methods have some limited improvements currently, which can be a direction for further exploration.

5.4 Business Intelligence Laboratory of Baidu Research Institute (百度研究院商业智能实验室)

They proposed the predefined PET-based templates with P-Tuning, namely Pattern-exploiting & Pattern-tuning Training (PPT) method. Such a combined method can prevent instability by using different PET templates and unexplainable p-tuning generated templates. Each iteration uses the predefined template to train the model to choose the best-performed model based on the validation performance. Then they used the previous chosen model to make the unlabeled data prediction and choose the high-confidence label as the pseudo-labeled training data. Finally, after several iterations, they use the final best model to make a prediction.

5.5 Team from Beijing University of Posts and Telecommunications (北京邮电大学)

Following the idea of the EFL method, the few-shot tasks are converted into textual implication tasks. Firstly, they use the public textual implication task (such as CMNLLI) for task training, and then use the task trained model for fine-tune. During the experiment, the team found that the choice of the model had a great impact on the results, so they used the Mask Language Task to choose the templates. They used the filtered templates to improve the effect of the model. At the same time, the ChID task is modified into a textual implication task composed of the two sentences after segmentation in the #idiom# position from the original sentence, which improves the effect of the original EFL method on Few-Shot learning task.

5.6 Team from Zhejiang University (浙江大学)

They converted all the task as Xin+T forms, which can use the template methods with learnable parameters. Specifically, based on the semantic matching idea, they used seven candidate sets in the sample and the [MASK]*4 to replace #idiom# in the original sentence, then used the converted samples to calculate the Cross Entropy loss.

6 Shared-Task Results

Table 3. Submission Results (Red indicates the last top 4 team and bold indicates the teams participated the final defense. Our baseline, PET, online score is 57.36.)

				Contest Defense Score		Semifinal Score		Total Score
Rank	Team Name	Team Member	Institution	Original	Normalized	Original	Normalized	Final score
1	篮网总冠军	李鹏、徐子云、李杨	阿里巴巴达摩院 & 计算平台 PAI	123.5	0.80	65.334	0.80	0.8
2	姜汁柠檬	曾嘉莉、姜雨帆	腾讯云小微教育	121.5	0.77	63.112	0.73	0.747
3	皮皮虾	周兴发、杨兰	长虹 AI 实验室	115	0.69	64.525	0.78	0.745
4	MLP fans	姜楠、王雅晴、熊昊一	百度研究院商业智能实验室	122	0.78	60.525	0.66	0.699
5	**姜汁可乐**	**王泽元、魏志宇**	**北京邮电大学**	**117**	**0.71**	**60.024**	**0.64**	**0.667**
6	**上山没老虎**	**陈湘楠、谢辛**	**浙江大学**	**93**	**0.40**	**52**	**0.40**	**0.4**
7	**卷心菜**	**从鑫、崔诗尧、盛囧伟、曹江峡**	**中国科学院信息工程研究所**	**0**	**0.00**	**55.24**	**-**	**-**
8	UPSIDEDOWN	苏江文	国网信通产业集团福建亿榕信息	0	0.00	55.233	-	-
9	paht_sjtu	朱威	平安科技	0	0.00	48.083	-	-

In this competition, the "篮球总冠军" team from Alibaba Dharma Academy and the Cloud Computing platform PAI won the double track championship with a parameter-restricted track and a non-parameter-restricted track. In the average score of the list, it won The results of the parameter-limited track 65.334 and the non-parameter limited track 66.128; the "姜汁柠檬 " team from Tencent Cloud Small and Micro Education and the "皮皮虾 " team from Changhong AI Lab achieved 63.112 and 64.525 respectively. The single result, followed by the second and third place after weighting the results of the rematch. "MLP_fans" from the Business Intelligence Laboratory of Baidu Research Institute and "姜汁可乐 " [4] from Beijing University of Posts and Telecommunications won the score of 60.525 and 60.024 respectively, and the semi-finals were also fourth and fifth. The results of other participating teams can refer to the specific list and official website[2].

[2] https://www.cluebenchmarks.com/NLPCC.html.

7 Conclusion

In this paper, we reported the results of the NLPCC 2021 shared-task 2, which evaluates the performance of different learning schemes under a data sparsity situation. Compared with the classic pretrain-finetune paradigm, few-shot learning methods such as PET, where an NLU task is reformulated as a cloze-test task, have achieved good results. In this shared-task, we provided 9 NLU tasks to systematically evaluate the few-shot learning ability of different methods. Several strong baselines based on PET, P-tuning, etc. have been presented, along with fine-tuning and zero-shot learning results. Teams from both the industry and academia participated in our shared-task. Several systems outperformed our baselines by a large margin. However, even the best few-shot learning method is still 15 points behind human score, suggesting that the tasks are challenging and that there is much room for improvement for the few-shot learning models and methods.

Acknowledgements. Many thanks to NLPCC for offering us this opportunity to organize this task and people who took part in this task.

References

1. Brown, T.B., et al.: Language models are few-shot learners (2020)
2. Cui, Y., Che, W., Liu, T., Qin, B., Wang, S., Hu, G.: Revisiting pre-trained models for Chinese natural language processing. In: Proceedings of the 2020 Conference on Empirical Methods in Natural Language Processing: Findings, pp. 657–668. Association for Computational Linguistics, Online, November 2020. https://www.aclweb.org/anthology/2020.findings-emnlp.58
3. Devlin, J., Chang, M.W., Lee, K., Toutanova, K.: BERT: pre-training of deep bidirectional transformers for language understanding. arXiv preprint arXiv:1810.04805 (2018)
4. Zhang, Y.W.E., Wang, Z.: Entailment method based on template selection for FewCLUE evaluation (2021)
5. Gao, T., Fisch, A., Chen, D.: Making pre-trained language models better few-shot learners. In: Association for Computational Linguistics (ACL) (2021)
6. Hu, H., Richardson, K., Liang, X., Lu, L., Kübler, S., Moss, L.: OCNLI: original Chinese natural language inference. In: Findings of Empirical Methods for Natural Language Processing (Findings of EMNLP) (2020)
7. IFLYTEK CO. L: Iflytek: a multiple categories Chinese text classifier. competition official website (2019). http://challenge.xfyun.cn/2019/gamelist
8. Zeng, J., Jiang, S.W.Y., Li, M.: Enhanced few-shot learning with multiple-pattern-exploiting training. In: CCF International Conference on Natural Language Processing and Chinese Computing (2021)
9. Lan, Z., Chen, M., Goodman, S., Gimpel, K., Sharma, P., Soricut, R.: ALBERT: a lite BERT for self-supervised learning of language representations. arXiv preprint arXiv:1909.11942 (2019)
10. Liu, X., et al.: GPT understands, too. arXiv preprint arXiv:2103.10385 (2021)
11. Liu, Y., et al.: RoBERTa: a robustly optimized BERT pretraining approach. arXiv preprint arXiv:1907.11692 (2019)

12. Conversational-AI Center of OPPO XiaoBu: BUSTM: OPPO Xiaobu dialogue short text matching dataset (2021). https://github.com/xiaobu-coai/BUSTM
13. Schick, T., Schütze, H.: Exploiting cloze-questions for few-shot text classification and natural language inference. In: Proceedings of the 16th Conference of the European Chapter of the Association for Computational Linguistics: Main Volume, pp. 255–269. Association for Computational Linguistics, Online, April 2021. https://aclanthology.org/2021.eacl-main.20
14. Schick, T., Schütze, H.: It's not just size that matters: small language models are also few-shot learners. In: Proceedings of the 2021 Conference of the North American Chapter of the Association for Computational Linguistics: Human Language Technologies, pp. 2339–2352. Association for Computational Linguistics, Online, June 2021. https://doi.org/10.18653/v1/2021.naacl-main.185, https://aclanthology.org/2021.naacl-main.185
15. Tam, D., Menon, R.R., Bansal, M., Srivastava, S., Raffel, C.: Improving and simplifying pattern exploiting training (2021)
16. Wang, S., Fang, H., Khabsa, M., Mao, H., Ma, H.: Entailment as few-shot learner (2021)
17. Wei, J., et al.: NEZHA: neural contextualized representation for Chinese language understanding. arXiv preprint arXiv:1909.00204 (2019)
18. Xu, L., et al.: CLUE: a Chinese language understanding evaluation benchmark. In: Proceedings of the 28th International Conference on Computational Linguistics, pp. 4762–4772. International Committee on Computational Linguistics, Barcelona, Spain (Online), December 2020. https://doi.org/10.18653/v1/2020.coling-main.419, https://aclanthology.org/2020.coling-main.419
19. Zheng, C., Huang, M., Sun, A.: ChID: a Large-scale Chinese IDiom dataset for Cloze test. In: Proceedings of the 57th Annual Meeting of the Association for Computational Linguistics, pp. 778–787 (2019)
20. Xu, Z., et al.: When few-shot learning meets large-scale knowledge-enhanced pre-training: Alibaba at FewCLUE. In: CCF International Conference on Natural Language Processing and Chinese Computing (2021)

When Few-Shot Learning Meets Large-Scale Knowledge-Enhanced Pre-training: Alibaba at FewCLUE

Ziyun Xu[1,2], Chengyu Wang[1], Peng Li[1], Yang Li[1], Ming Wang[1,3], Boyu Hou[1,4], Minghui Qiu[1](✉), Chengguang Tang[1], and Jun Huang[1]

[1] Alibaba Group, Hangzhou, Zhejiang 311121, China
{xuziyun.xzy,chengyu.wcy,jerry.lp,ly200170,jinpu.wm,houboyu.hby, minghui.qmh,chengguang.tcg,huangjun.hj}@alibaba-inc.com

[2] School of Computer Science, Carnegie Mellon University, Pittsburgh, PA 15213, USA

[3] School of Information Management and Engineering, Shanghai University of Finance and Economics, Shanghai 200433, China

[4] College of Computer Science, Chongqing University, Chongqing 400044, China

Abstract. With the wide popularity of Pre-trained Language Models (PLMs), it has been a hot research topic to improve the performance of PLMs in the few-shot learning setting. FewCLUE is a new benchmark to evaluate the few-shot learning ability of PLMs over nine challenging Chinese language understanding tasks, which poses significant challenges to the learning process of PLMs with very little training data available. In this paper, we present our solution to FewCLUE tasks by means of large-scale knowledge-enhanced pre-training over massive texts and knowledge triples, together with a new few-shot learning algorithm for downstream tasks. Experimental results show that the generated models achieve the best performance in both limited and unlimited tracks of FewCLUE. Our solution is developed upon the PyTorch version of the EasyTransfer toolkit and will be released to public.

Keywords: Pre-trained Language Model · Knowledge-enhanced Pre-trained Language Model · Knowledge Graph · Few-shot learning

1 Introduction

Recent years has witnessed the successful application of large-scale Pre-trained Language Models (PLMs) for solving various Natural Language Processing (NLP) tasks, based on the "pre-training and fine-tuning" paradigm [7,12,17]. To ensure high model accuracy for downstream tasks, it is necessary to obtain a sufficient amount of training data, which is often the bottleneck in low-resource scenarios.

In the literature, two types of approaches have been proposed to address the above-mentioned problem. The first approach is to inject factual knowledge

L. Wang et al. (Eds.): NLPCC 2021, LNAI 13029, pp. 422–433, 2021.
https://doi.org/10.1007/978-3-030-88483-3_34

from Knowledge Graphs (KGs) into PLMs. PLMs pre-trained on large-scale unstructured corpora pay little attention to semantic information of important entity mentions and their relations expressed in texts. In contrast, Knowledge-Enhanced PLMs (KEPLMs) can significantly improve the plain PLMs in terms of the language understanding abilities, by fusing both unstructured knowledge from texts and structured knowledge from KGs [6,14,26,31,32]. The other type of approaches is to leverage the few-shot learning abilities of PLMs. As ultra-large PLMs such as GPT-3 [3] are proved to have the abilities to solve an NLP task with very few training samples, it has become a hot research topic to design prompts for fine-tuning BERT-style PLMs in the few-shot learning setting [8, 11,20].

Motivated by this research trend, FewCLUE[1] is established as a new benchmark of few-shot learning for Chinese Language Understanding Evaluation. It contains nine challenging few-shot NLP tasks, covering a wide range of topics, such as sentiment analysis, natural language inference, keyword recognition and coreference resolution. To solve the FewCLUE tasks, we first train a large-scale PLM that digests the world knowledge from Knowledge Graphs (KGs), with the resulting model named KEBERT. After continual pre-training, KEBERT is then adapted to specific downstream tasks based on the proposed Fuzzy-PET few-shot learning algorithm. Specifically, Fuzzy-PET employs the Fuzzy Verbalizer Mapping (FVM) mechanism that gives the underlying PLM more generalization power during few-shot learning. The results show that our approach effectively solves FewCLUE tasks, producing the highest score among all the teams in the competition.

In summary, we make the following contributions in this paper:

- We introduce a novel knowledge-enhanced pre-trained model named KEBERT, which digests both the unstructured knowledge from massive text corpora and the structured knowledge from KGs.
- We propose a few-shot learning algorithm named Fuzzy-PET to improve the generalization abilities of PLMs for few-shot learning.
- Our solution achieves the best performance in both limited and unlimited tracks of FewCLUE, which will be released to public.

The rest of this paper is as follows. Section 2 gives a brief overview on related work. Section 3 presents our solution to FewCLUE tasks based on KEPLMs and few-shot learning. Experimental results are reported in Sect. 4. Finally, we give the concluding remarks and discuss possible extensions in Sect. 5.

2 Related Work

In this section, we summarize the related work on three aspects: PLMs, KEPLMs and few-shot learning for PLMs.

[1] https://github.com/CLUEbenchmark/FewCLUE.

2.1 Pre-trained Language Models

Recently, PLMs have achieved significant improvements on various NLP tasks based on the "pre-training and fine-tuning" paradigm [17]. Among these PLMs, BERT [7] is probably most influential, which learns bidirectional contextual representations by transformer encoders. RoBERTa [12] improves the pre-training process of BERT by several optimization techniques such as dynamic masking, larger sequence length and byte-level byte-pair encoding (BPE). Other PLMs based on transformer encoder architectures include ALBERT [10], Transformer-XL [6], XLNet [29], StructBERT [25] and many others. Apart from transformer encoders, the encoder-decoder architectures of transformers have also been exploited in PLMs for modeling generative NLP tasks. Typical PLMs of this type include T5 [18], UniLM [1], etc.

As pre-training large-scale PLMs is computationally expensive, a lot of efforts have also devoted into efficient distributed pre-training. Mixed precision [13] uses half-precision or mixed-precision representations of floating points for model training. This technique can be further improved through Quantization Aware Training [9], where the weights are quantized during training and the gradients are approximated with the straight-through estimator. Additionally, gradient checkpointing [4] is also frequently applied to save memory by extra computation. 3D parallelism [19] combines model parallelism (tensor slicing) and pipeline parallelism with data parallelism in complex ways to efficiently scale models by fully leveraging computing resources in clusters.

2.2 Knowledge-Enhanced Pre-trained Language Models

As shown in various studies, PLMs pre-trained on large-scale unstructured corpora only can capture the basic lexical and syntactical knowledge of languages. However, the lack of semantic information of important entity mentions in texts may affect the performance of these PLMs significantly [26]. Recently, KEPLMs are proposed to utilize the structured knowledge from KGs to enhance the language understanding abilities of PLMs. Here, we summarize the recent KEPLMs into the following two types.

The first type is knowledge enhancement by entity embeddings. In the literature, ERNIE-THU [32] injects entity embeddings into the deep language token representations via a knowledge-encoder and text-encoder modules. Entity embeddings are obtained by the existing knowledge embedding algorithms, such as TransE [2]. KnowBERT [14] uses the knowledge attention and re-contextualization technique (KAR) and entity-linking mechanisms to inject the knowledge embeddings to PLMs. The goal of entity linking here is to inject the knowledge into the PLMs with higher accuracy and less noise.

The other type of approaches can be categorized as knowledge enhancement by relation triple descriptions. These works encode knowledge description texts into PLMs, which refer to the texts converted from relation triples to replace the large-scale entity embeddings. For example, E-BERT [31] and KEPLER [26] encode entity description texts through the general text encoder such as the transformer encoder [22]. This method learns context-aware token representations and knowledge representations jointly into a unified semantic space.

2.3 Few-Shot Learning for Pre-trained Language Models

The emergence of the ultra-large PLM GPT-3 [3] shows that it has the few-shot learning abilities with prompting texts provided. For BERT-style models, Pattern-Exploiting Training (PET) [20] converts a variety of few-shot NLP tasks into cloze questions, with manually defined patterns (also called prompts in following works) as additional inputs. The fine-tuned PLMs generate predicted masked language tokens that are further mapped into class labels by pre-defined mappings. To construct prompts automatically, Gao et al. [8] generates prompts for PLMs from the T5 model [18]. AutoPrompt [21] is an automated approach to generate prompts using token-based gradient searching. These approaches focus on discrete prompts in the form of natural languages only. P-tuning [11] learns continuous prompt embeddings, which can be optimized with fully differentiable parameters. Our work on few-shot learning is extended from PET [20] and also considers fuzzy verbalizers to make the underlying PLMs more generalized to unseen data instances during training, hence improving the testing performance.

3 The Proposed Approach

In this section, we begin with a brief summary on FewCLUE tasks. After that, detailed techniques of pre-training and few-shot learning are elaborated.

3.1 Task Description

FewCLUE is a new benchmark of few-shot learning for Chinese Language Understanding Evaluation. It styles after CLUE [27] and SuperGLUE [23] with nine challenging tasks covering a wide range of language understanding topics including text classification, language inference, idiom comprehension and co-reference resolution, etc. Each task consists of five independent labeled subsets, each of which has 16 examples per class for training and the same amount of data for validation, as well as additional unlabeled examples. Models are evaluated based on the averaged test performance trained over five subsets for each task. Table 1 presents the summary of FewCLUE tasks.

3.2 Knowledge-Enhanced Pre-training

As reported in [26], PLMs pre-trained on texts may only "understand" the literal meanings of input texts, lacking the deep understanding of the background knowledge of entities and relations. Without additional knowledge, it is challenging for the underlying PLM to solve few-shot learning tasks with high accuracy, which is the case for FewCLUE tasks.

Table 1. A brief summary of FewCLUE tasks.

Dataset	Train	Dev	Test public	Test private	Labels	Unlabeled	Task
Single sentence classification tasks							
eprstmt	32	32	610	753	2	19565	Sentiment analysis
csldcp	536	536	1784	2999	67	18111	Long text classification
tnews	240	240	2010	1500	15	20000	Short text classification
ifytek	928	690	1749	2279	119	7558	Long text classification
Sentence pair classification tasks							
ocnli	32	32	2520	3000	3	20000	NLI
bustm	32	32	1772	2000	2	4251	Semantic similarity
Reading comprehension tasks							
chid	42	42	2002	2000	7	7585	Multiple choice (idiom)
csl	32	32	2828	3000	2	19841	Keyword recognition
cluewsc	32	32	976	290	2	0	Coreference resolution

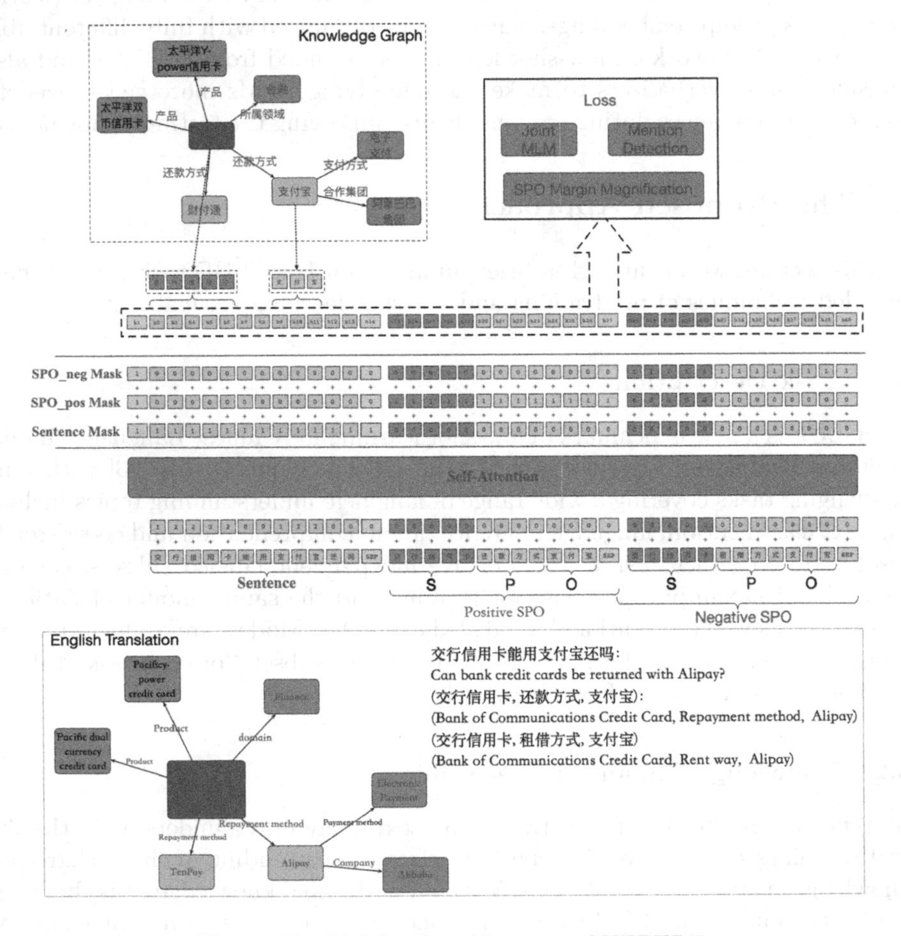

Fig. 1. The high-level architecture of KEBERT.

In this work, we introduce a new Knowledge-Enhanced BERT named KEBERT. The high-level architecture is presented in Fig. 1. To encode knowledge effectively, each pre-training instance is in the form of "Text-Positive SPO-Negative SPO" triples, where SPO refers to the Subjective-Property-Object relation triples stored in KGs. The knowledge of the positive SPO is semantically consistent with the input text, while the negative SPO is related to the input text but provides incorrect information. To facilitate knowledge understanding, three new pre-training tasks are proposed:

- Mention detection, which enhances the KEPLM's understanding abilities of entity mentions in the text;
- Joint MLM (Masked Language Modeling) of text and knowledge, which emphasizes the information sharing process between the structured knowledge and the unstructured texts, and improves the semantic understanding abilities of the model;
- SPO margin magnification, which is designed to widen the semantic gap of representations between the positive SPO and the negative SPO, and makes the underlying PLM be aware of the correctness of the knowledge.

In the future, we will release more technical details of KEBERT to public.

3.3 Continual Pre-training

We further design three variants of the MLM pre-training tasks to improve the robustness of the model over different few-shot tasks.

Whole-Word Masking. The whole-word masking strategy [5] is widely used for pre-training Chinese PLMs. Here, we use all unlabeled texts of each task as the pre-training corpus for whole-word masking. The purpose is to make the model more adaptive to the task-related data.

EFL-Based Pre-training. Previous work has shown the effectiveness of intermediate training on the GLUE benchmark [15], as well as few-shot matching tasks [30], which is a continual pre-training method that supplements PLMs with intermediate supervised tasks. Specifically, the Entailment as Few-Shot Learner (EFL) algorithm [24] is proposed to reformulate all language understanding tasks as entailment tasks. In our work, we follow the EFL baseline provided by the CLUE committee and use the CMNLI dataset [27] for continual pre-training. However, instead of directly using the "[CLS]" head to predict class labels, we create a prompt and a verbalizer (following PET [20]), to re-design the task as an MLM task, which is well aligned with other pre-training tasks.

MRC-Enhanced MLM. During the algorithm design process, we find that the PET algorithm [20] trained over vanilla KEPLMs may perform poorly over Machine Reading Comprehension (MRC) tasks. We argue that the problem is that PLMs are insensitive to the prompt formats of these tasks. Thus, we propose a variant of the MLM task to enhance the PLM's power for MRC tasks. For each sentence, we extract the keywords based on TF-IDF scores, mask one of the keywords and randomly create several false options with the same length of the masked word. Next, we create a special prompt to force the model to choose the correct word to fill in the masked position. An example is shown in Table 2.

Table 2. An example of MRC-enhanced MLM. English translations of Chinese texts are also provided.

Input Text	卓越的生命之花需要用美育浇灌。 (The exceptional flower of life needs to be watered by aesthetic education.)
Masked Word	浇灌 (watered)
Other Options	篮球，苹果，跳跃 (basketball, apple, jump)
Output Text	卓越的生命之花需要用美育[MASK][MASK]。候选词：篮球，苹果，浇灌，跳跃 (The exceptional flower of life needs to be [MASK] by aesthetic education. Candidates: basketball, apple, watered, jump)

3.4 Fuzzy-PET Algorithm for Few-Shot Fine-Tuning

For the few-shot fine-tuning process for a specific task, we propose the Fuzzy-PET algorithm by extending PET [20], in order to improve the model's generalization abilities during inference time. Specifically, Fuzzy-PET employs the *Fuzzy Verbalizer Mapping* (FVM) mechanism that allows multiple masked language tokens to be mapped into the same class label for one pattern. Let V be the full vocabulary set, L be the class label set, and $\mathbf{x}$ be an arbitrary input text sequence. We further define a pattern function P as a mapping from $\mathbf{x}$ to $P(\mathbf{x})$ such that $P(\mathbf{x})$ is a text sequence consisting of $\mathbf{x}$, the prompting texts and a masked language token.[2] Hence, the textual input to our PLM w.r.t. the text sequence $\mathbf{x}$ is $P(\mathbf{x})$.

To facilitate the many-to-one mappings from multiple masked language tokens to the same class label, in contrast to [20], we define the *Reverse Fuzzy Verbalizer* (RFV) f as a function: $f: v \rightarrow l$ such that $v \in V$ is the predicted result of the masked language token and $l \in L$. The RFV f gives a one-to-one mapping from a masked token to a class label. By handcrafting a collection of RFVs $F = \{f_1, f_2, \cdots\}$, it is straightforward to establish the multiple fuzzy mappings from masked tokens to class labels. Formally, denote $M(v|P(\mathbf{x}))$ as the un-normalized score that the underlying PLM assigns to $v \in V$ at the masked position. F_l is a subset of the RFV collection F such that the class label of each

[2] For the Chinese language, we can use multiple masked tokens to generate model outputs in the form of multiple Chinese characters. For simplicity, in the algorithm description, we assume there is only one masked token.

RFV $f \in F$ is $l \in L$. Given the input $P(\mathbf{x})$ and the class label l, the sum of un-normalized scores $s(\mathbf{x}, l)$ for the class label l is as follows:

$$s(\mathbf{x}, l) = \sum_{f \in F_l} M(v|P(\mathbf{x})). \tag{1}$$

Hence, the predicted probability distribution $\Pr(l|\mathbf{x})$ over all the class labels is then defined by the softmax function:

$$\Pr(l|\mathbf{x}) = \frac{e^{s(\mathbf{x},l)}}{\sum_{l' \in L} e^{s(\mathbf{x},l')}}. \tag{2}$$

During the training process of Fuzzy-PET, we use the cross-entropy loss of the true and predicted probability distributions as the loss function to fine-tune the model.

4 Experiments

In this section, we present the experimental results on the FewCLUE tasks. Detailed analysis of our approach is also provided.

4.1 Experimental Details

We conduct our experiments using the PyTorch version of the EasyTransfer toolkit [16]. In the pre-training steps, we use CLUECorpus2020 [28], which contains 100 GB Chinese raw texts retrieved from Common Crawl.[3] We also employ 100 million high-quality Chinese knowledge SPOs from our in-house KG as the structured knowledge source. We pre-train two KEBERT models for limited and unlimited tracks of FewCLUE, namely KEBERT_{large} and KEBERT_{xlarge}. The model configurations are shown in Table 3. All models are pre-trained with the described objectives. The training for all experiments are parallelized across 32 A100 GPUs, with the techniques of mixed precision, gradient checkpointing and 3D parallelism applied.

Table 3. Detailed model configurations.

Model	Layers	Attention head	Hidden size	Filter size	Total params
KEBERT_{large}	24	16	1024	4096	340M
KEBERT_{xlarge}	24	32	2048	8192	1.3B

During few-shot learning, we apply grid search to tune the best learning rate and the batch size for each individual task. The search space of the learning rate is from 1e−4 to 5e−5, and the search space of the batch size is from 4 to 32. The maximum learning epoch is 80 for all the tasks.

[3] https://commoncrawl.org/.

Table 4. The overall performance of KEBERT$_{xlarge}$ and KEBERT$_{large}$ over the private test set.

Task	eprstmt	csldcp	tnews	iflytek	ocnli	bustm	chid	csl	cluewsc	**Score**
Third place	**87.73**	60.26	**73.07**	45.25	66.18	70.40	57.30	56.94	60.76	63.11
Second place	85.50	59.17	72.84	44.04	67.77	**74.16**	58.05	60.75	**66.69**	64.53
KEBERT$_{large}$	85.92	57.60	72.69	44.78	70.72	61.45	**73.83**	**68.37**	59.93	65.33
KEBERT$_{xlarge}$	85.84	**60.78**	71.67	**46.07**	**71.86**	69.21	72.51	65.85	56.83	**66.13**

4.2 Experimental Results

In the competition, we win the first place in both limited and unlimited tracks of FewCLUE. Specifically, KEBERT$_{large}$ is used for the limited track submission, where the size of the PLM should be no larger than the RoBERTa-large model [12]. KEBERT$_{xlarge}$ is for the unlimited track submission. Table 5 lists the patterns that we use for each task. Please note that we apply variants of the PET algorithm on the *chid* and *cluewsc* tasks for better performance. For the *chid* task, four "[MASK]" tokens are used to replace the "#idiom" placeholders in the texts. The pattern and all idiom candidates are appended to the end of the text and the model is forced to predict the correct idiom in the masked position. For the *cluewsc* task, we replace the target possessive pronouns or adjectives with "[MASK]" tokens. The model should learn to predict the marked entity if the co-reference is true, and "[UNK]" tokens otherwise.

Table 5. Patterns for all FewCLUE tasks. English translations are also provided.

Task	Pattern
eprstmt	[MASK]满意，[TEXT] [MASK] satisfied, [TEXT]
csldcp	这篇论文阐述了[MASK][MASK]主题。[TEXT] This paper explains the [MASK][MASK] topic. [TEXT]
tnews	以下新闻的主题是[MASK][MASK]。[TEXT] The topic of the following news is [MASK][MASK]. [TEXT]
iflytek	作为一款[MASK][MASK]应用，[TEXT] As an application of [MASK][MASK], [TEXT]
ocnli	"[TEXT1]"和"[TEXT2]"的关系是[MASK][MASK]。 The relations between "[TEXT1]" and "[TEXT2]" is [MASK][MASK].
bustm	"[TEXT1]"和"[TEXT2]"意思[MASK]同。 The meanings of "[TEXT1]" and "[TEXT2]" are [MASK].
chid	[TEXT][MASK][MASK][MASK][MASK][TEXT] 候选词：[WORD1],[WORD2],[WORD3]... [TEXT][MASK][MASK][MASK][MASK][TEXT] Candidate words: [WORD1],[WORD2],[WORD3]...
csl	[TEXT]，关键词：[WORD1],[WORD2],[WORD3]...，答案: [MASK] [TEXT], Keywords: [WORD1],[WORD2],[WORD3]..., Answer: [MASK]
cluewsc	[TEXT][MASK][MASK][MASK][TEXT]

Our results (i.e., our final submissions) over the private test set are shown in Table 4, together with the performance of models submitted by the second and the third best candidates. Overall, our model KEBERT$_{large}$ outperforms the second best candidate in 5 out of 9 tasks. The large model KEBERT$_{xlarge}$ achieves the state-of-the-art performance in the FewCLUE benchmark, outperforming the second candidate by 1.6 point.

Table 6. Comparison of KEBERT with baseline algorithms over the public test set. The performance of baselines is reported by the CLUE committee.

Task	eprstmt	csldcp	tnews	iflytek	ocnli	bustm	chid	csl	cluewsc	Avg.
Fine-tuning [12]	63.2	35.7	49.3	32.8	33.5	55.5	15.7	50.0	49.6	42.8
PET [20]	87.2	56.9	51.2	35.1	43.9	64.0	61.3	55.0	50.0	56.1
P-tuning [11]	88.5	44.4	48.2	32.0	35.0	65.4	57.6	50.0	51.0	52.5
EFL [24]	85.6	46.7	53.5	44.0	67.5	67.6	28.2	61.6	54.2	56.5
KEBERT$_{large}$	88.72	58.14	61.79	49.96	71.05	61.95	73.86	67.35	**64.45**	66.36
KEBERT$_{xlarge}$	**89.27**	**60.18**	**64.30**	**51.01**	**72.32**	**67.63**	**73.89**	**67.92**	61.86	**67.60**

We also compare KEBERT against several baseline algorithms for few-shot learning, including RoBERTa fine-tuning [12], PET [20], P-tuning [11] and EFL [24]. The results over the public test set are presented in Table 6. As seen, the improvement gained by KEBERT is also consistent across all the tasks.

We further conduct an ablation study to test the effectiveness of the Fuzzy-PET algorithm, with results shown in Table 7. We compare the results of Fuzzy-PET against the vanilla PET algorithm [20]. The results show that our proposed Fuzzy-PET algorithm can both reduce the human effort to select best verbalizers and achieve higher performance in few-shot tasks.

Table 7. Ablation study of Fuzzy-PET over the public test set on KEBERT$_{large}$.

Task	eprstmt	csl	bustm
Vanilla PET [20]	87.87	66.92	60.70
Fuzzy-PET	**88.66**	**67.35**	**61.95**

5 Concluding Remarks

In this paper, we present our solution to the FewCLUE benchmark, which achieves the first place among all teams. Specifically, we propose the large-scale knowledge-enhanced pre-trained model named KEBERT to digest the relation triples from KGs, and the Fuzzy-PET algorithm for few-shot learning, together with continual pre-training techniques. The ablation studies have shown the importance of different integral parts of our solution. In the future, we will release more technical details to public and apply our approach to more scenarios and NLP tasks.

References

1. Bao, H., et al.: UniLMv2: pseudo-masked language models for unified language model pre-training. In: ICML, vol. 119, pp. 642–652 (2020)
2. Bordes, A., Usunier, N., García-Durán, A., Weston, J., Yakhnenko, O.: Translating embeddings for modeling multi-relational data. In: NIPS, pp. 2787–2795 (2013)
3. Brown, T.B., et al.: Language models are few-shot learners. In: NeurIPS (2020)
4. Chen, T., Xu, B., Zhang, C., Guestrin, C.: Training deep nets with sublinear memory cost. CoRR abs/1604.06174 (2016)
5. Cui, Y., Che, W., Liu, T., Qin, B., Yang, Z., Wang, S., Hu, G.: Pre-training with whole word masking for Chinese BERT. CoRR abs/1906.08101 (2019)
6. Dai, Z., Yang, Z., Yang, Y., Carbonell, J.G., Le, Q.V., Salakhutdinov, R.: Transformer-XL: attentive language models beyond a fixed-length context. In: ACL, pp. 2978–2988 (2019)
7. Devlin, J., Chang, M., Lee, K., Toutanova, K.: BERT: pre-training of deep bidirectional transformers for language understanding. In: NAACL-HLT, pp. 4171–4186 (2019)
8. Gao, T., Fisch, A., Chen, D.: Making pre-trained language models better few-shot learners. CoRR abs/2012.15723 (2020)
9. Jacob, B., et al.: Quantization and training of neural networks for efficient integer-arithmetic-only inference. In: CVPR, pp. 2704–2713 (2018)
10. Lan, Z., Chen, M., Goodman, S., Gimpel, K., Sharma, P., Soricut, R.: ALBERT: a lite BERT for self-supervised learning of language representations. In: ICLR (2020)
11. Liu, X., et al.: GPT understands, too. CoRR abs/2103.10385 (2021)
12. Liu, Y., et al.: RoBERTa: a robustly optimized BERT pretraining approach. CoRR abs/1907.11692 (2019)
13. Micikevicius, P., et al.: Mixed precision training. CoRR abs/1710.03740 (2017)
14. Peters, M.E., et al.: Knowledge enhanced contextual word representations. In: EMNLP, pp. 43–54 (2019)
15. Phang, J., Févry, T., Bowman, S.R.: Sentence encoders on stilts: supplementary training on intermediate labeled-data tasks. CoRR abs/1811.01088 (2018)
16. Qiu, M., et al.: EasyTransfer - a simple and scalable deep transfer learning platform for NLP applications. CIKM 2021 (2020). https://arxiv.org/abs/2011.09463
17. Qiu, X., Sun, T., Xu, Y., Shao, Y., Dai, N., Huang, X.: Pre-trained models for natural language processing: a survey. CoRR abs/2003.08271 (2020)
18. Raffel, C., et al.: Exploring the limits of transfer learning with a unified text-to-text transformer. J. Mach. Learn. Res. **21**, 140:1–140:67 (2020)
19. Rasley, J., Rajbhandari, S., Ruwase, O., He, Y.: DeepSpeed: system optimizations enable training deep learning models with over 100 billion parameters. In: SIGKDD, pp. 3505–3506. ACM (2020)
20. Schick, T., Schütze, H.: Exploiting cloze-questions for few-shot text classification and natural language inference. In: EACL, pp. 255–269 (2021)
21. Shin, T., Razeghi, Y., IV, R.L.L., Wallace, E., Singh, S.: AutoPrompt: eliciting knowledge from language models with automatically generated prompts. In: EMNLP, pp. 4222–4235 (2020)
22. Vaswani, A., et al.: Attention is all you need. In: NIPS, pp. 5998–6008 (2017)
23. Wang, A., et al.: SuperGLUE: a stickier benchmark for general-purpose language understanding systems. In: NeurIPS, pp. 3261–3275 (2019)
24. Wang, S., Fang, H., Khabsa, M., Mao, H., Ma, H.: Entailment as few-shot learner. CoRR abs/2104.14690 (2021)

25. Wang, W., et al.: StructBERT: incorporating language structures into pre-training for deep language understanding. In: ICLR (2020)
26. Wang, X., Gao, T., Zhu, Z., Liu, Z., Li, J., Tang, J.: KEPLER: a unified model for knowledge embedding and pre-trained language representation. CoRR abs/1911.06136 (2019)
27. Xu, L., et al.: CLUE: a Chinese language understanding evaluation benchmark. In: COLING, pp. 4762–4772 (2020)
28. Xu, L., Zhang, X., Dong, Q.: CLUECorpus 2020: a large-scale Chinese corpus for pre-training language model. CoRR abs/2003.01355 (2020)
29. Yang, Z., Dai, Z., Yang, Y., Carbonell, J.G., Salakhutdinov, R., Le, Q.V.: XLNet: generalized autoregressive pretraining for language understanding. In: NeurIPS, pp. 5754–5764 (2019)
30. Yin, W., Rajani, N.F., Radev, D.R., Socher, R., Xiong, C.: Universal natural language processing with limited annotations: try few-shot textual entailment as a start. In: EMNLP, pp. 8229–8239 (2020)
31. Zhang, D., et al.: E-BERT: a phrase and product knowledge enhanced language model for e-commerce. CoRR abs/2009.02835 (2020)
32. Zhang, Z., Han, X., Liu, Z., Jiang, X., Sun, M., Liu, Q.: ERNIE: enhanced language representation with informative entities. In: ACL, pp. 1441–1451 (2019)

TKB^{2}ert: Two-Stage Knowledge Infused Behavioral Fine-Tuned BERT

Jiahao Chen(✉), Zheyu He, Yujin Zhu, and Liang Xu

GammaLab, PingAn OneConnect, Shanghai, China
{chenjiahao637,hezheyu874,xuliang867}@pingan.com.cn,
zhuyujin204@ocft.com, 17212010084@fudan.edu.cn

Abstract. Machine Reading Comprehension (MRC) is a technique to make machine answer questions corresponding to the given passage. However, the existing MRC systems confront challenges to deal with the complex real-world fine-grained question answering tasks. To address the issue, we propose a novel method named TKB^{2}ert that generates prior knowledge data and utilizes it in behavioral fine-tuning at the first stage and trains itself through the dual attentive streams at the second stage. Subsequently, we take part in the MRC track of 2021 Language and Intelligence Challenge and win the first place. The result on the newly-built fine-grained MRC competition dataset validates the effectiveness of TKB^{2}ert.

Keywords: Machine Reading Comprehension · Question answering task · Prior knowledge data · Behavioral fine-tuning · Dual attentive streams

1 Introduction

The last five years have witnessed the prosperity of end-to-end machine reading and question answering models. As one of the fundamental and longstanding goal of Natural Language Processing (NLP), Machine Reading Comprehension (MRC) [3,7,9] is regarded as the key to improve the learning ability of an artificial intelligence system. However, the machine reading models might be deteriorated when implemented to the real-world scenes, such as search engine, community question answering, and domain-specific chatbot, because of their lack of prior-experience, complex reasoning capabilities, and robustness to informal expressions or incorrect grammar. What is worse, some MRC systems have the trouble to distinguish whether the given passage contains an answer for the question or not, as that they do not see samples without answers during their training process.

In this paper, we aim to solve the mentioned issues in a novel two-stage knowledge infused behavioral fine-tuned language model based on BERT [4], abbreviated as TKB^{2}ert. The proposed method is expected to not only transfer knowledge in advance for a better token-alignment between question and

L. Wang et al. (Eds.): NLPCC 2021, LNAI 13029, pp. 434–446, 2021.
https://doi.org/10.1007/978-3-030-88483-3_35

passage in attention mechanism, but also acquire certain behavior guidance to ensure its reasoning process according to the similar logic as human being. In detail, TKB^{2}ert generates samples with low-quality but a wide variety of prior knowledge to pre-train the model to learn more behavior patterns in the first stage, and adopts dual attentive streams to fine-tune the model respectively for classification and answer-location task in the second stage.

Finally, to test the effectiveness of TKB^{2}ert, we consider a new MRC dataset [15], named as Checklist. The dataset is proposed by China Computer Federation, Chinese Information Processing Society of China, and Baidu company, specially designe4d for complex MRC behavior evaluation. More details about the dataset will be given in Section III. Actually, we took part in the Machine Reading Comprehension track of 2021 Language and Intelligence Challenge (LIC 2021), which provides both Checklist and In-Domain MRC dataset, and achieved the first place[1]. In summary, the contributions of our work can be highlighted as follows:

- We propose a novel effective two-stage knowledge infused behavioral fine-tuned BERT model called **TKB^{2}ert**, in order to alleviate issues happened in existing MRC systems when dealing with complex real-world question-answering situations.
- We test TKB^{2}ert on a newly-proposed fine-grained MRC dataset **Checklist** and find that our model (f1 = 67.82%) outperforms the official baseline (f1 = 59.76%) nearly 8% in f1-score.
- We take part in **LIC 2021** and win the first place in machine reading comprehension track by using TKB^{2}ert.

2 Related Work

Reading comprehension models with attention mechanism are widely explored since The Stanford Question Answering Dataset (SQuAD 1.0) [14] and its 2.0 version [13] have been released. Confronting with these challenging tasks, MRC models with attention mechanism are categorized to two kinds according to their interaction paradigm: the one-dimensional or two-dimensional matching model.

The one-dimensional models, such as Attention Sum Reader [9] and AOA Reader [3], encode each word of the passage and the whole semantics of the question, then calculates the matching degree between each token of both of them through matching function.

On the other side, two-dimensional matching models, such as BiDAF [17], preserve both the semantics of each token of the passage and question at the same time. The matching matrix is obtained through the matching function, and then attention score is calculated by row or column to describe the semantic correlation between each word of the passage and the question.

In addition, the family of the pre-trained language models, such as Bert [4], RoBERTa [11], GPT [12], demonstrate their significant performance in

[1] https://aistudio.baidu.com/aistudio/competition/detail/66?isFromLuge=true.

many NLP tasks. For reading comprehension task, pre-trained models encode the passage and the question together, and then predict the answer boundary through pointer network [18]. Recently, these pre-trained models have continuously refreshed the best performance on SQuAD leaderboard, and even achieved results beyond human beings.

3 Data Description

We use Checklist [15] as one of datasets to test our model in this paper. The evaluation strategy for the existing MRC datasets are mostly focused on general and unexplainable tasks, leading to several over-fitting MRC methods that lack generalization in industrial question-answering scenarios. To boost the evaluation for a MRC system in more complex linguistic ability, Checklist is proposed to decompose potential real-world cases into specific behaviors and contains five fine-grained subtasks: vocabulary understanding, phrase understanding, semantic role understanding, fault tolerance, and negative reasoning. Detailed introductions are given as follows.

- **Vocabulary Understanding** task indirectly refers to keywords belonging to special types (i.e., nouns, adverbs, verbs, adjectives, quantifiers, units, and named entities) in passage by using synonyms, antonyms, and other related words, rather than repeating these words to the question. For instance, the keyword '蝉 (cicada)' from the passage is written as '知了 ' in the question. This task aims to investigate the understanding ability of a MRC system in vocabulary-level.
- **Phrase Understanding** task gives different descriptions for the sentence-level trunk from the passage and the question, respectively, to avoid the model making an inference relying on specific vocabulary matching forms. In this case, a MRC system should understand the precise intention of the question before offering the correct answer. For example, the passage says that '牛不能吃豆子 (cattle cannot eat beans)', but the question asks '牛肉是否能与豆子一同食用 (whether beef can be eaten with beans)'.
- **Semantic Role Understanding** task exchanges the order of crucial semantic roles (or the prefix of roles) in either passage or question, in order to inspect the discrimination ability of symmetric and asymmetric relationships in a MRC system. For instance, when the question asks that '山药能和胡萝卜一起吃吗 (can Chinese yam and carrot be eaten together)', the passage shows that '胡萝卜和山药不能一起吃 (carrot and Chinese yam cannot be eaten together)'.
- **Fault Tolerance** task is designed to test the stability of a MRC system by artificially introduce typos (i.e., wrongly written or mispronounced characters) in the question, considering the real community environment.
- **Negative Reasoning** task turn the meaning of the sentence to the negative side through introducing negative words. According to which part the negator is added to, this task could be divided into two sub-types: the question

negative reasoning and the passage negative reasoning. In this situation, a MRC system is expected to understand the opposite meaning between the question and the passage and not be misled.

Each sample in Checklist is given as a quadruples like (q, t, p, a), where given a question q, a passage p and its title t, the MRC system needs to judge whether p contains the answer a of q, and return a if so, output 'no answer' otherwise. Note that the distribution bias between Checklist and the conventional In-Domain MRC dataset is large. When dealing with samples from Checklist, the MRC model needs to consider more detailed behavior pattern discovering strategies for fine-grained cases mentioned above, implying that Checklist is challenging to the conventional MRC model.

4 Method

In this section, the introduction of TKB^{2}ert is arranged as follows: Stage 1: generating and bringing the prior knowledge data into behavioral fine-tuning process; Stage 2: running the dual attentive streams.

4.1 Stage 1: Behavioral Fine-Tuning

Behavioral Fine-tuning [16] is an effective pre-finetuning strategy to avoid semantic misunderstanding of model caused by the distribution deviation between the training in-domain dataset and the real-world out-domain situations, aiming to guide the model closer to the distribution of the target data.

Suppose a real-world environment consisting of a feature space $\mathcal{X}$, the target task with a label space $\mathcal{Y}$ is to fit a conditional probability distribution $P(Y|X)$ where $X = \{x_1, ..., x_n\} \in \mathcal{X}$, and $Y = \{Y_1, ..., Y_n\} \in \mathcal{Y}$, the process of behavioral fine-tuning is shown in Fig. 1. It can be seen that the strategy fine-tunes a model on relevant tasks to make it do well on the target task. By doing so, the model focuses on learning useful behaviours, that is how the strategy gets its name.

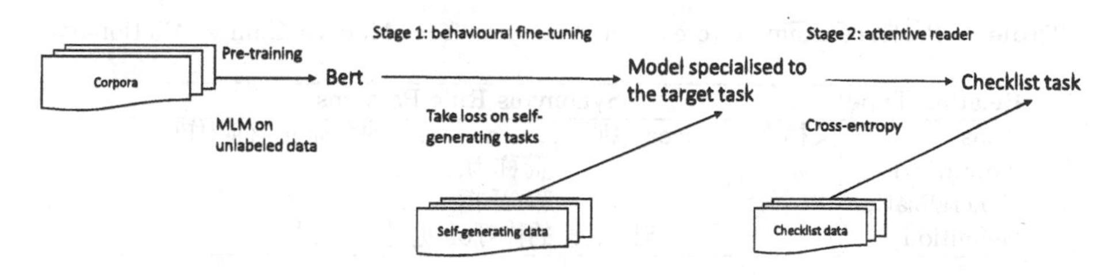

Fig. 1. Framework of TKB^{2}ert

The key for behavioral fine-tuning is to generate large amount of samples without high requirement for quality that bring a variety of prior knowledge as well as behavioral patterns into the model. For convenience, we call them prior knowledge data, distinguished from the augmentation samples used in the downstream task. In this paper, the prior knowledge data is generated by modifying the questions of samples from DuReader 2.0 [6] through methods mentioned in the next subsections. As a result, a new sample with/without answer can be generated alternatively depending on whether the modified question have the same meaning as the original one.

Words Replacement. As a typical analytic language, even tiny difference happened in character-level could change the meaning of the whole sentence in Chinese. Thus for vocabulary understanding subtask, we consider to infuse the character-level and word-level knowledge into the pre-trained language model during the behavioral fine-tuning through a variety of words replacement strategy.

Notional Words Replacement. In practice, we adopt the open source NLP library Language Technology Platform (LTP) [1] to identify nouns, verbs, adverbs, and adjectives in questions from DuReader 2.0, and then replace the found words by their synonyms or antonyms stored in our own built dictionary V, which takes Chinese commonly-used words as the indices while the synonyms or antonyms of indices as the contents. The replacements by synonyms does not change the meaning of the question, so that the answer is still suit for the newly-generated question. On the other hand, the antonyms turn a sample to 'no answer' one. The dictionary V is constructed from scratch as follows: (1) Extracting Modern Chinese Dictionary by rules to obtain a list $\mathcal{D}$ including nouns, verbs, adjectives, and adverbs. (2) Extracting synonyms and antonyms of words listed in $\mathcal{D}$ from Baiduhanyu[2] and storing them in V. (3) For the nouns in $\mathcal{D}$, extracting definition, interpretation, and examples of words from Modern Chinese Dictionary using rules to find synonyms from the extracted texts. Several rule pattern examples are given in Table 1. The dictionary contains 118,780 nouns, 8,378 verbs, 554 adverbs, and 3,512 adjectives in total.

Table 1. Patterns example to extract synonyms from Modern Chinese Dictionary

Relation Type	Synonyms Rule Patterns
Alias	又称为...... or 也叫...... or ......的别称or 又叫[作]......
Abbreviation	简称为......
Generalization	意/泛指......
Definition	对......的称呼or 见【......】

2 https://hanyu.baidu.com/.

Partially Repeated Words Replacement. Partially repeated words mean a pair of words with one or more same characters but having different meanings, e.g., '汽修站' (Auto Repair Station) & '汽车站' (Bus Station). The kind of these words could be very confused for the MRC model in matching process. Accordingly, we use open source question answering community corpus to build the maximum probability dictionary containing 100,000 words based on Expectation-Maximum (EM) algorithm [10], which maximizes the marginal likelihood of sentence probabilities. Finally, the words randomly selected from the question are mapped to their partially repeated words within a certain edit distance.

Entity Augmentation. We also find and replace entities from questions to their neighbors or hypernyms in an open source graph BigCilin[3], following the steps: (1) Obtaining all triples (head entity, relation, tail entity) from BigCilin. (2) Using a wide range of keywords, such as '名', '称', '同义', to search out synonymous relations from the triples. (3) Filtering the obviously incorrect relation types. (4) Retaining the triples with synonymous relationship, and generating a synonymous entity dictionary containing 42,500 entities with 1.53 synonyms for each one.

Numbers and Units Replacement. In a question, a number written in digits is randomly replaced to its Chinese form, e.g., '7' to '七', vice versa. For the quantifier or the measurement unit, they would be replaced randomly to a new quantifier/unit in the same measurement category but different written forms, e.g., '千克' (kilogram) to '公斤' (kilogram). Additionally, replacements between Chinese and English abbreviations are considered, e.g., '千克' to 'kg' (kilogram).

Phrase Understanding. We use UniLM [5] to generate variations of the input original questions from DuReader 2.0. UniLM is first trained on the open source sentence similarity datasets, and then generated positive or negative (i.e., 'no answer' type) questions that have similar or non-similar meaning with the input ones, respectively. Top 10 of samples with newly-generated questions from each side are selected as the prior knowledge data, while others will be filtered out through sentence probability checking process.

Negative Reasoning. We replace the core verbs and adjectives in questions of DuReader 2.0 with both synonyms and antonyms. Subsequently, we randomly insert negative words, such as '不 (no)', '没 (not)', '没有 (have not)', before the replaced words. Samples with the revised questions are added to the prior knowledge data. Note that the double negative process implies that the meaning of the question does not change.

[3] http://www.openkg.cn/dataset/hit.

Semantic Role Understanding. We analyze the sentence structure of a question, and if there are merely two semantic blocks in the sentence, the order of them will be exchanged. There are two ways to exchange: (1) Exchanging the whole semantic blocks, e.g., '上海迪斯尼和北京万达哪个好玩? (Which is fun, Shanghai Disney or Beijing Wanda?)' to '北京万达和上海迪斯尼哪个好玩? (Which is fun, Beijing Wanda or Shanghai Disney?)'. (2) Partially exchanging the blocks, e.g., '妈妈属虎，儿子属龙好不好? (Mom's zodiac is tiger, son's zodiac is loong, is that propitious?)' to '妈妈属龙，儿子属虎好不好? (Mom's zodiac is loong, son's zodiac is tiger, is that propitious?)'.

Candidates Filtering. When constructing the prior knowledge dataset, we create nearly 20 to 30 variations from each sample in DuReader 2.0 to ensure that the candidate pool is sufficient enough. Even though the behavioral fine-tuning does not requires high quality data, most of the generated samples are still quite semantically ambiguous. Consequently, we use GPT to further calculate the four kinds of probability for the rest candidates: the left-to-right language model sentence probability, the right-to-left language model sentence probability, the left-to-right token probability, and the right-to-left token probability. If the sum of probabilities is lower than a certain threshold, the current candidate sample should be discarded. Thereafter, the remaining samples are divided into 5% high quality ones and 95% low quality ones. The former is incorporated into the training set in stage 2 as augmentation data, while the latter is used for behavioral fine-tuning in stage 1.

4.2 Stage 2: Attentive Reader

In order to locate the correct answer, the model should consider information of not only the question but also the title. The reason is that in most real-world situations, such as community posts or search engines, the answer replies to the title directly without repeating the content of the title. Factually, a title is the summary of a passage and contains main clues to bridge the question and the passage. Furthermore, the model can determine whether a passage has an answer by comparing the meanings between question and title, implying that the deficiency of the title would cause serious information loss.

However, it is not suggested to simply concatenate all of the passage, question, title together as the input for the subsequent model, such as BERT:

$$[CLS]question[SEP]title[SEP]passage[SEP]. \tag{1}$$

Because in some cases, the title of the passage is given in the form of a question, so that the model would be confused about which question is the target when it sees two 'questions' appearing in the input text.

In order to sufficiently utilize information of both title and question while avoiding potential confusions mentioned above, we propose a novel model with dual attention streams following certain behavior guidance, whose framework is shown in Fig. 2.

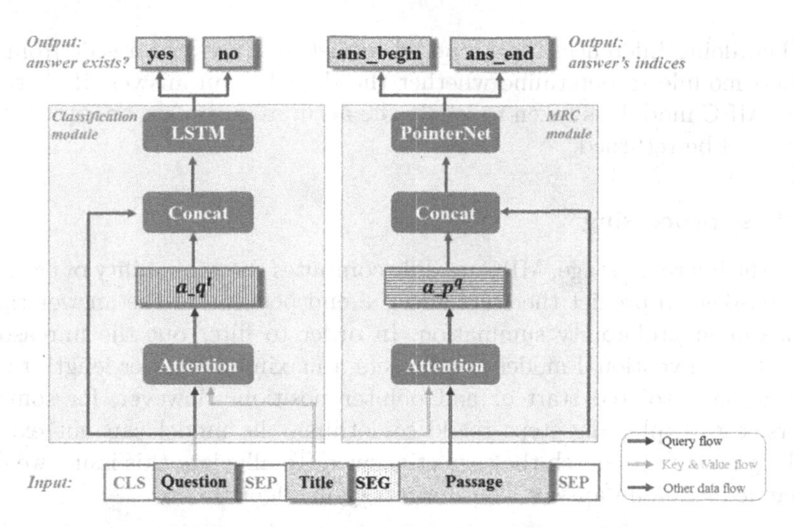

Fig. 2. Dual attentive streams of TKB²ert

According to the framework, we obtain the output from the last layer of RoBERTa [11] as the feature vector of each token. The attentive reader consists of two modules: Classification module and MRC module. As for Classification module, we first concatenate the weighted vector of the title a_q^t after attention matching with the question, and then get *output_question*, which is the question representation with title awareness. The calculation procedures are as follows:

$$a\_q^t = W_c E_{title} * Softmax(E_{title}^T W_{qt} E_{question}), \tag{2}$$

$$output\_question = concat(E_{question}, a\_q^t), \tag{3}$$

where $a\_q^t, E_{question}, output\_question \in \mathbb{R}^{d_{model} \times question\_length}$, while the title representation $E_{title} \in \mathbb{R}^{d_{model} \times title\_length}$, and the two attention weights for matching the question and title are: $W_c, W_{qt} \in \mathbb{R}^{d_{model} \times d_{model}}$. Then *output_question* is brought to the classifier BiLSTM [8] to judge whether there exists an answer in the passage.

Similarly, in MRC module, we concatenate the weighted vector of the question a_p^q after attention matching(query stream: passage, value&key stream: question) with the passage, and then get *output_passage*, which is the passage with question awareness. The calculation procedures are as follows:

$$a\_p^q = W_{mrc} E_{question} * Softmax(E_{question}^T W_{pq} E_{passage}), \tag{4}$$

$$output\_passage = concat(E_{passage}, a\_p^q), \tag{5}$$

where $a\_p^q, E_{passage}, output\_passage \in \mathbb{R}^{d_{model} \times passage\_length}$, and the two attention weights for matching the passage and question are: $W_{mrc}, W_{pq} \in \mathbb{R}^{d_{model} \times d_{model}}$. Then *output_passage* is brought into the subsequent pointer networks to infer the answer.

When doing inference, the proposed model first uses the result from Classification module to determine whether the data has an answer. If there is an answer, MRC module is taken to locate the accurate answer span, otherwise 'no answer' will be returned.

4.3 Post-processing

During the inference stage, MRC module computes the probability of each token in the passage to predict the start and the end position of the answer through the maximum probability summation. In order to filter out the unreasonable results, the conventional model usually sets a maximum answer length to avoid the occasional isolated start or end pointer position. However, for some long answers, e.g., explaining steps to do something, the model can not extract a span long enough to match the correct answer. To alleviate this issue, we design a dynamic maximum answer length strategy in Algorithm 1.

Algorithm 1: Dynamic Answer Length

```
Input data
For sample in data
    If keywords not in question
        If Passage_Length <250
            Max_Answer_Length = 100
        Else
            Max_Answer_Length = unlimited
        End If
    Else
        Max_Answer_Length = unlimited
    End If
End For
Return Max_Answer_Length
```

Concretely, we control the maximum answer length through the proposed algorithm: When no trigger words in question are identified, we control the maximum length of the answer according to the length of the passage. Otherwise, the length is automatically increased without limitation. Trigger words used in this paper include: '症状 (symptom)', '前兆 (forewarning)', '怎么回事 (What's going on ...)', '怎么办 (How to do ...)', '怎么 (How ...)', '什么意思 (What does something mean)', '如何 (How to ...)', '步骤 (steps)', '做法 (practice)', '菜 (cooking)'.

5 Experiment

5.1 Settings

The proposed method is tested by both Test Set 1 and Test Set 2 of LIC 2021 competition[4]. Test Set 1 is provided for online model verification, accepting

[4] https://www.luge.ai/.

repetitive submissions to post the evaluation scores online. On the other hand, the submission of Test Set 2 will be evaluated off-line by the Committee of the shared task and given the results on the final leaderboard. According to the task description, Test Set 2 consists of 2/3 Checklist and 1/3 general In-Domain dataset, while part of the data is randomly selected to add to Test Set 1. Micro Average F1 Score (F1) and Exact Match (EM) is used to evaluate the results.

The used pre-trained language model is RoBERTa-wwm-ext [2] (transformer structure with 768 hidden units, 24 layers, 24 heads, 300 million parameters). The learning rate is set as 5e-6, while batch size is fixed to 16. Each model selects two random seeds and do the evaluation in the second, fourth, sixth and eighth epoch, respectively. In pre-process, a fault tolerance module is used to correct wrongly written characters. The model is trained on 8 TESLA V100.

5.2 Ensemble Strategy

During the experiments, we consider the ensemble strategy, since a series of heterogeneous models were obtained by adopting different hyper-parameter settings and external algorithm modules. Consequently, we consider to pick models with higher f1-score among their counterparts and predictions, and then convert f1-scores into weights. Taking one sample as example, the current input question is denoted as q. 1 to K models are used to make the predictions for q, respectively. Suppose the predictions are defined as $pred$, the prediction-mean ensemble strategy is given as follows:

$$pred = \{P_m, \sum w_j | 1 \leq j \leq K, m, j \in N, 1 \leq m \leq s\}, \tag{6}$$

where w_j means the model confidence weight, and P_m denotes one of m non-repetitive predictions of q.

5.3 Results on Test Set 1

The experiment results on Test Set 1 are shown in Table 2. The first column lists the used methods, while the second column shows the evaluation results corresponding to methods. Since Test Set 1 consists of both Checklist and In-Domain data, their f1-scores are also listed in the third and fourth column, respectively. The symbol '+' or '–' means that the method in current row could improve or reduce how much scores compared to baseline. The last row shows that the proposed TKB^{2}ert significantly outperforms the baseline.

From Table 2, it can be seen that the behavioral fine-tuning and data augmentation are the most effective means to improve the performance. However, the In-Domain data has sharply declined after stage 1. The reason might be that most samples in Checklist and the prior knowledge data are labeled as 'no answer', leading the model to predict some In-Domain data, which actually have answers, to 'no answer'. Nevertheless, it is found that attentive reader, dynamic answer length, and ensemble strategy make improvements to both Checklist and In-domain data.

Table 2. Ablation results on test set 1 of LIC 2021

Method	Total-F1 (%)	Checklist-F1 (%)	In-Domain-F1 (%)
Baseline	64.42	59.76	72.93
Baseline + Behavioral Fine-tuning (Step 1)	+2.12	+3.16	−1.43
Baseline + Data augmentation (Step 1)	+1.65	+2.33	−1.21
Baseline + Attentive reader (Step 2)	+0.82	+0.89	+0.74
Baseline + Dynamic answer length (Step 2)	+0.73	+0.70	+0.89
Baseline + Ensemble strategy	+0.89	+0.98	+0.87
TKB^{2}ert (Baseline + All)	**70.63**	**67.82**	**72.79**

5.4 Results on Test Set 2

We cite the final results on Test Set 2 from LIC 2021 in Table 3. It could be found that TKB^{2}ert performs the best in 4/5 Checklist fine-grained classes while keeping its competitiveness in In-Domain data, validating its effectiveness.

Table 3. Final results on test set 2 of LIC 2021

Evaluation	TKB^{2}ert (Champion)	1st Runner-up	2nd Runner-up
Total-F1	**69.41**	68.849	68.442
EM	57.351	**58.559**	56.833
In-Domain-F1	70.555	**73.172**	71.587
Checklist-F1	**68.3474**	64.987	63.8708
Vocab-F1	66.246	64.764	**66.706**
Phrase-F1	**77.514**	75.478	73.867
Semantic-F2	**67.553**	59.093	57.901
Fault-F1	**59.763**	56.863	56.804
Reasoning-F1	**70.661**	68.737	64.076

6 Conclusion

In this paper, we introduce challenges for the existing MRC systems when dealing with real-world question answering situation. To address the issue, we apply a two-stage approach TKB^{2}ert that uses generated prior knowledge data in behavioral fine-tuning at the first stage, and runs its dual attention streams at the second stage. The subsequent experiment on the newly-proposed fine-grained Checklist demonstrates the effectiveness of TKB^{2}ert.

References

1. Che, W., Feng, Y., Qin, L., Liu, T.: N-LTP: a open-source neural Chinese language technology platform with pretrained models. arXiv preprint arXiv:2009.11616 (2020)
2. Cui, Y., Che, W., Liu, T., Qin, B., Wang, S., Hu, G.: Revisiting pre-trained models for Chinese natural language processing. In: Proceedings of the 2020 Conference on Empirical Methods in Natural Language Processing: Findings, pp. 657–668. Association for Computational Linguistics, Online, November 2020
3. Cui, Y., Chen, Z., Wei, S., Wang, S., Liu, T., Hu, G.: Attention-over-attention neural networks for reading comprehension. In: Proceedings of the 55th Annual Meeting of the Association for Computational Linguistics, vol. 1: Long Papers (2017). https://doi.org/10.18653/v1/p17-1055
4. Devlin, J., Chang, M.W., Lee, K., Toutanova, K.: BERT: pre-training of deep bidirectional transformers for language understanding. arXiv preprint arXiv:1810.04805 (2018)
5. Dong, L., et al.: Unified language model pre-training for natural language understanding and generation. arXiv preprint arXiv:1905.03197 (2019)
6. He, W., et al.: DuReader: a Chinese machine reading comprehension dataset from real-world applications (2018)
7. Hermann, K.M., et al.: Teaching machines to read and comprehend (2015)
8. Hochreiter, S., Schmidhuber, J.: Long short-term memory. Neural Comput. **9**(8), 1735–1780 (1997)
9. Kadlec, R., Schmid, M., Bajgar, O., Kleindienst, J.: Text understanding with the attention sum reader network (2016)
10. Kudo, T.: Subword regularization: improving neural network translation models with multiple subword candidates. In: Proceedings of the 56th Annual Meeting of the Association for Computational Linguistics (Volume 1: Long Papers), pp. 66–75. Association for Computational Linguistics, Melbourne, Australia, July 2018. https://doi.org/10.18653/v1/P18-1007
11. Liu, Y., et al.: RoBERTa: a robustly optimized BERT pretraining approach. arXiv preprint arXiv:1907.11692 (2019)
12. Radford, A., Narasimhan, K., Salimans, T., Sutskever, I.: Improving language understanding by generative pre-training. https://s3-us-west-2.amazonaws.com/openai-assets/researchcovers/languageunsupervised/languageunderstandingpaper.pdf (2018)
13. Rajpurkar, P., Jia, R., Liang, P.: Know what you don't know: unanswerable questions for SQuAD. In: Proceedings of the 56th Annual Meeting of the Association for Computational Linguistics (Volume 2: Short Papers), pp. 784–789. Association for Computational Linguistics, Melbourne, Australia, July 2018. https://doi.org/10.18653/v1/P18-2124
14. Rajpurkar, P., Zhang, J., Lopyrev, K., Liang, P.: SQUAD: 100,000+ questions for machine comprehension of text. arXiv preprint arXiv:1606.05250 (2016)
15. Ribeiro, M.T., Wu, T., Guestrin, C., Singh, S.: Beyond accuracy: behavioral testing of NLP models with checklist (2020)

16. Ruder, S.: Recent Advances in Language Model Fine-tuning. http://ruder.io/recent-advances-lm-fine-tuning (2021)
17. Seo, M., Kembhavi, A., Farhadi, A., Hajishirzi, H.: Bidirectional attention flow for machine comprehension (2018)
18. Vinyals, O., Fortunato, M., Jaitly, N.: Pointer networks. arXiv preprint arXiv:1506.03134 (2015)

A Unified Information Extraction System Based on Role Recognition and Combination

Yadong Zhang(✉) and Man Lan

East China Normal University, Shanghai, China
yadongzhang@stu.ecnu.edu.cn, mlan@cs.ecnu.edu.cn

Abstract. In this paper, we propose a unified information extraction system, which handles event extraction (EE) and relation extraction (RE) tasks. Given context and schema, event extraction aims to extract the events and the specific roles in the events, and relation extraction extracts all SPO triples. We formulate event extraction and relation extraction as one extraction schema, that is, role recognition and role combination. We use Multi-Label Pointer Network (MLPN) to recognize composite roles that contain both event/relation and role information and simultaneously train a Co-occurrence Matrix (CM) to determine the co-occurrence relationship of composite roles, i.e., whether two roles describe the same event/relation. Using such a **U**nified model based on **R**ole **R**ecognition and **C**ombination (URRC) and corresponding combination strategy, we implement three tasks: sentence-level event extraction, document-level event extraction, and relation extraction. In LIC 2021, our model achieved 6th in the Multi-format Information Extraction racing track with an average F_1 score of 77.44% in the final test dataset of three subtasks.

Keywords: Event extraction · Relation extraction · Document-level event extraction

1 Introduction

Information extraction (IE) distills structured data or knowledge from the unstructured text by identifying references to named entities as well as stated relationships between such entities [13]. Most current research focus on extracting information in a single format while lacking a unified evaluation platform for IE in different formats. Therefore, in the 2021 Language and Intelligence Challenge, the organizer set up a multi-format IE task designed to comprehensively evaluate IE from different dimensions. To be more specific, the multi-format IE task consists of three parts: sentence-level event extraction, document-level extraction, and relation extraction. In the following content of this section, we describe our understanding of these three tasks and the corresponding solution.

L. Wang et al. (Eds.): NLPCC 2021, LNAI 13029, pp. 447–459, 2021.
https://doi.org/10.1007/978-3-030-88483-3_36

For event extraction, traditional approaches to the task usually rely on sequential pipelines with multiple stages, including event triggers and arguments detection and identification [7]. However, the pipeline manner will bring error propagation [7], and in some cases, multiple events share the same trigger word. As a result, it is not ideal to use trigger words as the criteria for determining the number of events. For entity detection, sequence tagging is widely used, and it does show promising results in single-event extraction. However, it cannot solve the role overlapping problem in the multi-event case, where the same entity mention plays different roles in multiple events [1]. In this regard, we use Multi-Label Pointer Network (MLPN) based Pointer Network [15] to solve the above two problems simultaneously. The entities extracted by MLPN come with event type information, as event detection is not required, and each entity can have multiple role types, which solves the problem of entity overlap.

In sentence-level event extraction, the events described are relatively simple because of the text length, and multiple events of the same type appear in one sentence rarely. In contrast, at the document level, a large percentage of data has multiple events of the same type, e.g.,

"联建光电由于并购地雷爆发导致业绩持续亏损，2018 年巨亏 29 亿，2019 年巨亏 14 亿元。"

In the above example, there are two *loss* events, and if the roles of loss event type are put together to describe only one *loss* event is not accurate. Using MLPN can solve the problem of overlapping entities, but since it is not event-centric, how to combine the extracted entities to form complete events becomes an important issue. In the later sections, we introduce the proposed co-occurrence matrix module, which is designed to solve the assignment and combination of event roles.

The main objective of the relation extraction task is to extract the subject, object, and relation. As shown in Sect. 2.4, similar to document-level event extraction, a large percentage of data in relation extraction task also indicates multiple occurrences of the same type of relation. The event extraction approach introduced above can be summarized as role extraction and combination, and we find that this pattern can also be applied to the relation extraction task. Thus we use a unified model to solve the two major information extraction tasks, i.e., event extraction and relation extraction. The section of experiments also demonstrates that this approach works well on all three tasks.

2 Task Data and Analysis

Multi-format Information Extraction competition aims to extract structured knowledge such as entities, relations, and events from unstructured natural language texts. Generally, the competition has three tasks: sentence-level event extraction, document-level event extraction, and relation extraction. For each task, there is an individual dataset as well as evaluation criteria.

Table 1 shows the dataset composition of tree tasks.

Table 1. The dataset composition. *schema* refers to the number of event or relation types.

Task	Source	Schema	Train	Dev	Test1	Test2
Sentence-level event extraction	DuEE [10] 1.0	65	12k	1k	1.5k	3.5k
Document-level event extraction	DuEE-fin	13	7k	1.2k	1.2k	3.5k
Relation extraction	DuIE [8] 2.0	48	170k	20k	10k	20k

2.1 Sentence-Level Event Extraction

Given a sentence and predefined event types with corresponding argument roles, this task aims to identify all events of predefined event types mentioned in the sentence, and extract corresponding event arguments playing target roles. The data schema has 65 predefined event types, such as *parade* and *attack*, and each event type has corresponding event roles, for example, a *parade* event has three event roles: *organization, location*, and *time.*

The evaluation method uses the token level matching F_1 score, which is case insensitive.

2.2 Document-Level Event Extraction

This task shares almost the same definition as the previous task but restricts the event types in the financial field. Furthermore, unlike the previous task, this task considers input text fragments at document-level rather than sentence-level, which has a title and much longer context.

In terms of evaluation metrics, document-level event extraction is very different from sentence-level event extraction. First, although both use the F_1 score as the evaluation metric, the document-level is at the role-level rather than the token-level. Second, before evaluating the correctness of the roles, the extracted events are matched with the ground truth, and the roles for events that cannot be matched are directly considered wrong. In contrast, there is no event matching step for sentence-level event extraction.

2.3 Relation Extraction

The aim of this task is to extract all **SPO** (**S**ubject, **P**redicate, **O**bject) triples from a given sentence according to predefined schema. The schema defines relation P and the types of its corresponding subject S and object O. According to the complexity of object O, there are two types of schemas: simple object and complex object. Complex object is a structure composed of multiple slots. For example, in the relation of *play*, the object O gets two slots named *@value* and *inWork*, which respectively represent *what is the role the person played* and *in which film and television work the person played the role*. While simple object only has *@value* slot.

The relation extraction task uses the SPO-level F_1 score for evaluation metric that requires subject, object, and predicate all to be correct.

2.4 Data Analysis

A statistical examination of the training data shows that there are multiple occurrences of the same event/relation in all three tasks. The detailed statistics in Table 2 demonstrate that the proportion of the dataset with sentence-level event extraction is low (4.57%). In contrast, the proportion of document-level event extraction and relation extraction is much higher, which is why we believe that the problem of combination of roles cannot be ignored.

Table 2. Statistics of the number of events/relations on three datasets. *Multiple* represents the proportion of data in which there are multiple events in a single piece of data, where the type of event is not considered, whereas *Identical* requires that these multiple events contain at least two events of the same type.

Dataset	Context	Event/Relation	Multiple	Identical
DuEE 1.0 train	11,958	13,915	13.03%	4.57%
DuEE-fin train	2937	5403	69.46%	50.53%
DuIE 2.0 train	171,293	310,709	39.17%	18.05%

3 Model

Our proposed unified information extraction model is based on two main directions: role extraction and role combination. The role extraction is implemented by MLPN, and a co-occurrence matrix module is trained simultaneously to obtain the co-occurrence relationships between roles. Finally, the roles are combined using the corresponding matching strategy. These modules are described in detail below.

3.1 Multi-label Pointer Network

The MLPN module is designed to recognize all roles in the given context. With the encoded vector $\mathbf{x}$ produced by the pre-trained transformer encoder (BERT [4]), each token in the input text is assigned multiple binary tags (0/1) that indicates whether the current token corresponds to a start or end position of

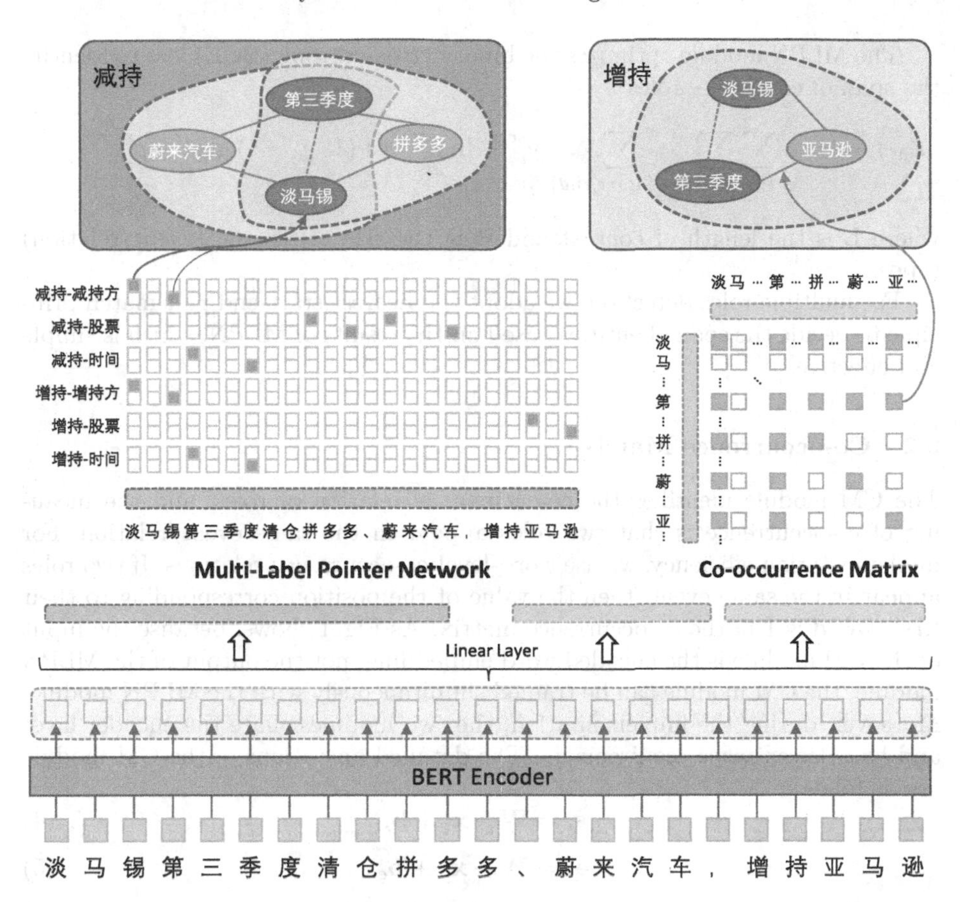

Fig. 1. The whole architecture of our proposed model. This framework consists of two modules: 1) Multi-Label Pointer Network; 2) Co-occurrence Matrix. The above two modules are trained jointly and share a BERT encoder.

a composite role that represents both event/relation and role information, e.g., 游行-时间. The detailed operations of the MLPN on each token are as follows:

$$p_i^{start\_r_j} = \sigma(\mathbf{W}_{start\_r_j} \cdot \mathbf{x}_i + \mathbf{b}_{start\_r_j}) \quad (1)$$

$$p_i^{end\_r_j} = \sigma(\mathbf{W}_{end\_r_j} \cdot \mathbf{x}_i + \mathbf{b}_{end\_r_j}) \quad (2)$$

where $p_i^{start\_r_j}$ and $p_i^{end\_r_j}$ represent the probability of the i-th toke in the given text as the start and end position of a composite role r_j, and $\mathbf{x}_i$ is the encoded representation of the $i-$th token. Additionally, $\mathbf{W}$ and $\mathbf{b}$ are the weights for a linear layer, and *sigma* is the sigmoid activation function.

The MLPN module optimizes the binary cross-entropy (BCE) loss to identify the span of composite roles :

$$\mathcal{L}_{MLPN} = -\sum_{i \in L}\sum_{j \in R}\sum_{se \in \{start, end\}} [y_i^{se\_r_j} \ln p_i^{se\_r_j} + (1 - y_i^{se\_r_j}) \ln (1 - p_i^{se\_r_j})] \quad (3)$$

where L is the length of context and R is the size of schema (event/relation) types.

For multiple roles detection, we adopt the nearest start-end pair match principle to decide the span of entities based on the results of MLPN, which is simple and effective.

3.2 Co-occurrence Matrix

The CM module identifies the co-occurrence relation of roles, and the meaning of co-occurrence is that two roles appear in the same event/relation. For implementation efficiency, we use word-level co-occurrence relations. If two roles appear in the same event, then the value of the position corresponding to their **first word** is 1 in the co-occurrence matrix. As Fig. 1 shows, because the input of the CM module is the encoded word embedding, not the output of the MLPN module, the CM module can be trained simultaneously with the MLPN module. Moreover, during the implementation, the two modules share the encoder layer and have fewer space requirements. The detailed operations of the CM module are as follows:

$$\mathbf{s}_i = \mathbf{W}_s \cdot \mathbf{x}_i + \mathbf{b}_s \quad (4)$$

$$\mathbf{e}_j = \mathbf{W}_e \cdot \mathbf{x}_j + \mathbf{b}_e \quad (5)$$

$$c_{ij} = \sigma(\mathbf{s}_i^{\mathsf{T}} \mathbf{e}_j) \quad (6)$$

where c_{ij} represents the probability of identifying the i-th token and j-th token pair has co-occurrence relation. More precisely, the target co-occurrence matrix is symmetric, and the diagonal value is one if it is the position corresponding to the first word of the entity. Similar to the adjacency matrix of an undirected graph where each word is a node in the graph. If two entities describe the same event, an edge exists between these two nodes (the first word corresponding to the entities), and each entity also has a self-loop.

The CM module optimizes the BCE loss to identify the co-occurrence relationship of roles:

$$\mathcal{L}_{CM} = -\sum_{i \in L}\sum_{j \in L}[y_{ij} \ln c_{ij} + (1 - y_{ij}) \ln (1 - c_{ij})] \quad (7)$$

where L is the length of text.

3.3 Roles Combination

After obtaining the composite roles and co-occurrence relation, the corresponding combination strategy is also needed to combine the composite roles to form an entire event or relation. For sentence-level event extraction, it is rare for a sentence to have multiple events of the same type, and the composite roles of the same event type can be put together with the CM module removed in the implementation. Document-level event extraction and relation extraction have their combination strategies.

Document-Level Event Extraction. For each document-level event type, we artificially identify an anchor role that best represents the event's uniqueness. For example, *loss value* was selected as the anchor role for the *loss* event because it is rare for two loss events to have identical loss values. Therefore, the number of anchor roles represents the number of events. All anchor role of the event types is listed in Table 3. For an anchor role, if a composite role has the same event type as it and they have a co-occurrence relationship, then they are put into a group, representing that they describe the same event. Alternatively, if roles of the same type do not have any anchor role, they are simply all placed in the same group. The detailed algorithm is summarized in Algorithm 1.

Table 3. Shape of a character in dependence on its position in a word

Event Type	Anchor Role
解除质押	质押股票/股份数量
股份回购	回购股份数量
股东减持	交易股票/股份数量
亏损	净亏损
中标	中标标的
高管变动	高管姓名
企业破产	破产公司
股东增持	交易股票/股份数量
被约谈	公司名称
企业收购	被收购方
公司上市	上市公司
企业融资	融资金额
质押	质押股票/股份数量

Algorithm 1. Document-level role combination algorithm

Input: all composite roles R of a document, co-occurrence matrix C
Output: *events* of the document
1: R grouped by event type, and each role r in R has their first word index r_{index} of the document context
2: *events*=List()
3: **for** each event type t in R **do**
4:   **if** $R[t]$ has anchor role **then**
5:     **for** each anchor role ar in $R[t]$ **do**
6:       *event*=List(ar)
7:       **for** all common role r in $R[t]$ **do**
8:         **if** $C[ar_{index}][r_{index}]$=1 **then**
9:           *event*.add(r)
10:       *events*.add(*event*)
11:   **else**
12:     *event*=$R[t]$
13:     *events*.add(*event*)
14: **return** *events*

Relation Extraction. Unlike document-level event extraction, relation extraction does not require identifying anchor roles because the subject and object pair is highly unique. For a simple object, a subject with the same relation type and co-occurrence will form a pair of SPO. For complex objects, we will first form the SPO pair using *@value* slot like simple object, and roles of the other slot that have co-occurrence relation with the subject or the object will be added to extend the object.

3.4 Enumeration Type Classification

There is a rather special role in document-level event extraction, where the values are not extracted from the text but selected from a list, called enumeration types. For example, the 上市环节 of 公司上市 event is an enumeration type and the values that can be selected are: 筹备上市, 暂停上市, 正式上市, and 终止上市. We formulate enumeration type classification as a document classification task, i.e., a multi-label classification of documents with the category being the values of all enumeration types. If a document is classified into a certain category, then the enumeration role uses the corresponding enumeration value. For this purpose, we trained a separate document classification model that does not share an encoder with the MLPN and CM module. The detailed operations are as follows:

$$\mathbf{p} = softmax(\mathbf{W} \cdot \mathbf{h} + \mathbf{b}) \tag{8}$$

where $\mathbf{h}$ is the CLS token of BERT, and $\mathbf{p}$ is the probability of each category. The training dataset only uses data from events with enumerated types, and only events with enumerated types are classified during testing.

4 Experiments and Results

4.1 Implement Details

All our model is implemented with PyTorch, and not use external corpus in this competition.

Sentence-Level Event Extraction. In the sentence-level extraction, we use Chinese RoBERTa-Base [2,3,12] released by HFL[1] as encoder and set the batch size to be 8, learning rate of Adam [6] optimizer to be 2.0×10^{-5}. With about 15 epochs, training process takes about 2 h on an 11G Nvidia 2080Ti. A dropout layer [14] follows the linear layer, and the dropout probability is 0.2.

Document-Level Event Extraction. For document-level event extraction, we share a BERT to jointly train the MLPN and CM, where the BERT version is Chinese RoBERTa-Large released by HFL[2]. The overall loss is $\mathcal{L} = \mathcal{L}_{MLPN} + \alpha\mathcal{L}_{CM}$, where the value of α is 1.0. The size of the linear layer of the CM is 256. All linear layers are followed by a dropout layer with 0.2 dropout probability. The threshold value of MLPN and CM are all 0.5. We set the batch size to be 2, learning rate of Adam optimizer to be 2.0×10^{-5}. With 12 epochs, the training process takes about 4 h on an 11G Nvidia 2080Ti.

Relation Extraction. As with document-level event extraction, we share the same encoder and jointly train the MLPN and CM module for relation extraction using Chinese RoBERTa-base. The threshold values of MLPN and CM are 0.3 and 0.5, respectively. All other parameters are set the same as document-level event extraction. We set the batch size to be 16, learning rate of Adam optimizer to be 2.0×10^{-5}. With 4 epochs, the training process takes about 4 h on an 11G Nvidia 2080Ti.

4.2 Ensemble

To make better use of the dataset and improve the stability of the model, we combine the training set with the development set for cross-validation [5] and then average the output probability for MLPN and CM. Among them, sentence-level event extraction and document-level event extraction use 5-fold cross-validation. The relation extraction itself has sufficient training data, so only 3-fold cross-validation is performed, where the size of development sets is the same as the original setting.

[1] https://huggingface.co/hfl/chinese-roberta-wwm-ext.

[2] https://huggingface.co/hfl/chinese-roberta-wwm-ext-large.

4.3 Results

Sentence-Level Event Extraction. Table 4 is the comparison of different models and settings on sentence-level event extraction test1. It can be seen that Bert-CRF [11] has a considerable improvement over Baseline, partly because CRF simulates label dependencies in the output. In addition, the approach using machine reading comprehension [9] is competitive with our proposed method. However, the MRC-based system will autonomously scale up the amount of data, and the training efficiency using pipeline manner is lower than our joint approach.

Table 4. Performance comparison on sentence-level event extraction test1. Baseline is from Baidu official submission; Bert-CRF is the method of sequence tagging like Baseline, but with the added restriction of CRF; URRC is our proposed approach and only has the MLPN module since the role combination is ignored in the sentence-level event extraction task; ensemble setting is introduced in Sect. 4.2.

Model	Precision	Recall	F_1
Baseline (Official)	78.85	75.39	77.09
Bert-CRF	80.06	79.18	79.62
MRC	81.94	**84.29**	83.10
URRC$_{-CM}$	**85.11**	82.12	**83.59**
MRC$_{+ensemble}$	81.39	**85.56**	83.42
URRC$_{-CM+ensemble}$	**86.47**	83.80	**85.11**

Document-Level Event Extraction. Table 5 is the comparison of different models and settings on document-level event extraction test1. The comparison of Baseline and Bert-CRF is similar to sentence-level event extraction. The simple replication mechanism (MLPN-Copy) produces huge fluctuations in both precision and recall, and the final F_1 is 1% lower than MLPN, which shows that the combination of roles has a large impact on document-level event extraction and needs to be addressed using better strategies. URRC is our proposed method, and the addition of the CM module brings a notable enhancement (5.68%) to the whole system. Moreover, enumerated type classification and the use of Chinese RoBERTa-large have a slight boosting effect.

Table 5. Performance comparison of different models and settings on document-level event extraction test1. Baseline is from Baidu official submission; Bert-CRF is the method of sequence tagging like Baseline, but with the added restriction of conditional random fields. MLPN-Copy is a simple combination strategy of putting all other composite roles of the same event type together based on anchor roles, where composite roles and anchor roles are extracted by the MLPN. URRC is our proposed method, with the markers *enum* and *large* representing the addition of enumeration type classification and the use of Chinese RoBERTa-large, respectively.

Model	Precision/Recall/F_1
Baseline (Official)	36.42/58.80/44.98
Bert-CRF	65.81/55.76/60.37
MLPN	68.42/60.06/63.97
MLPN-Copy	55.35/**72.95**/62.95
URRC	74.06/65.73/69.65
$URRC_{+enum}$	74.64/66.20/70.17
$URRC_{+enum+large}$	**75.07**/66.51/**70.53**
$URRC_{+enum+large+ensemble}$	**80.59/66.30/72.75**

Relation Extraction. Table 6 is the comparison of different models and settings on relation extraction test1. Baseline is from Baidu official submission; Pipeline is a model implemented referring to the idea of *Godzilla* team in the 2020 LIC relation extraction task, which first extracts subject, then extracts object based on subject, and then predict relation based on subject and object; URRC is our proposed model. Compared with the two strong baselines, the overall score of URRC is greatly improved. In addition, compared to Pipeline, URRC uses a joint training approach, which is simpler and has higher training efficiency.

Table 6. Performance compared with two strong baselines for relation extraction.

Model	Precision/Recall/F_1
Baseline (Official)	76.81/65.63/70.78
Pipeline	75.94/68.97/72.29
URRC	**81.06/72.33/76.44**
$URRC_{+ensemble}$	**79.94/75.09/77.44**

Final Test. Table 7 shows the models we used on test2 of three tasks and their performance. Test2 is the final test set of the Multi-format Information Extraction Competition, and the average F_1 score of the three tasks is used as the basis for the ranking. The score of our models on the final test is 77.44.

Table 7. Performance and models used for three tasks on the final test set.

Task	Model	Precision/Recall/F_1
Sentence-level event extraction	$\text{URRC}_{+ensemble}$	87.41/80.18/83.64
Document-level event extraction	$\text{URRC}_{+enum+large+ensemble}$	79.46/64.60/71.27
Relation extraction	$\text{URRC}_{+ensemble}$	79.84/75.16/77.43

5 Conclusion

In this work, we propose a model that adapts to both event extraction and relation extraction. Based on the use of Multi-Label Pointer Network to solve the overlap problem in entity recognition, we further discover the problem of combining composite roles in the multi-event case, which is effectively solved by the Co-occurrence Matrix module jointly trained with MLPN. This pattern is also applied to relation extraction by using a different combination strategy. Through experiments, we demonstrate that our model achieves good results in two level event extraction tasks as well as relation extraction, with good space utilization and high training efficiency. Our model also helps us win the 6th place in the Multi-format Information Extraction Competition of the 2021 Language and Intelligence Challenge.

References

1. Chen, Y., Xu, L., Liu, K., Zeng, D., Zhao, J.: Event extraction via dynamic multi-pooling convolutional neural networks. In: Proceedings of the 53rd Annual Meeting of the Association for Computational Linguistics and the 7th International Joint Conference on Natural Language Processing (Volume 1: Long Papers), pp. 167–176 (2015)
2. Cui, Y., Che, W., Liu, T., Qin, B., Wang, S., Hu, G.: Revisiting pre-trained models for Chinese natural language processing. In: Proceedings of the 2020 Conference on Empirical Methods in Natural Language Processing: Findings, pp. 657–668. Association for Computational Linguistics, Online, November 2020. https://www.aclweb.org/anthology/2020.findings-emnlp.58
3. Cui, Y., et al.: Pre-training with whole word masking for Chinese BERT. arXiv preprint arXiv:1906.08101 (2019)
4. Devlin, J., Chang, M.W., Lee, K., Toutanova, K.: BERT: pre-training of deep bidirectional transformers for language understanding. arXiv preprint arXiv:1810.04805 (2018)
5. Kearns, M., Ron, D.: Algorithmic stability and sanity-check bounds for leave-one-out cross-validation. Neural Comput. **11**(6), 1427–1453 (1999)
6. Kingma, D., Ba, J.: Adam: a method for stochastic optimization. Comput. Sci. (2014)
7. Li, Q., Ji, H., Huang, L.: Joint event extraction via structured prediction with global features. In: Proceedings of the 51st Annual Meeting of the Association for Computational Linguistics (Volume 1: Long Papers), pp. 73–82 (2013)

8. Li, S., et al.: DuIE: a large-scale chinese dataset for information extraction. In: Tang, J., Kan, M.-Y., Zhao, D., Li, S., Zan, H. (eds.) NLPCC 2019. LNCS (LNAI), vol. 11839, pp. 791–800. Springer, Cham (2019). https://doi.org/10.1007/978-3-030-32236-6_72
9. Li, X., Feng, J., Meng, Y., Han, Q., Wu, F., Li, J.: A unified MRC framework for named entity recognition. arXiv preprint arXiv:1910.11476 (2019)
10. Li, X., et al.: DuEE: a large-scale dataset for Chinese event extraction in real-world scenarios. In: Zhu, X., Zhang, M., Hong, Yu., He, R. (eds.) NLPCC 2020. LNCS (LNAI), vol. 12431, pp. 534–545. Springer, Cham (2020). https://doi.org/10.1007/978-3-030-60457-8_44
11. Liu, M., Tu, Z., Wang, Z., Xu, X.: LTP: a new active learning strategy for BERT-CRF based named entity recognition. arXiv preprint arXiv:2001.02524 (2020)
12. Liu, Y., et al.: RoBERTa: a robustly optimized BERT pretraining approach. arXiv preprint arXiv:1907.11692 (2019)
13. Mooney, R., Bunescu, R.: Mining knowledge from text using information extraction. SIGKDD Explor. **7**, 3–10 (2005). https://doi.org/10.1145/1089815.1089817
14. Srivastava, N., Hinton, G., Krizhevsky, A., Sutskever, I., Salakhutdinov, R.: Dropout: a simple way to prevent neural networks from overfitting. J. Mach. Learn. Res. **15**(1), 1929–1958 (2014)
15. Vinyals, O., Fortunato, M., Jaitly, N.: Pointer networks. arXiv preprint arXiv: 1506.03134 (2015)

An Effective System for Multi-format Information Extraction

Yaduo Liu, Longhui Zhang, Shujuan Yin, Xiaofeng Zhao, and Feiliang Ren(✉)

School of Computer Science and Engineering, Northeastern University, Shenyang 110169, China
renfeiliang@cse.neu.edu.cn

Abstract. The multi-format information extraction task in the 2021 Language and Intelligence Challenge is designed to comprehensively evaluate information extraction from different dimensions. It consists of an multiple slots relation extraction subtask and two event extraction subtasks that extract events from both sentence-level and document-level. Here we describe our system for this multi-format information extraction competition task. Specifically, for the relation extraction subtask, we convert it to a traditional triple extraction task and design a voting based method that makes full use of existing models. For the sentence-level event extraction subtask, we convert it to a NER task and use a pointer labeling based method for extraction. Furthermore, considering the annotated trigger information may be helpful for event extraction, we design an auxiliary trigger recognition model and use the multi-task learning mechanism to integrate the trigger features into the event extraction model. For the document-level event extraction subtask, we design an Encoder-Decoder based method and propose a Transformer-alike decoder. Finally, our system ranks No.4 on the test set leader-board of this multi-format information extraction task, and its F1 scores for the subtasks of relation extraction, event extractions of sentence-level and document-level are 79.887%, 85.179%, and 70.828% respectively. The codes of our model are available at https://github.com/neukg/MultiIE.

Keywords: Multi-format information extraction · Relation extraction · Sentence-level event extraction · Document-level event extraction

1 Introduction

Information extraction (IE) aims to extract structured knowledge from unstructured texts. Named entity recognition (NER), relation extraction (RE) and event extraction (EE) are some fundamental information extraction tasks that focus on extracting entities, relations and events respectively. However, most researches only focus on extracting information in a single format, while lacking a unified

Y. Liu and L. Zhang—Contribute equally to this research and are listed randomly.

L. Wang et al. (Eds.): NLPCC 2021, LNAI 13029, pp. 460–471, 2021.
https://doi.org/10.1007/978-3-030-88483-3_37

evaluation platform for IE in different formats. Thus the 2021 Language and Intelligence Challenge (LIC-2021) sets up a multi-format IE competition task which is designed to comprehensively evaluate IE from different dimensions. The task consists of a relation extraction subtask and two event extraction subtasks that extract events from both sentence-level and document-level. The definitions of these subtasks are as follows[1].

Relation Extraction is a task that aims to extract all SPO triples from a given sentence according to a predefined schema set. The schema set defines relation P and the types of its corresponding subject S and object O. According to the complexity of object O, there are two types of schemas. The first one is the single-O-value schema in which the object O gets only a single slot and value. The second one is the schema in which the object O is a structure composed of multiple slots and their corresponding values.

Event Extraction consists of two subtasks. The first one is the sentence-level event (SEE) extraction whose aim is that given a sentence and predefined event types with corresponding argument roles, to identify all events of target types mentioned in the sentence, and extract corresponding event arguments playing target roles. The predefined event types and argument roles restrict the scope of extraction. The second one is the document-level event (DEE) extraction task, which shares almost the same definition with the previous one, but considers input text fragments in document-level rather than sentence-level, and restricts the event types to be extracted in the financial field.

Table 1 demonstrates some schema examples for these subtasks. For example, in the given multiple-O-values schema RE example, the object consists of two items: *inWork* and *@value*, but traditional RE may only get *@value* item in this example. Compared with traditional single-format IE task, there are following challenges in the multi-format IE task designed in LIC-2021.

First, despite the success of existing joint entity and relation extraction methods, they cannot be directly applied to the multiple slots RE task in LIC-2021 because of the multiple-O-values schema relations.

Second, the event extraction is more challenging due to the *overlapping event role* issue: an argument may have multiple roles. Besides, there are annotated triggers provided in the given datasets, and how to effectively use this kind of trigger information is still an open issue.

Third, it is difficult to extract different events that have the same event type in the DEE subtask. For example, for the "*be interviewed*" event type shown in Table 1, different people maybe interviewed at different time. For this case, in the SEE subtask, two different time arguments and two interviewed people are regarded as arguments for the same event. However, these two arguments need to be classified into arguments of two different events in the DEE subtask.

In our system, we use some effective methods to overcome these challenges. Specifically, for the first one, we design a schema disintegration module to convert each multiple-O-values relation into several single-O-value relations, then use a voting based module to obtain the final relations. For the second one, we convert

[1] https://aistudio.baidu.com/aistudio/competition/detail/65.

Table 1. Schema examples of three information extraction subtasks in LIC-2021.

Task	Schema example
RE (multiple-O-values schema)	object_type: {inWork: film and television work, @value: role} predicate: play subject_type: entertainer
SEE	event_type: finance/trading-limit down role_list: [role: time, role: stocks fell to limit]
DEE	event_type: be interviewed role_list: [role: the name of company, role: disclosure time, role: time of being interviewed, role: interviewing institutions]

the SEE subtask into a NER task and use a pointer labelling based method to address the mentioned *overlapping* issue. To use the trigger information, we design an auxiliary trigger recognition module and use the multi-task learning mechanism to integrate the trigger information. For the third one, we design an Encoder-Decoder based method to *generate* different events.

2 Related Work

RE is a long-standing natural language processing task whose aim is to mine factual triple knowledge from free texts. These triples have the form of (S, P, O), where both S and O are entities and P is a kind of semantic relation that connects S and O. For example, a triple (Beijing, capital_of, China) can express the knowledge that *Beijing is the capital of China*. Currently, the methods that extract entities and relations jointly in an end to end way are dominant. According to the extraction route taken, these joint extraction methods can be roughly classified into following three groups. (i) Table filling based methods ([3,12]), that maintain a table for each relation, and each item in such a table is used to indicate whether the row word and the column word possess the corresponding relation. (ii) Tagging based methods ([4,16]) that use the sequence labeling based methods to tag entities and relations. (iii) Seq2seq based methods ([18,19]), that use the sequence generation based methods to *generate* the items in each triple with a predefined order, for example, first *generates* S, then P, then O, etc.

Event extraction (EE) can be categorized into sentence-level EE (SEE) and document-level EE (DEE) according to the input text. The majority of existing EE researches (such as [1] and [13]) focus on SEE. Usually, they use a pipeline based framework that predicts triggers firstly, and then predict arguments. Recently, DEE is attracting more and more attentions. Usually there are two main challenges in DEE, which are: (i) arguments of one event may scatter across some long-distance sentences of a document or even different documents, and (ii) each document is likely to contain multiple events. A representative Chinese DEE work is [17], whose DEE model contains two main components: a SEE

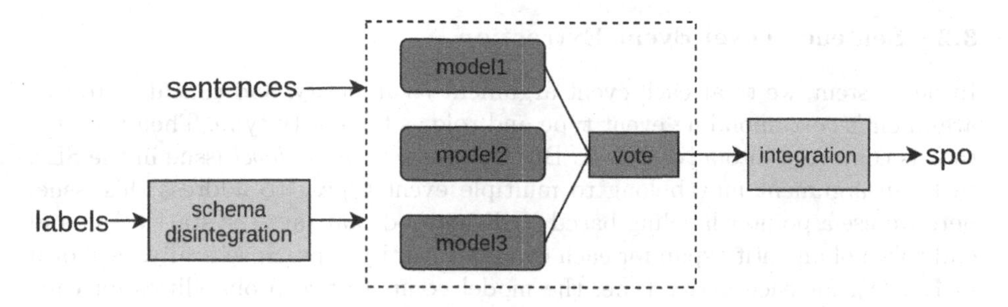

Fig. 1. The architecture of our RE model.

model that extracts event arguments and event triggers from a sentence; and a DEE model that extracts event arguments from the whole document based on a key event detection model and an arguments-completion strategy. Another representative Chinese DEE work is [20], which uses an entity-based directed acyclic graph to fulfill the DEE task effectively. Moreover, they redefine a DEE task with the no-trigger-words design to ease document-level event labeling. Both of these two DEE work are evaluated on Chinese financial field.

3 Methodology

3.1 Relation Extraction

The architecture of our RE model is shown in Fig. 1. We can see that it consists of two main modules: a schema disintegration module and a voting module.

Given a multiple-O-values triple[2] {s, p, $[(o_{k_1}, o_{v_1}), ..., (o_{k_m}, o_{v_m})]$}, the schema disintegration module would transform it into *2m-1* single-O-value triples. The concrete transformation process is as follows. First, o_{v_1} will be taken as an object to form a triple {s, p, o_{v_1}}. Then from $i = 2$ on, each (o_{k_i}, o_{v_i}) will form following two triples: {s, p-o_{k_i}, o_{v_i}} and {o_{v_1}, o_{k_i}, o_{v_i}}. Here both s and o_{v_1} will be repeatedly taken as subjects in the formed triples, and p-o_{k_i} is a new generated predicate. Accordingly, the given example will generate following triples, which are {s, p, o_{v_1}}, {s, p-o_{k_2}, o_{v_2}}, {o_{v_1}, o_{k_2}, o_{v_2}}, ..., {s, p-o_{k_m}, o_{v_m}}, {o_{v_1}, o_{k_m}, o_{v_m}}. If there are some single-O-value triples, they can be formed into a multi-O-value triple with a reverse process.

In the voting module, three existing state-of-the-art triple extraction models are used to extract single-O-value triples separately. Then their results would be voted to output the triples that receive more votes. Next, these obtained triples are converted into some multiple-O-values triples[3] as final results of this RE task. In our system, following three existing state-of-the-art models are used for voting: *TPLinker* [14], *SPN* [10] and *CasRel* [16].

[2] Although a multiple-O-values relation contains multiple objects, for simplicity, we still call it as a triple.

[3] Note the single-O-value triples can also be viewed as a specifical kind of multi-O-value triples.

3.2 Sentence-Level Event Extraction

In our system, we treat each event argument as an entity, and concatenate this argument's corresponding event type and role as the entity type. Then the SEE task is converted into a NER task. But there is a *multiple label* issue in the SEE task: an argument may belong to multiple event types. To address this issue, here we use a pointer labeling based NER method that tags the start token and end token of an entity span for each candidate entity type. Specifically, as shown in Eq. (1), for each entity type, the model compute two probabilities for each word to indicate the possibilities of this word being the start and end tokens of an entity that possesses this entity type.

$$\begin{aligned} p_{s_{ij}} &= sigmoid(W_i^S h_j + b_i^S), \\ p_{e_{ij}} &= sigmoid(W_i^E h_j + b_i^E) \end{aligned} \tag{1}$$

where $\{W_i^{(.)} \in \mathbb{R}^d\}_{i=1}^r$, $\{b_i^{(.)} \in \mathbb{R}\}_{i=1}^r$ are learnable parameters for the i-th entity type, $h_j \in \mathbb{R}^d$ is the token representation for the j-th word and is obtained by a pretrained language model, r is the number of entity types, $p_{s_{ij}}$ and $p_{e_{ij}}$ is the probabilities of the j-th word being the start token and the end token of an entiy that should be labeled with the i-th entity type.

Furthermore, to make full use of the features from the annotated triggers, we design an auxiliary trigger recognition module that also recognizes triggers in the same NER manner as used above. This auxiliary module are trained jointly with the above argument recognition module in a multi-task learning manner.

During training, the trigger recognition module will generate a representation for each trigger. Then these trigger representations will be merged into the token representations used by the argument recognition module with an attention mechanism for the argument recognition of next iteration. But during inferencing, our model will first recognize all triggers (the results are denoted as $T = \{t_1, t_2, ...\}$). Then these triggers' representations will be fused into a unified representation (denoted as t) by a max-pooling operation. Next, an additive-attention based method is used to obtained a new token representation sequence (denoted as $\hat{H}$), which will then be used as input for the argument recognition module. Specifically, $\hat{H}$ is computed with following Eq. (2).

$$\begin{aligned} \alpha &= softmax(V^{\mathrm{T}} tanh(W_1 H + W_2 t)), \\ \hat{H} &= \alpha \cdot H \end{aligned} \tag{2}$$

where H is the original token representations, V, W_1, and W_2 are learnable parameters, and the superscript T denotes a matrix transpose operation.

3.3 Document-Level Event Extraction

The architecture of our DEE model is shown in Fig. 2. We can see it contains three main modules, which are *Encoder*, Decoder, and *Output*.

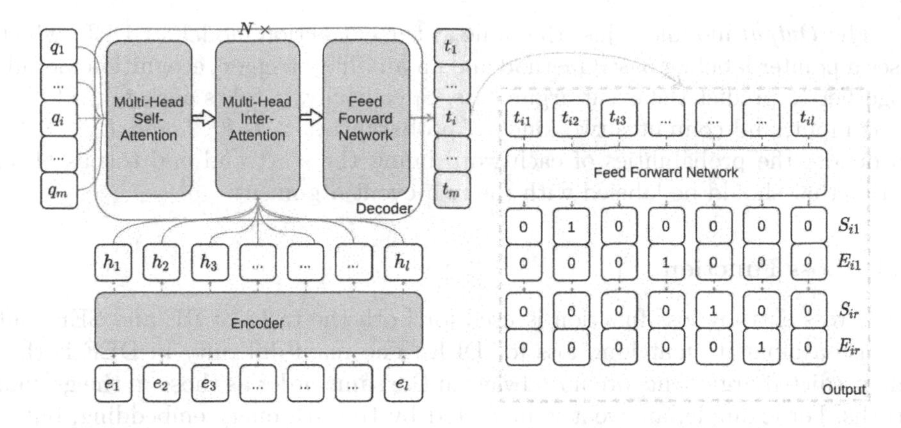

Fig. 2. The architecture of our DEE model. N is the number of transformer decoder.

BERT [2] and some of its variants are used as *Encoder* to get a context-aware representation for each token in a document. Note that both BERT and its variants have a length restriction for the input text (usually 512 TOKENs), so we use the *sliding window* based method to split a long document into several segments. Then each segment is taken as an input to be processed by our DEE model and the results of these segments are combined as final results. Given a segment $S = \{e_0, e_1, ..., e_l\}$, the output of *Encoder* is a token embedding sequence, we denote it as $H \in \mathbb{R}^{l \times d}$, where l is the number of tokens in the given segment, and d is the dimension of the token embedding.

In the *Decoder* module, some learnable query embeddings (as shown in Fig. 2, denoted as $(q_0, q_1, ..., q_m)$) along with the token embedding sequence outputted by the *Encoder* module are taken as input to predict events. Each query embedding corresponds to a specific event. Although different documents may have different numbers of events, we set a fixed number (denoted as m, and is set to 16 here) for the used queries, which is obtained according to the maximal event number in a document of the training dataset.

Here our designed *Decoder* is similar to the one used in Transformer [11], but we delete the components of masking and positional encoding because there is no correlations and position relations between event's query embeddings in the DEE task. Our *Decoder* module consists of N stacked layers, and each layer consists of a multi-head self-attention module, a multi-head inter-attention module, and a feed forward network. It will generate a refined embedding representation sequence (denoted as $T \in \mathbb{R}^{m \times l \times d}$) for the input queries. Specifically, the multi-head self-attention module is first performed to highlight some queries. Then, the multi-head inter-attention operation is performed to highlight the correlations between this input queries and the input sentence. Next, a feed forward network is used to generate a new embedding representation sequence for the input query. The above three steps will be performed N times to obtained T.

The *Output* module is just the same as the extraction model for DEE, which uses a pointer labeling based method and an auxiliary trigger recognition module together to predict the event arguments. Specifically, it takes each $t_i \in \mathbb{R}^{l \times d}$ in T as input, and computes two kinds of probabilities, $S_i \in \mathbb{R}^{l \times r}$ and $E_i \in \mathbb{R}^{l \times r}$, to denote the probabilities of each word being the start and end tokens of an entity that should be labeled with the i-th event argument.

3.4 Loss Function

The cross entropy loss function is used for both the tasks of RE and SEE, but we use a bipartite matching loss for DEE. The main difficulty in DEE is that the predicted arguments are not always in the same order as those in the ground truths. For example, an event is predicted by the i-th query embedding, but it may be in the j-th position of the ground truths. Thus, we do not apply the cross-entropy loss function directly because this loss is sensitive to the permutation of the predictions. Instead, we use the bipartite matching loss proposed in *SPN* to produce an optimal matching between m predicted events and m ground truth events. Specifically, we concatenate S and E (see above, generated in DEE), into $\hat{Y} \in \mathbb{R}^{m \times l \times 2 \times r}$, that can represent the predicted events. The ground truths is denoted as $Y \in \mathbb{R}^{m \times l \times 2 \times r}$. The process of computing bipartite matching loss is divided into two steps: finding an optimal matching and computing the loss function. To find optimal matching between Y and $\hat{Y}$, we search for a permutation of elements π^* with the lowest cost l_{match}, as shown in Eq. (3).

$$\pi^* = \underset{\pi \in S(m)}{\arg\min} \sum_{i=1}^{m} l_{match}(Y_i, \hat{Y}_{\pi(i)}) \tag{3}$$

where $S(m)$ is the space of all m-length permutations. $Y_i \in \mathbb{R}^{l \times 2 \times r}$ and $\hat{Y}_{\pi(i)} \in \mathbb{R}^{l \times 2 \times r}$ are outputs of the i-th predicated event and the i-th ground truth. $l_{match}(Y_i, \hat{Y}_{\pi(i)})$ is the pair-wise matching cost between the ground truth Y_i and the prediction event with index $\pi(i)$, and it is be computed as follow:

$$l_{match} = \hat{Y}_{\pi(i)} \circ Y_i \tag{4}$$

where $\circ$ is Hadamard product, and the optimal π^* can be be computed in polynomial time ($O(m_3)$) via the Hungarian algorithm[4]. Readers can find more detailed introduction about this loss in *SPN*.

After π^* is obtained, we change the event order of the predicated $\hat{Y}_{\pi^*}$ to be in line with π^*, and the re-ordered result is denoted as Y^*. Then we use the cross entropy loss function to compute the loss between Y^* and Y.

3.5 Model Enhancement Techniques

In our system, some model enhancement techiques are used to further improve the performance of each subtask.

[4] https://en.wikipedia.org/wiki/Hungarianalgorithm.

Table 2. Statistics of dataset.

Dataset	Training Set	Development Set	Test Set
DuIE2.0	171,293	20,674	21,080
DuEE1.0	11,958	1,498	3,500
DuEE-fin	7,047	1,174	3,524

Adversarial Training. To train a robust model, we use the FGM adversarial training mechanism [9] to disturb the direction of gradients, which makes the loss increase toward the maximum direction. Then the model will be pushed to find more robust parameters during the training process to weaken the impact of this aleatoric uncertainty. Assume the input text's embedding representation sequence is x, then its disturbance embedding r_{adv} is computed with Eq. (5).

$$\begin{aligned} g &= \nabla_x L(\theta, x, y), \\ r_{adv} &= \epsilon \cdot g/\|g\|_2 \end{aligned} \tag{5}$$

where ϵ is a hyperparameter to control the disturbance degree.

During training, r_{adv} will be updated with above equation after the loss of a subtask is computed. Then it will be taken as a new input of each task for the training of next iteration.

Data Augmentation. We use following two kinds of data augmentation strategies to enhance the performance of our models for different tasks. The first strategy is *Synonyms Replace*, which first randomly selects some words from each input text, and then these words will be replaced by their synonyms that are selected from a synonym dictionary. The second strategy is *Randomly Delete*, under which every input word is randomly deleted with a predefined probability. But if a word to be deleted is an entity, it will be remained.

Model Ensemble. We use the bagging strategy to ensemble multiple base models. It simply average or vote the weighted results of some base models. Concretely, we split all data in the training set into 10 folds, and train with 9 folds and validate with the remained 1 fold. By this way, we can obtain ten different base models. Besides, we also replace the *Encoder* of different models with different pretrained language models, including BERT, RoBERTa-wwm-ext [7], and NEZHA [15]. Accordingly, another kinds of base models can be trained. Finally, all the base models of a task are ensembled into one model. In our system, we simply vote on the predicted results of all base models, and select the outputs that get more votes.

4 Experiments

4.1 Basic Settings

Datasets. The LIC-2021 competition uses three large-scale Chinese information extraction datasets, including DuIE2.0 [5], DuEE1.0 [6] and DuEE-fin [6].

Table 3. Models' performance in relation extraction, sentence-level event extraction and document-level event extraction.

RE	F1 score	SEE	F1 score	DEE	F1 score
CasRel	0.7568	Backbone	0.8254	Backbone	0.6630
TPLinker	0.7605	Backbone+TR	0.8402	Backbone+TR	0.6843
SPN	0.7324			Backbone+BML	0.6702
Ens	0.7894	Ens	0.8509	Ens	0.7089
Ens+AT	0.7905	Ens+AT	0.8535	Ens+AT	0.7134
Ens+AT+PLM	0.8051	Ens+AT+PLM	0.8594	Ens+AT+PLM	0.7235

DuIE2.0 is the largest schema-based Chinese relation extraction dataset, which contains more than 430,000 triples over 210,000 real-world Chinese sentences, restricted by a pre-defined schema of 48 relation types. In this dataset, there are 43 single-O-value relations and 5 multiple-O-values relations. DuEE1.0 is a Chinese event extraction dataset, which consists of 17,000 sentences containing 20,000 events of 65 event types. DuEE-fin is a document-level event extraction dataset in the financial domain. The dataset consists of 11,700 documents across 13 event types, in which there are negative samples that do not contain any target events. All of these datasets are built by Baidu. Some basic statistics of these datasets are shown in Table 2.

Evaluation Metrics. According to the settings of LIC-2021, the scores of a participating system on DuIE2.0, DuEE1.0 and DuEE-fin are given respectively, and the macro average is used as the final score of the system. F1 score is used as the evaluation metrics for all three subtasks. For the RE subtask, a predicted relation with multiple-O-values schema would be regarded as correct only when all its slots are exactly matched with a manually annotated golden relation. For the SEE subtask, a predicted event argument is evaluated according to a token level matching F1 score. If an event argument has multiple annotated mentions, the one with the highest matching F1 will be used. For the DEE subtask, for each document, the evaluation would will first select a most similar predicted event for every annotated event (Event-Level-Matching). The matched predicted event must have the same event type as the annotated event and predict the most correct arguments. Each predicted event is matched only once.

Implementation Details. AdamW [8] is used to train all the models . The learning rates for RE, SEE, and DEE are 3e–5, 5e–5, and 5e–5 respectively. And the epochs for these three subtasks are 10, 40, and 30 respectively.

4.2 Results

Main Results. The main experimental results are shown in Table 3, in which Backbone, TR, BML, Ens, AT, and PLM to denote the backbone models, the trigger recognition module, the bipartite matching loss, the ensemble model, the adversarial training, and the pretrained language models respectively. In all of our experiments, Backbone refers to the model in which the transformer decoder and the bipartite matching loss are removed.

From Table 3 we can see that our RE model achieves far better results than all the compared state-of-the-art triple extraction models like *CasRel*, *TPLinker*, and *SPN*. Here we donot compare our two EE models with other state-of-the-art models because most of existing EE models cannot be used here directly.

Ablation Results. From the ablation results in Table 3 we can see that the pretrained language models are much helpful for all subtasks and they always bring a significant performance gain for a subtask. Besides, the adversarial training is also helpful and it consistently improves the performance of a subtask.

From the results of both SEE and DEE we can see that the designed trigger recognition module plays a very important role to the performance, and it improves nearly 1.5% point for the F1 score of SEE, and more than 2.1% point for the F1 score of DEE. In fact, this module plays even more roles than the pretrained language models. Based on these results we can conclude that the triggers do contain some important cues for both kinds of EE subtasks, and making full of these cues are much helpful for improving the performance.

From the results of DEE we can see that the bipartite matching loss also plays a helpful role to improve the performance. And it brings more performance gain than the adversarial training.

Besides, we also conduct experiments on the DEE subtask to evaluate: (i) the impact of the layer number of the transformer-based decoder (the number N in Fig. 2) , and (ii) the impact of different sizes of sliding windows. For the first one, we test different numbers in the range of 0 to 5. Our experiments show that when the number of decoder layer is set to 3, the model achieves the best results. We think this is mainly because that a moderate layer number will lead to more complete integration of input information into event queries, while a larger or smaller number will make the model be overfit or underfit. For the second one, we set the size of sliding windows to 128, 256, and 512. And their F1 scores are 64.8%, 66.9%, and 69.2% respectively when the number of the decoder's layer is set to 3 and the bipartite matching loss is used. These results show that usually, the performance would increase as the size of the sliding window increases. This is because that more context features can be taken into consideration with a larger sliding windows size.

5 Conclusions

This paper describes our system to the LIC-2021 multi-format IE task. We use different methods to overcome different challenges in this competition task, including the schema disintegration method for the multiple-O-values schema issue in the RE subtask, the multi-task learning method for the SEE subtask, and the Transform-alike decoder for the DEE subtask. Experimental results show that our system is effective and it ranks No.4 on the final test set leader-board of this competition. Its F1 scores are 79.887% on DuIE2.0, 85.179% on DuEE1.0, and 70.828% on DuEE-fin respectively.

However, there is still plenty of room for improvement, and lots of work should be further explored. First, in the RE subtask, many triples are not annotated. These triples gives model wrong supervisions, which is very harmful for the performance. But this *missing annotation* issue is still an open issue, and should be further explored. Second, in the DEE subtask, how to process long text is still a challenge that is worthy being further studied. In addition, if two arguments of one event are far away in the given text (either a sentence or a document), it would be difficult to extract them correctly. This issue also should be well studied in the future.

Acknowledgments. This work is supported by the National Key R&D Program of China (No.2018YF C0830701), the National Natural Science Foundation of China (No.61572120), the Fundamental Research Funds for the Central Universities (No.N181602013 and No.N171602003), Ten Thousand Talent Program (No.ZX20200035), and Liaoning Distinguished Professor (No.XLYC1902057).

References

1. Chen, Y., Xu, L., Liu, K., Zeng, D., Zhao, J.: Event extraction via dynamic multi-pooling convolutional neural networks. In: Proceedings of the 53rd Annual Meeting of the Association for Computational Linguistics and the 7th International Joint Conference on Natural Language Processing (volume 1: Long Papers), pp. 167–176. Association for Computational Linguistics, Beijing, China (July 2015)
2. Devlin, J., Chang, M.W., Lee, K., Toutanova, K.: BERT: pre-training of deep bidirectional transformers for language understanding. In: Proceedings of the 2019 Conference of the North American Chapter of the Association for Computational Linguistics: Human Language Technologies, vol. 1 (Long and Short Papers), pp. 4171–4186. Association for Computational Linguistics, Minneapolis, Minnesota (June 2019)
3. Gupta, P., Schütze, H., Andrassy, B.: Table filling multi-task recurrent neural network for joint entity and relation extraction. In: Proceedings of COLING 2016, the 26th International Conference on Computational Linguistics: Technical Papers, pp. 2537–2547. The COLING 2016 Organizing Committee, Osaka, Japan (December 2016)
4. Hang, T., Feng, J., Wu, Y., Yan, L., Wang, Y.: Joint extraction of entities and overlapping relations using source-target entity labeling. Exp. Syst. Appl. **177**, 114853 (2021)
5. Li, S., et al.: Duie: a large-scale Chinese dataset for information extraction. In: CCF International Conference on Natural Language Processing and Chinese Computing, pp. 791–800 (2019)
6. Li, X., et al.: Duee: A large-scale dataset for Chinese event extraction in real-world scenarios. In: CCF International Conference on Natural Language Processing and Chinese Computing, pp. 534–545 (2020)
7. Liu, Y., et al.: RoBERTa: A Robustly Optimized BERT Pretraining Approach. arXiv e-prints arXiv:1907.11692 (July 2019)

8. Loshchilov, I., Hutter, F.: Decoupled Weight Decay Regularization. arXiv e-prints arXiv:1711.05101 (November 2017)
9. Miyato, T., Dai, A.M., Goodfellow, I.J.: Adversarial training methods for semi-supervised text classification. In: 5th International Conference on Learning Representations, ICLR 2017, Toulon, France, April 24–26, 2017, Conference Track Proceedings. OpenReview.net (2017)
10. Sui, D., Chen, Y., Liu, K., Zhao, J., Zeng, X., Liu, S.: Joint Entity and Relation Extraction with Set Prediction Networks. arXiv e-prints arXiv:2011.01675 (November 2020)
11. Vaswani, A., et al.: Attention is all you need. In: Guyon, I., et al. (eds.) Advances in Neural Information Processing Systems, vol. 30, Curran Associates, Inc. (2017)
12. Wang, J., Lu, W.: Two are better than one: Joint entity and relation extraction with table-sequence encoders. In: Proceedings of the 2020 Conference on Empirical Methods in Natural Language Processing (EMNLP), pp. 1706–1721. Association for Computational Linguistics, Online (November 2020)
13. Wang, X., Han, X., Liu, Z., Sun, M., Li, P.: Adversarial training for weakly supervised event detection. In: Proceedings of the 2019 Conference of the North American Chapter of the Association for Computational Linguistics: Human Language Technologies, Volume 1 (Long and Short Papers), pp. 998–1008. Association for Computational Linguistics, Minneapolis, Minnesota (June 2019)
14. Wang, Y., Yu, B., Zhang, Y., Liu, T., Zhu, H., Sun, L.: TPLinker: single-stage joint extraction of entities and relations through token pair linking. In: Proceedings of the 28th International Conference on Computational Linguistics, pp. 1572–1582. International Committee on Computational Linguistics, Barcelona, Spain (Online) (December 2020)
15. Wei, J., et al.: NEZHA: Neural Contextualized Representation for Chinese Language Understanding. arXiv e-prints arXiv:1909.00204 (August2019)
16. Wei, Z., Su, J., Wang, Y., Tian, Y., Chang, Y.: A novel cascade binary tagging framework for relational triple extraction. In: Proceedings of the 58th Annual Meeting of the Association for Computational Linguistics, pp. 1476–1488. Association for Computational Linguistics, Online (July 2020)
17. Yang, H., Chen, Y., Liu, K., Xiao, Y., Zhao, J.: Dcfee: A document-level chinese financial event extraction system based on automatically labeled training data, pp. 50–55 (01 2018)
18. Zeng, D., Zhang, H., Liu, Q.: Copymtl: copy mechanism for joint extraction of entities and relations with multi-task learning. In: The Thirty-Fourth AAAI Conference on Artificial Intelligence, AAAI 2020. AAAI Press (2020)
19. Zeng, X., Zeng, D., He, S., Liu, K., Zhao, J.: Extracting relational facts by an end-to-end neural model with copy mechanism. In: Proceedings of the 56th Annual Meeting of the Association for Computational Linguistics (Volume 1: Long Papers), pp. 506–514. Association for Computational Linguistics, Melbourne, Australia (July 2018)
20. Zheng, S., Cao, W., Xu, W., Bian, J.: Doc2EDAG: An end-to-end document-level framework for Chinese financial event extraction. In: Proceedings of the 2019 Conference on Empirical Methods in Natural Language Processing and the 9th International Joint Conference on Natural Language Processing (EMNLP-IJCNLP), pp. 337–346. Association for Computational Linguistics, Hong Kong, China (November 2019)

A Hierarchical Sequence Labeling Model for Argument Pair Extraction

Jingyi Sun, Qinglin Zhu, Jianzhu Bao, Jipeng Wu, Caihua Yang, Rui Wang, and Ruifeng Xu(✉)

School of Computer Science and Technology, Harbin Institute of Technology, Shenzhen, China
{zhuqinglin,wujipeng}@stu.hit.edu.cn,xuruifeng@hit.edu.cn

Abstract. Argument pair extraction (APE) is a new task in the field of argument mining, aiming at mining interactive argument pairs from peer review and rebuttal. Previous work framed APE as a combination of a sequence labeling task and a sentence pair classification task via multi-task learning. However, this method lacks explicit modeling of the relations between arguments, and the two subtasks may not cooperate well. Towards these issues, we propose a hierarchical sequence labeling model that can efficiently represent arguments and explicitly capture the correlations between them. Our method matches argument pairs from the perspective of review and rebuttal respectively, and merges the results, enabling a more comprehensive extraction of argument pairs. Also, we propose a series of improved models and fuse them using weighted voting. Our method achieved first place in the NLPCC 2021 Shared Task on APE, proving the effectiveness of our method.

Keywords: Argument pair extraction · Hierarchical sequence labeling · Argument relation

1 Introduction

Argument mining aims to identify the semantic and logical structure in argumentative texts, which has important academic value and broad application background. Argument pair extraction (APE), aiming to extract interactive argument pairs from two argumentative passages of a discussion, is a new task in the field of argument mining.

Cheng et al. [4] constructed a review-rebuttal (RR) dataset from the peer review and rebuttal domain to facilitate research on this task. Peer review and rebuttal contain rich arguments and involve strong correspondences, as the rebuttal arguments often respond to the review arguments case by case. Figure 1 is an example of APE including a review passage and its corresponding rebuttal passage, divided into arguments and non-arguments at the sentence level. Two arguments from review passage and rebuttal passage separately discussing the same view can form an argument pair, where each argument contains one

L. Wang et al. (Eds.): NLPCC 2021, LNAI 13029, pp. 472–483, 2021.
https://doi.org/10.1007/978-3-030-88483-3_38

or more sentences. The APE task is challenging for accurately identifying the arguments on both sides and further extracting the complex relations between the review and rebuttal arguments.

[This paper proposes an optimization principle that is called gamma-optimization principle for stochastic algorithms (SA) in nonconvex and over-parametrized optimization.][1] ... NON-ARGU

[Could you please explain how you could achieve the value of the common global minimizer for the loss on all individual component functions?][5]... [It is unclear to me to understand how you could obtain it.][7] Review-Arg-1

[You have not mentioned the loss function that you are using for your numerical experiments.][8] [Can you show that Fact 1 is true for Softmax cross-entropy loss?][9] Review-Arg-2

Pair-1

Pair-2

[We thank the reviewer for providing valuable feedback.][1] [Below is our response.][2] [Any further comment is very welcome.] [3] NON-ARGU ...

[We assume the (neural network) model has enough expressive power to interpolate all the training data samples.][4] [We used the non-negative cross-entropy loss function to achieve a small value.][5] Reply-Arg-1

[We use the cross-entropy loss in the following experiments.] [6]... [we will adopt the MSE loss under the teacher-student setting to guarantee the existence of a known common global minimum (Please see our general response).][8] Reply-Arg-2

Fig. 1. A sample of the review-rebuttal (RR) dataset. The left and right tables show the review and the corresponding rebuttal passage, respectively. The number i at the end of each sentence indicates the i-th sentence in the review/rebuttal passage. These passages are divided into argument and non-argument at the sentence level, and each argument may contain one or more sentences. The two argument pairs are marked in yellow and blue, respectively.

To automatically detect arguments and match argument pairs, cheng et al. [4] proposed a multi-task learning framework based on multi-level LSTM [9]. The arguments in the review and rebuttal passages are extracted separately by a sequence labeling task, and then a classification task is used to determine the relation between each sentence pair. However, the method is suboptimal since the two tasks may not cooperate well and lack explicit modeling of the relations between arguments.

To address these issues, we propose a hierarchical sequence labeling model. Firstly, the arguments in the review and rebuttal are labeled by Bi-LSTM-CRF respectively. Then, we utilize GCN to gain the structural representation of sentences and acquire the semantic representation of arguments by mean pooling. Finally, we match argument pairs from the perspective of review and rebuttal respectively and merge the results. Compared with previous work, our method can explicitly mine the semantic information between arguments and thus extract argument pairs more accurately. To summarize, our contributions are as follows:

1. A hierarchical sequence labeling model is proposed for solving the APE task.
2. We propose a series of improved models and ensemble them by weighted voting to further improve the performance.
3. The official evaluation results show that our model achieves the best performance (1/15) in the review-rebuttal dataset, indicating the effectiveness of our method.

2 Relative Work

Argument mining has gradually become one of the hot spots of academic research in recent years, such as legal decision support [15,21], argument quality assessment [7,8,18], argument structure parsing [1,13,17], automated essay scoring [11,16,20], and so on. Argument mining has been applied in many different types of texts, such as student essays [11,16], public speeches [14], online debates and discussions [3,5], etc. Recently, several studies focus on the analysis of interactive argument relations. Ji et al. [10] proposed the task of identifying interactive argument pairs in online debate platforms such as ChangeMyView (CMV). Cheng et al. [4] collected texts from the peer review and rebuttal process and proposed a new task to extract the contained argument structure. This task is more challenging as it requires not only to mine the arguments in review and rebuttal passages but also to match these arguments into argument pairs.

3 Approach

We tackle this task with a hierarchical sequential tagging model, named H-BLC (Hierarchical Bidirectional LSTM-CRF model), as shown in Fig. 2. The model consists three parts, where the argument tagger and the argument representer are designed to identify and represent all possible arguments in the review and rebuttal passages. Afterward, these arguments are utilized by two argument pair taggers to tag the corresponding arguments in the rebuttal passages and review passages, respectively. The argument tagger, argument representer, and argument pair tagger are described below.

3.1 Task Defination

According to Cheng's [4] work, we need to detect the argument pairs from the pairs of review passages and rebuttal passages. The arguments in review and rebuttal are composed of one or more sentences with no overlapping with other arguments. Then, we extract argument pairs by matching individual review arguments and rebuttal arguments, such as {Review-Arg-1, Reply-Arg-1} and {Review-Arg-2, Reply-Arg-2} in Fig. 1.

3.2 Argument Tagger

The argument tagger is used to get the argument spans in the review and rebuttal passages. The pre-trained model BERT [6] is used to make the preliminary representation of the sentences in a passage and the i-th sentence embedding $s_i^v/s_i^b \in \mathbb{R}^{d_b}$ can be obtained by averaging its corresponding token embeddings, where d_b is the vector dimension of the last layer of BERT. Thus, the review passage with m sentences and rebuttal passage with n sentences can be represented as $\mathcal{P}^v = [s_1^v, s_2^v, \ldots, s_m^v]$ and $\mathcal{P}^b = [s_1^b,\ s_2^b, \ldots, s_n^b]$. To further model the contextual information, we feed preliminary sentence representations into the

Bi-LSTM, and get contextual sentence embeddings. $\mathcal{H}^v = [h_1^v, h_2^v, \ldots, h_m^v]$ and $\mathcal{H}^b = \left[h_1^b, h_2^b, \ldots, h_n^b\right]$ denote the sentence representations of the review passage and rebuttal passage respectively, where $h_i^v/h_i^b \in \mathbb{R}^{2d_l}$ is the i-th sentence representation and d_l is the hidden size of LSTM.

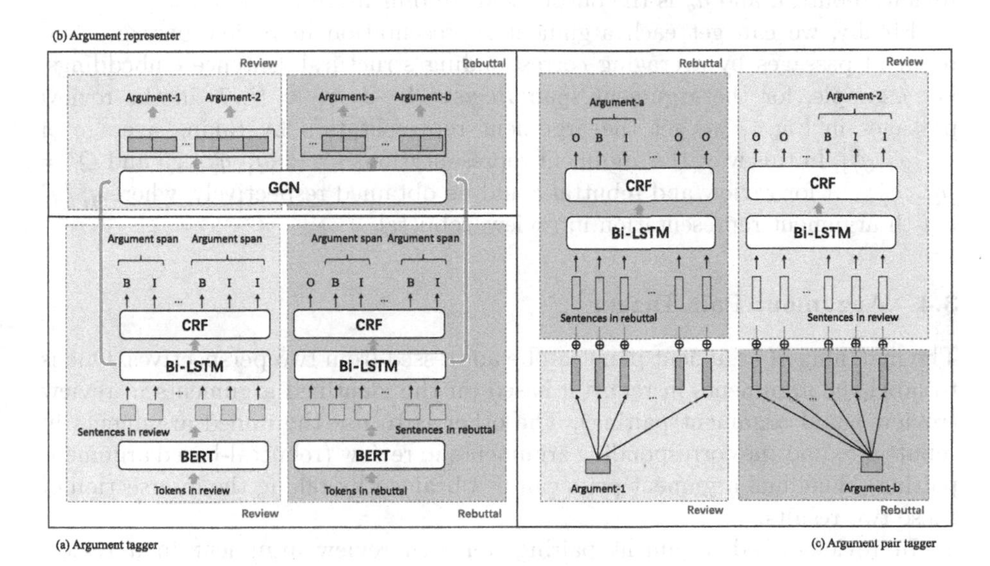

Fig. 2. (a) The argument tagger identifies argument spans in review and rebuttal passages separately. (b) The argument representer takes the sentence representations from Bi-LSTM, enhances them with the structure information obtained from GCN, and then outputs the argument representations through mean pooling. (c) The argument pair tagger extracts argument pairs from both review and rebuttal perspectives. For Argument-1 in the review passage, we concatenate it to the representations of sentences in the rebuttal to find its paired Argument-a. Similarly, we can find Argument-2 in review to pair with Argument-b. The complete pairing process has been omitted for brevity.

We use a CRF to obtain the argument spans $\mathcal{R}^v = [r_1^v, r_2^v, \ldots]$ in the review passage and $\mathcal{R}^b = [r_1^b, r_2^b, \ldots]$ in the rebuttal passage by sequence labeling, where r_i^v/r_i^b is the i-th argument span in review/rebuttal.

3.3 Argument Representer

The argument representer is designed to get a representation of each argument. First, we enhance the sentence representation by introducing structural information. Intuitively, when two sentences share some words (except stop words), they are probably related to the same topic. To model these features, we consider sentences as nodes, linking all sentences with co-occurring words. A graph is constructed for each pair of review passages and rebuttal passages. Using the sentence embeddings H^v and H^b from the argument tagger as initial node representations, we use a graph convolutional network (GCN) to model this graph. We keep the node vectors of the last layer of the GCN and get $\mathbf{G}^v = [g_1^v, g_2^v, \ldots, g_m^v]$ for review passage and $\mathbf{G}^b = [g_1^b, g_2^b, \ldots, g_m^b]$ for corresponding rebuttal passage, where $\mathbf{g}_i^v / \mathbf{g}_i^b \in \mathbb{R}^{d_g}$ is the structural representation for the i-th sentence in review/rebuttal, and d_g is the output feature dimension of GCN.

Finally, we can get each argument representation in review passages and rebuttal passages by averaging corresponding structural sentence embeddings. For example, for the argument span *Review-Arg-1*: $r_1^v = (5, 7)$ in the review passages in Fig. 1, we get the argument representation by taking average of $[g_5^v, g_6^v, g_7^v]$. In this way, the argument representations $\mathcal{Q}^v = [q_1^v, q_2^v, \ldots]$ and $\mathcal{Q}^b = [q_1^b, q_2^b, \ldots]$ for review and rebuttal could be obtained respectively, where q_i^v / q_i^b is i-th argument representation in review/rebuttal.

3.4 Argument Pair Tagger

The matching of argument pairs can be addressed from two perspectives. One is to label the arguments in rebuttal based on the identified arguments in review (review-based argument pairing). The other is to use the mined arguments in rebuttal to find the corresponding arguments in review (rebuttal-based argument pairing). The final argument pairs can be obtained by taking the intersection of these two results.

In review-based argument pairing, for i-th review argument in a review passage, we concatenate its argument representation q_i^v to each structural sentence embedding g_j^b in rebuttal respectively. These combined embeddings are fed into a Bi-LSTM to get argument-related representation of sentences $\mathcal{H}^{v,b} = [h_1^{vb}, h_2^{vb}, \ldots, h_n^{vb}]$ in the rebuttal passage, where $h_j^{vb} \in \mathbb{R}^{2d_l}$ can be obtained by:

$$h_j^{vb} = BiLSTM\left(q_i^v \oplus g_j^b, h_{j-1}^{vb}\right) \quad (1)$$

Then we use a CRF to identify sentences highly related to the target argument i. The labeling result can be represented as $\mathcal{Y}^{vpair} = [y_1^{vb}, y_2^{vb}, \ldots, y_n^{vb}]$, and the tagged arguments can form pairs with the target argument i in review. In this way, we can find all review-based argument pairs $\mathcal{S}^{vpair} = [(r_1^v, r_1^{vp}), (r_2^v, r_2^{vp}), \ldots]$, where (r_i^v, r_i^{vp}) represents the argument pair found by i-th argument in review.

Similarly, we can perform rebuttal-based argument pairing to get $\mathcal{S}^{bpair}$, in which case each rebuttal argument is used to guide the identification of its paired review arguments. Finally, by merging the result of review-based pairing and rebuttal-based pairing, we can get the set of all argument pairs $\mathcal{S}^{pair} = \mathcal{S}^{vpair} \cup \mathcal{S}^{bpair}$.

3.5 Training

We maximize the log-likelihood of the argument tagger and the argument pair tagger, and sum the loss for training:

$$\begin{aligned}\mathcal{L} = &\log p\left(\hat{Y}^v \mid \mathcal{P}^v\right) + \log p\left(\hat{Y}^b \mid \mathcal{P}^b\right) + \\ &\sum_i \log p\left(\hat{Y}_i^{vpair} \mid \mathcal{P}^b, \mathcal{R}^v\right) + \sum_i \log p\left(\hat{Y}_i^{bpair} \mid \mathcal{P}^v, \mathcal{R}^b\right)\end{aligned} \tag{2}$$

where $\hat{Y}^v$ and $\hat{Y}^b$ are the ground-truth argument labels of the review and rebuttal, $\hat{Y}_i^{vpair}$ and $\hat{Y}_i^{bpair}$ are the i-th argument-related ground-truth sequence labels of the review and rebuttal.

4 Experimental Setup

4.1 Dataset

Cheng et al. [4] collected peer reviews and rebuttals of ICLR from openreview.net. Each pair of review and rebuttal passage is annotated with sentence-level labels containing: the review/reply tag, the argument tag, the argument pair tag, and the paper number. The training set, validation set, and test set cover 3818, 474, and 474 samples, respectively.

4.2 Evaluation Metric

The performance of the model is evaluated by the overlap between the predicted set of argument pairs and the golden set of argument pairs, and the evaluation metric F1 is calculated as follows:

$$\text{Precision}(P) = TP/(TP + FP) \tag{3}$$

$$\text{Recall}(R) = TP/(TP + FN) \tag{4}$$

$$F1 = 2 * P * R/(P + R) \tag{5}$$

where TP represents true positives, FP represents false positives, TN represents true negatives, and FN represents false negatives.

To validate the model performance, we first extract the golden argument pairs in the review and rebuttal passages. For example, for the argument pair {Review-Arg-1, Reply-Arg-1} in Fig. 1, we represent it as the argument span pair {(5,7), (4,5)} to show that the sentences [5, 6, 7] in the review passage form an argument and are paired with the argument consist of sentences [4, 5] in the rebuttal passage. In this way, we can get all sets of golden argument span pairs for each pair of review and rebuttal passages, and then we compare these argument span pairs with those model predicts to calculate the F1 score for APE.

4.3 Implementation Details

We use BERT$_{Base}$ as the basic sentence encoder and fine-tune BERT during training. In addition, we adopt two-layer LSTMs for each Bi-LSTM with the hidden size of 512. The Adam [12] optimizer is utilized for parameter optimization, with the initial learning rate of 1e–5 for the BERT model and 1e–3 for the other models. The batch size is 2 and the dropout probability is 0.5. We set the maximum iteration number to 10, and train our model on the officially given training set with an early stop strategy. When the F1 score for argument pair extraction does not increase on the validation set over 5 consecutive times, we stop training and save the model as the best model.

4.4 Compared Models

To evaluate the performance of our proposed model H-BLC, we compare it with the following models.

- **MT-H-LSTM-CRF** It is the first model (proposed by Cheng et al. [4]) to solve the argument pair extraction task. The bert-as-service [22] and a Bi-LSTM are used to represent each sentence, followed by another Bi-LSTM to mine arguments and pair sentences simultaneously. Finally, the argument pairs can be inferred from the sentence pairs.
- **Non-FT-H-BLC** Our proposed model H-BLC fine-tunes BERT for sentence encoding. To compare with MT-H-LSTM-CRF, we remove the fine-tuning process of BERT and design the Non-FT-H-BLC.

Besides, we try several methods based on H-BLC to improve its performance on argument pairing. On the one hand, we introduce the attention mechanisms to capture essential contextual information of argument-sentence pairs. For brevity, we take the process of review-based pairing for description.

- **H-BLQAC** When pairing the arguments, we adopt the attention mechanism proposed by Bahdanau et al. [2] and use the given review argument as a query to interact with sentences in rebuttal. In this way, we get the attention representation of argument-sentence pairs $\mathcal{O}^{vb} = [o_1^{vb}, o_2^{vb}, \ldots, o_n^{vb}]$. Then we concatenate $\mathcal{O}^{vb}$ with the original argument-sentence pair representation $\mathcal{H}^{vb} = [h_1^{vb}, h_2^{vb}, \ldots, h_n^{vb}]$ to predict the relatedness between the given review argument and the rebuttal sentences.
- **H-BLAC** We implement the self-attention [19] mechanism in this model to capture the key information in the rebuttal context. Specifically, according to the representation of argument-sentence pairs $\mathcal{H}^{vb} = [h_1^{vb}, h_2^{vb}, \ldots, h_n^{vb}]$ got from Bi-LSTM, new representation $\mathcal{A}^{vb} = [a_1^{vb}, a_2^{vb}, \ldots, a_n^{vb}]$ is obtained using the self-attention mechanism, and then $\mathcal{H}^{vb}$ and $\mathcal{A}^{vb}$ are concatenated to predict the relevance of each sentence in rebuttal to the given argument in review.

On the other hand, we use the sentence/argument embeddings from the argument tagger to assist argument pairing.

- **H-BLAC-L** This model adds the Bi-LSTM representation of sentences in rebuttal $\mathcal{H}^b = [h_1^b, h_2^b, \ldots, h_n^b]$ from the argument tagger for argument pairing. Specifically, $\mathcal{H}^b$ is concatenated with $\mathcal{H}^{vb}$ and $\mathcal{A}^{vb}$ to predict the relevance of sentences to the given argument.
- **H-BLAC-G** When pairing arguments, this model uses the GCN representation of sentences in rebuttal $\mathbf{G}^b = [g_1^b, g_2^b, \ldots, g_n^b]$, together with $\mathcal{H}^{vb}$ and $\mathcal{A}^{vb}$, to match the argument pairs.
- **H-BLAC-LArg** This model averages the sentence representations $\mathcal{H}^v = [h_1^v, h_2^v, \ldots, h_m^v]$ got from the argument tagger to represent the given argument q^v, and concatenates it with $\mathcal{H}^{vb}$, $\mathcal{A}^{vb}$, $\mathcal{H}^b$ to find the arguments which are highly related to the given argument in the corresponding passage.

5 Results

5.1 Results of Basic Models

We first compare our basic model H-BLC with the baseline model, and the model performances are shown in the first part of the Table 1. Compared to MT-H-LSTM-CRF model proposed by Cheng et al. [4], our basic model H-BLC obtains an increase in F1 score of 13.04%, showing the effectiveness of our basic framework. In addition, our Non-FT-H-BLC model with fixed BERT parameters still outperforms the MT-H-LSTM-CRF model by 11.90%, indicating that the performance improvement of our model does not rely on fine-tuning BERT.

Table 1. The performance of the baseline and our proposed models on RR dataset

Method	P(%)	R(%)	F1(%)
MT-H-LSTM-CRF [4]	26.69	26.24	26.46
Non-FT-H-BLC	38.44	38.29	38.36
H-BLC	40.71	38.36	39.50
H-BLQAC	43.13	38.94	40.93
H-BLAC	42.96	39.63	41.22
H-BLAC-G	44.44	38.83	41.44
H-BLAC-LArg	**49.82**	36.07	41.84
H-BLAC-L	43.06	**40.80**	**41.89**

5.2 Results of Improved Models

To validate the effectiveness of our improvement methods, we also train the improved models on the same train set and list their performance in Table 1.

As shown in the middle row of the table, the models introduce different attention mechanisms to improve our basic model H-BLC. H-BLQAC and H-BLAC, using the attention mechanism [2] and self-attention mechanism [19] respectively, can achieve 1.42% and 1.72% increase in F1 score compared to H-BLC. It also shows that the self-attention mechanism is more effective in capturing the relevance between arguments, so the subsequent models are based on H-BLAC.

The last part of the table includes the results of models further improving H-BLAC by introducing sentence representations/argument representations to the argument pair tagger. H-BLAC-L achieves the highest F1 score, increasing about 0.67% compared to H-BLAC, which indicates that the sentence representations obtained by Bi-LSTM in argument tagger can effectively help for argument-sentence pair identification. H-BLAC-LArg achieves a similar but slightly lower F1 score as H-BLAC-L. It seems that concatenating argument representations with other argument-sentence pair features used in H-BLAC-L can not increase the improvement. Comparably, H-BLAC-G, which adds the sentence representation from GCN directly to argument-sentence pair representation, improves H-BLAC by 0.22%. One possible reason making the lower F1 of H-BLAC-G than H-BLAC-L is that the structure information captured by GCN might not be suitable to be concatenated with other features directly. Nevertheless, the sentence representations obtained from GCN still work for improving H-BLAC.

In conclusion, our proposed models all outperform the baseline model MT-H-LSTM-CRF, and the best model H-BLAC-L achieves a significant improvement of 15.43% in F1 score.

5.3 Ensemble Approach

To improve the overall performance on the APE task, we combine the results of different models mentioned in Sect. 4.4. Specifically, argument pairs predicted by distinct models are considered and retained if half of the models agree.

For model selection, we discover that the more diverse the models used for fusion, the better the fusion performance, implying that the strengths and limitations of our proposed model frameworks might complement each other. As a result, we choose the best models from each of our six proposed frameworks.

We finally adopt a weighted voting strategy and assign different weights to the chosen models. The specific weights used in the best voting result on the validation set are shown in Table 2.

Table 2. The weights of the fusion models

Model	H-BLC	H-BLQAC	H-BLAC	H-BLAC-G	H-BLAC-LArg	H-BLAC-L
Weight	2	1	1	1	1	2

For NLPCC 2021 shared task on APE, we submitted the result of a single model in phase 1 and the fusion models in phase 2. As shown in Table 3 and

Table 3. Phase 1 Leaderboard.

Rank	F1-score
Rank1(ours)	**0.3790**
Rank2	0.3579
Rank3	0.2803

Table 4. Phase 2 Leaderboard.

Rank	F1-score
Rank1(ours)	**0.3231**
Rank2	0.2629
Rank3	0.2625

Table 4, we achieve the highest rankings in both phase 1 and phase 2, demonstrating the effectiveness of our proposed models and the weighted voting strategy.

6 Conclusion

In this paper, we propose a hierarchical sequence labeling model that focuses on capturing the argument-level interactions to extract argument pairs. We also propose a series of improved models and utilize a weighted voting method to merge these models. Official evaluation results of NLPCC 2021 shared task on APE show that our system yields the best performance in 15 participants, which proves the effectiveness of our work.

Considering that argumentation requires background knowledge, especially in peer review and rebuttal domains, we intend to introduce external knowledge to help capture the semantic relatedness between arguments in the future.

Acknowledgments. This work was partially supported by the National Natural Science Foundation of China (61632011, 61876053, 62006062), the Guangdong Province Covid-19 Pandemic Control Research Funding (2020KZDZX1224), the Shenzhen Foundational Research Funding (JCYJ20180507183527919 and JCYJ20180507183608379), and China Postdoctoral Science Foundation (2020M670912).

References

1. Afantenos, S., Peldszus, A., Stede, M.: Comparing decoding mechanisms for parsing argumentative structures. Argument Comput. **9**(3), 177–192 (2018)
2. Bahdanau, D., Cho, K., Bengio, Y.: Neural machine translation by jointly learning to align and translate (2016). https://arxiv.org/abs/1409.0473
3. Chakrabarty, T., Hidey, C., Muresan, S., McKeown, K., Hwang, A.: AMPERSAND: argument mining for PERSuAsive oNline discussions. In: Proceedings of the 2019 Conference on Empirical Methods in Natural Language Processing and the 9th International Joint Conference on Natural Language Processing (EMNLP-IJCNLP), pp. 2933–2943. Association for Computational Linguistics, Hong Kong, China (November 2019). https://doi.org/10.18653/v1/D19-1291, https://aclanthology.org/D19-1291

4. Cheng, L., Bing, L., Yu, Q., Lu, W., Si, L.: APE: argument pair extraction from peer review and rebuttal via multi-task learning. In: Proceedings of the 2020 Conference on Empirical Methods in Natural Language Processing (EMNLP), pp. 7000–7011. Association for Computational Linguistics, Online (November 2020). https://doi.org/10.18653/v1/2020.emnlp-main.569, https://aclanthology.org/2020.emnlp-main.569
5. Chowanda, A.D., Sanyoto, A.R., Suhartono, D., Setiadi, C.J.: Automatic debate text summarization in online debate forum. Procedia Comput. Sci. **116**, 11–19 (2017). https://doi.org/10.1016/j.procs.2017.10.003, https://www.sciencedirect.com/science/article/pii/S1877050917320409, discovery and innovation of computer science technology in artificial intelligence era: The 2nd International Conference on Computer Science and Computational Intelligence (ICCSCI 2017)
6. Devlin, J., Chang, M.W., Lee, K., Toutanova, K.: BERT: pre-training of deep bidirectional transformers for language understanding. In: Proceedings of the 2019 Conference of the North American Chapter of the Association for Computational Linguistics: Human Language Technologies, Volume 1 (Long and Short Papers), pp. 4171–4186. Association for Computational Linguistics, Minneapolis, Minnesota (June 2019). https://doi.org/10.18653/v1/N19-1423, https://aclanthology.org/N19-1423
7. Gleize, M., et al.: Are you convinced? choosing the more convincing evidence with a Siamese network. In: Proceedings of the 57th Annual Meeting of the Association for Computational Linguistics, pp. 967–976. Association for Computational Linguistics, Florence, Italy (July 2019). https://doi.org/10.18653/v1/P19-1093, https://aclanthology.org/P19-1093
8. Gretz, S., et al.: A large-scale dataset for argument quality ranking: Construction and analysis. In: Proceedings of the AAAI Conference on Artificial Intelligence, vol. 34, no. 05, pp. 7805–7813 (April 2020)
9. Hochreiter, S., Schmidhuber, J.: Long short-term memory. Neural Comput. **9**(8), 1735–1780 (1997). https://doi.org/10.1162/neco.1997.9.8.1735
10. Ji, L., Wei, Z., Li, J., Zhang, Q., Huang, X.: Discrete argument representation learning for interactive argument pair identification. In: Proceedings of the 2021 Conference of the North American Chapter of the Association for Computational Linguistics: Human Language Technologies, pp. 5467–5478. Association for Computational Linguistics, Online (June 2021). https://doi.org/10.18653/v1/2021.naacl-main.431, https://aclanthology.org/2021.naacl-main.431
11. Ke, Z., Carlile, W., Gurrapadi, N., Ng, V.: Learning to give feedback: modeling attributes affecting argument persuasiveness in student essays. In: Proceedings of the Twenty-Seventh International Joint Conference on Artificial Intelligence, IJCAI-18, pp. 4130–4136. International Joint Conferences on Artificial Intelligence Organization (7 2018). https://doi.org/10.24963/ijcai.2018/574, https://www.ijcai.org/Proceedings/2018/0574.pdf
12. Kingma, D.P., Ba, J.: Adam: A method for stochastic optimization. arXiv preprint arXiv:1412.6980 (2014), https://arxiv.org/abs/1412.6980
13. Kuribayashi, T., et al.: An empirical study of span representations in argumentation structure parsing. In: Proceedings of the 57th Annual Meeting of the Association for Computational Linguistics, pp. 4691–4698. Association for Computational Linguistics, Florence, Italy (July 2019). https://doi.org/10.18653/v1/P19-1464, https://aclanthology.org/P19-1464

14. Lippi, M., Torroni, P.: Argument mining from speech: Detecting claims in political debates. In: Proceedings of the AAAI Conference on Artificial Intelligence, vol. 30, no. 1 (March 2016). https://ojs.aaai.org/index.php/AAAI/article/view/10384
15. Palau, R.M., Moens, M.: Argumentation mining: the detection, classification and structure of arguments in text. In: The 12th International Conference on Artificial Intelligence and Law, Proceedings of the Conference, June 8–12, 2009, Barcelona, Spain. pp. 98–107. ACM (2009). https://doi.org/10.1145/1568234.1568246
16. Song, W., Song, Z., Liu, L., Fu, R.: Hierarchical multi-task learning for organization evaluation of argumentative student essays. In: Bessiere, C. (ed.) Proceedings of the Twenty-Ninth International Joint Conference on Artificial Intelligence, IJCAI-20. pp. 3875–3881. International Joint Conferences on Artificial Intelligence Organization (7 2020). https://doi.org/10.24963/ijcai.2020/536
17. Stab, C., Gurevych, I.: Parsing argumentation structures in persuasive essays. Comput. Linguist. **43**(3), 619–659 (2017). https://doi.org/10.1162/COLI_a_00295
18. Toledo, A., et al.: Automatic argument quality assessment - new datasets and methods. In: Proceedings of the 2019 Conference on Empirical Methods in Natural Language Processing and the 9th International Joint Conference on Natural Language Processing (EMNLP-IJCNLP), pp. 5625–5635. Association for Computational Linguistics, Hong Kong, China (November 2019). https://doi.org/10.18653/v1/D19-1564, https://aclanthology.org/D19-1564
19. Vaswani, A., et al.: Attention is all you need. In: Guyon, I., et al. (eds.) Advances in Neural Information Processing Systems, vol. 30. Curran Associates, Inc. (2017). https://proceedings.neurips.cc/paper/2017/file/3f5ee243547dee91fbd053c1c4a845aa-Paper.pdf
20. Wachsmuth, H., Al-Khatib, K., Stein, B.: Using argument mining to assess the argumentation quality of essays. In: Proceedings of COLING 2016, the 26th International Conference on Computational Linguistics: Technical Papers, pp. 1680–1691. The COLING 2016 Organizing Committee, Osaka, Japan (December 2016). https://aclanthology.org/C16-1158
21. Walker, V.R., Foerster, D., Ponce, J.M., Rosen, M.: Evidence types, credibility factors, and patterns or soft rules for weighing conflicting evidence: argument mining in the context of legal rules governing evidence assessment. In: Slonim, N., Aharonov, R. (eds.) Proceedings of the 5th Workshop on Argument Mining, ArgMining@EMNLP 2018, Brussels, Belgium, November 1, 2018, pp. 68–78. Association for Computational Linguistics (2018). https://doi.org/10.18653/v1/w18-5209
22. Xiao, H.: bert-as-service. https://github.com/hanxiao/bert-as-service (2018)

Distant Finetuning with Discourse Relations for Stance Classification

Lifeng Jin(✉), Kun Xu, Linfeng Song, and Dong Yu

Tencent AI Lab, Bellevue, WA 98003, USA
{lifengjin,kxkunxu,lfsong,dyu}@tencent.com

Abstract. Approaches for the stance classification task, an important task for understanding argumentation in debates and detecting fake news, have been relying on models which deal with individual debate topics. In this paper, in order to train a system independent from topics, we propose a new method to extract data with silver labels from raw text to finetune a model for stance classification. The extraction relies on specific discourse relation information, which is shown as a reliable and accurate source for providing stance information. We also propose a 3-stage training framework where the noisy level in the data used for fine-tuning decreases over different stages going from the most noisy to the least noisy. Detailed experiments show that the automatically annotated dataset as well as the 3-stage training help improve model performance in stance classification. Our approach ranks 1st among 26 competing teams in the stance classification track of the NLPCC 2021 shared task Argumentative Text Understanding for AI Debater, which confirms the effectiveness of our approach.

Keywords: Stance classification · Distant finetuning · Noisy data

1 Introduction

In natural language understanding, it is important to understand how sentences are used in order to argue for or against particular topics in conversations and articles. This not only relates to automatic debates [17] but also is useful in detecting fake news in media and allowing colorful persona in robots [12]. Stance detection or stance classification [3,12] is the task where one has to decide whether a given claim is in support of or against a given topic, or the two are unrelated in terms of argumentation. The support and against relations between topics and claims are usually more abstract than such relations in opinion mining, because in instead of directly taking or refuting a topic, a claim is usually a piece of evidence or a logical consequence following a stance towards some topic, which makes detecting the stance of such claims difficult and knowledge-intensive. This problem is partially tackled by approaches where topic-specific models are used. Obviously it is difficult to generalize to new topics with these models, because new models have to be trained with annotated data for the new

L. Wang et al. (Eds.): NLPCC 2021, LNAI 13029, pp. 484–495, 2021.
https://doi.org/10.1007/978-3-030-88483-3_39

topics, and possible topics in real life scenarios are numerous. Generalizability is also a problem for machine learning models from the stand point of training data, because common stance detection datasets have only a couple hundred topics but thousands of claims, allowing such models to easily overfit to the topics in training data.

We propose to address the generalizability issue as well as the knowledge-intensive nature of the task with knowledge-rich pretrained models. Pretrained models have shown good performance in a variety of natural language understanding tasks which require both linguistic and commonsense knowledge. Such knowledge is invaluable to the stance detection task. In order to further improve model performance, we extract a noisy training dataset from large quantities of unlabeled text, following the intuition that discourse relations are indicative of stance in general. For example, the relationship between a topic, such as "大数据带来了更多的好处 (big data brings more good than bad)", and a supporting argument, such as "生成的大数据可作为预测工具和预防策略 (the generated big data can be used as a predictive tool and preventive strategy)", may be rewritten as a causal relation:

1. 因为生成的大数据可作为预测工具和预防策略，所以大数据带来了更多的好处。(Because the generated big data can be used as a predictive tool and preventive strategy, big data brings more good than bad.), and the same topic and an against argument, such as "大数据的准确性难以确保 (the accuracy of big data is hard to be sure of)", may be rewritten as a contradiction relation:
2. 虽然大数据带来了更多的好处，大数据的准确性难以确保。(Although big data brings more good than bad, the accuracy of big data is hard to be sure of.),

which suggests that raw sentences in such relations may be in turn used as noisy training instances for the stance detection task.

Training neural network models with such noisy datasets improves robustness of the model, reduces greatly the chance of overfitting, allows the model to acclimate to task-specific data format, and provides chances to learn more knowledge for the stance detection task. Experiments on development data show large improvements over baselines where such noisy data is not used. Amongst the 26 teams participating in the Claim Stance Classification for Debating track of the Argumentative Text Understanding for AI Debater shared task, our approach ranks 1st, with 2.3% absolute performance improvement over the runner-up approach.

2 Related Work

Stance classification has been a subject of research in many different environments, such as congressional debates [21], online debates on social media [4,19] and company-internal discussions [14]. Previous approaches focus on learning topic-specific models to classify stances of related claims with machine learning models [2,10,18] as well as deep learning models [7,8,15,16,20,24]. Previous work has also looked at doing stance classification at challenging situations such as zero-shot [1] and unsupervised settings [9,11,19]. Since stance classification has been thought of as a subtask of sentiment analysis [12], the use of sentiment lexicon is popular in previous work. Compared to previous work, our approach does not rely on any sentiment lexicon, which is a linguistic resource difficult to construct. Our approach also does not require topic-specific model training, which improves generalizability of a trained model to unseen topics and claims.

3 Unsupervised Data Preparation

We follow the intuition that the Support relation in stance classification between claims and topics can be categorized as a causal or conditional relation, because one should be able to deduce the topic from the claim if the claim supports the topic. Similarly, the Against relation between claims and topics can be categorized as a contraction relation where the claim does not naturally follow a topic. If a claim and a topic are to be connected by discourse connectives, connectives of corresponding discourse relations need to be used in order to preserve discourse coherence. Sentences with such discourse relations could better prepare the pretrained language models for finetuning with gold data and help the language models fight against overfitting. We first present a few different sets of data we extract from raw text with no supervision, and then explain how they are used in our finetuning framework.

3.1 Data D_1 Extraction for Distant Finetuning

A dataset for unsupervised distant finetuning is extracted from a large text corpus CLUE [23][1] based on discourse relations. Table 1 shows examples of discourse connectives used for extracting sentences with particular discourse relations. A pair of sentences are kept when the second sentence starts with a multiple line connective, which follows this pattern "S_1。 $c_1 S_2$。" where S_i is a sentence or a sentence fragment, and c_i is a discourse connective. For Support, S_1 is a topic and S_2 is a claim, where the opposite is adopted for Against. For single sentences, one sentence is kept if it contains a pair of single line connectives where the second connective is in a sentence fragment directly after a comma, which follows this pattern "$S_1 c_1 S_2$， $S_3 c_2 S_4$。". In the case of single sentences, for Support, $S_1 c_1 S_2$ is a topic and $S_3 c_2 S_4$ is a claim, where the opposite is adopted for Against. Candidate sentences are discarded when they contain non-Chinese characters, exceed 100 characters, or contain pronouns. The discourse connectives are deleted from the sentences to remove obvious and easy cues to the relation classes. The sentence pairs with the Neutral label are selected randomly from sentences in the same article which are close to the topic sentence. The

[1] https://github.com/CLUEbenchmark/CLUE.

final D_1 dataset includes 1.2 million data points labeled as Support, 0.7 million labeled as Against, and 1.9 million labeled as Neutral. Table 2 shows examples of extracted sentences with different silver labels.

Table 1. Example Chinese discourse connectives used in extraction.

Relation	Type	Connectives
Support	Multiple line	因此, 因而, 所以
	Single line	因为... 所以..., 只要... 就..., 要是... 就..., 之所以... 是因为...
Against	Multiple line	但是, 然而, 可是
	Single line	虽然... 但是..., 虽然... 可是..., 尽管... 但是...

Table 2. Examples of extracted sentence pairs from raw text.

Relation	Type	Connectives
Support	Topic	常将弹性工时与变形工时相互混淆
	Claim	国内学界对于弹性工时概念未有统一解释
Against	Topic	选择不同作用机制的癫痫药物，才可能获得疗效的叠加
	Claim	如果两种癫痫药物有相同的不良反应，就不能联合使用
Neutral	Topic	其中的人数是最基本的数据
	Claim	人口数据是一个国家和地区的基本数据

3.2 Low-Noise Finetuning Data D_2 Extraction

Although the distant finetuning data prepared in Sect. 3.1 can provide training signal to further pretrain language models, it may be too noisy for final finetuning purposes. The Conditional relation does not always equal to Support, as portrayed in this example "只要小明去，小张就会去。(If Xiao Ming goes, Xiao Zhang goes too.)" in which the condition has only an arbitrary connection to the result. Similarly the Contradiction relation is not always Against, shown in this example "虽然兔毛可以抵御严寒，但是兔子也怕热。(Although rabbit fur can be good for rigid cold, rabbits are also prone to overheating)" where the two facts are more supplementary than contradictory to each other. Further filtering is needed to reduce the noise level within the extracted pretrain dataset.

A list of high frequency topic indicators is used to find sentences that are most likely to be statements of positions on certain issues, which are the best candidates for topcis. The list includes words such as "应 (should)" and "最 (most)". More importantly, we consider Entailment and Contradiction relations from the natural language inference (NLI) task very close to the Support and Against relations in stance detection, therefore we employ an NLI model for data selection. Specifically, a Chinese BERT with a classification layer is finetuned with the XNLI dataset on all available languages and the best model is chosen based on evaluation on the Chinese NLI portion of the XNLI evaluation dataset. This model is then used to make predictions of NLI labels on all data points in D_1. Finally, 30,000 data points which are either labeled Support by the connectives and Entailment by the XNLI model, or Against and Contradiction, or Neutral and No Entailment are randomly sampled from D_1, resulting in a low-noise finetuning dataset D_2 with 30,000 data points in total, which is about 5 times the size of the gold training set.

3.3 Stance Detection Data in Other Languages

Datasets for stance detection also exist in other languages such as English. With a pretrained language model able to take multilingual input, we expect such datasets help the model learn the concept of Support and Against more robustly. The multilingual stance detection dataset XArgMining [22] from the IBM Debater project contains human-authored data points for stance detection in English, as well as such data points translated into 5 other languages: Spanish, French, Italian, German and Dutch. With both human authored and machine translated data points combined, the dataset used for training has 400,000 data points. The dataset D_x is the concatenation of these datasets.

4 Staged Training with Noisy Finetuning

Our model used for the task follows the standard pretraining-finetuning paradigm. A base transformer-based language model pretrained on large quantities of unlabeled data is used as an encoder to encode the topic and the claim. The contextualized embedding of the [CLS] token is used for classification, which goes through a linear layer to generate the logits for the three labels.

In order to utilize the large amount of noisy data to help our model get better results, a novel training process where datasets with different noise levels are used in different stages to finetune the model, which is shown in Fig. 1. There are three stages in the whole finetuning process. The first stage is to use D_1 and D_x for distant finetuning, and the second stage is to use the low-noise refined dataset and the back-translated gold dataset for noisy finetuning, and the final stage is to use the gold data with a small portion of noisy data for final finetuning.

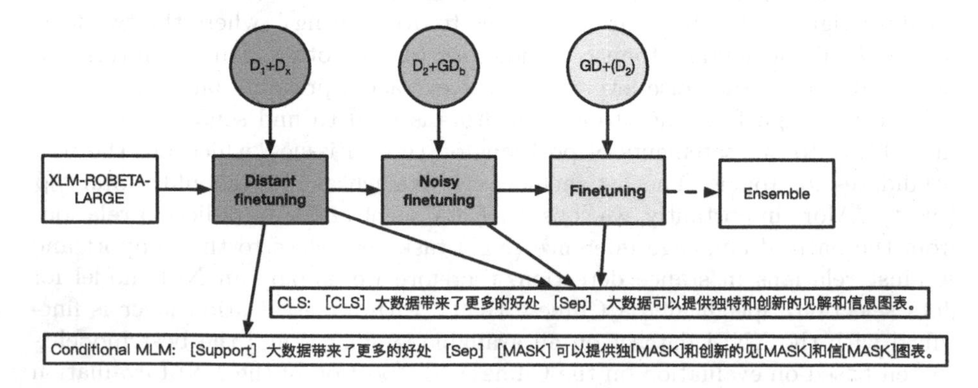

Fig. 1. The training process with noisy datasets.

4.1 Distant Finetuning

In this finetuning stage, datasets with high noise level D_1 or with data points in other languages D_x are used as training data. There are two training objectives used in this stage: conditional masked language modeling and classification. For each batch of training data points, one training objective is randomly chosen. For the conditional masked language modeling objective, the topic sentence and the claim sentence are first concatenated and tokenized by a tokenizer from a pretrained language model, and then the [CLS] token at the beginning of the tokenized sequence is replaced by a special token indicating the label of the pair. Part of either the topic or the claim, chosen randomly, will be masked with a special [MASK] token and predicted by the language model. For the classification objective, the concatenated sequence without any modification is encoded by the language model, and the [CLS] token is used for classifying the pair. The classification objective is identical to the one used in a common clean finetuning setup for a classification task. In a noise-free scenario, using the classification objective may be enough for finetuning the language model. However, the conditional masked language modeling objective is able to allow the model to learn how a topic and a claim interacts conditioned on a noisy relation without forgetting how to do language modeling. Preliminary experiments show that this objective is very important in ensuring model performance. Shown in Sect. 3.1, the D_1 dataset is imbalanced with a large number of data points labeled as Neutral or Support. Random sampling with small weights on Support and Neutral is performed on this dataset such that there are 0.7 million data points for all classes, ensuring balanced training of all labels.

4.2 Noisy and Clean Finetuning

After distant finetuning, the encoder from the pretrained language model is ready for a finetuning stage where training data is less noisy and more similar to data used in the downstream task. At this stage, the refined noisy dataset described in Sect. 3.2, combined with the original gold dataset and a dataset with gold data points back-translated from English, is used for training. Only the classification objective is used in this stage, resembling the common finetuning process. After two epochs, the encoder is ready for clean finetuning with the gold training set. In order to increase robustness of the model and regularize learning, a small portion of D_2 equal to 8% of the gold training set is added into the gold training set for the final clean finetuning.

4.3 Ensembling

Due to the small size of gold training data, different random seeds yield models with varying performances. Randomness caused by the noisy data sampling process also causes models to be trained with different training sets thus having different performances. We propose to ensemble best-performing models trained with different configurations together, which leads to a final composite model with high robustness. The final prediction probabilities are calculated as the product of the prediction probabilities from all the models:

$$p_{\text{final}}(\mathbf{y}|\mathbf{x}_i) = \prod_j p_j(\mathbf{y}|\mathbf{x}_i) \tag{1}$$

where i is the index of an input $\mathbf{x}$, and $\mathbf{y}$ is the output probabilities and j is the index of a model in the ensemble.

5 Experiments

The datasets provided in the shared task include a training set with 6,416 data points and a development set with 990 data points, which are used for model development and hyperparameter tuning. For hyperparameters, we use the XLM RoBERTa large model [5] as the base pretrained language model encoder in our classifier, which has 24 hidden layers, 16 attention heads with 4096 as intermediate embedding size and 1024 as the size of the final hidden embeddings. Dropout for all layers is set to be 0.1.

The classifier is first trained with the distant finetuning setup with D_1 and D_x datasets for 58,500 steps with a batch size of 8 per step. A gradient update is performed every 4 steps, making the effective batch size to be 32. The learning rate for this stage is set to be 8×10^{-6}. The mix ratio between D_1 and D_x is 4:1, meaning that 80% of the time, a batch is sampled from D_1. After a batch is sampled from a dataset, a training objective is chosen randomly between classification and language modeling. For the noisy and clean finetuning, the number of epochs is chosen to be 2. The learning rate is 6×10^{-6} and the batch size is 32.

AdamW [13] is used as the optimizer at all stages. The top classification layer is re-initialized between stages. Performances of different experiment setups are reported in accuracy on the development set, because the test set is not released.

5.1 Encoders

We first examine performances of different pretrained languages models as the base encoder in clean finetuning. The goal of this experiment is to measure model performances when finetuned with gold data only. Table 3 shows the results of finetuning with a number of popular Chinese pretrained models [6] as well as the XLM-RoBERTa model. Interestingly, the only model that is not trained entirely on Chinese data, XLM-RoBERTa large, is the best performing model of all. This indicates that multilingual training is helpful even when the downstream task is in a specific language only. The Electra model, which has been reported to reach state-of-the-art performances on many language understanding tasks, is not able to outperform both RoBERTa large and XLM-RoBERTa. Finally, there is a substantial performance gap between smaller BERT base models and larger XLM-RoBERTa models, showcasing the importance of training data size for pretraining as well as objectives used in pretraining.

Table 3. Performance of different encoders with finetuning on the development set

Encoder type	Development accuracy
Chinese BERT wwm base	76.22
Chinese BERT wwm ext	78.78
Chinese Electra 180g large	80.80
Chinese RoBERTa wwm ext large	82.61
XLM-RoBERTa large	**85.24**

5.2 Distance Finetuning

With XLM-RoBERTa large chosen as the encoder of the model, we explore the number of steps needed for the best performance with distant finetuning. Table 4 shows the model performance on the development set with only distant finetuning with no gold training set used at all. Model performance increases steadily as the number of steps increases. A pretrained encoder with a randomly initialized classification layer gets 32.72 accuracy, but when distant finetuning is used, the model is able to reach 70.49, which is close to how Chinese BERT base performs with finetuning. This shows that the training signal in the dataset used in distant finetuning is very strong, and the model is able to learn robustly to detect stances of sentences, despite the fact that it has not seen any gold training data and there exists a style difference between the noisy dataset from the internet and human-authored gold training data. Finally, the model trained with 58500 steps is used for clean finetuning because of time constraints in the shared task, but it is likely that further improvement may be acquired with even more training steps.

Table 4. Performance of the model in distant finetuning on the development set with XLM-RoBERTa large

Number of distant finetuning steps	Development accuracy
0	32.72
16500	68.28
38500	70.20
58500	**70.49**

5.3 Stages of Finetuning

Good performance from distant finetuning can be further improved by finetuning the model with gold training data. As described in Sect. 4.2, two finetuning stages follow the distant finetuning, which both involve gold training data. Table 5 shows how different combinations of finetuning stages affect model performance.

The 3-stage finetuning is most effective in improving model performance and robustness, as it further increases model accuracy by 1.52 points compared to directly using clean finetuning after distant finetuning. Although a large amount of automatically generated data is used in noisy finetuning, model performance is only slightly lower than clean finetuning, showing both the high quality of the noisy data and high robustness of the model.

Table 5. Model performance with different combinations of finetuning stages.

Stage	Development accuracy
Distant finetuning	70.49
Distant + Noisy finetuning	89.80
Distant + Clean finetuning	90.20
Distant + Noisy + Clean finetuning	**91.72**

5.4 Added Noisy Samples in Finetuning

We also look at if adding noisy samples into the clean training set in clean finetuning is able to help the model improve its performance, most likely by regularizing model training. Different numbers of noisy training data points from D_2 are randomly sampled and added to the gold training set, as shown in Table 6. Model performance averaged across 50 seeds is reported here. Using no noisy data in final clean finetuning yields lowest performance in general, and adding a small amount of noisy data does help model performance. Comparing to the whole gold training set with more than 6000 training instances, adding 500 noisy data points does not introduce too much noise but the regularizing effect from the noisy data points helps the model to be more robust to test items not found in training.

Table 6. Model performance with different number of noisy data points added into the gold training set in the final clean finetuning. Performance numbers are average accuracy over 50 random seeds.

Number of samples from D_2	Development accuracy
0	89.25
250	89.49
500	**89.51**
1000	89.35

6 Conclusion

A new method to extract data with silver labels from raw text to finetune a system for stance classification has been proposed in this paper. The reliance on specific discourse relations in the data extraction has ensured that the extracted silver topic and claim pairs are of high quality and the relations between the extracted pairs are relevant to the stance classification task. In order to use such silver data, a 3-stage training scheme where the noisy level in the training data decreases over different stages going from most noisy to least noisy is also proposed in the paper. We show through detailed experiments that the automatically annotated dataset as well as the 3-stage training help improve model performance in stance classification. Our approach ranks 1st among 26 competing teams in the stance classification track of the NLPCC 2021 shared task Argumentative Text Understanding for AI Debater, which confirms the effectiveness of our approach.

References

1. Allaway, E., McKeown, K.: Zero-shot stance detection: a dataset and model using generalized topic representations. In: Proceedings of the 2020 Conference on Empirical Methods in Natural Language Processing (EMNLP), pp. 8913–8931. Association for Computational Linguistics (November 2020)
2. Anand, P., et al.: Cats rule and dogs drool!: classifying stance in online debate. In: Proceedings of the 2nd Workshop on Computational Approaches to Subjectivity and Sentiment Analysis (WASSA 2.011), pp. 1–9. Association for Computational Linguistics, Portland, Oregon (June 2011)
3. Bar-Haim, R., Bhattacharya, I., Dinuzzo, F., Saha, A., Slonim, N.: Stance classification of context-dependent claims. In: Proceedings of the 15th Conference of the European Chapter of the Association for Computational Linguistics: vol. 1, Long Paper, pp. 251–261. Association for Computational Linguistics, Valencia, Spain (April 2017)
4. Conforti, C., Berndt, J., Pilehvar, M.T., Giannitsarou, C., Toxvaerd, F., Collier, N.: Will-They-won't-they: a very large dataset for stance detection on Twitter. In: Proceedings of the 58th Annual Meeting of the Association for Computational Linguistics, pp. 1715–1724. Association for Computational Linguistics (July 2020)
5. Conneau, A., et al.: Unsupervised cross-lingual representation learning at scale. In: Proceedings of the 58th Annual Meeting of the Association for Computational Linguistics, pp. 8440–8451. Association for Computational Linguistics (July 2020)
6. Cui, Y., Che, W., Liu, T., Qin, B., Wang, S., Hu, G.: Revisiting Pre-Trained models for Chinese natural language processing. In: Findings of the Association for Computational Linguistics: EMNLP 2020, pp. 657–668. Association for Computational Linguistics (November 2020)
7. Dey, K., Shrivastava, R., Kaushik, S.: Topical stance detection for twitter: a two-phase LSTM model using attention. In: Pasi, G., Piwowarski, B., Azzopardi, L., Hanbury, A. (eds.) ECIR 2018. LNCS, vol. 10772, pp. 529–536. Springer, Cham (2018). https://doi.org/10.1007/978-3-319-76941-7_40

8. Ghosh, S., Singhania, P., Singh, S., Rudra, K., Ghosh, S.: Stance detection in web and social media: a comparative study. In: Crestani, F., et al. (eds.) CLEF 2019. LNCS, vol. 11696, pp. 75–87. Springer, Cham (2019). https://doi.org/10.1007/978-3-030-28577-7_4
9. Ghosh, S., Anand, K., Rajanala, S., Reddy, A.B., Singh, M.: Unsupervised stance classification in online debates. In: Proceedings of the ACM India Joint International Conference on Data Science and Management of Data, CoDS-COMAD 2018, pp. 30–36. Association for Computing Machinery, New York (January 2018)
10. Hasan, K.S., Ng, V.: Extra-Linguistic constraints on stance recognition in ideological debates. In: Proceedings of the 51st Annual Meeting of the Association for Computational Linguistics (vol. 2: Short Papers), pp. 816–821. Association for Computational Linguistics, Sofia, Bulgaria (August 2013)
11. Kobbe, J., Hulpuş, I., Stuckenschmidt, H.: Unsupervised stance detection for arguments from consequences. In: Proceedings of the 2020 Conference on Empirical Methods in Natural Language Processing (EMNLP), pp. 50–60. Association for Computational Linguistics (November 2020)
12. Küçük, D., Can, F.: Stance detection: a survey. ACM Comput. Surv. **53**(1), 1–37 (2020)
13. Loshchilov, I., Hutter, F.: Decoupled weight decay regularization (November 2017)
14. Murakami, A., Raymond, R.: Support or oppose? classifying positions in online debates from reply activities and opinion expressions. In: Coling 2010: Posters, pp. 869–875. Coling 2010 Organizing Committee, Beijing, China (August 2010)
15. Popat, K., Mukherjee, S., Yates, A., Weikum, G.: STANCY: stance classification based on consistency cues. In: Proceedings of the 2019 Conference on Empirical Methods in Natural Language Processing and the 9th International Joint Conference on Natural Language Processing (EMNLP-IJCNLP), pp. 6413–6418. Association for Computational Linguistics, Hong Kong, China (November 2019)
16. Sirrianni, J., Liu, X., Adams, D.: Agreement prediction of arguments in cyber argumentation for detecting stance polarity and intensity. In: Proceedings of the 58th Annual Meeting of the Association for Computational Linguistics, pp. 5746–5758. Association for Computational Linguistics (July 2020)
17. Slonim, N., et al.: An autonomous debating system. Nature **591**(7850), 379–384 (2021)
18. Sobhani, P., Mohammad, S., Kiritchenko, S.: Detecting stance in tweets and analyzing its interaction with sentiment. In: Proceedings of the Fifth Joint Conference on Lexical and Computational Semantics, pp. 159–169. Association for Computational Linguistics, Berlin, Germany (August 2016)
19. Somasundaran, S., Wiebe, J.: Recognizing stances in online debates. In: Proceedings of the Joint Conference of the 47th Annual Meeting of the ACL and the 4th International Joint Conference on Natural Language Processing of the AFNLP, pp. 226–234. Association for Computational Linguistics, Suntec, Singapore (August 2009)
20. Sun, Q., Wang, Z., Zhu, Q., Zhou, G.: Stance detection with hierarchical attention network. In: Proceedings of the 27th International Conference on Computational Linguistics, pp. 2399–2409. Association for Computational Linguistics, Santa Fe, New Mexico, USA (August 2018)
21. Thomas, M., Pang, B., Lee, L.: Get out the vote: Determining support or opposition from congressional floor-debate transcripts. In: Proceedings of the 2006 Conference on Empirical Methods in Natural Language Processing, pp. 327–335. Association for Computational Linguistics, Sydney, Australia (July 2006)

22. Toledo-Ronen, O., Orbach, M., Bilu, Y., Spector, A., Slonim, N.: Multilingual argument mining: Datasets and analysis. In: Findings of the Association for Computational Linguistics: EMNLP 2020, pp. 303–317. Association for Computational Linguistics (November 2020)
23. Xu, L., et al.: CLUE: a Chinese language understanding evaluation benchmark. In: Proceedings of the 28th International Conference on Computational Linguistics, pp. 4762–4772. International Committee on Computational Linguistics, Barcelona, Spain (December 2020). https://doi.org/10.18653/v1/2020.coling-main.419, https://aclanthology.org/2020.coling-main.419
24. Yu, J., Jiang, J., Khoo, L.M.S., Chieu, H.L., Xia, R.: Coupled hierarchical transformer for stance-aware rumor verification in social media conversations. In: Proceedings of the 2020 Conference on Empirical Methods in Natural Language Processing (EMNLP), pp. 1392–1401. Association for Computational Linguistics (November 2020)

The Solution of Xiaomi AI Lab to the 2021 Language and Intelligence Challenge: Multi-format Information Extraction Task

Wen Dai, Xinyu Hua(✉), Rongrong Lv, Ruipeng Bo, and Shuai Chen

Xiaomi Corporation, AI Lab, Beijing, China
{daiwen,huaxinyu,lvrongrong,boruipeng,chenshuai3}@xiaomi.com

Abstract. Information Extraction is a challenging task in Natural Language Processing. There are several formats of information extraction that researchers mainly focused on. Typically, relation extraction and event extraction are two type of information extraction tasks that could facilitate many related business. In this work, we will introduce our solutions for the 2021 Language and Intelligence Challenge. We propose our methods for the relation extraction and event extraction tasks, respectively. The event extraction task is further divided into sentence level and document level. In relation extraction task, we propose a knowledge-based relation extraction method; in event extraction task, we propose a hybrid method which utilize event types and event triggers separately to extract event roles and event arguments. Finally, our solution ranked the second in the private leaderboard.

Keywords: Relation extraction · Event extraction

1 Introduction

Information Extraction is a challenging task in Natural Language Processing, and also plays a crucial role in many NLP applications, like knowledge base construction, question answering, document analysis, etc. There are several formats of information extraction that researchers focused on. Typically, relation extraction and event extraction are two type of information extraction tasks that could facilitate many related business.

Relation extraction aims to extract entities and their relations from unstructured text. It is a fundamental problem in information extraction. For instance, in the triplet (Albert Einstein, ancestralHome, Germany), "Albert Einstein" is subject and "Germany is object", "ancestralHome" is the relation between these two entities. Event extraction can help to automatically extract valuable information from large amount of corpus. This would accelerate the efficiency of document analysis and decision making, especially in the financial field. Thus, it is quite valuable to improve the performance of information extraction. In this paper, we organize the solutions as two main sections. The first section is the solution to relation extraction task, and the second section is the solution to event extraction, including sentence level and document level.

L. Wang et al. (Eds.): NLPCC 2021, LNAI 13029, pp. 496–508, 2021.
https://doi.org/10.1007/978-3-030-88483-3_40

In relation extraction section, there are two existing methods: the pipeline model and the joint model, and both of them have shortcomings. The pipeline model regard relation extraction as two separate tasks, and propagate classification errors will through the tasks. Joint model considers the subtask interaction, but some need artificial features, which will rely on many NLP tools. In this work, we propose a Knowledge-based relation extraction method (KREM) to jointly extract entities and their relationships. Specifically, we adopt the pre-trained RoBERTa-large-wwm as backbone network and use the knowledge base established from remote supervision to obtain the corresponding entity category information, which is integrated into the network as additional information. Compared with existing works, our method has two advantages: Firstly, it can take full advantage of entity's category information, which can improve the accuracy of the model's recognition of entities. Secondly, modeling entity's category and entity simultaneously can improve computation efficiency.

In event extraction section, we propose an event type and event trigger hybrid model. In the task of event extraction, we usually have to retrieve trigger words firstly before extracting event roles and event arguments. In fact, trigger words are indications of event types. From another perspective, we can judge the type of event in the corpus firstly, and then extract relevant event roles and event arguments. As the number of pre-defined event types is specific, the judgement of event types can be formulated as a multi-label classification problem. These two pipeline can be viewed as two perspective of the same problem, both of which could extract valuable information and complement each other. Thus, we propose a novel hybrid method which utilize event types and event triggers separately to extract event roles and event arguments. The proposed method is composed of three main procedures. In the first step, we use a joint model to predict the event type and event trigger, as there is interrelationship between event type and event trigger. In the second step, we use two models to extract the event arguments. In detail, we use predicted event type to extract event roles and event arguments. Meanwhile, we use predicted event trigger to extract event roles and event arguments. It should be noted that these two operations are two parallel event extraction model and have no order. In the third step, we combine the results of these two event extraction model into one by voting.

2 Related Work

In relation extraction, recent works employ neural networks and achieve promising results in previous datasets. Compared with the pipline method and the joint method, Giannis et al. model the relation extraction task as a multi-head selection problem, potentially identifying multiple relationships for each entity [1]. Markus et al. extract all possible entities, since each selected segment is independent, the span-level feature can be extracted directly to solve the overlapping entity problem [2]. For a text with T tokens, there are theoretically N = T(T + 1)/2 kinds of fragment arrangements. However, they did not use the entity category feature. According to [3] and [4], the category of entity is very important to the relational model.

In event extraction, there are three main strategies. The first strategy is pattern matching. Template matching is performed to extract event arguments from raw text [5]. Pattern construction is a time-consuming job in this strategy, and some automatic tools are utilized to improve the efficiency. The second strategy is using traditional machine learning algorithms for event extraction. In this strategy, we train machine learning models for trigger detection and argument detection [6, 7]. Feature engineering plays an important role in this strategy. The third strategy is deep learning based event extraction. This strategy can get rid of exhausting feature engineering which requires linguistic knowledge and domain expertise. In this strategy, raw text are usually taken as the input for event extraction. Convolutional neural network, recurrent neural network and graph neural network are used to extract more abstract representation for input tokens. Recently, large-scale pre-trained language models are more and more frequently used in event extraction tasks to extract representations [8]. In this strategy, a named entity recognition model is firstly used to extract event triggers. Then, another named entity recognition model is used to extract the event arguments corresponding to the predefined event roles [9]. Event trigger detection and event argument detection can also be undertaken jointly in the task of event extraction [10].

3 Methods

3.1 Task of Relation Extraction

In this section, we first introduce our proposed relation extraction model. Then we detail the model structure.

3.1.1 Knowledge-Based Relation Extraction Method

Recently, Bidirectional Encoder Representations from Transformers (BERT) achieves great success by pre-training a language representation model on large scale corpora then fine-tuning on downstream tasks.

Firstly, we use the training data to build a knowledge base which can directly get SPOs from the sentence by using Aho-Corasick automation. Then we use the established knowledge base to get the corresponding spo in the sentence. Using the form of {[cls] + subject + subject type + object + object type + [sep]} as sentence b. If it is not retrieved, then sentence b is empty. Sentence A is the original input text. For instance, if input text is "Jackie Chan played in Rush Hour and Crime Squad", then sentence b is {[cls] + Jackie Chan + Person + Rush Hour + inWork + [sep]}.

For the case of multiple O, they must be decoded together, but to prevent over-fitting and only use, the knowledge of one O builds sentence B.

The encoder module extracts feature information from sentence A and sentence B, which will be fed into subsequent tagging modules. Similar to the paper proposed by [11], we use the double pointer extraction method to extract entities. Specifically, we firstly extract the subject and then extract the corresponding relationship and object. As illustrated in Fig. 1, after using BERT to encode the sentence information, it adopts two binary classifiers to detect the start and end position of subjects by assigning each token a binary tag (0/1) that indicates whether the current token corresponds to a start or end position of a subject.

In this competition, there are two types, one is a simple O value, and the other is complex O. For simple O, we can directly decode it; For complex O, there are multiple combinations between it and Sp. Here we introduce the distance variance algorithm to select the nearest SP and the corresponding O.

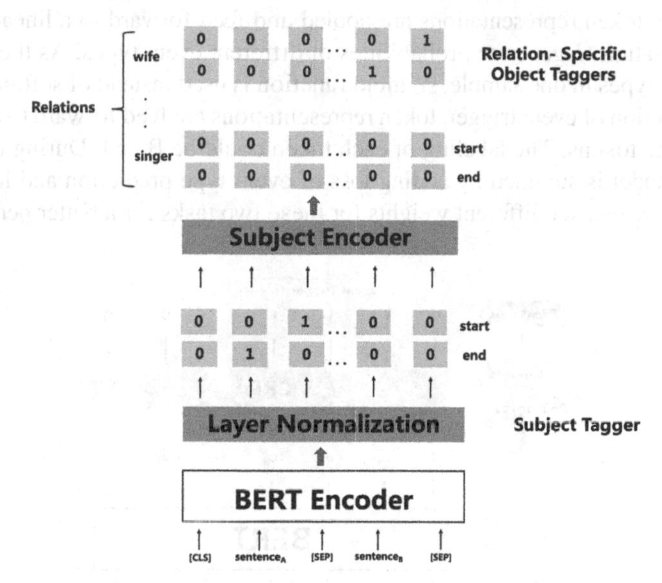

Fig. 1. The framework of Knowledge-based relation extraction method

3.1.2 Voting Strategy

In order to improve the robustness of the model, we divide the training data into k parts and train k models separately. Then we propose ensemble processes: k models vote for each SPO in the sequence and choose a spo greater than the voting threshold T;

3.2 Task of Event Extraction

In this work, we propose a hybrid method which utilize event types and event triggers separately to extract event roles and event arguments. In most existing approaches, event triggers are extracted as the first step. In fact, event type is also a valuable indicator for event argument extraction. Thus, we propose a novel method which utilize event type information for the task of event argument extraction. As the prediction of event type and event trigger influence each other in a specific instance, we use a joint model to predict the event type and event trigger simultaneously.

3.2.1 Joint Model of Event Type and Event Trigger

The prediction of event type and event trigger influence each other. For example, in the event type of "life-be engaged", the trigger words is impossible to be "extramarital affairs"; on the contrary, when the trigger words are "date of wedding", the event type can't be "life- be derailed". Thus, we construct a joint model of event type and event trigger. As illustrated in Fig. 2, the event type prediction model and event trigger prediction model shares the same representation layer, which is a pre-trained language model like BERT, etc. Then different output layers are utilized for different tasks. For the prediction of event type, token representations are pooled and feed forward to a linear layer with sigmoid activation to generate probabilities of different event types. As there might be several event types in one sample, sigmoid function is used instead of softmax function. For the prediction of event trigger, token representations are feed forward to a CRF layer to label all the tokens. The labeling of each token could be B or I. During training, the loss of the model is summed by adding loss of event type prediction and loss of event trigger. And we can set different weights for these two tasks for a better performance.

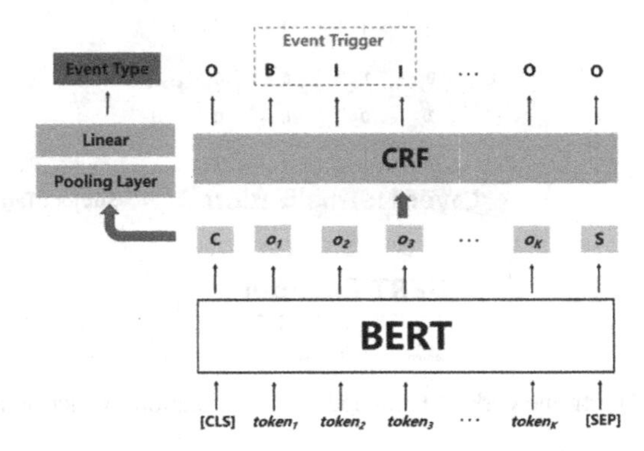

Fig. 2. The framework of joint model of event type and event trigger

3.2.2 Parallel Sequence Labeling Models

Traditionally, role extraction and argument extraction are two procedures as a pipeline. Firstly, we have to detect the possible argument spans. Then, we assign a role to each detected argument. Here, we integrated these two steps into one. We regard the all the event roles in different event types as different entities, and each entity has two tags, namely 'B' and 'I'. Each tag consists of three parts, namely event type, event role and B/I. For example, for the tag "life-be engaged_time_B(人生-订婚_时间_B)", it means that event type is life-be engaged(人生-订婚), event role is time(时间), and position is beginning. Thus, we only have to assign a tag to each token, and this is a typical sequence labeling task. In the task of sentence level event extraction, there are 217 kind of event roles in total corresponding to 65 event types. For the 217 event roles, each event role has 'B' and 'I' tags. Apart from that, there is one more 'O' tag for tokens which doesn't belong to any of the 217 event roles. Thus, we have 435 different tags in total, and event extraction task is converted to assign a tag from the 435 tags to each token. Similarly, in the task of document level event extraction, we have 185 different tags in total.

In this part, we propose two sequence labeling models for event roles and event arguments extraction. These two models are based on the predicted event type and event trigger in Sect. 3.1, respectively. With the assistance of event type and event trigger information, we believe that the performance of extraction would be much better.

The first sequence labeling model is based on event trigger. As illustrated in Fig. 3, we use BERT + CRF to encode the context and extract all possible entities. To utilize event trigger information from Sect. 3.1, we concatenate trigger distance features with token representations. The trigger distance feature is measured by the shortest distance between a token and trigger words. The distance values are converted to trainable embedding vectors. When we have obtained the tags of each token, we have to decode the labeling result. As previously mentioned, each tag consists of three parts. We can obtain the event type from the first part, and event role from the second part. Besides, the whole token span is the extracted event argument corresponding to the event role.

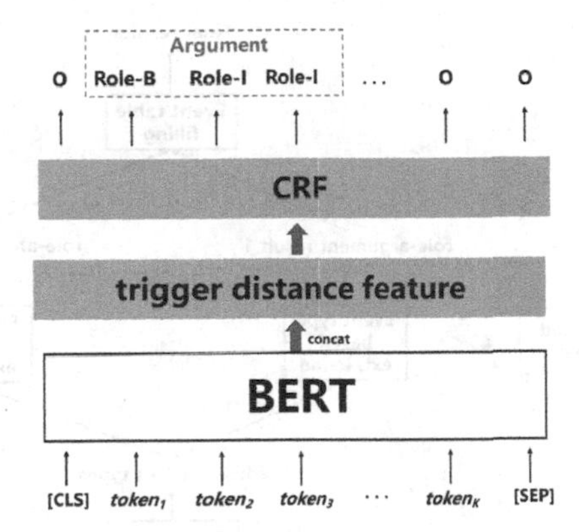

Fig. 3. Event trigger based role and argument extraction

The second sequence labeling model is based on event type. As illustrated in Fig. 4, we also use BERT + CRF to encode the context and extract all possible entities. To utilize event type information from Sect. 3.1, we regard the event type tokens as the first sentence, and the original input as the second sentence. In this way, we reorganize the input as a sentence pair. In decoding, we use the predicted event type to filter out event arguments of the other event types. If there are several event types for one sample, we input these event types one by one and accordingly separate this sample into several.

3.2.3 Voting Strategy

We use 10-fold cross validation to train 10 models for event trigger based extraction and event type based extraction, respectively. In total, we got 20 extraction results for each sample, and we use voting strategy to combine these results. For a (event type, event role, event argument) triplet, if it occurred in 10 or more than 10 results, we keep it, otherwise, we abandon it.

3.2.4 Document Level

In document level event extraction, our method is based on the sentence level. Nevertheless, there are several challenges in document-level event extraction. Firstly, there are no

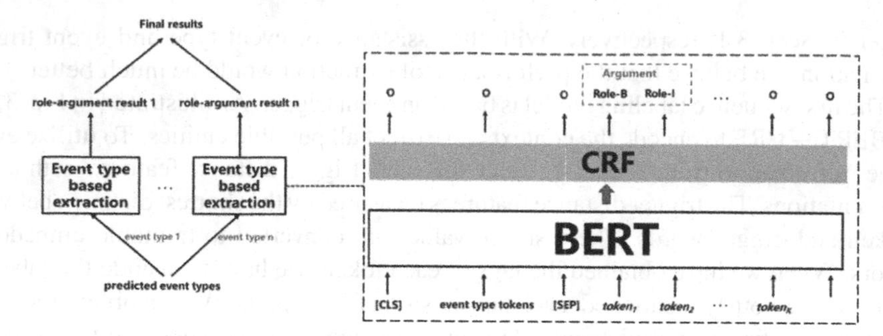

Fig. 4. Event type based role and argument extraction

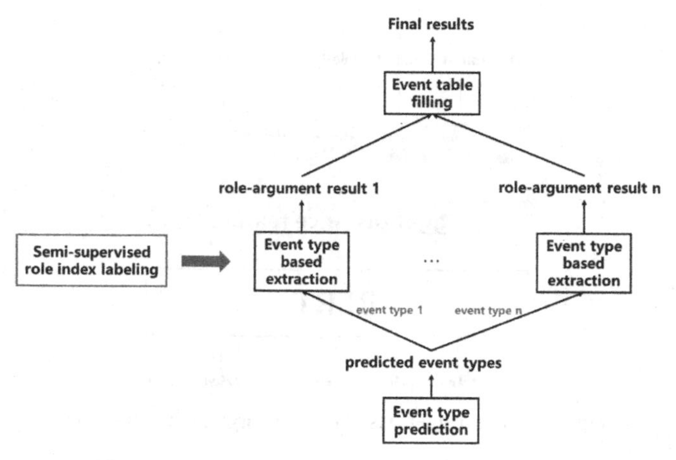

Fig. 5. Pipeline of document-level event extraction

index information for the labeled roles, and there are no trigger information. Secondly, document-level samples is much longer than those of sentence-level, which means that the contained events information might be more scattered. Thirdly, there is an enum role in document level task, whose argument might not exist in the document text. Fourthly, we need to assemble the extracted roles and arguments in the form of event, whereas in the sentence-level task we can present the result in the form of (event type, event role, event argument) triplet.

As to the first challenge, we need to find a way to label the role index information before training the event extraction model. But we don't label the trigger index information, for the reason that the trigger words usually occur many times in a document, and most of these occurrences could act as the correct trigger. It is confusing to choose a single trigger for index labeling. Thus, we only consider event type information in document level event extraction task.

As to the second challenge, we try several different strategies, for example, splitting the documents into blocks of 512 tokens, or splitting the documents into paragraphs. Then we take each block or each paragraph as the input for event extraction, and combine the results. But the performance is no better than just using the 512 tokens in the beginning of the document. There might be two reasons for the phenomenon: the authors usually convey their message in the beginning parts when writing an information; the labeling workers might mainly focus on the beginning parts and slightly ignore the following parts. Thus, we take the beginning 512 tokens as the input for out event extraction model.

As to the third challenge, we can train a classification model to judge the value of the enum role. For example, the enum role of "phase(环节)" have four values, thus, we train a classification model of four labels to predict the argument.

As to the fourth challenge, we have to assemble the (event type, event role, event argument) triplets to organize them in the correct form.

Finally, the pipeline of document-level event extraction is illustrated as in Fig. 5. And the major adjustments to the method of sentence level are detailed as below:

Adjustment 1: semi-supervised role index labeling

In the task of document level event extraction, there are no start index information for argument. When there are several occurrence for an argument, using first occurrence might influence the model training, and further affect the performance in the period of inference. Thus, here we use semi-supervised learning in the phase of model training. The pseudo code is shown as in Fig. 6. The symbols with overline means event arguments which have start index information, while those without it means arguments which doesn't have start index information. In the first phase (line 2–5), we extract event arguments which occurs only once in the text of the corresponding sample. There is no uncertainty on the start index of these arguments. We construct the training samples with these arguments. Then in the second phase (line 2–5), we train a label sequencing model based on these training samples with start index information, and use the trained model to extract event arguments and obtain corresponding start indexes for the original samples. If an extracted event argument is also manually labeled in the original sample, we add it to the training sample, as illustrated in line 11 in Fig. 6. Then we repeat the second phase for three times and obtain the final training samples, in which all the event arguments have start index information. We carried on an empirical study and found that when we repeat the second phase for three times, the number of event arguments is tending towards stability. Compared with using first occurrence as start index, semi-supervised learning could improve the F1 score by about 0.5%.

Adjustment 2: event table filling

In sentence level event extraction, the result unit is (event type, event role, event argument) triplet. Nevertheless, the result unit is event table in document level, and we have to decide which triplets belong to the same event table. The schema of document level event table has been defined by the competition sponsor. In fact, there are three types of event roles in each event type. The first are indispensable roles, which definitely will occur in one event table, and different event tables can't have reduplicated event arguments in same event role. The second are roles that must be reduplicated in different event tables. The third are the remaining roles. To decide the property of all the event roles, we firstly retrieve all the samples that have more than one event table. Then, for each event role in each event type, we count the number of different event arguments. For example, if a role occurs in three event tables, and the number of different arguments is also three, then this role is the first kind; if a role occurs in three event tables, and the number of different arguments is one, then this role is the second kind. Take the event type of "pledge(质押)" as an example, the first kind of roles are "Number of pledged shares (质押股票/股份数量)", the second kind of roles are "Disclosure time(披露时间)", "Company that the pledge belonging to(质押物所属公司)", the third kind of roles are the other remaining roles.

During event table filling, we decide the number of event tables by the number of different event arguments of the first kind of event roles. As to the second event roles, we duplicate their event arguments in each event table. Finally, we have to choose an event table to fill in for the third kind of event roles. As to each event argument in the third kind, we calculate the distance between it and the indispensable roles of each event table, and choose the nearest one to be filled in with this event role and event argument.

Algorithm: Semi-Supervised Sequence Labeling Model Training

1: Initialization: n = 0 ; samples S ; event arguments A_i ,
 and $\overline{A_i} = \varnothing$ for $s_i \in S$

2: for $s_i \in S$

3: for $a_j^i \in A_i$ do

4: if occurrence(a_j^i , text(s_i)) == 1 then

5: $\overline{A_i} \leftarrow \overline{A_i} \cup \{\overline{a_j^i}\}$

6: **repeat**

7: $n \leftarrow n + 1$

8: M = training({ $\overline{A_i}$ })

9: R = inference(M , S)

10: for $\overline{r_i} \in R$

11: $\overline{A_i} \leftarrow \overline{A_i} \cup \{\overline{r_i} \cap a_j^i\}$

12: **until** $n == 3$

Output: samples $\overline{A_i}$

Fig. 6. Psuedo code of semi-supervised sequence labeling model training

4 Experiments and Results

4.1 Task of Relation Extraction

4.1.1 Dataset

DuIE2.0 [12] is the largest schema-based Chinese relation extraction dataset in the industry, which contains more than 430,000 triples over 210,000 real-world Chinese sentences, restricted by a pre-defined schema of 48 relation types. The sentences are extracted from Baidu Baike, Baidu Tieba, and Baidu News Feeds. This task will match output of a participating system with the manually annotated SPO triples, and score the system according to the F1 score.

4.1.2 Experimental Settings

We use pre-trained RoBERTa-large-wwm as backbone network, then fine-tune it on downstream relation extraction subtask with our proposed Knowledge-based relation extraction method. The RoBERTa-large-wwm has 24 layers of transformer blocks, and each block has 16 self-attention heads. The dimension of hidden size is 1024. We apply Adam optimizer to update model parameters. The initial learning rate is set to 1e-5, and the mini-batch size is 32. We use the official validation set for hyper-parameters tuning.

4.1.3 Compared Methods

To evaluate the performance of different methods, we compare our method with the following baselines.

PSO: The PSO model firstly classifies the relationship of the sentence, and then uses a double pointer network to extract the subject and object of the relationship.

SPO: The SPO model firstly identifies the subject, and then uses the double pointer network to extract the corresponding object.

PSO - KREM: The PSO - KREM model integrates the relationship category feature in the sentence on the basis of the spo.

SPO - KREM: The SPO - KREM model integrates the relationship category feature in the sentence on the basis of the spo.

SPO-KREM-E1: The SPO-KREM-E1 uses model ensemble to get the final spo.

PSO-KREM-E1: The PSO-KREM-E1 uses model ensemble to get the final spo.

4.1.4 Results

Table 1. Result

Model	F1
PSO	77.21
SPO	78.65
PSO – KREM	78.33
SPO – KREM	79.21
PSO - KREM - E1	80.18
SPO - KREM - E1	81.79

Compared with the baseline spo model, our proposed KREM has achieved significant results on the challenging DUIE data set.

The spo F1 scores have increased by 1.93%. Finally, the method of model fusion is used to achieve the best results, because after the data set is divided, the risk of overfitting is reduced, and it is also better to obtain spo with higher confidence.

4.2 Task of Event Extraction

4.2.1 Dataset

The sentence-level event extraction task uses the DuEE1.0 [13] dataset, which contains 65 event types. The training set has 13456 samples, the public test set has 15000 samples. The document-level event extraction task uses the DuEE-fin dataset, which contains 13 event types. The training set has 8221 samples, the public test set has 30000 samples, of which some are confounding samples.

4.2.2 Experimental Settings

We use RoBERTa-large-wwm as the backbone model. We conducted a 10-fold cross validation experiment on the training samples, and use each single trained single model for event extraction on the test set. Finally, we use voting strategy to select the (event type, event role, event argument) triplets whose frequency is above threshold. In the document level, after voting for (event type, event role, event argument) triplets, we undertake event table filling based on the selected triplets to obtain the final results.

4.2.3 Results

In sentence level event extraction, we firstly trained a joint model to predict the event type and event trigger at the same time. Then based on the predicted event trigger and event type, we trained two event extraction models, respectively. We test our performance on public test set. In comparison, we also trained a model without event trigger and event type information. As illustrated in Table 2, without event trigger and event type information, the event extraction model obtains a F1 score of 0.8105. When we exploit event trigger information, the event extraction model obtains 0.8396. And when we exploit event type information, the event extraction model obtains 0.8469. Finally, we ensemble these two models and obtain a F1 score of 0.8599.

Table 2. F1 score of sentence level event extraction models

Model	F1
Without event trigger and event type	0.8105
Event trigger based event extraction model	0.8396
Event type based event extraction model	0.8469
Model ensemble	0.8599

In document level event extraction, we firstly use semi-supervised learning to prepare event arguments with start index information. Then we conducted a 10-fold cross validation with event type based event extraction model, and use voting strategy to select the (event type, event role, event argument) triplets. Here, we try a naive event table filling method by simply merge all the triplets that have the same event type. As illustrated in Table 3, this results in a F1 score of 0.6397. Compared with that, after using our event table filling method describe above, the F1 score increases to 0.7227, which validates the effectiveness of our event table filling method.

Finally, our proposed method got 0.8503 and 0.7039 in the private test set.

Table 3. F1 score of document level event extraction models

Event table filling method	F1
Naive merge all the triplets that have the same event type	0.6397
Our event table filling method	0.7227

5 Conclusion

In the task of relation extraction, we propose a new Transformer-based Knowledge-based relation extraction method, which can make full use of the features of the entity category, so as to achieve a better extraction effect. Finally, we got the first ranking in the task of relation extraction.

In the task of event extraction, we introduce a novel hybrid method which utilize event types and event triggers separately to extract event roles and event arguments. The proposed method consists of three main steps: joint model of event type and event trigger, parallel sequence labeling models and voting strategy. In the document level event extraction, we make several adjustments to adapt to the situation in document level task. Finally, our method obtained an F1 score of 0.8503 in sentence level and 0.7039 in document level in the private test set. Our total score ranked second in the final private leaderboard.

References

1. Bekoulis, G., et al.: Joint entity recognition and relation extraction as a multi-head selection problem. Expert Syst. Appl. **114**, 34–45 (2018)
2. Eberts, M., Ulges, A.: Span-based joint entity and relation extraction with transformer pre-training. arXiv preprint arXiv:1909.07755 (2019)
3. Zhong, Z., Chen, D.: A frustratingly easy approach for joint entity and relation extraction. arXiv preprint arXiv:2010.12812 (2020)
4. Peng, H., et al.: Learning from context or names? An empirical study on neural relation extraction. arXiv preprint arXiv:2010.01923 (2020)
5. Riloff, E.: Automatically constructing a dictionary for information extraction tasks. In: AAAI. Citeseer (1993)
6. Chieu, H.L., Ng, H.T.: A maximum entropy approach to information extraction from semi-structured and free text. AAAI/IAAI **2002**, 786–791 (2002)
7. Ahn, D.: The stages of event extraction. In: Proceedings of the Workshop on Annotating and Reasoning about Time and Events (2006)
8. Yang, S., et al.: Exploring pre-trained language models for event extraction and generation. In: Proceedings of the 57th Annual Meeting of the Association for Computational Linguistics (2019)

9. Li, D., et al.: Biomedical event extraction based on knowledge-driven tree-LSTM. In: Proceedings of the 2019 Conference of the North American Chapter of the Association for Computational Linguistics: Human Language Technologies, vol. 1 (Long and Short Papers) (2019)
10. Sha, L., et al.: Jointly extracting event triggers and arguments by dependency-bridge RNN and tensor-based argument interaction. In: Proceedings of the AAAI Conference on Artificial Intelligence (2018)
11. Wei, Z., et al.: A novel cascade binary tagging framework for relational triple extraction. In: Proceedings of the 58th Annual Meeting of the Association for Computational Linguistics (2020)
12. Li, S., et al.: Duie: A large-scale Chinese dataset for information extraction. In: Tang, J., Kan, M.-Y., Zhao, D., Li, S., Zan, H. (eds.) NLPCC 2019. LNCS (LNAI), vol. 11839, pp. 791–800. Springer, Cham (2019). https://doi.org/10.1007/978-3-030-32236-6_72
13. Li, X., et al.: DuEE: A large-scale dataset for Chinese event extraction in real-world scenarios. In: Zhu, X., Zhang, M., Hong, Yu., He, R. (eds.) NLPCC 2020. LNCS (LNAI), vol. 12431, pp. 534–545. Springer, Cham (2020). https://doi.org/10.1007/978-3-030-60457-8_44

A Unified Platform for Information Extraction with Two-Stage Process

Chongshuai Zhao[1], Dongjie Guo[2], Xudong Dai[2], Chengmin Gu[3], Lingling Fa[4], and Peng Liu[1](✉)

[1] National Joint Engineering Laboratory of Internet Applied Technology of Mines, China University of Mining and Technology, Xuzhou, Jiangsu, China
{zhaochs,liupeng}@cumt.edu.cn

[2] University of Science and Technology of China, Hefei, Anhui, China

[3] Xi'an Jiaotong University, Xi'an, Shaanxi, China
guchengmin@stu.xjtu.edu.cn

[4] Nanjing University of Science and Technology, Nanjing, Jiangsu, China

Abstract. The multi-format Information Extraction (IE) task in Language and Intelligence Challenge 2021 (LIC2021) consists of three subtasks: Relation Extraction (RE), Sentence-level Event Extraction (SentEE) and Document-level Event Extraction (DocEE). Deep learning methods have made great progress in each subtask these years. However, most of them cannot solve these subtasks by a unified platform. In this paper, we develop a unified neural model with two-stage process, which adopt the Enhanced NER module in stage one to obtained the **ELEMENTs** and corresponding **LABELs**. In stage two, we designed the customized manoeuvres to solve challenges in different subtasks. The submission shows that our model achieves competitive performance, which ranks 3rd on the final leaderboard.

Keywords: Multi-format information extraction · Unified platform · Overlapping triples · Multi-value arguments

1 Introduction

Information extraction (IE) aims to extract structured knowledge such as entities, relations, and events from unstructured natural language texts. Most current researches focus on extraction of information in a single format, while lacking a unified evaluation platform for IE in different formats. The LIC2021 competition setup a multi-format IE task, which is designed to comprehensively evaluate IE from different dimensions. Multi-format IE consists of three subtasks, the definitions and challenges of these subtasks are as follows.

1.1 Relation Extraction (RE)

Definition. The RE task is to extract all (s, r, o) triples from a given sentence according to predefined schema, which defines the relation r and the types of its corresponding subject entity s and object entity o.

L. Wang et al. (Eds.): NLPCC 2021, LNAI 13029, pp. 509–518, 2021.
https://doi.org/10.1007/978-3-030-88483-3_41

Challenges. (1) **Overlapping Triples**: Zeng [1] is the first to consider the problem of overlapping triple, which is categorized into No Overlap (Normal), Single Entity Overlap (SEO) and Entity Pair Overlap (EPO). A plenty of overlapping triples have been observed in DuIE2.0 [2], which are the difficulties of this task. (2) **Entity Position Uncertainty**: Many triples in DuIE2.0 contain entities that are repeated multiple times in a sentence, but the exact position is not given. Since the semantics of the same entity in different positions may not be the same, this problem will induces errors into the model.

1.2 Sentence-Level Event Extraction (SentEE)

Definition. Given a sentence and predefined event types with corresponding argument roles, the aim of this task is to identify all events of target types mentioned in the sentence, and extract corresponding event arguments. The predefined event types and argument roles restrict the scope of extraction.

Table 1. Challenges in event extraction task.

Challenges	Data example	Event type: (Argument role, Argument)
Role Sharing	伦纳德压哨绝杀，猛龙淘汰76人闯进东部决赛	竞赛行为-胜负：（胜者，猛龙）
		竞赛行为-胜负：（晋级方，猛龙）
Nested Arguments	注意！苹果MacBook部分型号召回	召回：（召回方，苹果）
		召回：（召回内容，苹果MacBook部分型号）
MVA	北京市监约谈美团、饿了么	约谈：（公司名称，美团）
		约谈：（公司名称，饿了么）

Challenges. (1) **Role Sharing**: As shown in Table 1, there are two different event types in the sentence, but they share the same argument [猛龙]. The proportion of such samples in the training and validation sets reached 6.8% and 7.3% respectively. (2) **Nested Arguments**: In the second row of Table 1, the arguments [苹果] and [苹果MacBook部分型号] in different events are partially nested. The proportion of such examples in the training and validation sets are about 6.4% and 6.9% respectively. (3) **Multi-value Arguments (MVA)**: The last row in Table 1 shows that multiple arguments can belong to the same argument role. Such samples account for 5.2% and 4.8% of the training and validation sets respectively.

1.3 Document-Level Event Extraction (DocEE)

Definition. This task is also an event extraction task but considers input text fragments in document-level rather than sentence-level. And there are other challenges need to be dealt with.

Challenges. (1) **Longer Inputs**: Compared to sentence-level event extraction, document-level event extraction poses longer input sequences. Nearly half of the sequences in the dataset are longer than 512, which makes it difficult to use most pretrained models directly. (2) **More MVA**: Notably, 24% of the samples suffer from MVA problem. It requires the proposed model to identify potential arguments and the relevancy between them.

2 Model Description

The RE task is defined as entity extraction, while SentEE and DocEE are defined as arguments extraction. Although the definitions are different, all of them can be modeled as a named entity recognition (NER) task. Empirically, solving these three tasks only with NER model is not satisfying. To achieve better performance, we designed a unified platform with two-stage extraction. A unified Enhanced NER module is applied as the first stage, followed by a customized second stage module. The details of the two stages are as follows.

2.1 Enhanced NER Module

Figure 1 summarizes the proposed Enhanced NER module. It is designed to distinguish all elements and the corresponding labels by directly decoding the encoded vector $\mathbf{h}_N$, which is produced by the BERT [3] encoder. As shown in Fig. 1, the size of the vector produced by Enhanced NER module is $R \times L \times 2$. The detailed operations are as following:

$$p_{l_j,i}^{start_e} = \sigma(\mathbf{W}_{start}^{l_j}\mathbf{x}_i + \mathbf{b}_{start}^{l_j}) \tag{1}$$

$$p_{l_j,i}^{end_e} = \sigma(\mathbf{W}_{end}^{l_j}\mathbf{x}_i + \mathbf{b}_{end}^{l_j}) \tag{2}$$

where $p_{l_j,i}^{start_e}$ and $p_{l_j,i}^{end_e}$ represent the probability of identifying the i-th token as the start and end position of a element in specific label l_j, label id $j \in (1, 2, ..., R)$. The corresponding token will be assigned with tag 1 if the probability exceeds a certain threshold or with a tag 0 otherwise. $\mathbf{x}_i$ is the encoder representation of the i-th token in the input sequence, where $\mathbf{W}_{(.)}^{l}$ represents the learnable weight, $\mathbf{b}_{(.)}^{l_j}$ is the bias and σ is the sigmoid activation function.

In this paper, elements and labels have different references in different subtasks. In RE task, **ELEMENT** refers to [subject] and **LABEL** refer to corresponding [relation type] of subject. In SentEE and DocEE task, we can treat [trigger word/argument] as **ELEMENT** and [event type/argument role] as **LABEL**.

2.2 Customized Manoeuvres

Relation Extraction. The first stage provides all possible subjects (elements) in the input sentence and their corresponding relation types (labels), which are marked as (s, r) pairs. The second stage in RE aims to obtain the related objects according to the (s, r) pairs to construct multiple (s, r, o) triples. The details are shown in Fig. 2.

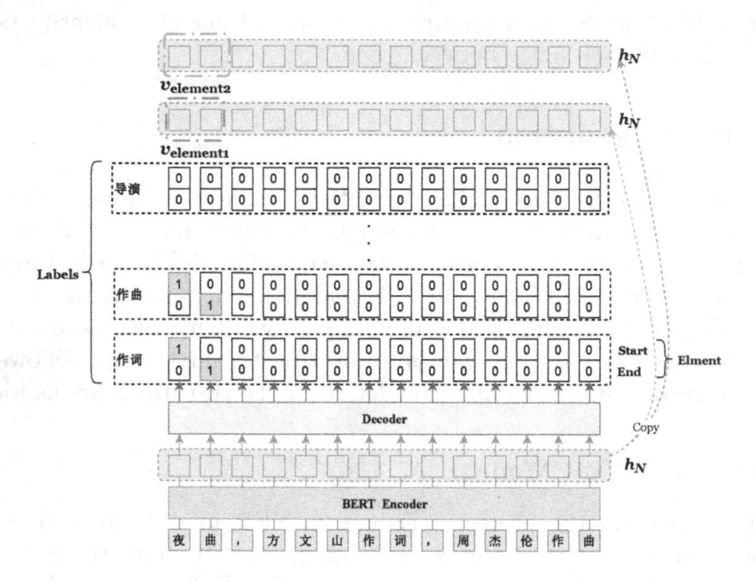

Fig. 1. A framework of the proposed Enhance NER module. Taking the RE task as an example, there is a candidate element(subject) detected at the low level. The presented 0/1 binary tags are specific to the elements in different labels(relation types).

We first create a simple lookup table that stores all relation embeddings. Notably, the parameters of these relation embeddings are randomly initialized and updated in the training progress. Secondly, we retrieve the corresponding relation embedding by a specific relation id, which can be expressed by the following formula:

$$\mathbf{l}_{r_j} = \mathbf{RET}(rel_j) \tag{3}$$

where rel_j is the relation id j, $\mathbf{l}_{r_j}$ is the vector of j-th in relation embeddings table (RET). The embedding module contains R vectors with size h, where R is the number of relation types and h is the dimension of encoded hidden states.

Through the above two steps, we obtain the start and end positions of all subjects in the input sentence and their token embeddings, denoted as $\mathbf{v}_{sub}^{start}$ and $\mathbf{v}_{sub}^{end}$ respectively. The representation of the subject $\mathbf{v}_{sub}$ is the sum of the start and end embeddings. As Fig. 2 shown, given $\mathbf{v}_{sub}$ ($\mathbf{v}_{element}$) and its corresponding relation embedding $\mathbf{l}_{r_j}$ (label embedding). we utilize binary classification to determine whether a token is the start/end of an object. The detailed operations of the model on each token are as follows.

$$p_i^{start_{obj}} = \sigma(\mathbf{W}_{start}^{r_j}(\mathbf{x}_i + \mathbf{v}_{sub} + \mathbf{l}_{r_j}) + \mathbf{b}_{start}^{r_j}) \tag{4}$$

$$p_i^{end_{obj}} = \sigma(\mathbf{W}_{end}^{r_j}(\mathbf{x}_i + \mathbf{v}_{sub} + \mathbf{l}_{r_j}) + \mathbf{b}_{end}^{r_j}) \quad (5)$$

where $p_i^{start_{obj}}$ and $p_i^{end_{obj}}$ represent the probability of the i-th token in the input sequence being the start and end of an object respectively. For each (s, r) pair, we iteratively apply the same process to it.

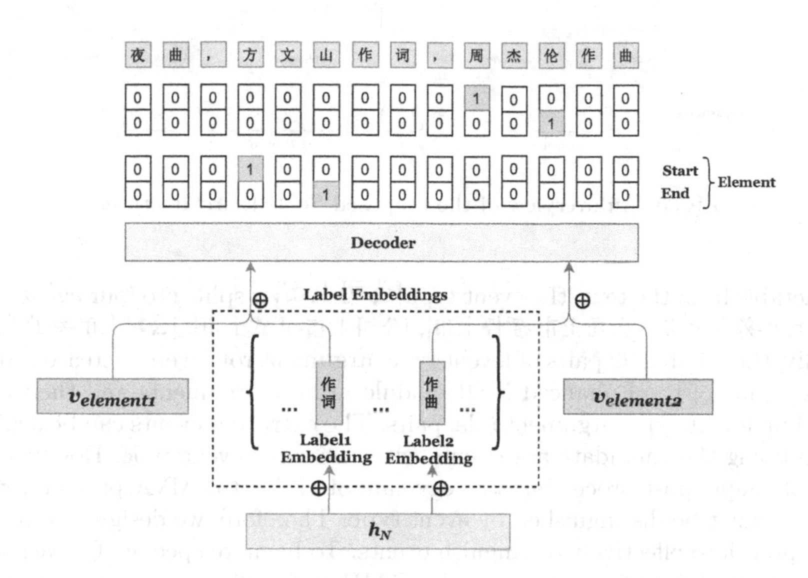

Fig. 2. The model input is the addition of three vectors $\mathbf{h}_N, \mathbf{v}_{element}$, and label embedding. Purple/red blocks with tag 1 reflecting the triple (夜曲,作词,方文山)/(夜曲,作曲, 周杰伦) led by the input pairs of (夜曲,作词)/(夜曲,作曲). (Color figure online)

Sentence-Level Event Extraction. Different from DocEE, the SentEE task provides the exact positions of trigger words and each of them points to a single event. We leverage the trigger position and the event type information obtained from first stage to guide the extraction of the corresponding arguments. As shown in Fig. 3, the input sequence is in the format of MRC, where we concatenate a specific trigger word and its event type as the query and the context is the original sentence. To extract the event arguments corresponding to this specific trigger word, we set the segment ids of the trigger word positions to zero. If a single sentence contains multiple events, there must be multiple trigger words as well. In such case, we iteratively apply the same process but using different trigger word.

Document-Level Event Extraction. DuEE-Fin dataset consists of 11,700 documents across 13 event types, each of which contains multiple argument roles. We group them together into 92 pairs of [event type, argument role], such as [质押, 质押股票/股份数量]. Specifically, to make all arguments directly

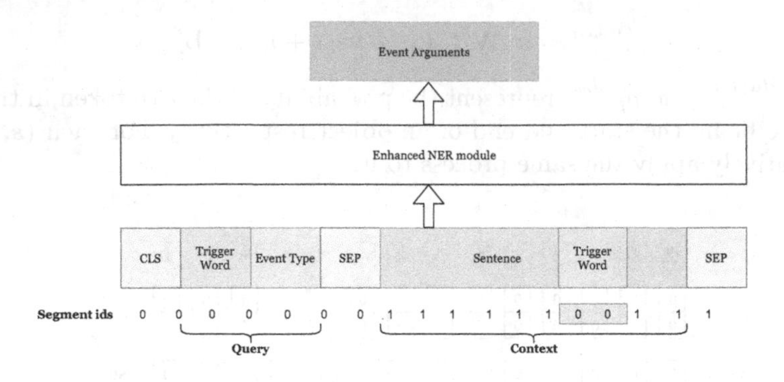

Fig. 3. An overview of the proposed SentEE architecture.

extractable from the text, the event type [公司上市] is split into four event types: [公司上市-筹备上市], [公司上市-暂停上市], [公司上市-正式上市], [公司上市-终止上市]. Finally, the original 92 pairs of [event type, argument role] were increased to 112. The outputs of the Enhanced NER module are the arguments and their corresponding [event-type, argument role] pairs. The extracted events can be achieved by grouping the candidate arguments with the same event type. However, this kind of simple post-processing strategy cannot tackle the MVA problem, where events cannot be distinguished by event type. Therefore, we design a novel neural approach to effectively distinguish events. To be more specific, for each event type, we select a center argument role (CAR) as the discriminative criterion for different events. For example, multiple events may have the same event type of [质押], but their arguments with the role of [质押股票/股份数量] must be different. In this case, as shown in Fig. 4, the CAR is [质押股票/股份数量]. After archiving all the candidate arguments, we first conduct event detection by picking out the arguments with the role of CAR, which are the core arguments (CA) of each event. Then we reuse the model architecture of Fig. 2 and fit remaining arguments (RAs) to every single event by predicting whether it matches the corresponding CA or not. The detailed operations are as follows:

$$p_i^{start_{arg}} = \sigma(\mathbf{W}^l_{start}(\mathbf{x}_i + \mathbf{v}_{CA} + \mathbf{v}_{RA_j} + \mathbf{l}_{CA} + \mathbf{l}_{RA_j}) + \mathbf{b}^l_{start}) \tag{6}$$

$$p_i^{end_{arg}} = \sigma(\mathbf{W}^l_{end}(\mathbf{x}_i + \mathbf{v}_{CA} + \mathbf{v}_{RA_j} + \mathbf{l}_{CA} + \mathbf{l}_{RA_j}) + \mathbf{b}^l_{end}) \tag{7}$$

where $p_i^{start_{arg}}$ and $p_i^{end_{arg}}$ represent the probability of the i-th token in the input sequence being the start and end of an argument respectively. The representation of the core argument $\mathbf{v}_{CA}$ is the sum of the start and end embeddings. Similarity, $\mathbf{v}_{RA_j}$ is the sum of the start and end embeddings of j-th RA. $j \in (1, 2, ..., n)$ and n is the number of RAs. $\mathbf{l}_{CA}$ and $\mathbf{l}_{RA_j}$ are the embeddings of the [Event type-Argument role] corresponding to CA and RA$_j$ respectively. These embeddings are initialized randomly and updated during the training process. Notably, if the argument predicted by the model are the same as the input RA, it means that CA matches RA.

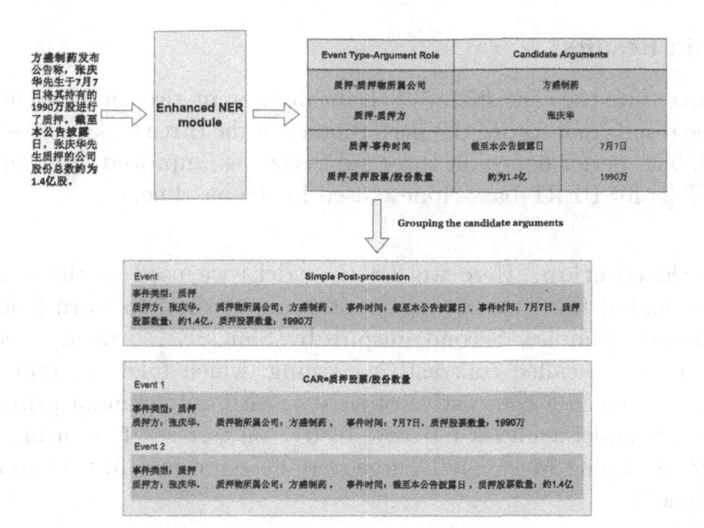

Fig. 4. Different strategies of grouping the candidate arguments. The example contains two events. Only one confusing event could be obtained by simple post-processing. Adopting CAR, on the contrary, picks up two distinctive events.

3 Experiments

3.1 Experimental Settings

Datasets. We evaluate the proposed model on three large-scale Chinese information extraction datasets[1] built by Baidu, including DuIE2.0, DuEE1.0 [4] and DuEE-fin.

Preprocessing. (1) **Sliding window.** In RE and SentEE tasks, the maximum length of the input sequence is 512 and the exceeding part is truncated. In DocEE task, we split articles into overlapping passages by sliding window. We set the window size as 512 words, and the stride as 256 words. (2) **Relation simplified.** The DuIE2.0 dataset provides 48 predefined schema, including 5 multiple-O-values(MOV) schema. The object O in MOV schema is a structure composed of multiple slots and values. For example, in the relation of [play], the object O gets two slots named [@value] and [inWork], which respectively represent [what is the role the person played] and [in which film and television work the person played the role]. We divide the object into multiple relation types according to the number of slots corresponding to the MOV schema, such as [play-@value] and [play-inWork]. (3) **Case-insensitive letters** All uppercase letters are converted to lowercase letters.

[1] For details of the datasets, please refer to https://aistudio.baidu.com.

3.2 Main Results

In order to ensure the consistency of the evaluation metircs in different tasks, we submit the results and record the performance of the three tasks on test set 1 and test set 2. The performance on three subtasks was improved after substituting MacBERT [5] for BERT-base-chinese used in the baseline.

Relation Extraction. Here we list the tricks we used in the experiments. First, we adopted the PULearning [6] training method to learn from positive and unlabeled examples. Second, inspire by SimCSE [7], we applied a unsupervised approach, called contrastive learning, which takes an input sentence and predicts itself in a contrastive objective, with only standard dropout used as noise. The model achieves F1-score 0.8077 on test set 1 by using these two tricks. By employing MacBERT Large, the F1-score was further improved by 0.92% (Table 2).

Table 2. Relation extraction results on test set 1

Model	Prec	Recall	F1
Baseline	0.8105	0.7482	0.7781
Baseline + PULearning	0.8111	0.7782	0.7943
Baseline + PULearning + Contrastive Learning	0.8293	0.7872	0.8077
MacBERT Large + PULearning + Contrastive Learning	0.8396	0.7954	**0.8169**

Sentence-Level Event Extraction. Considering the input sequence contains multiple events in this task, we verified the effects of three different strategies including adding trigger word and event type, and marking the position of trigger word in the segment ids. As shown in Table 3, the model achieves F1-score 0.8559 on test set 1 by using these three strategies.

Table 3. Sentence-level event extraction results on test set 1

Strategies	Prec	Recall	F1
Baseline	0.8242	0.8399	0.8320
Baseline + Event Type	0.8391	0.8474	0.8396
Baseline + Trigger Words	0.8305	0.8649	0.8473
Baseline + Segment Id	0.8475	0.8516	0.8495
Baseline + Event Type + Trigger Word + Segment id	0.8454	0.8557	0.8505
MacBERT Large + Event Type + Trigger Word + Segment id	0.8477	0.8642	**0.8559**

Document-Level Event Extraction. The baseline model only adopts Enhanced NER module with simply post-procession to identify all arguments. In this experiment, we select one CAR for each event type to distinguish events. The bold text in Table 4 indicates that the F1-score of the CAR is higher than that of the Baseline. These CARs are positive CARs and the others are negative CARs. The final performance is obtained by fusing the results of all positive CARs processing. The sample in Fig. 4 illustrates that for the event containing MVA, more events can be obtained through grouping candidate arguments by the CAR. Because the performance of model is calculated under the Event-level-Matching condition, the recall score can be dramatically improved aided by the further events.

Table 4. Document-level event extraction results on test set 1.

CAR	**Prec**	**Recall**	**F1**
Baseline	0.6949	0.6582	0.6760
Baseline + 亏损-净亏损	0.7063	0.6663	**0.6857**
Baseline + 解除质押-质押股票/股份数量	0.7018	0.6591	**0.6798**
Baseline + 质押-质押股票/股份数量	0.7023	0.6650	**0.6831**
Baseline + 高管变动-高管姓名	0.7086	0.6786	**0.6933**
Baseline + 企业融资-融资金额	0.6971	0.6543	0.6751
Baseline + 股东减持-交易股票/股份数量	0.7012	0.6618	**0.6809**
Baseline + 企业收购-收购标的	0.6993	0.6533	0.6755
Baseline + 企业破产-破产时间	0.7018	0.6516	0.6758
Baseline + 被约谈-公司名称	0.7035	0.6559	**0.6789**
Baseline + 中标-中标金额	0.6993	0.6558	**0.6768**
Baseline + all positive CAR	0.7147	0.7423	**0.7282**
MacBERT Large + all positive CAR	0.7138	0.7464	**0.7297**

The final performance on test set 2 on three tasks has been shown in Table 5.

Table 5. Performance on test set 2

Task	Prec	Recall	F1
Relation Extraction	0.8377	0.7919	0.8142
Sentence-level Event Extraction	0.8539	0.8250	0.8392
Document-level Event Extraction	0.7005	0.7266	0.7133

4 Conclusion

In this article, we report our solution to the multi-format information extraction task in LIC2021. We analyze the challenges of each task and design a unified platform for IE in different formats. The proposed platform can be summarized into two stages. Firstly, the Enhanced NER module is utilized to extract all elements and labels. Secondly, the strategies are customized in terms of different IE tasks and datasets. The proposed method has obtained competitive performance and ranked 2nd on test set 1 and 3rd on test set 2.

References

1. Zeng, X., Zeng, D., He, S., Kang, L., Zhao, J.: Extracting relational facts by an end-to-end neural model with copy mechanism. In: Proceedings of the 56th Annual Meeting of the Association for Computational Linguistics, vol. 1: Long Papers (2018)
2. Li, S., He, W., Shi, Y., Jiang, W., Zhu, Y.: DuIE: a large-scale Chinese dataset for information extraction. In: Natural Language Processing and Chinese Computing (2019)
3. Devlin, J., Chang, M.W., Lee, K., Toutanova, K.: Pre-training of deep bidirectional transformers for language understanding. Bert (2018)
4. Li, X., Li, F., Pan, L., Chen, Y., Zhu, Y.: DuEE: a large-scale dataset for Chinese event extraction in real-world scenarios. In: 9th CCF International Conference on Natural Language Processing and Chinese Computing, NLPCC, Zhengzhou, China, 14–18 October 2020, Proceedings. Part II (2020)
5. Cui, Y., Che, W., Liu, T., Qin, B., Wang, S., Hu, G.: Revisiting pre-trained models for chinese natural language processing (2020)
6. Kiryo, R., Niu, G., du Plessis, M.C., Sugiyama, M.: Positive-unlabeled learning with non-negative risk estimator. arXiv preprint arXiv:1703.00593 (2017)
7. Gao, T., Yao, X., Chen, D.: Simcse: simple contrastive learning of sentence embeddings. arXiv preprint arXiv:2104.08821 (2021)

Overview of the NLPCC 2021 Shared Task: AutoIE2

Weigang Guo[1], Xuefeng Yang[1](✉), Xingyu Bai[2], Taiqiang Wu[2], Weijie Liu[1], Zhe Zhao[1], Qi Ju[1], and Yujiu Yang[2]

[1] Tencent Technology, Shenzhen, China
{jimwgguo,ryanxfyang,jagerliu,nlpzhezhao,damonju}@tencent.com

[2] Tsinghua University, Shenzhen International Graduate School, Shenzhen, China
{bxy20,wtq20}@mails.tsinghua.edu.cn,yang.yujiu@sz.tsinghua.edu.cn

Abstract. This is an overview paper of the NLPCC 2021 shared task on AutoIE2, which aims to evaluate the sub-event identification systems with limited annotated data. Given definitions of specific sub-events, 100K unannotated samples and 300 annotated seed samples, participants are required to build a sub-event identification system. 30 teams registered and 14 of them submitted results. The top system achieves 8.43% and 8.25% accuracy score improvement upon the baseline system with or without extra annotated data respectively. The evaluation result indicates that it is possible to use less human annotation and large unlabeled corpora for the sub-event identification system. ALL information about this task can be found at https://github.com/IIGROUP/AutoIE2.

Keywords: Sub-event identification · Low resource natural language processing

1 Introduction

Information Extraction(IE) [5,20] aims to automatically extract structured information such as entities, relationships between entities, event attributes from unstructured sources. The extracted structure knowledge may be used as an individual application or supporting downstream applications like dialogue system [4] and information retrieval [11]. The important role IE played in language intelligence makes it a hot topic [24], and many IE systems (e.g., sub-events identification) have been developed in the last decades.

Given continuous and large stream of social media information, sub-event identification aims to automatically indicate and keep track of the sub-parts of an event [3]. It is challenging due to the following two characteristics. First, data confusion and imbalance. Events usually evolve rapidly and successive sub-events occur. Only a few target sub-event data need to be identified from the large volume of related events and noisy data. Second, low resource scenarios. Usually limited annotated seed data is given for learning and more annotated data is expensive and time-consuming, especially in emergencies (e.g., terrorist attacks) [16].

L. Wang et al. (Eds.): NLPCC 2021, LNAI 13029, pp. 519–529, 2021.
https://doi.org/10.1007/978-3-030-88483-3_42

Seed Dataset

数据	标签
摘下口罩再戴时，手的洁净至关重要。建议洗手后再戴口罩	1
美国商务部18日宣布将59家中国实体列入所谓出口管制"实体清单"。	3
原定在东京举行的夏季奥运会及残奥运会重新安排到2020年以后	2
最可怕的是，如果另外病毒肆虐，人类怎么办?	0
到底病毒来自哪里还没有答案。最可怕的是，如果另外病毒肆虐，人类怎么办	0
如果美方愿意谈，我们愿在平等磋商、相互尊重的基础上磋商，解决分歧。	0

Unlabeled Dataset

数据
戴口罩，勤洗手，不聚餐，不聚集。坚持就是胜利，不要让无数人的努力功亏一篑!
病毒可以战胜。但是奥运会。可以推迟!
美国好像不看好结果，特朗普昨天刚对中国44亿美元家具加征关税，国内都没报道
等着吧，中国迟早要背锅，毕竟很多口罩made in China
到6月，已经抛了1000多亿
可能是给汽车打广告

Fig. 1. Examples of seed dataset and unlabeled dataset

It is challenging for participants to build a robust sub-event identification system with 100K unannotated samples and totally 300 seed samples in this evaluation task. The task setting is non-trivial in practical application scenarios since a few annotated seed samples are given.

We choose three recent events and obtain a corpus from Sina Weibo[1] comments (generally 8 to 120 characters long). The corpus is processed into three parts, i.e., unlabeled dataset(100K samples for 3 events), seed dataset(100 samples per event) and test dataset(2K per event). Figure 1 shows some examples of the dataset.

30 teams signed in this shared task, and 14 submitted solutions. The good news is that most submitted systems perform better than baseline, and the top-ranked system achieves at most 8.43% accuracy score improvement without extra labeled samples and 8.25% with extra samples.

This overview is organized as follows. In Sect. 2, we will review some important related works. Details of this evaluation task are provided in Sect. 3. After introducing the task settings, some important factors and the proposed solutions are analyzed in Sect. 4. The conclusion is given in Sect. 5.

2 Related Work

Firstly, we will review some widely used sub-event identification benchmarks. Meladianos [17] proposed a dataset including 185 sub-events extracted from 20 football matches during 2010 to 2014 FIFA World Cups. It is composed of totaling over 2M pre-processed tweets filtered from 6.1M collected ones. [21] built a dataset utilizing 808661 tweets about the election of Australia during August 2013 to September 2013. 115 keywords such as Ausvotes and Tony Abbott

[1] http://www.weibo.com/.

were considered as the sub-event labels. For Chinese datasets, Brand-Social-Net dataset [9] was constructed upon 3 million micro-blogs from Sina Weibo during June 2012 to July 2012, and divided into 20 saga events.

In recent years, sub-event identification in social media platforms has attracted extensive research attention and several methods have been proposed. Given new data, the similarity between the new data and the existing events is computed for the nearest neighbor classification. Features like TF/IDF [1,25], word appearance [8], Wavelet [23] are investigated to measure the similarity. Unsupervised methods such as clustering aim to group similar information streams to detect specific sub-events. Becker [2] introduces the learning similarity metrics under a clustering task to identify events in social media streams. Reuter and Cimiano [19] proposed a trained Support Vector Machine classifier to deal with incremental data in social media streams. In their method, temporal information, geographical information and textual information are used for feature representation. Furtherly, Jie [12] frames sub-event identification as a sequence labeling problem by exploiting the chronological relation between consecutive tweets. It is noted that most of the supervised classifier-based methods i.e., CLASS-LR, CLASS-SVM and neural network architectures, show the best performance in experimental results given enough labeled data [10]. However, it is hard to fully meet the requirements in real-world scenarios, and thus better few-shot learning and data selection models for low resource sub-event identification are crucial.

From the perspective of low-resource, two directions are strongly related to this task. Firstly, to fully unlock the potential of limited labeled data, unsupervised pre-trained language models are widely developed. BERT and its variants (e.g., RoBERTa [14], ALBERT [13] and T5 [18]) have achieved state-of-the-art performance in many few-shot learning tasks. Secondly, to augment the labeled data, data selection by various active learning strategies [7] and pseudo labels generation by semi-supervised training [6,15,22] are proposed.

3 Evaluation Task

The goal of this task is to build an IE (Information Extraction) system that can quickly adapt to emerging sub-events. The task is strongly related to two research directions, namely few-shot learning and data selection. Generally, the directions are very important for data-driven machine learning applications.

3.1 Setting

There are two challenges in sub-event identification applications. Firstly, available labeled data is rare and valuable in the beginning period of sub-events. How to train high-quality models basing on small training samples remains a big problem. Secondly, events usually evolve rapidly and a large number of related messages appear. How to select the most vital samples from the abundant unlabeled data and minimize the labeling burden is quite crucial.

Specifically, there are two settings in this task, noted as task 1 and task 2 respectively. Task 1 aims to build an IE system to identify the target sub-events with a few labeled seed data. Task 2 shares the same goal while allowing extra annotated samples selected from the unlabeled dataset. The extra data may not exceed 100 per sub-event and can be acquired by model or human annotation.

3.2 Dataset

The corpus used in this task is originally obtained from Weibo comments and cleaned by lots of techniques to support this evaluation. Three hot event topics are chosen and focused in the task, including Trade War, Tokyo Olympic Games, COVID-19.

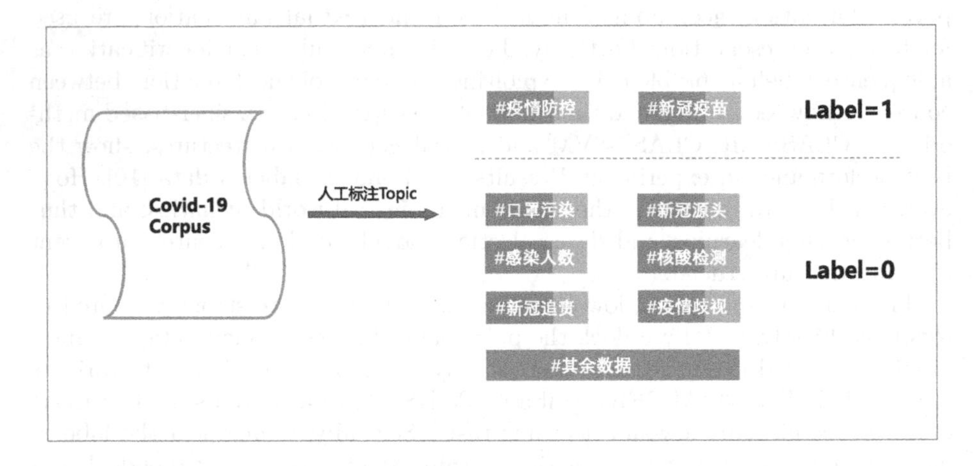

Fig. 2. Process of selecting positive and negative samples

The whole corpora are split into three parts, i.e., unlabeled dataset, seed dataset and test dataset. In the unlabeled dataset, there are 100K samples involving the three chosen event topics without annotations. Seed dataset(100 samples per sub-event) are fully labeled data according to the definitions of the three target sub-events. As shown in Figure 2, we describe the definition of COVID-19 sub-event. It means various preventive control measures taken by governments, hospitals and other organizations during the period of COVID-19, e.g., policies and regulations about masks, vaccines and home quarantine. Trade War sub-event means US actions for reducing the trade deficit with China during the trade war, e.g., imposing tariffs on Chinese imports and restricting the ability of Chinese companies to trade with US firms. The sub-event for Tokyo Olympic Games means holding the view that it will be postponed or held as scheduled.

The unlabeled dataset and labeled seed dataset can be combined to construct the training set and developing set. The test dataset(around 2k per sub-event) is reserved for testing the submissions. The statistic details of the datasets are shown in Table 1. The task is challenging since the size of test dataset is far larger than seed dataset.

Table 1. Statistic of proposed dataset

Part	Size	Avg. len	Size of sub-event samples			
			COVID-19	Olympic	Trade war	Negative
Seed	300	49.73	43	57	46	154
Test	6000	46.24	515	1356	747	3382
All	6300	46.41	558	1413	793	3536
Unlabeled	100000	29.55	Null	Null	Null	Null

3.3 Baseline

The UER[2] framework supports comprehensive downstream NLU(Natural Language Understanding) solutions (e.g. classification) and achieves state-of-the-art performance for many Chinese NLP tasks. It is employed to build the baseline system for our evaluation.

There are two models in the baseline system, corresponding to the two task settings separately. The first one is a fine-tuned classification model basing on pre-trained BERT base. We can obtain the pseudo-labels of the unlabeled dataset predicted by that model and select 300 high confidence samples as the extra annotated data for task 2. The second model can then be trained. The baseline code is released in our github repository[3].

4 Task Analysis

In this part, empirical studies about the influence of the pre-trained models, labeled data size, data selection strategy and data annotation quality are conducted to explore the AutoIE2 evaluation task. After the factors analysis, submitted systems are reviewed and evaluation results are provided.

4.1 Factor Analysis

To understand the effect caused by pre-trained models, labeled data size, data selection strategy and data annotation quality, we study the performances of baseline system with different settings.

---

[2] https://github.com/dbiir/UER-py.
[3] https://github.com/IIGROUP/AutoIE2/tree/main/baseline.

Effect of Pre-trained Models. Pre-trained models are widely used and show promising performance in various NLP tasks. We study the performance effect of different pre-trained models on the AutoIE2 task, including BERT, RoBERTa.

As shown in Table 2, RoBERTa performs nearly 2.5% better than BERT for both settings. It indicates that the tasks are sensitive to pre-trained models and more suitable pre-trained models may bring better performance. Additionally, we also found that both models perform better in task 2 than in task 1. It meets our expectation for task 2 setting that more labeled training data within a certain range can bring better performance.

Table 2. Accuracy on test dataset for pre-trained models

Pretrained model	Accuracy in task 1	Accuracy in task 2
BERT	70.52	70.85
RoBERTa	72.73	73.33

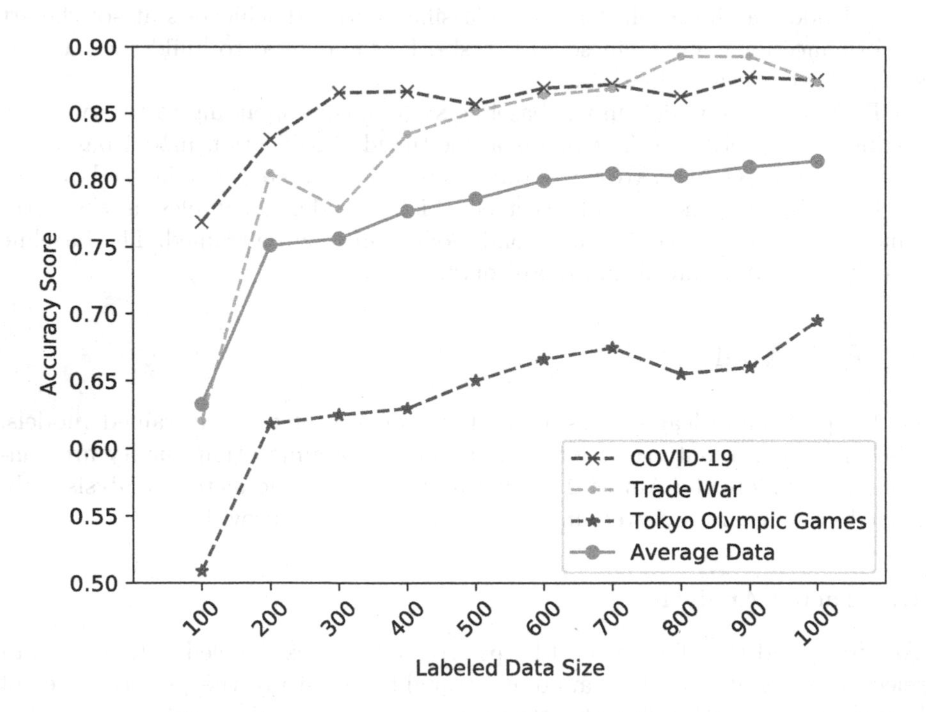

Fig. 3. Accuracy score with different labeled data size

Dataset Size. In this section, we study the effect of different sizes of labeled data in the baseline system by conducting binary classification on the three sub-events separately. We increase the size of training samples from 100 to 1000 in steps of 100 by randomly sampling while fixing all other hyper-parameters for a fair comparison.

As shown in Fig. 3, the accuracy score increases most significantly by 11.87% on average, with the size of training data ranging from 100 to 200. After the size exceeds 200, the accuracy score increases only by 6.33% on average and stabilizes gradually. The distribution illustrates that setting the labeled dataset size as 200 can maximize the value of labeling effort in this task. Specifically, we set the size of seed dataset as 100 per sub-event for task 1 and max size of extra annotated data as 100 per sub-event for task 2.

Data Selection. There are two factors that may affect the performance in task 2, i.e., data selection strategy and data annotation quality. To simulate different data selection strategies under the task 2 setting, we randomly selcct 100 from 700 labeled samples as the extra annotated data. The random trials were repeated 20 times for each sub-event and the experiment results are displayed in the form of boxplot.

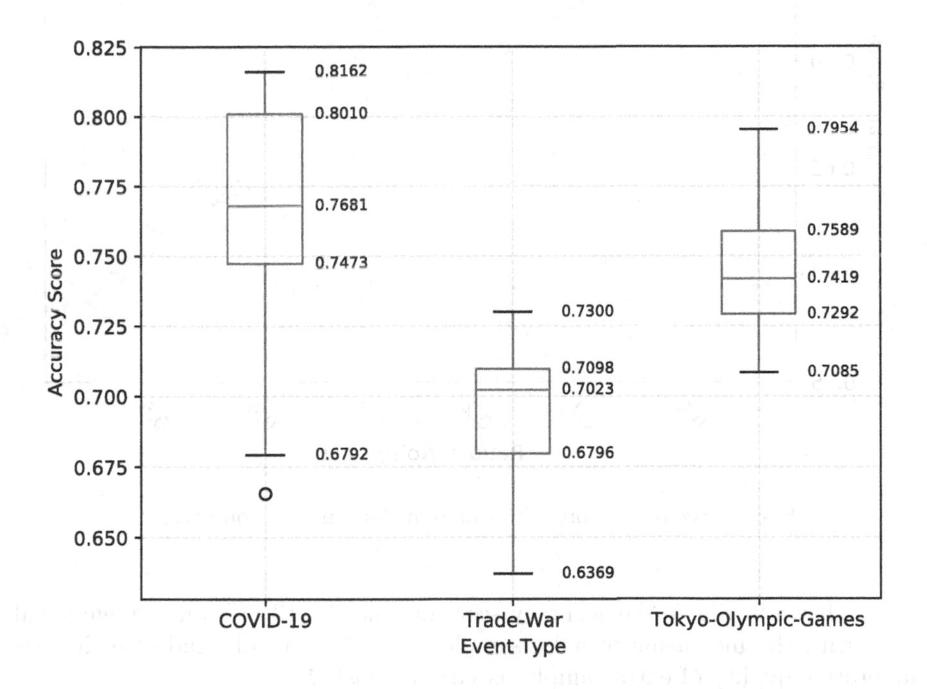

Fig. 4. Accuracy score with different data selection

As shown in Fig. 4, the performance scores vary dramatically with different data combinations. The difference between the maximum scores and minimum scores is close to 10% on average, which is far more than the improvement brought by pre-trained models and comparable with the improvement brought by the size of training samples. However, the size of training samples contributes mostly when it is less than 200 thus the data selection strategies may have the greatest potential to leverage the limited annotation resource.

Data Annotation Quality. To study the effect of annotation quality, we add different ratios of noise samples in the 100 extra annotated samples to simulate different levels of annotation quality. The noise data is generated by deliberately mislabeling. We increase the ratio of noisy samples from 0 to 0.5 in steps of 0.1 for observation.

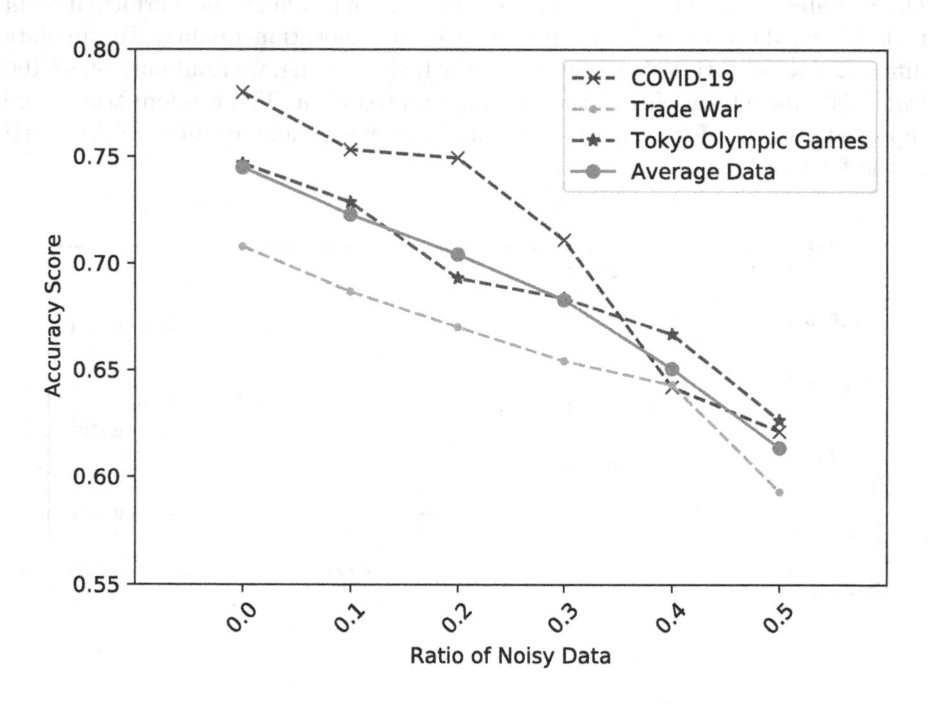

Fig. 5. Accuracy score with different data annotation quality

As shown in Fig. 5, the accuracy score declines by 13.11% on average for all tasks with the increasing of noise samples ratio. The results indicate that the annotation quality of extra samples is vital for task 2.

4.2 Submission Analysis

13 of 14 teams perform better than the baseline system for task 1, and 11 of 12 teams perform better for task 2 with extra annotated data. This promising result indicates that algorithm development plays an important role for sub-event identification problem under low resource.

Evaluation Results. The evaluation results of the top 3 submissions and baseline are given in Table 3. Apparently, all three systems significantly increase the performance with or without extra annotated data, and the best solution makes an improvement over 8.61% and 8.25% accuracy score. Another interesting conclusion is that the top 3 systems perform worse in task 2 with more annotated samples than in task 1.

Table 3. Evaluation leaderboard for two tasks

Rank	Team name	Accuracy for task 1	Accuracy for task 2	Overall accuracy
1	BIT-Event	79.13	79.10	79.12
2	Happy	77.98	76.50	77.24
3	Alphonse	76.22	76.05	76.14
15	Baseline System	70.52	70.85	70.69

System Review. The top 3 ranking systems are reviewed in the shared task. The team *BIT-Event* proposes a hybrid system to enhance the performance through a large amount of unlabeled corpus. In the hybrid model, the pre-trained model is fine-tuned to adapt to the linguistics style of the training data. Besides, Adversarial Training and Virtual Adversarial Training are mixed to enhance the effect of single model with unlabeled domain data. Active Learning was also applied as an iterative process to screening the worthy-labeling data. The team *happy* chooses RoBERTa[14] as the pre-trained model and fine-tune on the seed dataset to generate pseudo labels of the unlabeled dataset. Then relative confidence scores instead of absolute confidence scores are calculated to select the most valuable samples for annotation in task 2. The team *Alphonse* focuses on task 1 with abundant combination attempts of hyper parameters and also achieves impressive results. Different pre-trained models are applied in the tasks, and RoBERTa[14] may provide the best features. For fine-tuning skills, semi-supervised and Adversarial Training may contribute most under low-resource setting of sub-event identification task.

Data Selection. It is interesting to find that the top 3 systems perform worse in task 2 than in task 1, which is contrary to the baseline system and experiment results. Data selection by various active learning strategies[7] has been proved meaningful for many academic tasks and corpora, however, it is more challenging for task 2. Because our setting and dataset are closer to the practical application scenarios and suffer from serious semantic confusion. We are excited to expecting more effective solutions.

5 Conclusion

This paper briefly presents an overview of the AutoIE2 evaluation task in NLPCC 2021. The evaluation results, especially the setting without extra labeled data, are quite exciting and promising. Meanwhile, it's also worth noting that the evaluation results with extra labeled data may perform slightly worse than the other setting. It may reveal our weakness in data selection, which would excite us to explore deeper and broader around the technology. We believe these proposed solutions may help in practical sub-event identification applications.

References

1. Allan, J., Papka, R., Lavrenko, V.: On-line new event detection and tracking. In: Proceedings of the 21st Annual International ACM SIGIR Conference on Research and Development in Information Retrieval, pp. 37–45 (1998)
2. Becker, H., Naaman, M., Gravano, L.: Learning similarity metrics for event identification in social media. In: Davison, B.D., Suel, T., Craswell, N., Liu, B. (eds.) Proceedings of the Third International Conference on Web Search and Web Data Mining, WSDM 2010, New York, NY, USA, 4–6 February 2010, pp. 291–300. ACM (2010). https://doi.org/10.1145/1718487.1718524
3. Bekoulis, G., Deleu, J., Demeester, T., Develder, C.: Sub-event detection from twitter streams as a sequence labeling problem. In: Proceedings of the 2019 Conference of the North American Chapter of the Association for Computational Linguistics: Human Language Technologies, vol. 1 (Long and Short Papers), pp. 745–750. Association for Computational Linguistics, Minneapolis, Minnesota (2019). https://doi.org/10.18653/v1/N19-1081, https://www.aclweb.org/anthology/N19-1081
4. Chen, H., Liu, X., Yin, D., Tang, J.: A survey on dialogue systems: recent advances and new frontiers. SIGKDD Explor. Newsl. **19**(2), 25–35 (2017)
5. Cowie, J., Lehnert, W.: Information extraction. Commun. ACM **39**(1), 80–91 (1996)
6. Du, J., Grave, E., Gunel, B., Chaudhary, V., Celebi, O., Auli, M., Stoyanov, V., Conneau, A.: Self-training improves pre-training for natural language understanding. In: Proceedings of the 2021 Conference of the North American Chapter of the Association for Computational Linguistics: Human Language Technologies. pp. 5408–5418. Association for Computational Linguistics, Online (2021), https://www.aclweb.org/anthology/2021.naacl-main.426
7. Ein-Dor, L., et al.: Active Learning for BERT: An Empirical Study. In: Proceedings of the 2020 Conference on Empirical Methods in Natural Language Processing (EMNLP), pp. 7949–7962. Association for Computational Linguistics, Online (2020). https://doi.org/10.18653/v1/2020.emnlp-main.638, https://www.aclweb.org/anthology/2020.emnlp-main.638
8. Fung, G.P.C., Yu, J.X., Yu, P.S., Lu, H.: Parameter free bursty events detection in text streams. In: Proceedings of the 31st International Conference on Very Large Data Bases, pp. 181–192. Citeseer (2005)
9. Gao, Y., Wang, F., Luan, H., Chua, T.S.: Brand data gathering from live social media streams. In: Proceedings of International Conference on Multimedia Retrieval, pp. 169–176 (2014)

10. Gao, Y., Zhao, S., Yang, Y., Chua, T.-S.: Multimedia social event detection in microblog. In: He, X., Luo, S., Tao, D., Xu, C., Yang, J., Hasan, M.A. (eds.) MMM 2015. LNCS, vol. 8935, pp. 269–281. Springer, Cham (2015). https://doi.org/10.1007/978-3-319-14445-0_24
11. Greengrass, E.: Information retrieval: a survey (2000)
12. Jie, Z., Xie, P., Lu, W., Ding, R., Li, L.: Better modeling of incomplete annotations for named entity recognition. In: Proceedings of the 2019 Conference of the North American Chapter of the Association for Computational Linguistics: Human Language Technologies, vol. 1 (Long and Short Papers), pp. 729–734. Association for Computational Linguistics, Minneapolis (2019). https://doi.org/10.18653/v1/N19-1079, https://www.aclweb.org/anthology/N19-1079
13. Lan, Z., Chen, M., Goodman, S., Gimpel, K., Sharma, P., Soricut, R.: ALBERT: a lite BERT for self-supervised learning of language representations. In: 8th International Conference on Learning Representations, ICLR 2020, Addis Ababa, Ethiopia, 26–30 April 2020. OpenReview.net (2020). https://openreview.net/forum?id=H1eA7AEtvS
14. Liu, Y., et al.: Roberta: a robustly optimized bert pretraining approach. arXiv preprint arXiv:1907.11692 (2019)
15. Mann, G.S., McCallum, A.: Generalized expectation criteria for semi-supervised learning with weakly labeled data. J. Mach. Learn. Res. **11**(2), 955–984 (2010)
16. Meladianos, P., Nikolentzos, G., Rousseau, F., Stavrakas, Y., Vazirgiannis, M.: Degeneracy-based real-time sub-event detection in twitter stream. In: Proceedings of the International AAAI Conference on Web and Social Media, vol. 9 (2015)
17. Meladianos, P., Xypolopoulos, C., Nikolentzos, G., Vazirgiannis, M.: An optimization approach for sub-event detection and summarization in twitter. In: Pasi, G., Piwowarski, B., Azzopardi, L., Hanbury, A. (eds.) ECIR 2018. LNCS, vol. 10772, pp. 481–493. Springer, Cham (2018). https://doi.org/10.1007/978-3-319-76941-7_36
18. Raffel, C., et al.: Exploring the limits of transfer learning with a unified text-to-text transformer. arXiv preprint arXiv:1910.10683 (2019)
19. Reuter, T., Cimiano, P.: Event-based classification of social media streams. In: Proceedings of the 2nd ACM International Conference on Multimedia Retrieval, pp. 1–8 (2012)
20. Sarawagi, S.: Information Extraction. Now Publishers Inc., Norwell (2008)
21. Unankard, S., Li, X., Sharaf, M., Zhong, J., Li, X.: Predicting elections from social networks based on sub-event detection and sentiment analysis. In: Benatallah, B., Bestavros, A., Manolopoulos, Y., Vakali, A., Zhang, Y. (eds.) WISE 2014. LNCS, vol. 8787, pp. 1–16. Springer, Cham (2014). https://doi.org/10.1007/978-3-319-11746-1_1
22. Wang, Y., et al.: Adaptive self-training for few-shot neural sequence labeling. arXiv preprint arXiv:2010.03680 (2020)
23. Weng, J., Lee, B.S.: Event detection in twitter. In: Proceedings of the International AAAI Conference on Web and Social Media, vol. 5 (2011)
24. Yang, X., Wu, B., Jie, Z., Liu, Y.: Overview of the NLPCC 2020 shared task: AutoIE. In: Zhu, X., Zhang, M., Hong, Yu., He, R. (eds.) NLPCC 2020. LNCS (LNAI), vol. 12431, pp. 558–566. Springer, Cham (2020). https://doi.org/10.1007/978-3-030-60457-8_46
25. Yang, Y., Pierce, T., Carbonell, J.: A study of retrospective and on-line event detection. In: Proceedings of the 21st Annual International ACM SIGIR Conference on Research and Development in Information Retrieval, pp. 28–36 (1998)

Task 1 - Argumentative Text Understanding for AI Debater (AIDebater)

Yuming Li[1,2](✉), Maojin Xia[1](✉), and Yidong Wang[1](✉)

[1] Qingbo Intelligence Technology Co. LTD, Hefei, China
{liyuming,xiamaojin,wangyidong}@gsdata.cn
[2] University of Auckland, 12 Grafton Road, Auckland, New Zealand

Abstract. Debate refers to the behavior of humans using certain reasons to explain their views on things or issues, exposing their contradictions, and finally obtaining a common understanding and opinions. In the current era of massive information and misleading culture, we expect that an AI system that realizes fully autonomous debate can promote the development of intelligent debate, help establish more reasonable arguments, and make more informed decisions. Based on artificial intelligence technology, this task uses deep learning-based algorithms to identify the attributes of input debate topics and claims, including support, opposition, and neutrality. In this task, we transform the Argumentative Text Understanding problem into a classification problem, firstly compared roberta-large, macbert-large, nezha-large, and nezha-large-wwm, and finally chose to use nezha-large with the highest accuracy rate, and achieve 90.2% accuracy in the second stage.

Keywords: AIDebater · Text Understanding · Classification

1 Introduction

1.1 Background

Debate refers to an act in which humans use certain reasons to explain their own views on things or problems, expose each other's contradictions, and finally get a common understanding and opinions. Artificial Intelligence is defined as the ability and possibility of computer science algorithms or machines to perform tasks that associated with intelligent creatures. Argument and debate are fundamental capabilities of human intelligence, essential for a wide range of human activities, and common to all human societies.

In 2020, IBM's strongest AI debater Project Debater [1] represents the culmination of current computing debate research. In the current era of massive information and misleading culture, we expect that an AI system that realizes fully autonomous debate can promote the development of intelligent debate, help establish more reasonable arguments, and make more informed decisions. The research on debate can be traced back to ancient Greece, when ancient Greek philosophers such as Socrates discussed politics and truth with people in the market, and the content of the debate was all-encompassing. One of the current challenges in artificial intelligence research is how to make machines understand the arguments in the natural language debate.

L. Wang et al. (Eds.): NLPCC 2021, LNAI 13029, pp. 530–537, 2021.
https://doi.org/10.1007/978-3-030-88483-3_43

1.2 Task Description

In this task, given a pair of topic and claim, participants are required to classify the stance of the claim towards the topic into either Support, Against or Neutral.

The data is the debating topics collected from online forums and topic-related articles from Wikipedia (for English) and Baidu Encyclopedia (for Chinese). Human annotators were recruited to annotate claims from the collected articles and identify their stances with regards to the given topics. Human translators were recruited to translate the English claims into Chinese.

The format of data file is as follows: The data is in TXT format, and each line includes three items: topic, claim, and the label ({support, against, neutral}), separated by tab. Below are some examples:

大学教育应该免费<tab>高等教育免费导致了较低的学术水平<tab>against
应该放弃独生子女政策<tab>独生子女很难独自照顾年迈的父母<tab>support
人工智能最终会取代人类<tab>人工智能在计算机领域内得到了愈加广泛的重视<tab>neutral

The overall process can be summarized as Fig. 1:

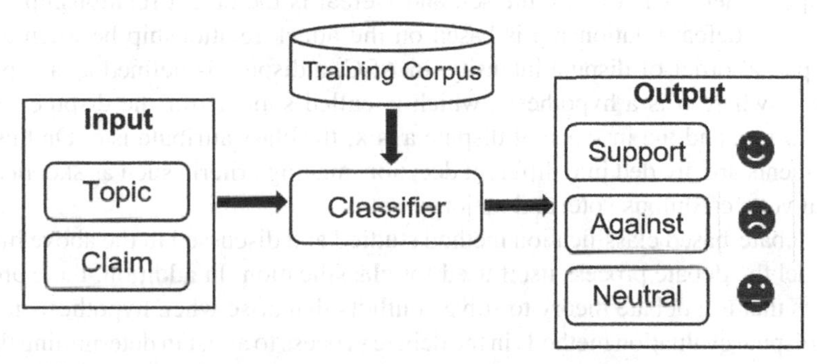

Fig. 1. The overall process of Argumentative Text Understanding for AI Debater (AIDebater) task.

Finally, accuracy is used as the evaluation metric.

2 Related Work

The research on debate theory originated in the fields of philosophy, logic and law, and its main purpose is to model the process of daily debate. Representative studies in this area mainly include Wigmore model [2] and Toulmin model [3]. Among them, the Wigmore model uses graphic elements such as boxes and arrows to describe the debate process. It is generally used in court debates. However, due to its high complexity, it is not easy to implement in algorithms, and its versatility is not strong. The Toulmin model decomposes disputes into six parts: claim, premise, warranty, backing, modality, and rebuttal, which realizes the structured representation of disputes. Conducive to the processing of disputed information by computers, it has been widely recognized in the fields of artificial intelligence and law. However, the Toulmin model mainly focuses on the dispute itself, lacks a formal description of the dialogue process, and does not analyze the state of the dispute. With the development of non-monotonic logic, Dung [4] proposed Abstract Argumentation Framework (AAF), which ignores the internal structure of disputes and focuses on the relationship between disputes and disputes in order to evaluate the acceptability of disputes. He proposed four extension definitions: preferred extension, stable extension, complete extension and grounded extension. However, this model only considers the attack relationship between disputes, while ignoring the possible support relationship between disputes. Many subsequent studies have extended Dung's abstract debate framework. Including preference-based argumentation framework (PAF) [5, 6], Value-based Argumentation Framework (VAF), Bipolar Argumentation Framework (BAF) and so on.

Classification is a kind of supervised learning, which can predict the class attributes of new instances based on the information contained in existing samples. The purpose of classification is to derive a general description of given data from historical data records, so as to predict the type of unknown data. Amgoud et al. [7] based on Dung's abstract debate framework, using Agent to generate and explain the classification process, and proposed an argumentation system for classification (ASC). ASC = <A, Defeat> is a two-tuple, where A is the dispute set, and Defeat is the defeat relationship between disputes. The defeat relationship is based on the attack relationship between disputes and the partial order of dispute intensity. In ASC, a dispute is defined as a triplet a = <h, x, c>, where h is a hypothesis, which is called support for the dispute, (x, c) is the conclusion, and the instance of dispute a is x, the class attribute is c. On this basis, the arguments are divided into different decision-making criteria such as skeptical vote, universal vote, credulous vote, and majority vote.

The debate-based classification method studied and discussed in the above literature is to model the debate process itself used for classification. In addition, there are some literatures that use debate theory to solve conflicts that arise when hypothesis matches, and use dispute evaluation methods in the debate process to assist in determining the classification results. Representative studies in this area include: Gomez et al. [8] proposed a classification method based on repealable debate and neural network. This method solves the problem of classification conflicts caused by the overlap of class intervals obtained by clustering technology. In its classification method, dispute is defined as a set of rules in revocable logic procedures. Dispute A attacks dispute B if and only if A refutes B's conclusion or A attacks B's premise. On this basis, the time stamp, the scope of the class, and the partial order of the modular evaluation of the class are introduced to determine the defeating relationship between the disputes. The process of dialogue between disputes is carried out in accordance with the timestamp, and the auxiliary classification of the debate process is graphically expressed as the structure of the argument tree, and the dispute evaluation is based on the judgment of the acceptability of the nodes in the argument tree.

3 Methodology

In this article, we transform the Argumentative Text Understanding problem into a classification problem, and conduct comparative experiments based on the four models: roberta-large, macbert-large, nezha-large, and nezha-large-wwm.

3.1 Roberta

Pre-training models have become an indispensable resource in the field of natural language processing at this stage, such as ELMo [9], GPT [10], BERT [11], XLM [12], and XLNet [13]. Recently, the BERT-based RoBERTa model [14] proposed by Facebook has further refreshed the best results of multiple data sets and has become one of the most popular pre-training models. The Roberta Masked language model is shown in Figure below. The Roberta Masked Language model is composed of a Language Model head on top of the base language model (Fig. 2).

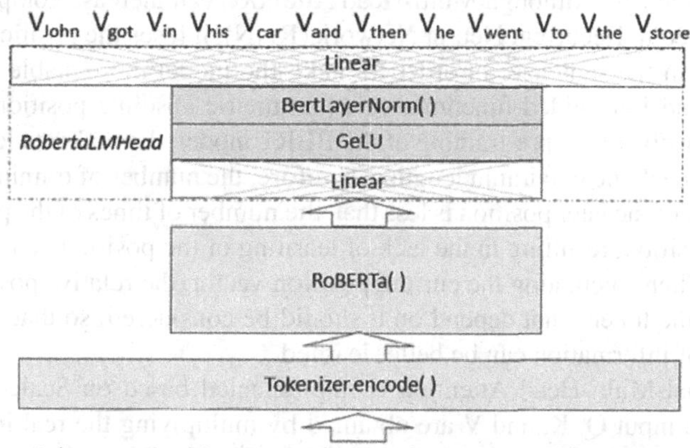

Fig. 2. The framework of RoBERTa masked language model

In order to further improve the effect of Chinese natural language processing tasks, the Harbin Institute of Technology IFLYTEK Joint Laboratory and the State Key Laboratory of Cognitive Intelligence released the RoBERTa-large Chinese pre-training model. And the F1 of this model exceeded 60% in the CMRC 2018 reading comprehension challenge set, indicating that the Chinese pre-training model exceeded the "pass line" for the first time on difficult problems.

3.2 MacBERT

MacBERT [15] replaces bert's original Masked Language Model (MLM) task with a Mac (MLM as correction) task, and reduces the difference between the pre-training and fine-tuning stages. Compared with traditional BERT, MacBERT's innovations can be summarized as follows:

- Extensive empirical studies are carried out to revisit the performance of Chinese pre-trained models on various tasks with careful analyses.
- MacBERT mitigate the gap between the pre-training and fine-tuning stage by masking the word with its similar word, which has proven to be effective on down-streamtasks.
- It create and release the Chinese pre-trained model series to the community to further accelerate future research on Chinese NLP

3.3 Nezha

NEZHA (Neural Contextualized Representation for Chinese Language Understanding) [16] is a Transformer-based pre-training model. It improves the BERT model in four points, as follows:

3.3.1 Functional Relative Positional Encoding

In order to increase the parallel efficiency of the model, Transformer uses the Multi-Head Attention mechanism. Although Multi-Head Attention can increase computational efficiency compared to Recurrent Neural Network (RNN), it loses the position information of each token in the sentence. In order to make the model more stable and effective, Transformer and Bert added functional and parametric absolute position codes to the model respectively. In the pre-training of the BERT model, the real data length of many data does not reach the maximum length. Therefore, the number of training times of the position vector of the later position is less than the number of times of the position vector of the front position, resulting in the lack of learning of the position coding of the later parameters. When calculating the current position vector, the relative position relationship between the tokens that depend on it should be considered, so that the interactive transmission of information can be better learned.

The original Multi-Head Attention is implemented based on Scaled Dot-Product Attention. The input Q, K, and V are obtained by multiplying the real input sequence $x = (x_1, x_2, ..., x_n)$ by different weights $W^Q W^K W^V$. The output is the sequence $x = (x_1, x_2, ..., x_n)$, the length of which is consistent with the input sequence. The calculation formula for output z_1 is as follows:

$$z_i = \sum_{j=i}^{n} a_{ij}(x_j W^V)$$

Among them, a_{ij} is obtained by seeking softmax from the hidden states of position i and position j, as follows:

$$a_{ij} = \frac{exp e_{ij}}{\sum_k exp e_{ik}}$$

In which e_{ij} is obtained by scaling the dot product of the input element through W^Q and W^K transformation:

$$e_{ij} = \frac{(x_i W^Q)(x_j W^K)^T}{\sqrt{d_z}}$$

In the relative position coding scheme, the output z is added to the parameter of the relative distance between the two positions. In the above formula 1 and formula 3, the relative position information of the two tokens are added respectively, and the modification is obtained as follows:

$$z_i = \sum_{j=i}^{n} a_{ij}\left(x_j W^V + a_{ij}^V\right) \qquad e_{ij} = \frac{\left(x_i W^Q\right)\left(x_j W^K + a_{ij}^K\right)^T}{\sqrt{d_z}}$$

Among them, a_{ij}^V and a_{ij}^K are the relative position codes of position i and position j. Define the position code of a_{ij} as follows:

$$a_{ij}[2k] = \sin\left((j-i)/\left(10000^{\frac{2k}{d_z}}\right)\right) \qquad a_{ij}[2k+1] = \cos\left((j-i)/\left(10000^{\frac{2k}{d_z}}\right)\right)$$

3.3.2 Whole Word Masking

The BERT model implements a two-way Transformer by means of word masks. In the BERT model, the masked words are randomly selected. Research shows that replacing random masked words with full word masks can effectively improve the effect of the pre-training model, that is, if a Chinese character is masked, other Chinese characters belonging to the same Chinese character are masked together. In WWM implementation process in NEZHA, a tokenization tool Jieba2 was used to perform Chinese word segmentation and find the boundaries of Chinese words.

3.3.3 Mixed Precision Training

When implementing mixed precision training, every step in the training process maintains a copy of FP32 for all weights of the model, called Master Weights; in the process of forward and backward propagation, Master Weights will be converted to FP16 (Half-precision floating point number) format, where weights, activation functions and gradients are all expressed in FP16, and finally the gradient will be converted to FP32 format to update Master Weights. The ultimate goal is to improve the training speed, calculating float16 is faster than float32.

3.3.4 LAMB Optimizer

Usually when the batch size of deep neural network training is very large (exceeding a certain threshold), it will have a negative impact on the generalization ability of the model, and the Layer-wise Adaptive Moments optimizer for Batching training (LAMB) optimizer adjusts the learning rate for each parameter in an adaptive way. Ensure that the effect of the model is not lost when the Batch Size is large, so that the model training can use a large Batch Size, which greatly improves the training speed. The batch size of each GPU in the NEZHA base model is 180, and the batch size of each GPU in the large model is 64.

4 Result

In the experiment, our parameter configuration is shown in Table 1:

Table 1. Parameter configuration.

Parameter	Value
Max_len	60
Verbose_step	1000
Batch_size	32
Backword_step	1
K_fold	5
Ckpt_step	200
N_epochs	3
Lr	1e−5
SEED	42

In this task, accuracy is used as the evaluation metric. It is the ratio of the number of samples correctly classified to the number of all samples. The calculation formula is:

$$\text{Accuracy} = \frac{\text{TP} + \text{TN}}{\text{TP} + \text{TN} + \text{FP} + \text{FN}}$$

In which TP (True Positive) refers to samples that actually belong to the category and are predicted to be in the category, FP (False Positive) refers to samples that do not belong to the category but are predicted to be in the category, and FN (False Negative) refers to samples that should belong to this category but are predicted to be in other categories.

Table 2. Result of comparative experiment.

Model	Accuracy (%)
roberta-large	82.31
macbert-large	82.33
nezha-large	84.6
nezha-large-wwm	83.8

In the first stage, we compared four SOTA models: roberta-large, nacbert-large, nezha-large and nezha-large-wwm. The final results are shown in Table 2. It can be seen that nezha-large has the best effect, with an accuracy rate of 84.6%, which is higher than the other three models. At the same time, we also introduced the FGM (Fast Gradient Method) in the adversarial training, which is a method of introducing noise. The purpose is to regularize the parameters and improve the robustness and generalization ability of the model.

Finally, we chose to use the nezha large model for this task, and the accuracy reached 90.2% in the second stage.

References

1. Slonim, N.: Project Debater. In: COMMA, p. 4. (2018)
2. Rowe, G., Reed, C.: Translating wigmore diagrams. Front. Artif. Intell. Appl. **144**, 171 (2006)
3. Toulmin, S.E.: The Uses of Argument. Cambridge Univ. Press, Cambridge (1958)
4. Dung, P.M.: On the acceptability of arguments and its fundamental role in nonmonotonic reasoning. logic programming and n-person games. Artif. Intell. **77**(2), 321–357 (1995)
5. Amgoud, L., Cayrol, C.: On the acceptability of arguments in preference-based argumentation. In: Proceedings of the 14th Annual Conference on Uncertainty in Artificial Intelligence (UAI-98), San Francisco, CA, USA (1998)
6. Amgoud, L., Cayrol, C.: A reasoning model based on the production of acceptable arguments. Ann. Math. Artif. Intell. **34**(1), 197–215 (2002)
7. Amgoud, L., Serrurier, M.: Agents that argue and explain classifications. Auton. Agent. Multi-Agent Syst. **16**(2), 187–209 (2008)
8. Gómez, S.A., Chesnevar, C.I.: Integrating defeasible argumentation with fuzzy art neural networks for pattern classification. J. Comput. Sci. Technol. **4**(1), 45–51 (2004)
9. Peters, M.E., et al.: Deep contextualized word representations. arXiv preprint arXiv:1802.05365 (2018)
10. Radford, A., Narasimhan, K., Salimans, T., Sutskever, I.: Improving language understanding by generative pre-training (2018)
11. Devlin, J., Chang, M.W., Lee, K., Toutanova, K.: Bert: pre-training of deep bidirectional transformers for language understanding. arXiv preprint arXiv:1810.04805 (2018)
12. Lample, G., Conneau, A.: Cross-lingual language model pretraining. arXiv preprint arXiv:1901.07291 (2019)
13. Yang, Z., Dai, Z., Yang, Y., Carbonell, J., Salakhutdinov, R.R., Le, Q.V.: Xlnet: generalized autoregressive pretraining for language understanding. Adv. Neural Inf. Process. Syst. **32** (2019)
14. Liu, Y., et al.: Roberta: a robustly optimized bert pretraining approach. arXiv preprint arXiv:1907.11692 (2019)
15. Cui, Y., Che, W., Liu, T., Qin, B., Wang, S., Hu, G.: Revisiting pre-trained models for chinese natural language processing. arXiv preprint arXiv:2004.13922 (2020)
16. Wei, J., et al.: Nezha: neural contextualized representation for Chinese language understanding. arXiv preprint arXiv:1909.00204 (2019)

Two Stage Learning for Argument Pairs Extraction

Shuheng Wang[1], Zimo Yin[2], Wei Zhang[2](✉), Derong Zheng[2], and Xuan Li[2]

[1] School of Computer Science and Engineering, Nanjing University of Science and Technology, Nanjing 210094, China
wsh@njust.edu.cn

[2] Pingan Life Insurance Company of China, Shenzhen 518000, China
{yinzimo618,zhangweif95,zhengderong704,lixuan208}@pingan.com.cn

Abstract. Argument pair extraction (APE) is an important research area for real-world applications such as online debates and persuasive essays. To facilitate the research of APE, NLPCC 2021 shared Task1 releases the RR-dataset, which aims to extract argument pairs from the peer reviews and its rebuttals. We propose a two-stage learning strategy for this task which divides it into two sub-tasks: arguments mining and arguments pairing. In arguments pairing task, instead of modeling arguments pairing task in sentence level, we cast it as a paragraph-level pairing task, which can alleviate the mismatch between the definition of task and training. And we apply transfer learning and fine-tuning strategy on all sub-tasks, which exploits large scale pre-trained semantic knowledge to benefit downstream APE task. Experiment results show that our method achieves significant improvements compared with strong baseline BERT-based multi-task learning framework, and finally ranks the 3rd in NLPCC2021 shared task 1 track 3 evaluation phrase.

Keywords: Argument pair extraction · Pre-trained language model · Arguments mining · Arguments pairing

1 Introduction

Argument pair extraction (APE) is an information extraction task, which aims to extract argument pairs from the reviews and rebuttals [3]. Given a peer review and its rebuttal, argument pairs denotes the arguments in review and the replies to the argument in the rebuttal. In a way, APE can be seen as a special case of arguments mining [10]. In recent years, arguments mining has attracted many attentions from researchers. However, at present, many researchers focus on mining arguments from online debates or discussions [1,2]. Compared with the online discussion or debates, peer reviews and rebuttals contains rich arguments. And

S. Wang—Author contributed equally.

L. Wang et al. (Eds.): NLPCC 2021, LNAI 13029, pp. 538–547, 2021.
https://doi.org/10.1007/978-3-030-88483-3_44

there are strong interactions and relations between the peer review and its corresponding rebuttal. Meantime, due to its better structure, peer reviews and rebuttals are more suitable for argument pairs extraction than online debates and discussion.

Table 1. An example of Argument pair in review-rebuttal. *ARG1* and *Reply_ARG1* denote the first argument and its reply.

Reviews:
$\{$Comment:$\}_{Non}$
$\{$I am not at all familiar with quantization methods, therefore no knowledge of relevant related works. If, however, the authors did a thorough job of surveying related works and chose sensible baselines, I think the experiments demonstrate the usefulness of the new DualPrecision technique.$\}_{ARG1}$
Rebuttal:
$\{$We thank the reviewer for your feedback and for appreciating our work.$\}_{Non}$
$\{$We have updated our manuscript so that it states our contributions explicitly. While our proposed DualPrecision is simple yet effective, we would like to emphasize that it is only one of our contributions... $\}_{Reply\_ARG1}$

To demonstrate the annotation explicitly, We show an example in the Table 1. The review and its reply are two passages, and there are some arguments in the review passage and its replies in the rebuttal passage. And indices are labeled on the arguments to show the alignment between the review and rebuttal. As shown in the example, APE can be split two sub-tasks: 1, Arguments mining: detecting whether the sentences include the arguments. 2, Arguments pairing: pairing the arguments in rebuttal with the arguments in review. While there are some works on reviews [8,19], they only focus on the reviews but the rebuttals are ignored. To extract argument pairs, [3] proposed the RR-dataset to facilitate the study of argument pair extraction. Meantime, to promote the study of argument pair extraction, NLPCC 2021 Shared task 1 released the dataset.

Benefiting from the developments of pre-training language models [4,11], in this work, we introduce a pre-training model based method to extract argument pairs. Specially, we divide APE task into two sub tasks, and we design the model for the two sub tasks separately. For Arguments mining, we cast it as the sequence-level sequence label task. And for arguments pairing, we cast it as a paragraph-level binary classification task. Due to the difficulty of paragraph extraction, we use the pipeline method to combine two sub tasks. Compared with the existing methods, there are two advantages in this work:

- Unlike previous works modeling arguments pairing task in sentence level, our method casts arguments pairing as a paragraph-level pairing task, which can alleviate the mismatch between the definition of task and training.
- Our method applies transfer learning and fine-tuning strategy, which exploits large scale pre-trained semantic knowledge to benefit downstream APE task.

To evaluate the performance of our method, we conduct experiments on the RR-dataset and our method achieves significant improvements than baseline model. And our method ranks the third place in the NLPCC 2021 Shard task 1 Track 3 Evaluation.

2 Related Works

Pre-training Language Models. With the developments of corpus and compute resource, pre-training technology has been widely-used in natural language processing (NLP) [4]. At present, pre-train mainly focuses on language models. And pre-trained language models aim to learn universal and powerful contextual representations from a large corpus by self-supervised learning [4,5,11,20]. Based on the pre-trained language models, several NLP tasks have achieved new state-of-the-art levels. Among those pre-trained models, BERT is the first and popular pre-trained model. There are two tasks in the BERT model, masked language model (MLM) and next sentence prediction (NSP). Given a sentence $X = \{x_1, x_2, ..., x_n\}$, BERT randomly selects a part of tokens and replaces them with the special token "[MASK]". And then, BERT generates the masked tokens based on the left tokens. This process can be written as:

$$L_{MLM} = -\sum_{t=1}^{N} logP(x_t|X_L;\theta), \tag{1}$$

where N denotes the number of masked tokens.

While BERT boosts the performance of NLP tasks, BERT has some weaknesses, for example, the NSP task may decrease the performance, and BERT adapts the static masking method. To address this issue, RoBERTa [11] has been proposed to optimize BERT model. So, in this work, our method mainly uses the RoBERTa model to extract the sentence representations.

Arguments Mining. Arguments mining is an important field that attracts growing attention in recent years [10]. And in recent years, a series of works have been done to mine argument [15,17] and argument relation detection [7, 14]. And compared the works that model two tasks separately, [13] proposed a pipeline method to model two tasks in one model. However, their method mainly focus on one single passages, and review-rebuttal argument pairs extraction has two passages. So, [3] uses multi-task learning model to extract arguments and detect the relations. When the arguments are extracted, they detect whether the sentence in rebuttal matches the sentence in review. However, we think this method is not suitable. Arguments in review and rebuttal consist of multi sentences, and sentence-level pairing can not reflect the relation of arguments. So, in this work, we directly detect the arguments relations in review and rebuttals.

3 Method

We propose a two-stage learning framework to accomplish this task. The first stage is arguments mining, in which we use a combination of pre-trained language model and bi-directional LSTM [6] model to extract multiple continual sentence blocks from both review and rebuttal as arguments. The second stage is arguments pairing, in which we use a pre-trained language model to predict the probability of an argument pairs being correlated.

We define the first stage of the task as a sentence-level sequence labeling task. Despite the role of each sentence is given as known information in this task, we reckon it is beneficial to use the whole review-rebuttal paragraph as input for model training. Given a sequence of review-rebuttal sentences $S = \{s_1^1, s_1^2, \ldots, s_1^m, s_2^1, s_2^2, \ldots, s_2^n\}$, ur task is to predict a label sequence Y where each element y_i belongs to the label set $\{B, I, O\}$, where B represents the start sentence of an argument, I represents the rest part of an argument, O represents non-argument. For the model architecture, we choose RoBERTa [11] as the sentence encoder. For every input sentence feed into RoBERTa model, a special "[CLS]" token is added in front of the beginning word. We use the embedding of "[CLS]" token as the representation of the sentence. After having a sequence of representations for each sentence from review and rebuttal, we use it as input for a bi-directional LSTM layer. The bi-LSTM layer takes the order information of the sentences into account. Then the output of bi-LSTM layer is fed into a linear layer to predict the probability of each sentence being a label of $\{B, I, O\}$ (Fig. 1).

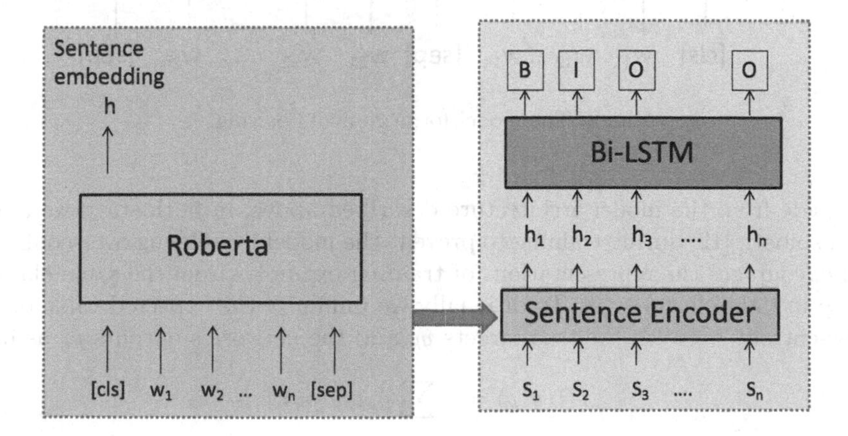

Fig. 1. The model for arguments mining.

The second stage is an arguments pairing task where we need to find argument pairs from a set of review arguments and rebuttal arguments. In the training process, we define the task as a paragraph-pair binary classification task. During sampling, we randomly choose one argument from the review and one argument from rebuttal, we use it as positive sample if they are labeled as pair, and negative otherwise. Noted that using such a strategy will lead to severely unbalanced dataset with too many negative samples, we use hyper-parameter α to control the ratio between positive and negative samples. We use RoBERTa model as the classifier, the model weights are not shared with the RoBERTa model we use in first stage. The input argument pairs is formed as: $\{[CLS], x^1_{review}, x^2_{review}, \ldots, x^n_{review}, [sep], x^1_{rebuttal}, x^2_{rebuttal}, \ldots, x^m_{rebuttal}, [sep]\}$. "[CLS]" token embedding is used as the paragraph-pair embedding and flows into a liner classifier. In the prediction phase, for every review argument extracted in the first stage, we bind it with all rebuttal arguments and use the model to select the one with the highest probability of being positive (Fig. 2).

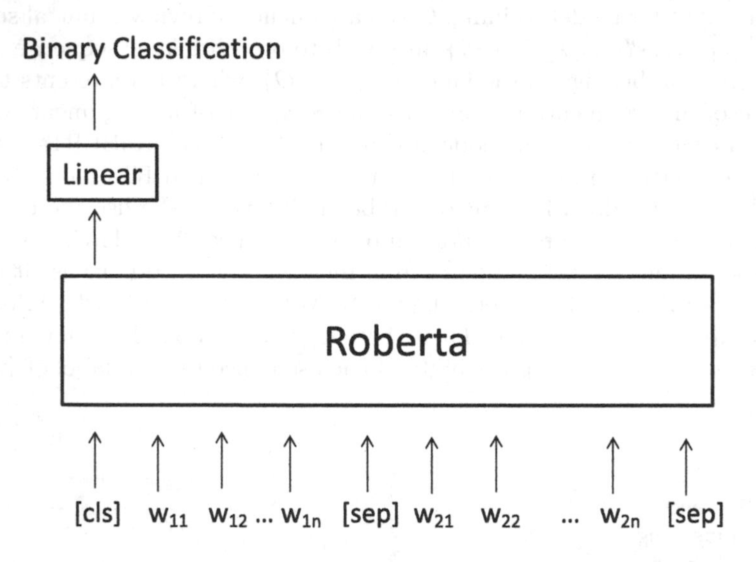

Fig. 2. The model for arguments pairing.

Aside from the model architecture described above, in both stage we apply label smooth [12] during training to prevent the model from being over-confident and encourages the representations of training examples from the same class to group in tight clusters [16]. Traditionally we minimize the expected value of the cross-entropy between the true targets y_k and the network's outputs p_k as in

$$H(y,p) = -\sum y_k log(p_k), \tag{2}$$

where y_k is "1" for the correct class and "0" for the rest. For a network trained with a label smoothing of parameter α, we minimize instead the cross-entropy between the modified targets y_k^{LS} and the networks' outputs p_k, where y_k^{LS} is calculated as:

$$y_k^{LS} = y_k(1-\alpha) + \alpha/K. \tag{3}$$

4 Experiment and Results

4.1 Dataset and Metrics

We use the RR-submission dataset released by NLPCC 2021 Shard task 1 Track 3 to evaluate the performance of our method. Each sample in the dataset consists of two passages, and BIO tags are used to point out which sentences are arguments and its replies. And the statistics of RR-submission dataset are shown in the Table 2. NLPCC 2021 Shard task 1 Track 3 uses Macro-F1 as the evaluation metric, which is calculated as:

Table 2. The statistics of dataset.

Passage	Typies	Train set	Valid set	Test set
Review	Sentences	90853	11349	11892
	Arguments	18642	2310	2396
	Avg arguments sentences	2.82	2.86	2.93
Rebuttal	Sentences	86910	11264	10301
	Arguments	14315	1811	1756
	Avg arguments sentences	4.15	4.35	4.15

$$P = TP/(TP + FP), \tag{4}$$

$$R = TP/(TP + FN), \tag{5}$$

$$F1 = 2 * P * R/(P + R), \tag{6}$$

where TP, FP and FP denote true positives, false positives and false negatives respectively. The final result is reported on the test set reserved by the organizer.

4.2 Settings

We implement our method based on the framework PyTorch and Transformers [18]. All models are trained on a Nvidia Tesla V100 GPU. The default dimension of pre-trained model is used as the dimension of sentence embedding. The dimension of LSTM is set as 512. The number of LSTM layers is set as 3. And we use Adam [9] with learning rate $1e-5$ to optimize our model. We use the last checkpoint to create the final model. Noted that we use each review-rebuttal pair as a sample, and due to the memory limitation, the batch size is set as 1.

4.3 Baselines

To evaluate the performance of our method, we choose some strong baselines:

- LSTM+LSTM: We use LSTM to extract the sentence embeddings and use LSTM to label sentence-level sequence and arguments pairing.
- BERT_{base}+LSTM: We use the BERT_{base} model to extract the sentence embeddings and then use LSTM to label sentence-level sequence and arguments pairing.
- RoBERTa_{base}+LSTM: This is our submitted method. We use RoBERTa_{base} to extract the sentence embeddings and use LSTM to label sentence-level sequence and arguments pairing.

4.4 Main Results

In the inference stage, our method is applied as two modules to conduct the subtasks of argument mining and arguments pairing sequentially. So, in this section, we evaluate the performance of our method on argument mining, arguments pairing, and the whole argument pairs extraction.

The Performance on Argument Mining. We first show the performance of our method on argument mining task in the Table 3. As the table shown, with the different methods to extract the sentence embeddings, the models achieve different results. As we know, pre-trained model improve the performance of many models. And it can be seen in our experiments. BERT_{base}+LSTM achieves 1.6 improvements on F1 compared with LSTM+LSTM. In addition, RoBERTa_{base}+LSTM, which uses RoBERTa_{base} as the pre-trained model, achieves better results. The reason is that RoBERTa_{base} uses dynamic masking, which strengthens the model's capability learning the semantic representation.

Table 3. The comparison on Arguments Mining

Model	F1	P	R
LSTM+LSTM	61.080	60.528	61.643
BERT_{base}+LSTM	64.609	63.095	66.198
RoBERTa_{base}+LSTM	65.899	65.939	65.859

The Performance on Arguments Pairing. We also show the results of our method on arguments pairing task in the Table 4. And we can see that on a difficult task, LSTM+LSTM achieves a much worse result. In contrast, pre-trained based methods both achieve a better result. And it denotes that the better representations will achieve better performance on difficult task. Similarly, we also observe the model RoBERTa_{base}+LSTM, with a better pre-training method, obtains better results than BERT_{base}+LSTM.

Table 4. The performance of Arguments pairing.

Model	ACC	F1
LSTM+LSTM	58.295	61.876
BERT$_{base}$+LSTM	85.507	83.766
RoBERTa$_{base}$+LSTM	87.785	86.985

The Overall Performance of Our Method. Table 5 shows the overall extraction performance on RR-passage dataset. We can observe that our method achieves significant improvement compared to the organizer's baseline, which denotes the effects of our method. Compared with baseline, using sentence-level pairing in arguments pairing, the model, which achieves better result on arguments pairing, will achieves better performance on the overall argument pairs extraction. And with pre-trained model, our method continues getting better results. We submit RoBERTa$_{base}$+LSTM as our final model, and our model obtains 26.25 F1 scores and ranks the third place.

Table 5. Main results on RR-passage dataset. "/" denotes no values in the item.

Model	P	R	F1
Baseline	/	/	26
LSTM+LSTM	14.941	6.035	8.597
BERT$_{base}$+LSTM	41.654	29.592	34.602
RoBERTa$_{base}$+LSTM	43.764	31.392	36.560
Final	/	/	26.25

5 Conclusions

In this work, we introduce a new method to extracting argument pairs from review and rebuttal passages. We divide the APE task into two sub-task, and model them sequentially. For argument mining, we treat the task as a sentence-level label task. And for argument pairing, we model it as a paragraph-level pairing task. The experiments show that our method achieves significant improvement than baseline. And finally, our method achieves the third place in NLCC 2021 Shard Task 1 Track 3 Evaluation.

Acknowledgements. We thank the reviewers for their careful reviewing and valuable advises, which are important for us to improve our work.

References

1. Boltužić, F., Šnajder, J.: Identifying prominent arguments in online debates using semantic textual similarity. In: Proceedings of the 2nd Workshop on Argumentation Mining, pp. 110–115 (2015)

2. Chakrabarty, T., Hidey, C., Muresan, S., McKeown, K., Hwang, A.: Ampersand: argument mining for persuasive online discussions. arXiv preprint arXiv:2004.14677 (2020)
3. Cheng, L., Bing, L., Yu, Q., Lu, W., Si, L.: Argument pair extraction from peer review and rebuttal via multi-task learning. In: Proceedings of the 2020 Conference on Empirical Methods in Natural Language Processing (EMNLP), pp. 7000–7011 (2020)
4. Devlin, J., Chang, M.W., Lee, K., Toutanova, K.: Bert: pre-training of deep bidirectional transformers for language understanding. arXiv preprint arXiv:1810.04805 (2018)
5. Dong, L., et al.: Unified language model pre-training for natural language understanding and generation. arXiv preprint arXiv:1905.03197 (2019)
6. Hochreiter, S., Schmidhuber, J.: Long short-term memory. Neural Comput. **9**(8), 1735–1780 (1997)
7. Hou, Y., Jochim, C.: Argument relation classification using a joint inference model. In: Proceedings of the 4th Workshop on Argument Mining, pp. 60–66 (2017)
8. Hua, X., Nikolov, M., Badugu, N., Wang, L.: Argument mining for understanding peer reviews. arXiv preprint arXiv:1903.10104 (2019)
9. Kingma, D.P., Ba, J.: Adam: a method for stochastic optimization. arXiv preprint arXiv:1412.6980 (2014)
10. Lawrence, J., Reed, C.: Argument mining: a survey. Comput. Linguist. **45**(4), 765–818 (2020)
11. Liu, Y., et al.: Roberta: a robustly optimized bert pretraining approach. arXiv preprint arXiv:1907.11692 (2019)
12. Müller, R., Kornblith, S., Hinton, G.: When does label smoothing help? arXiv preprint arXiv:1906.02629 (2019)
13. Persing, I., Ng, V.: End-to-end argumentation mining in student essays. In: Proceedings of the 2016 Conference of the North American Chapter of the Association for Computational Linguistics: Human Language Technologies, pp. 1384–1394 (2016)
14. Rocha, G., Stab, C., Cardoso, H.L., Gurevych, I.: Cross-lingual argumentative relation identification: from English to Portuguese. In: Proceedings of the 5th Workshop on Argument Mining, pp. 144–154 (2018)
15. Shnarch, E., et al.: Will it blend? blending weak and strong labeled data in a neural network for argumentation mining. In: Proceedings of the 56th Annual Meeting of the Association for Computational Linguistics, vol. 2: Short Papers, pp. 599–605 (2018)
16. Szegedy, C., Vanhoucke, V., Ioffe, S., Shlens, J., Wojna, Z.: Rethinking the inception architecture for computer vision. In: Proceedings of the IEEE Conference on Computer Vision and Pattern Recognition, pp. 2818–2826 (2016)
17. Trautmann, D., Daxenberger, J., Stab, C., Schütze, H., Gurevych, I.: Fine-grained argument unit recognition and classification. In: Proceedings of the AAAI Conference on Artificial Intelligence, vol. 34, pp. 9048–9056 (2020)
18. Wolf, T., et al.: Transformers: state-of-the-art natural language processing. In: Proceedings of the 2020 Conference on Empirical Methods in Natural Language Processing: System Demonstrations, pp. 38–45. Association for Computational Linguistics, Online (2020). https://www.aclweb.org/anthology/2020.emnlp-demos.6

19. Xiong, W., Litman, D.: Automatically predicting peer-review helpfulness. In: Proceedings of the 49th Annual Meeting of the Association for Computational Linguistics: Human Language Technologies, pp. 502–507 (2011)
20. Yang, Z., Dai, Z., Yang, Y., Carbonell, J., Salakhutdinov, R.R., Le, Q.V.: Xlnet: generalized autoregressive pretraining for language understanding. In: Advances in Neural Information Processing Systems, vol. 32 (2019)

Overview of Argumentative Text Understanding for AI Debater Challenge

Jian Yuan[1], Liying Cheng[2,3], Ruidan He[2], Yinzi Li[1], Lidong Bing[2(✉)], Zhongyu Wei[1,4(✉)], Qin Liu[5], Chenhui Shen[2], Shuonan Zhang[2], Changlong Sun[2], Luo Si[2], Changjian Jiang[6], and Xuanjing Huang[5]

[1] School of Data Science, Fudan University, Shanghai, China
{19210980107,20210980077}@fudan.edu.cn
[2] DAMO Academy, Alibaba Group, Beijing, China
{liying.cheng,ruidanhe,l.bing,chenhui.shen,shuonan.zhang, changlong.sun,luo.si}@alibaba-inc.com
[3] Singapore University of Technology and Design, Singapore, Singapore
[4] Research Institute of Intelligent and Complex Systems, Fudan University, Shanghai, China
[5] School of Computer Science, Fudan University, Shanghai, China
{liuq19,xjhuang}@fudan.edu.cn
[6] School of International Relations and Public Affairs, Fudan University, Shanghai, China
changjian@fudan.edu.cn

Abstract. In this paper we present the results of the Argumentative Text Understanding for AI Debater Challenge held by the 10th CCF International Conference on **N**atural **L**anguage **P**rocessing and **C**hinese **C**omputing (NLPCC2021), and introduce the related datasets. We organize three tracks to handle the argumentative texts in different scenarios, namely, supporting material identification (track 1), argument pair identification from online forum (track 2) and argument pair extraction from peer review and rebuttal (track 3). Each track is equipped with its distinct dataset and baseline model respectively. In total, 110 competing teams register for the challenge, from which we received 54 successful submissions. In this paper, we will present the results of the challenge and a summary of the systems, highlighting commonalities and innovations among participating systems. Datasets and baseline models of the Argumentative Text Understanding for AI Debater Challenge have been already released and can be accessed through the official website (http://www.fudan-disc.com/sharedtask/AIDebater21/index.html) of the challenge.

Keywords: Argumentation mining · AI Debater · Natural Language Processing

J. Yuan, L. Cheng and R. He—contributed equally

L. Wang et al. (Eds.): NLPCC 2021, LNAI 13029, pp. 548–568, 2021.
https://doi.org/10.1007/978-3-030-88483-3_45

1 Introduction

Argument and debate are fundamental capabilities of human intelligence, essential for a wide range of human activities, and common to all human societies. Striving to enable models to automatically understand and generate argument texts, computational argumentation, a newly emerging research field, is obtaining increasing attention from the research community [30]. According to the number of involved participants in the given argumentative text, computational argumentation can be categorized into two aspects, monological argumentation and dialogical argumentation.

Monological argumentation, by the definition, mainly focus on the monological contexts where only one participant is involved, existing research includes argument structure prediction [26,29], persuasiveness evaluation [4,17] and argument summarization [8,9]. Specifically, as for an autonomous AI debater, understanding the argumentative texts, to a large extent, requires the capability of identifying whether a piece of evidence is for or against the given argument, i.e. the capability of supporting material identification.

In contrast, dialogical argumentation usually handles the contexts such as debates and online forums, in which two or more parties take turns to either state their own arguments or refute the other's ones. With the rapid development of modern technology, online forums such as idebate[1] and changemyview[2] which enable people to exchange opinions on some specific topics freely become suitable data sources for argumentation mining, especially for the design of an AI debater since online forums highly resembles real world debates. Initial researches in this filed focused on analyzing the ChangeMyView data [36,38] to summarize the key factors of persuasive arguments. Furthermore, Ji et al. [22] propose the task of identifying interactive arguments in such online debating forums. An example of such interactive argument pair is as shown in Fig. 1. As can be seen, opinions from both sides are voiced with multiple arguments and the reply post B is organized in-line with post A's arguments. Formally, an interactive argument pair is formed with two arguments from both sides (with the same underline), which focuses on the same perspective of the discussion topic. The automatic identification of these pairs will be a fundamental step towards the understanding of dialogical argumentative structure.

In addition, peer review and rebuttal in evaluation process of scientific works, contain rich arguments and are thus another worth-studying domain for dialogical argumentation. In practice, a review and its rebuttal have strong interactions and relations, as a rebuttal usually contains coherent segments that argue with specific arguments in review. Compared to online debates and discussions, peer reviews and rebuttals on scientific works are a data source of rich structures and long passages for studying the argument pair extraction (APE). This study can also be applied to the study of online debates and discussions on how to respond to opponents' arguments. Cheng et al. [12] proposed such novel task in 2020.

---

[1] https://idebate.org/.

[2] https://www.reddit.com/r/changemyview/.

Table 1 is an example of review-rebuttal passage pair, where the two passages are segmented into arguments and non-arguments. For arguments, indices are also labeled to show the alignment between review and rebuttal, i.e., the rebuttal argument that attacks the review argument.

CMV: The position of vice president of the USA should be eliminated from our government.
Post A: a1: [If the president is either killed or resigns, the vice president is a horrible choice to take over office.] **a2:** The speaker of the House would be more qualified for the position. **a3:** [I'm willing to bet that John Boehner would have an easier time dealing with congress as president than Joe Biden would due to his constant interaction with it.] **a4:** If Boehner took office, as a republican, would he do something to veto bills Obama supported?
Post B: b1:[Seriously, stop this hyperbole.] **b2:** [Do you think that have anything to do with the fact that Boehner is a republican, and republicans control congress?] **b3:** That argument has much less to do with the individuals than it does with the current party in control.

Fig. 1. An example of dialogical argumentation consists of two posts from change my view, a sub-forum of Reddit.com. Different types of underlines are used to highlight the interactive argument pairs.

Table 1. An example of review-rebuttal passage pair. REVIEW-1 and REPLY-1 indicate the first argument in the review and its reply in the rebuttal.

Review:
[The authors introduce an extension of Continuous Ranked Probability Scores (CRPS) to the time-to-event setting termed Survival-CRPS for both right censored and interval-censored event data.]NON-ARGU ... [The claim that the proposed approach constitutes the first time a scoring rule other than maximum likelihood seems too strong, unnecessary and irrelevant to the value of the presented work.]REVIEW-1 [It is not clear how did the authors handle the irregularity (in time) of EHR encounters in the context of an RNN specification. Also, if the RNN specification considered is similar to Martinsson, 2016, why this wasn't considered as a competing model in the experiments?]REVIEW-2 ...
Rebuttal:
[Dear AnonReviewer3, we thank you for the time you spent reviewing and for the thoughtful comments.]NON-ARGU [You point out that our claim that the approach constitutes the first time a scoring rule other than maximum likelihood has been used seems too strong. ... However we have recently come across works such as adversarial time to event models, and therefore will remove the claim from the paper.]REPLY-1 [You also mention that it is unclear how irregularity (in time) of EHR encounters was handled by the RNN model. ... This approach naturally handles the irregularity of time between EHR encounters.]REPLY-2 ... [Thank you again for your time spent reviewing and constructive feedback on our work.]NON-ARGU

Therefore, with an aim of developing an autonomous debating system, we hosted the Argumentative Text Understanding for AI Debater Challenge[3], including three tracks each representing one aforementioned valuable subtask, namely, supporting material identification (track 1), argument pair identification from online forum (track 2) and argument pair extraction from peer review and rebuttal (track 3). In track 1, we present the fundamental scenario of supporting material identification for a given debating topic. We then move to the understanding of dialogical argumentative text in two domains, i.e., online debating forum and scientific paper review process. We provide three datasets in this task, one for each track.

In total, 111 teams from over 35 colleges and corporates enter for the challenge, 54 of which successfully submit their models and obtain their model's performance. We hope that research and practice in argumentation mining and autonomous debating AI will be stimulated by the challenge.

In this paper, we present a detailed description for the subtasks and their datasets, along with a summary of the submissions, and discuss the possible future research directions of the task.

2 Related Works

2.1 Supporting Material Identification

Detecting the stance of an argument is essential towards building a Debating system. This task is related to opinion mining. However, opinion mining focuses on detecting the sentiment polarity expressed by a text while stance detection aims to determine the position that the text holds with regards to a topic that is generally more abstract. Therefore, in stance classification, texts can transmit a negative sentiment or opinion, but be in favor of the targeted topic.

The majority of previous approaches learn topic-specific models using supervised machine learning algorithms [1,5,19,20,31–33]. Under this formulation, new topics require new models, which is not feasible for real-life applications. Another direction makes use of various context information such as the relevant posts and the network of authors [1,7,20]. In our formulation, we only target claim-topic pairs in isolation. We consider a topic independent setting to stance detection for short claims without considering the context. Our problem formulation is mostly related to [6] and [25]. Since in our formation, we are given a topic-argument pair as input and aim to detect the stance label, this problem is also related to other sentence-pair classification tasks such as the detection of semantic text similarity (STS) [2], paraphrase recognition [10], and textual entailment [15].

[3] This event is an NLPCC 2021 task sponsored by Fudan University and Alibaba Group.

2.2 Argument Pair Identification from Online Forums

As mentioned before, computational argumentation (or equivalently, argumentation mining) can be divided into two categories, monological argumentation and dialogical argumentation. Among recent researches in the dialogical argumentation aspect, Swanson et al. [35] aim to extract arguments from a large corpus of posts from online forums, they frame their problem as two separate subtasks: argument extraction and argument facet similarity; Chakrabarty et al. [11] focus on argument mining in online persuasive discussion forums based on the CMV dataset; Roxanne et al. [17] compare content- and style-oriented classifiers on editorials to explore the effect of the writing style of editorials to the audience of different parties; Ji et al. [22] propose the task of identifying interactive argument pairs in online debate forum such as ChangeMyView (CMV). Cheng et al. [12] collects the text data from peer review and rebuttal process to mine the argumentative relationship entailed in such discussion; Al-Khatib et al. [24] constructs a monological argumentation graph by extracting knowledge from *Debatepedia.org* and use human annotation to further improve the quality of their knowledge graph.

2.3 Argument Pair Extraction from Peer Review and Rebuttal

To understand the structure and the content of peer reviews, researchers have introduced several new datasets. PeerRead [23] is collected from venues including ACL, NIPS and ICLR. It consists of 14.7K paper drafts with their corresponding decisions (i.e. accept or reject) and 10.7K peer reviews for a subset of papers. This dataset is also applied on other tasks, such as predicting the paper's decision. Hua et al. [21] present a partially annotated dataset named AMPERE collected from four venues including ICLR, UAI, ACL and NeurIPS. They aim to detect the propositions and to classify them into different types to understand peer reviews. There are two key differences between our APE task and the two previous works. First, APE is a task to explore the reviews and rebuttals jointly, instead of only studying peer reviews. Second, the dataset for APE is fully annotated with 4,764 pairs of reviews and rebuttals, while only 400 reviews are annotated in AMPERE. As mentioned, few works explore the rebuttal process, which is the other important element of the peer review process. Gao et al. [18] propose a task to predict after-rebuttal scores from initial reviews and author responses. Their dataset is collected from ACL, which includes around 4K reviews and 1.2K author responses. However, APE focuses more on the internal structure and relations between reviews and rebuttals. In addition, the size of our dataset is much larger than theirs.

3 Task Description

In this section, we formally define the specific task, introduce the construction of the corresponding dataset, scoring metrics as well as the baseline model for each track respectively.

3.1 Track1: Supporting Material Identification

Task Formulation. Our problem formulation is similar to [6,25]. Formally, given an argumentative claim and a topic, our task is to detect the stance that the claim has with respect to the topic into either *Support*, *Against* or *Neutral*. The labels represent whether the claim is in favor of, against, or unrelated to the given topic.

Table 2. Data statistics of supporting material identification.

	Support	Against	Neutral
Train	2505	2011	1900
Dev	358	346	286
Test	380	356	264

Dataset Construction. We collected 131 debating topics from online forums. The collected topics are either in English or Chinese. We follow the claim annotation guideline from [3] to collect the set of claims for each topic. Specifically, given a topic, we first collect 10 English Wikipedia or Chinese Baidu Encyclopedia articles with promising content. Then, human annotators were recruited to annotate claims from the collected articles and identify their stances with regards to the given topic. Finally, human translators were recruited to translate the English topics and claims into Chinese.

In total we constructed 8406 topic-claim pairs. The data is splited into training, dev, and test sets with 6416, 990, and 1000 numbers of instances, respectively. We consider a cross-topic setting where the set of topics in the training set is not overlapped with those in the dev and test sets. Detailed statistics are given in Table 2

Scoring Metric. We use classification accuracy as the evaluation metrics.

Baseline Model. We train a sentence-pair classification model based on *bert-base-chinese* as our baseline. Specifically, the topic text and the claim text are concatenated and fed into the BERT model. We take the hidden state of the "[CLS]" token into a linear classifier to output the final prediction. Cross-entropy is used as the loss function.

3.2 Track2: Argument Pair Identification from Online Forum

Task Formulation. We formulate our task according to Ji et al.'s setting [22]. Formally, given an argument q from the original post, a candidate set of replies $r_1^- \sim r_5^-$, which consist of one positive reply and four negative replies, and their corresponding argumentative contexts $c_q, c_1 \sim c_5$, our goal is to automatically identify which candidate reply has interaction relationship with the quotation q.

The task is then formulated as a pairwise ranking problem. In practice, we calculate the matching score $S(q, r)$ for each candidate reply r_i with quotation q and treat the one with the highest matching score as the proposed answer.

Dataset Construction. In the challenge, we adopt Ji et al.'s dataset and further augment it following their construction method. Their data collection is built on the CMV dataset released by Tan et al. [36]. The original dataset is crawled using Reddit API. Discussion threads from the period between January 2013 and September 2015 are collected.

Original Post: ... Strong family values in society lead to great results. *I want society to take positive aspects of the early Americans and implement that into society.* This would be a huge improvement than what we have now. ...
User Post: *> I want society to take positive aspects of the early Americans and implement that into society.* What do you believe those aspects to be? ...

Fig. 2. An example illustrating the formation process of a quotation-reply pair in CMV.

Table 3. Overview statistics of the constructed dataset (mean and standard deviation). *arg.*, q, r^+, r^- represent *argument*, *quotation*, *positive reply* and *negative reply* respectively. $q - r^+$ represents the quotation-reply pair between posts.

	Training set	Test set
# of arg. per post	11.8 ± 6.6	11.4 ± 6.2
# of token per post	209.7 ± 117.2	205.9 ± 114.6
# of token per q	20.0 ± 8.6	20.0 ± 8.6
# of token per r^+	16.9 ± 8.1	17.3 ± 8.4
# of token per r^-	19.0 ± 8.1	17.3 ± 8.4
max # of $q - r^+$ **pairs**	12	9
avg. # of $q - r^+$ **pairs**	1.5 ± 0.9	1.4 ± 0.9

An observation shown in Fig. 2 on CMV shows that when users reply to a certain argument in the original post, they quote the argument first using the special token ">" (or "*>*" in raw data) and generate their own responsive argument directly, thus forming a quotation-reply pair. Therefore, they extracted all the arguments following the ">" token as the quotation argument and the first sentence right after it as the positive reply argument. After the collection of all the quotation-reply pairs, they randomly select four negative replies from other posts that are also related to the original post to pair with the quotation as negative samples. In addition, they also collect the contexts of all the quotation and reply posts, but remove the quoted sentences in the reply contexts for fairness. Finally, each entry in the dataset includes the quotation, one positive reply, four negative replies along with their corresponding posts. In order to assure the quality of quotation-reply pairs, they only keep the instances where the number of words in the quotation and replies range from 7 to 45. Instances extracted out of the discussion threads from the period between January 2013 and May 2015 are collected as training set, while instances extracted out of threads between May 2015 and September 2015 are considered as test set. In total, the number of instances in training and test set is 11,565 and 1,481, respectively. The statistic information of their dataset is shown in Table 3.

Scoring Metric. For the released multiple choice task, we take accuracy as the evaluation metric. Specifically, if the ground truth of the ith problem is y_i, and the system predicts the answer to be $\hat{y_i}$, then the average accuracy on the test dataset of size n is calculated as below:

$$accuracy = \frac{\sum_{i=1}^{n} y_i = \hat{y}_i}{n}$$

Baseline Model. The baseline model for the subtask is based on the sentence-par classification model of BERT [16]. As is shown in Fig. 3, the baseline model first concatenate the quotation argument q with candidate reply arguments r_i separately. In this way, we convert the 5-way multiple choice into 5 sentence-pair classification problems. For each sentence pair, it is fed into a BERT classification model and then we take the hidden state of the "[CLS]" token into a linear classifier to output the final prediction.

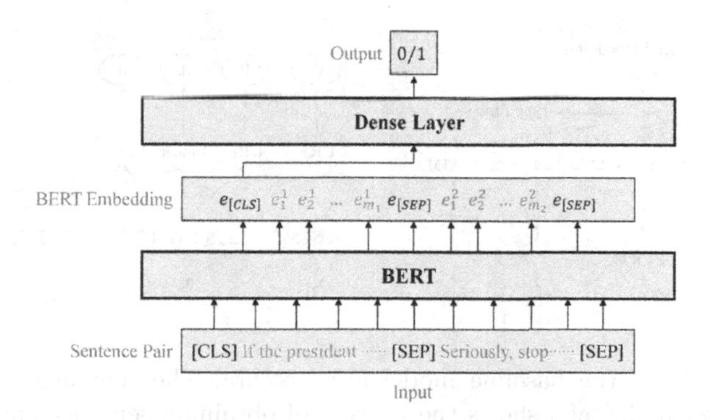

Fig. 3. The overall framework of the BERT-based baseline model for track 2.

3.3 Track3: Argument Pair Extraction from Peer Review and Rebuttal

Task Formulation. Our problem formulation is to identify the argument pairs from each passage pair of review and rebuttal. Specifically, given the two passages, a review passage of m sentences $R_1 = [s_1^1, s_1^2, \cdots, s_1^m]$ and a rebuttal passage of n sentences $R_2 = [s_2^1, s_2^2, \cdots, s_2^n]$, we first perform argument mining subtask to get the review arguments and rebuttal arguments. We then match individual review arguments with the rebuttal arguments, such as {REVIEW-1, REPLY-1} and {REVIEW-2, REPLY-2}, to extract the argument pairs.

Dataset Construction. We collect the dataset from ICLR 2013–2020 (except for 2015 that is unavailable) on http://openreview.net. After excluding those reviews receiving no reply, we extract 4,764 review-rebuttal passage pairs for data annotation. In total, 5 professional data annotators are hired to annotate the dataset based on the unified guidelines. Firstly, for the argument mining part, the annotators segment reviews and rebuttals by labeling the arguments and non-arguments. Given the arguments of both passages, the second step is to pair the review arguments with the rebuttal arguments.

We label 40,831 arguments in total, including 23,150 review arguments and 17,681 rebuttal arguments. 5% of the original review-rebuttal passage pairs (252 pairs) are randomly sampled from the full dataset for quality assessment. The annotation accuracy of the dataset is 98.4%. We split our dataset on submission level randomly by a ratio of 8:1:1 for training, development and testing.

Scoring Metric. We use F_1 score on argument pair extraction as the evaluation metrics.

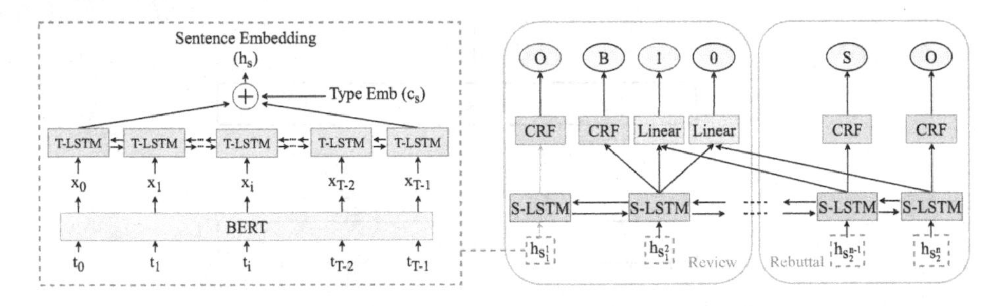

Fig. 4. Overview of the baseline model architecture. The sentence encoder (in the red dotted box on the left) shows the process of obtaining sentence embeddings from pre-trained BERT token embeddings with T-LSTM. The multi-task learning based framework (on the right) shows the process of generating labels for both subtasks with the shared S-LSTM.

Baseline Model. As shown in Fig. 4, our baseline model is a multi-task learning framework with hierarchical LSTM networks, to tackle the two subtasks (i.e., argument mining and sentence pairing). During the training phase, the red dotted box on the left shows our sentence encoder which uses the pre-trained BERT [16] token embeddings as the input of a token-level biLSTM (T-LSTM) to get sentence embeddings. The encoded sentence embeddings are then input to a sentence-level biLSTM (S-LSTM) to encode the sequence information of the passages. Finally, two types of labels, i.e., IOBES and 1/0, are predicted simultaneously from the shared features generated from the S-LSTM. During the inference, the trained multi-task model will be decoupled into two sub-modules to perform the two subtasks in a pipeline manner to get the final argument pairs.

4 Challenge Details

In the challenge, we divide the whole process into two phases. In the first phase, the input data as well as the labels of test sets are released to all contestants, which allows them to check their model's performance immediately, and they are required to submit the result file as well as their model. We calculate their score mainly on their result files, but also check their reproducibility by running their model after the due of phase 1. In phase 2, we do not release any extra datasets and contestants are only allowed to submit their trained models, we run their models in our local environment on the black-box test data. Finally, all teams are ranked according to their performance in the second phase for fairness.

There are over 110 teams from various universities as well as enterprises who have registered for the AI Debater Challenge, 24 teams who have submitted their models in the first phase, and 41 teams who have submitted their final models. The final accuracy shows that neural models can achieve considerable results on the task. We have collected the technical reports of the award-winning teams. In the following parts, we summarize their methods and tricks according to their reports. The performance of all participants on the challenge will be found in Appendix.

4.1 Track1: Supporting Material Identification

Participants and Results. There are 47 and 26 teams submitted their final models for Track 1 in phase 1 and 2, respectively. Table 4 presents the accuracy of top-10 participated teams of the task.

Submitted Models. In this subsection, we summarize the major components and techniques adopted by the participated teams:

- **Pretrained Language Model** Large pretrained language model has become an indispensable component for most of the NLP tasks at this stage. In particular, for Chinese tasks like track 1, XLMR [13], Chinese BERT [16] and

Table 4. Performance of participants on Track 1. All teams are sorted according to their final performance in Phase 2.

Team	Score 1	Score 2
Arrival	0.841	0.928
清博 AI	0.822	0.905
OVERFIT	0.838	0.895
Again	–	0.868
Error	–	0.865
OnceAgain	–	0.864
朝天椒	0.821	0.859
B318	–	0.859
IntellGame	–	0.857
注册号太难了+1	–	0.856

RoBERTa [28] are the common language models adopted among all participated teams. Besides, MacBERT [14] and NEZHA [37] are also popular options since they are specifically trained and optimised for Chinese tasks.

- **Tuning Methods** Most of the teams formulated the problem as a conventional sentence-pair classification task and performed fine-tuning on the pretrained language model. A few teams instead transformed this problem into a MLM (Masked Language Model) task and adopted prompt-based tuning [27]. In this setting, the classification labels are transformed into different relation patterns. For example, Support/Against/Neutral can be transformed into "因为", "但是", and "并且". The model is trained to predict the masked relation pattern inserted between the topic and claim.
- **Data Augmentation** Among the top-ranked teams, various data augmentation methods have been proposed. There are two common approaches. One is to exchange the positions of topics and claims to double the number of training instances. Another approach constructs claim-claim pairs under the same topic as the augmented data. Figure 5 illustrates the rules to assign labels to those augmented pairs which is based on the implicit logical relationship between claims.

  Except the above data augmentation methods, the 1st-ranked team "Arrival" proposed a novel approach to construct training datasets with different noise levels from large quantities of unlabeled text. Specifically, their idea is based on the intuition that discourse relations are indicative of stance. For each classification label, they constructed a list of discourse patterns and extracted sentences from a large Chinese text corpus CLUE [39] based on the predefined discourse patterns. The constructed noisy dataset contains over 3 million training instances. Furthermore, the authors employed a natural language inference (NLI) model for data selection to extract a subset of 30k low-noise data. In their experiments, they demonstrated that even without using the

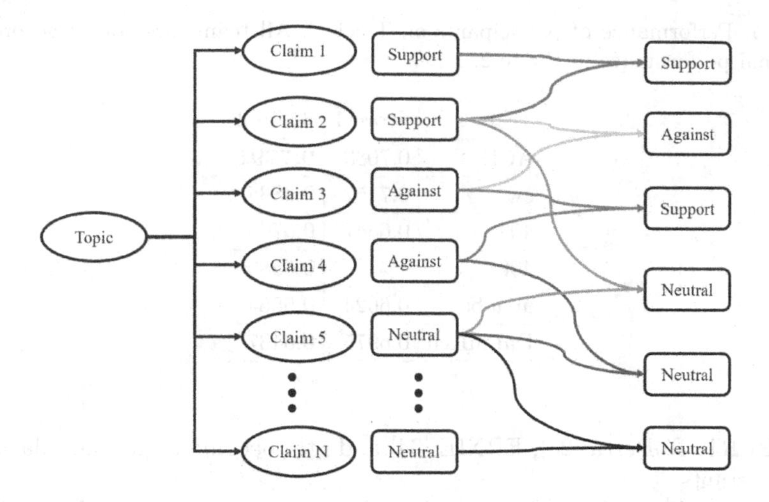

Fig. 5. Track 1 data augmentation.

gold-labeled data, XLMR with two-stage fine-tuning on the two augmented datasets can achieve 89.8% accuracy on the dev set which is more than 4% higher than the result obtained by fine-tuning XLMR directly on the gold training set.

- **Model Ensemble** Model ensemble is adopted by most of the top-ranked teams where majority voting or weighted average are used to combine the predicted results of multiple models. There are three major ensemble strategies. One is to use the same model architecture and obtain multiple models with different parameter initialization. The second approach also adopts the same model architecture but train it on different splits of train and dev data to obtain multiple models. Another approach combines the results of models with different model architecture.

4.2 Track2: Argument Pair Identification from Online Forum

Participants and Results. In total, there are 57 teams who have registered for Track 2, in which 9 and 13 teams submit their final models respectively in phase 1 and 2. The final accuracy shows that neural models can achieve considerable results on the task, though their performance decreases when given a larger test set. In Table 5, we list the scores of top-6 participants of the task.

Submitted Models. Among all submitted models from the contestants, the general framework contains:

- **Pretrained Language Model** Ever since BERT [16] was publicly proposed, the whole NLP area has been pushed into a new era, with almost all tasks improved in performance. Therefore, the pretrained language models such

Table 5. Performance of participants on Track 2. All teams are sorted according to their final performance in Phase 2.

Team	Score 1	Score 2
ACE	0.7083	**0.7291**
cw	0.7157	0.6834
404	0.6847	0.6709
400	–	0.6589
abaaba	0.6624	0.6554
DaLiuRen	0.6975	0.6437

as BERT, RoBerta [28], ERNIE [34] and etc. become highly popular among contestants.
- **Sentence Pair Classification** After leveraging the pretrained models mentioned above to obtain embedding for tokens and sentences, nearly all teams try to convert the quotation and each reply argument into a sentence pair, and further feed it into the classification model to calculate the final logit probability for the given argument pair to be interactive.

Other than the standard "pretrained model and fine-tuning(i.e. classification)" mode, there are some useful tricks which can address the issues met in the task and improve the models' performance significantly. We summarize them as follows:

- **Fine-tuning with external corpus** Some teams tried to fine-tune their pretrained model by adding external corpus in online debate forums. Such method helps improve the model since external debating corpus enables the pretrained language models to learn more topic-specific language knowledge and therefore performs better in argumentation settings.
- **Data Augmentation and Data Balancing** Among the winning teams, various data augmentation methods are utilized to improve the generalization capability and robustness of the model. For instance, the "ACE" team uses the Mixup method to generate a new synthetic sample by linearly interpolating a pair of training samples and their training targets; and the "abaaba" team generate more training instances by symmetrically swapping the argument pairs and dropout masks to further develop the online test result. According to those teams, the augmentation improves their model's performance by over 2%.
- **Ensemble** Some participants train several different classification models over different samples from the whole dataset, and finally combine them with majority voting or weighted average strategies to combine their predicting results. Among all the participants using such method, the "abaaba" team trains ten classification models based on Roberta on the 10-fold cross validation dataset and adopts the majority voting method among the top 4 models to reduce the variance of a single model and therefore improve the robustness

of the model, which finally helps their model to achieve the third prize of the competition.

4.3 Track3: Argument Pair Extraction from Peer Review and Rebuttal

Participants and Results. In total, there are 67 teams who have registered for Track 3, in which 13 and 16 teams submit their final models respectively in phase 1 and 2. Table 6 shows the scores of top-6 participants of this task.

Table 6. Performance of participants on Track 3. All teams are sorted according to their final performance in Phase 2.

Team	Score 1	Score 2
HLT-HITSZ	0.379	0.3231
nju_pumps	–	0.2629
PAL	0.296	0.2625
404	0.358	0.2563
ARGUABLY	0.270	0.2087
abaaba	0.244	0.205

Submitted Models. Most of the submitted models split the task into two subtasks:

- **Argument mining** The first stage of the task is defined as a sentence-level sequence labeling problem. A combination of pre-trained language model (i.e. BERT, RoBerta [28], ERNIE [34]) and bi-directional LSTM is used to extract multiple continual sentence blocks from both review and rebuttal as arguments.
- **Argument pairing** The second stage aims to find argument pairs from a set of review arguments and rebuttal arguments. It is then defined as a binary classification problem.

On top of this framework, there are some innovative ideas and methods that are able to further improve the performance.

- **Enhance argument representation** Intuitively, the shared words between two sentences are probably linked to the same topic. Team "HLT-HITSZ" incorporated such structure information to enhance the argument representation. To model these features, they considered sentences as nodes, linked all sentences with co-occurring words, and constructed a graph for each pair of review passage and rebuttal passage. A graph convolutional network (GCN) is then used to model this graph.
- **Paragraph-level argument pairing** Most of the teams modeled the argument pairing subtask on the sentence level. In order to improve the argument pairing subtask performance, team "PAL" casts argument pairing part as a paragraph-level pairing problem. In addition, they used the RoBerta model as the classifier, where the weights were not shared with the model in the first subtask.
- **double attention mechanism** In order to capture long-range relations, team "ARGUABLY" adopted double attention mechanism after obtaining the sentence representations. The final F_1 score increased 1.1 when using one attention layer, while it further increased 2.3 when incorporating double attention layers. They also observed that the use of the attention layer improved the convergence of CRF-loss.

5 Conclusion

In Argumentative Text Understanding for AI Debater Challenge, we make an initial step to understand argumentative text of debating in this shared task, including three tracks of supporting material identification, argument pair identification from online forum and argument pair extraction from peer review and rebuttal. In this challenge, we construct and release a brand new dataset for extracting the focus of dispute in the judgement documents. The performance on the task was significantly raised with the efforts of over 100 participants. In this paper, we summarize the general architecture and promising tricks they employed, which are expected to benefit further researches on argumentation mining. However, there is still a long way to go to fully achieve the goal of an autonomous AI debater that is capable of conducting reasoning and debating just like a human being. We hope our effort can encourage more researchers to pay attention to such a vibrant and interesting research field (Tables 7, 8, 9).

Appendix: Full Rank of All Participants

Table 7. Performance of all participants on Track 1. All teams are sorted according to their final performance in Phase 2. The best performance in each phase is in **bold**.

Rank	Team	Score 1	Score 2
1	Arrival	**0.841**	**0.928**
2	QingBoAI	0.822	0.905
3	OVERFIT	0.838	0.895
4	Again	–	0.868
5	Error	–	0.865
6	OnceAgain	–	0.864
7	ChaoTianJiao	0.821	0.859
8	B318	–	0.859
9	IntellGame	–	0.857
10	1	–	0.856
11	PuPuTao	–	0.851
12	CTJ	–	0.850
13	HuoLaLa	–	0.849
14	PAL	0.833	0.847
15	404	0.783	0.833
16	HLT-HITSZ	0.823	0.825
17	400	–	0.823
18	abaaba	0.804	0.822
19	UPSIDEDOWN	0.835	0.819
20	dawanqu	–	0.815
21	FanZuiFengZi	–	0.768
22	hunzi	–	0.710
23	DaBuDaJiangYou	–	0.606
24	SXU_CS	–	0.606
25	ARGUABLY	0.803	–
26	ZhenLiZhiCai	0.799	–
27	adam000	0.799	–
28	bjtu_yuhx	0.799	–
29	ACE	0.796	–
30	XinXiJingJiXue	0.788	–

Table 8. Performance of all participants on Track 2. All teams are sorted according to their final performance in Phase 2. The best performance in each phase is in **bold**.

Rank	Team	Score 1	Score 2
1	ACE	0.708	**0.729**
2	cw	0.716	0.683
3	404	0.685	0.671
4	400	–	0.659
5	abaaba	0.662	0.655
6	DaLiuRen	0.698	0.644
7	QingBoAI	**0.742**	0.621
8	HLT-HITSZ	0.725	0.498
9	YongShiZongGuanJun	–	0.461
10	AI Debater	–	0.203
11	Rush B	–	0.197
12	PAL	0.718	0.193
13	NOCLUE	–	0.000
14	EDZuHe	0.680	–

Table 9. Performance of all participants on Track 3. All teams are sorted according to their final performance in Phase 2. The best performance in each phase is in **bold**.

Rank	Team	Score 1	Score 2
1	HLT-HITSZ	0.379	**0.3231**
2	nju_pumps	–	0.2629
3	PAL	0.296	0.2625
4	404	0.358	0.2563
5	abaaba	0.243	0.205
6	QingBoAI	0.283	0.1755
7	WW_XX	0.280	0.1753
8	SXU_CS	–	0.1692
9	LieShaShiKe	0.284	0.1539
10	kk	–	0.1304
11	DaBuDaJiangYou	–	0.0642
12	NOCLUE	0.237	–
13	ARGUABLY	0.270	–
14	GoodMornin	0.239	–

References

1. Addawood, A., Schneider, J., Bashir, M.: Stance classification of Twitter debates: the encryption debate as a use case. In: Proceedings of the 8th International Conference on Social Media and Society. Association for Computing Machinery (2017)
2. Agirre, E., Cer, D., Diab, M., Gonzalez-Agirre, A.: SemEval-2012 task 6: a pilot on semantic textual similarity. In: *SEM 2012: The First Joint Conference on Lexical and Computational Semantics - Volume 1: Proceedings of the main conference and the shared task, and Volume 2: Proceedings of the Sixth International Workshop on Semantic Evaluation (SemEval 2012), pp. 385–393. Association for Computational Linguistics, Montréal, Canada, 7–8 June 2012. https://aclanthology.org/S12-1051
3. Aharoni, E., et al.: A benchmark dataset for automatic detection of claims and evidence in the context of controversial topics. In: Proceedings of the First Workshop on Argumentation Mining, pp. 64–68. Association for Computational Linguistics, Baltimore, Maryland, June 2014. https://doi.org/10.3115/v1/W14-2109, https://aclanthology.org/W14-2109
4. Al Khatib, K., Völske, M., Syed, S., Kolyada, N., Stein, B.: Exploiting personal characteristics of debaters for predicting persuasiveness. In: Proceedings of the 58th Annual Meeting of the Association for Computational Linguistics, pp. 7067–7072. Association for Computational Linguistics, Online, July 2020. https://doi.org/10.18653/v1/2020.acl-main.632, https://www.aclweb.org/anthology/2020.acl-main.632
5. Anand, P., Walker, M., Abbott, R., Fox Tree, J.E., Bowmani, R., Minor, M.: Cats rule and dogs drool!: classifying stance in online debate. In: Proceedings of the 2nd Workshop on Computational Approaches to Subjectivity and Sentiment Analysis (WASSA 2.011), pp. 1–9. Association for Computational Linguistics, Portland, Oregon, June 2011. https://aclanthology.org/W11-1701
6. Bar-Haim, R., Bhattacharya, I., Dinuzzo, F., Saha, A., Slonim, N.: Stance classification of context-dependent claims. In: Proceedings of the 15th Conference of the European Chapter of the Association for Computational Linguistics: Volume 1, Long Papers, pp. 251–261. Association for Computational Linguistics, Valencia, Spain, April 2017. https://aclanthology.org/E17-1024
7. Bar-Haim, R., Edelstein, L., Jochim, C., Slonim, N.: Improving claim stance classification with lexical knowledge expansion and context utilization. In: Proceedings of the 4th Workshop on Argument Mining, pp. 32–38. Association for Computational Linguistics, Copenhagen, Denmark, September 2017. https://doi.org/10.18653/v1/W17-5104, https://aclanthology.org/W17-5104
8. Bar-Haim, R., Eden, L., Friedman, R., Kantor, Y., Lahav, D., Slonim, N.: From arguments to key points: towards automatic argument summarization. In: Proceedings of the 58th Annual Meeting of the Association for Computational Linguistics, pp. 4029–4039. Association for Computational Linguistics, Online, July 2020. https://doi.org/10.18653/v1/2020.acl-main.371, https://www.aclweb.org/anthology/2020.acl-main.371
9. Bar-Haim, R., Kantor, Y., Eden, L., Friedman, R., Lahav, D., Slonim, N.: Quantitative argument summarization and beyond: Cross-domain key point analysis. In: Proceedings of the 2020 Conference on Empirical Methods in Natural Language Processing (EMNLP), pp. 39–49. Association for Computational Linguistics, Online, November 2020. https://doi.org/10.18653/v1/2020.emnlp-main.3, https://www.aclweb.org/anthology/2020.emnlp-main.3
10. Bhagat, R., Hovy, E.: What is a paraphrase? Computational Linguistics (2013)

11. Chakrabarty, T., Hidey, C., Muresan, S., McKeown, K., Hwang, A.: Ampersand: argument mining for persuasive online discussions. arXiv preprint arXiv:2004.14677 (2020)
12. Cheng, L., Bing, L., Yu, Q., Lu, W., Si, L.: Argument pair extraction from peer review and rebuttal via multi-task learning. In: Proceedings of the 2020 Conference on Empirical Methods in Natural Language Processing (EMNLP), pp. 7000–7011 (2020)
13. Conneau, A., et al.: Unsupervised cross-lingual representation learning at scale. In: Proceedings of the 58th Annual Meeting of the Association for Computational Linguistics, pp. 8440–8451. Association for Computational Linguistics, Online, July 2020. https://doi.org/10.18653/v1/2020.acl-main.747, https://aclanthology.org/2020.acl-main.747
14. Cui, Y., Che, W., Liu, T., Qin, B., Wang, S., Hu, G.: Revisiting pre-trained models for Chinese natural language processing. In: Findings of the Association for Computational Linguistics: EMNLP 2020, pp. 657–668. Association for Computational Linguistics, Online, November 2020. https://doi.org/10.18653/v1/2020.findings-emnlp.58, https://aclanthology.org/2020.findings-emnlp.58
15. Dagan, I., Dolan, B., Magnini, B., Roth, D.: Recognizing textual entailment: rational, evaluation and approaches - erratum. Nat. Lang. Eng. **16**(1), 105 (2010)
16. Devlin, J., Chang, M.W., Lee, K., Toutanova, K.: BERT: pre-training of deep bidirectional transformers for language understanding. In: Proceedings of NAACL (2019)
17. El Baff, R., Wachsmuth, H., Al Khatib, K., Stein, B.: Analyzing the persuasive effect of style in news editorial argumentation. In: Proceedings of the 58th Annual Meeting of the Association for Computational Linguistics, pp. 3154–3160. Association for Computational Linguistics, Online, July 2020. https://doi.org/10.18653/v1/2020.acl-main.287, https://www.aclweb.org/anthology/2020.acl-main.287
18. Gao, Y., Eger, S., Kuznetsov, I., Gurevych, I., Miyao, Y.: Does my rebuttal matter? Insights from a major NLP conference. In: Proceedings of NAACL (2019)
19. Ghosh, S., Singhania, P., Singh, S., Rudra, K., Ghosh, S.: Stance detection in web and social media: a comparative study. CoRR abs/2007.05976 (2020). https://arxiv.org/abs/2007.05976
20. Hasan, K.S., Ng, V.: Extra-linguistic constraints on stance recognition in ideological debates. In: Proceedings of the 51st Annual Meeting of the Association for Computational Linguistics, ACL 2013, 4–9 August 2013, Sofia, Bulgaria, Volume 2: Short Papers, pp. 816–821. The Association for Computer Linguistics (2013). https://www.aclweb.org/anthology/P13-2142/
21. Hua, X., Nikolov, M., Badugu, N., Wang, L.: Argument mining for understanding peer reviews. In: Proceedings of NAACL (2019)
22. Ji, L., Wei, Z., Li, J., Zhang, Q., Huang, X.: Discrete argument representation learning for interactive argument pair identification. arXiv preprint arXiv:1911.01621 (2019)
23. Kang, D., et al.: A dataset of peer reviews (PeerRead): collection, insights and NLP applications. In: Proceedings of NAACL (2018)
24. Khatib, K.A., Hou, Y., Wachsmuth, H., Jochim, C., Bonin, F., Stein, B.: End-to-end argumentation knowledge graph construction. In: The Thirty-Fourth AAAI Conference on Artificial Intelligence, AAAI 2020, The Thirty-Second Innovative Applications of Artificial Intelligence Conference, IAAI 2020, The Tenth AAAI Symposium on Educational Advances in Artificial Intelligence, EAAI 2020, 7–12 February 2020, New York, NY, USA, pp. 7367–7374. AAAI Press (2020). https://aaai.org/ojs/index.php/AAAI/article/view/6231

25. Kobbe, J., Hulpus, I., Stuckenschmidt, H.: Unsupervised stance detection for arguments from consequences. In: Proceedings of the 2020 Conference on Empirical Methods in Natural Language Processing (EMNLP), pp. 50–60. Association for Computational Linguistics, Online, November 2020. https://doi.org/10.18653/v1/2020.emnlp-main.4, https://aclanthology.org/2020.emnlp-main.4
26. Li, J., Durmus, E., Cardie, C.: Exploring the role of argument structure in online debate persuasion. In: Proceedings of the 2020 Conference on Empirical Methods in Natural Language Processing (EMNLP), pp. 8905–8912. Association for Computational Linguistics, Online, November 2020. https://doi.org/10.18653/v1/2020.emnlp-main.716, https://www.aclweb.org/anthology/2020.emnlp-main.716
27. Liu, X., et al.: GPT understands, too. CoRR abs/2103.10385 (2021). https://arxiv.org/abs/2103.10385
28. Liu, Y., et al.: RoBERTa: a robustly optimized BERT pretraining approach. arXiv preprint arXiv:1907.11692 (2019)
29. Morio, G., Ozaki, H., Morishita, T., Koreeda, Y., Yanai, K.: Towards better non-tree argument mining: proposition-level biaffine parsing with task-specific parameterization. In: Proceedings of the 58th Annual Meeting of the Association for Computational Linguistics, pp. 3259–3266. Association for Computational Linguistics, Online, July 2020. https://doi.org/10.18653/v1/2020.acl-main.298, https://www.aclweb.org/anthology/2020.acl-main.298
30. Slonim, N., et al.: An autonomous debating system. Nature **591**(7850), 379–384 (2021)
31. Sobhani, P., Mohammad, S., Kiritchenko, S.: Detecting stance in tweets and analyzing its interaction with sentiment. In: Proceedings of the Fifth Joint Conference on Lexical and Computational Semantics, pp. 159–169. Association for Computational Linguistics, Berlin, Germany, August 2016. https://doi.org/10.18653/v1/S16-2021, https://aclanthology.org/S16-2021
32. Somasundaran, S., Wiebe, J.: Recognizing stances in ideological on-line debates. In: Proceedings of the NAACL HLT 2010 Workshop on Computational Approaches to Analysis and Generation of Emotion in Text, pp. 116–124. Association for Computational Linguistics, Los Angeles, CA, June 2010. https://aclanthology.org/W10-0214
33. Sun, Q., Wang, Z., Zhu, Q., Zhou, G.: Stance detection with hierarchical attention network. In: Proceedings of the 27th International Conference on Computational Linguistics, pp. 2399–2409. Association for Computational Linguistics, Santa Fe, New Mexico, USA, August 2018. https://aclanthology.org/C18-1203
34. Sun, Y., et al.: ERNIE: enhanced representation through knowledge integration. arXiv preprint arXiv:1904.09223 (2019)
35. Swanson, R., Ecker, B., Walker, M.: Argument mining: extracting arguments from online dialogue. In: Proceedings of the 16th Annual Meeting of the Special Interest Group on Discourse and Dialogue, pp. 217–226 (2015)
36. Tan, C., Niculae, V., Danescu-Niculescu-Mizil, C., Lee, L.: Winning arguments: interaction dynamics and persuasion strategies in good-faith online discussions. In: Bourdeau, J., Hendler, J., Nkambou, R., Horrocks, I., Zhao, B.Y. (eds.) Proceedings of the 25th International Conference on World Wide Web, WWW 2016, Montreal, Canada, 11–15 April 2016, pp. 613–624. ACM (2016). https://doi.org/10.1145/2872427.2883081
37. Wei, J., et al.: NEZHA: neural contextualized representation for Chinese language understanding. CoRR abs/1909.00204 (2019). http://arxiv.org/abs/1909.00204

38. Wei, Z., Liu, Y., Li, Y.: Is this post persuasive? ranking argumentative comments in online forum. In: Proceedings of the 54th Annual Meeting of the Association for Computational Linguistics (Volume 2: Short Papers), pp. 195–200. Association for Computational Linguistics, Berlin, Germany, August 2016. https://doi.org/10.18653/v1/P16-2032, https://www.aclweb.org/anthology/P16-2032
39. Xu, L., et al.: CLUE: a Chinese language understanding evaluation benchmark. In: Proceedings of the 28th International Conference on Computational Linguistics, pp. 4762–4772. International Committee on Computational Linguistics, Barcelona, Spain (Online), December 2020. https://doi.org/10.18653/v1/2020.coling-main.419, https://aclanthology.org/2020.coling-main.419

ACE: A Context-Enhanced Model for Interactive Argument Pair Identification

Yi Wu[1,2] and Pengyuan Liu[1,2(✉)]

[1] School of Information Science, Beijing Language and Culture University, Beijing, China
202021198387@stu.blcu.edu.cn, liupengyuan@pku.edu.cn

[2] Chinese National Language Monitoring and Research Center (Print Media), Beijing, China

Abstract. In recent years, Interactive Argument Pair Identification has attracted widespread attention as an important task in argument mining. Existing methods for the task usually explicitly model the matching score between the argument pairs. However, identifying whether two arguments have an interactive relationship not only depends on the argument itself, but also usually requires its context to better grasp the views expressed by the two arguments. In the NLPCC-2021 shared task Argumentative Text Understanding for AI Debater, we participated in track 2 Interactive Argument Pair Identification in Online Forum and proposed ACE, A Context-Enhanced model which makes good use of the context information of the argument. In the end, we ranked 1st in this task, indicating the effectiveness of our model.

Keywords: Interactive argument pair identification · Pre-trained language model · Mixup

1 Introduction

Argument mining [14] aims to detect the argumentative discourse structure in text. While most work has focused on well-structured texts [8,9,16] (e.g., essay and peer reviews), user-generated content such as online debates and product reviews have also attracted widespread attention [5,6,17,21]. Compared with a well-structured argument text, users' comments will be more noisy, which makes the work more difficult.

In online forums, people can freely express their opinions on a topic, and people can exchange their opinions with each other, or put forward their own opinions in response to others' opinions. However, when people respond to a certain comment, the opinions may contain multiple arguments. and it's important to figure out which arguments are relevant to the arguments in the original post, i.e., which argument pairs have an interactive relationship. Moreover, automatically identifying these interactive argument pairs can be beneficial to downstream tasks.

L. Wang et al. (Eds.): NLPCC 2021, LNAI 13029, pp. 569–578, 2021.
https://doi.org/10.1007/978-3-030-88483-3_46

Therefore, Interactive Argument Pair Identification [11] is proposed as an important task of argument mining. It aims to identify whether two arguments have an interactive relationship. Table 1 shows the details of a sample in the task. It is difficult to identity the relationship between two arguments without rich contextual information, since an argument may contain only a few words, and information that can be provided to the model is very limited. Therefore, this task provides additional contextual information for each argument. It is worth considering how this contextual information can be used to help identify the relationship between argument pairs.

In this paper, we present a context-enhanced model to solve this problem. We introduce our work from the aspects of model structure, implementation details, experimental results, etc., and summarize the article in the last part.

Table 1. An instance of a sample in the dataset. Each sample includes a quotation and its corresponding 5 candidate replies. Each argument has context information, and the position of the argument in the context is marked in italics.

	Argument	Context
Quotation	i have never seen him not looking generally flustered and awkward	...or that they do not matter: *i have never seen him not looking generally flustered and awkward.* working on international issues requires developing personal relationships...
Reply-pos	working on international issues requires developing personal relationships, thus being good at making friends.	...here ill try and explain why. *working on international issues requires developing personal relationships, thus being good at making friends.* i do not think he is like that. this is pretty subjective...
Reply-neg1	that said, i do not know enough about sanders to say whether or not he would have foreign policy chops.	...but foreign policy is one area where he deserves some praise. *that said, i do not know enough about sanders to say whether or not he would have foreign policy chops.* just pointing out that looking generally flustered and awkward...
...	...	...
Reply-neg4	that is part of becoming the representative of a country.	...both for her looks and her body language. *that is part of becoming the representative of a country.* and second, on the international stage it is much different qualities that count...

2 Proposed Model

2.1 Task Definition

This task is designed to identify whether two arguments extracted from online forum have an interactive relationship. Given an argument q and it's five candidate replies which include one positive reply r^+ and four negative replies $r_1^- - r_4^-$, and the context corresponding to each argument, we need to find the correct one from the 5 candidates.

Different from the previous work that regards this as a pairwise ranking problem, we treat it as a multiple choice task and use context to get a better sentence representation.

2.2 Model Structure

Our work is mainly based on the pre-trained language models. BERT [7] is based on a multi-layer bidirectional Transformer [19] and is trained on large-scale corpus for masked word prediction and next sentence prediction tasks. The representations learned by BERT have been shown to generalize well to downstream tasks. The input of BERT is a sequence of no more than 512 tokens, the first token of the sequence is always [CLS] which is a special token used for classification and there is another special token [SEP] for separating segments.

Our model is mainly composed of two parts, the first part is BERT-based Multiple Choice with no context information added, and the second part is an enhanced representation of the argument using the context information. In addition, we also used data augmentation methods to improve model performance. The overall structure of our model can be found in Fig. 1.

BERT-based Multiple Choice. In this part, we do not use context information, the input of the model is the original arguments. First, quotation and its five candidate replies are concatenated together to form a sequence, namely {[CLS]; quotation; [SEP]; reply; [SEP]}. In this way, 5 options form 5 sequences. Then send these 5 sequences into the BERT to get the scores of the final 5 options.

Context-Enhanced Argument Representation. In previous work, the embedding of the context is calculated separately and then concatenated with the argument to provide additional information. In this case, the context information is not well used. Inspired by BERT's input, our proposed model can make full use of context information and generate a better representation of argument based on context information. Given an argument a and its context $ctxt$, we add two special token [S] and [E] to indicate where the argument starts and ends in its context, respectively. The enhanced argument's representation, R_{ca}, is shaped by the argument itself and its surrounded context. Namely, { [CLS]; $ctxt_l$; [S]; a; [E]; $ctxt_r$; [SEP] }. Through the new input sequences, the context of the argument can be well integrated into the representation of the argument.

$$R = BERT(ctxt_l; a; ctxt_r) \quad (1)$$

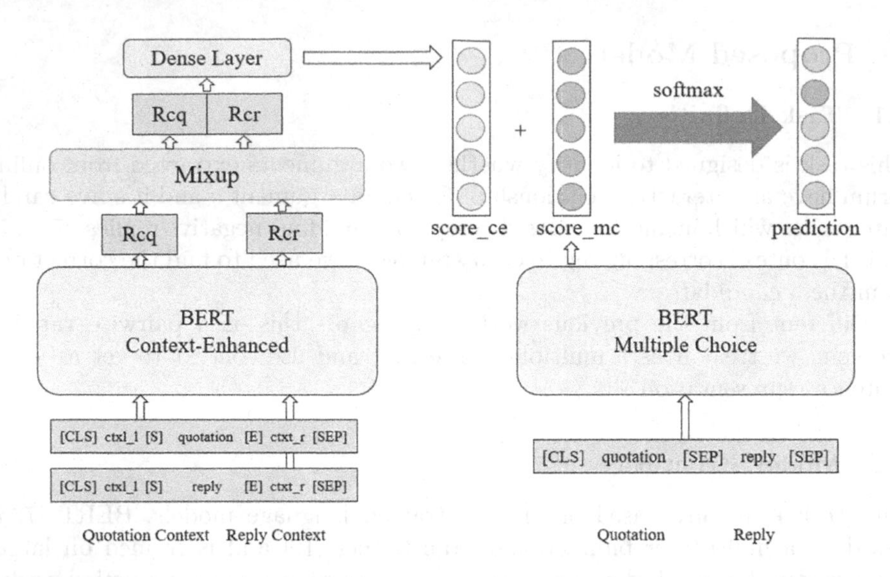

Fig. 1. A Context-Enhanced model for Interactive Argument Pair Identification.

After that, we used the average-pooling strategy to get the final context-based argument representing R_{cq} and R_{cr}.

$$avg = \sum_{i=0}^{n-1}(R_{ca}[i])/n \tag{2}$$

Finally, we concatenated R_{cq} and its corresponding 5 candidate options R_{cr} together to get the scores of the 5 options.

$$score = DenseLayer(R_{cq}; R_{cr}) \tag{3}$$

At last, the calculated scores of the two parts are simply added together, and then go through a softmax to get the final prediction result.

$$pred = Softmax(score_{ce} + score_{mc}) \tag{4}$$

2.3 Data Augmentation

Since high-quality annotation data is limited, we use data augmentation methods to improve the generalization and robustness of the model.

Some common data augmentation methods, e.g., random insertion [20], back translation [22], rely on specific data and specialized knowledge. The Mixup method is a simple and data-independent data augmentation method that expands the training distribution by constructing virtual training samples. Specifically, the method is to generate a new synthetic sample by linearly interpolating a pair of training samples and their training targets. Consider a sample

(x^i, y^i), we randomly select another sample (x^j, y^j) from the remaining samples to pair with it, and the new synthetic sample is generated by the following formula

$$\tilde{x}^{ij} = \lambda x^i + (1 - \lambda)x^j \tag{5}$$

$$\tilde{y}^{ij} = \lambda y^i + (1 - \lambda)y^j \tag{6}$$

where λ is the mixing policy or mixing-ratio for the sample pair. λ is sampled from a Beta(α, α) distribution with a hyper-parameter α.

3 Experiments

3.1 Datasets

The original raw dataset was collected from an online forum ChangeMyView in reddit.com. The experimental dataset is the argument pairs extracted from the CMV dataset [18]. For a quotation in the original post, the user's reply to the current post is regarded as an argument pair with interactive relationship, and 4 unrelated replies are randomly selected as negative samples from other posts. For these 6 arguments, including a quotation and its corresponding 5 candidate replies, each of them provides context information. The detailed statistics of dataset can be found in Table 2.

Table 2. Statistics of dataset

	Train	Dev	Test
Num	9252	2313	1481

3.2 Experiment Settings

For both two modules we use pre-trained BERT as the backbone networks, which are implemented in the Huggingface Transformers package(4.6.1) with Pytorch(1.8.1). We adopt two different input forms mentioned in part 3.2, and then fine-tune the model on the training set. The batch size is set to 1. We use an Adam optimizer to tune the parameters with a learning rate of 8e–6. In the data augment part, the hyperparameter α is set to 1. For reproducibility, we fixed all seeds and set them to 2021. We train for 5 epochs to get the final result.

3.3 Results and Analysis

In order to verify that our model utilizes the context information of the argument in a proper way, we have compared different methods of using context information and make a detailed analysis. The metric used in the experiment is accuracy.

- **MC w/o context** Only use the original argument, with no context information added. Quotations and their corresponding replies are concatenated respectively to get the final score for each option.
- **MC + SBERT** For the context of each argument, we use sentence-BERT to encode each sentence in the context, and then use BiLSTM to get the overall representation of the context.
- **MC + C** The difference from the ACE model we proposed is that the input part of the model does not add additional special tokens [S] and [E] to indicate the starting and ending positions of the argument in the context.
- **MC + CE** The model we finally adopted for this task but without data augmentation part. It uses the method demonstrated in part 3.2 to add context information.

Table 3. The results of models that incorporate context information in different ways

Methods	Dev	Test(Phase1)
MC w/o context	0.6667	0.6583
MC + SBERT	0.7160	0.7205
MC + C	0.7216	0.7245
MC + CE	**0.7717**	**0.7780**

Based on the results of the Table 3, we made the following analysis. Compared with the other three methods that incorporate context information, the first method is obviously inferior. This shows that it is difficult to identify the relationship between pairs just by relying on the argument itself, and the context can provide more auxiliary information. The last three methods show that using different methods to incorporate context information has a great impact on model performance. The SBERT method encodes each sentence in the context, and then uses BiLSTM to obtain the overall context representation. Although compared with the methods do not add context, it has made a great improvement, it still has a big gap with our model. The third method is to input the context as a whole into the BERT. Compared with ACE, it lacks [S] and [E] to indicate the position of the argument in the context, indicating that the ACE method can well capture the key information in the context and enhance the representation of the argument by using the left and right context.

3.4 Ablation Study

Our model consists of several important modules. As shown in Fig. 1, the MC module models the original argument matching task into a Multiple Choice task. The CE module provides Context-Enhanced information. As a method of data augmentation, the Mixup module generates a corresponding synthetic sample

Table 4. Ablation study results

Methods	Dev	Test(Phase1)	Test(Phase2[a])
ACE w/o MC	0.6891	0.6792	–
ACE w/o CE	0.6667	0.6583	–
ACE w/o Mixup	**0.7717**	**0.7780**	0.72
ACE	0.7665	0.7683	**0.7291**

[a] Since the test data of phase2 has not been released, we get the results based on the two submission records in phase2.

for each sample. In order to verify the effectiveness of each module, we further conducted experiments on each module.

We tested the performance of the model with the three modules removed respectively. The experimental results are shown in Table 4. When the MC or CE modules are removed, the model does not perform well, indicating that a single part is not sufficient to handle the task. The CE module makes the model pay attention to the left and right context of the argument, and provides more auxiliary information, so that when the meaning of argument is not clear, it can provide extra help. When our model removes the MC module, its performance is not as good as when the two coexist. If we only rely on the context to model the argument representation, it is likely to be affected by irrelevant information in the context, causing the model to make wrong choices. Therefore, it is necessary to add the original argument to the calculation. It is worth noting that when we added Mixup, the data augmentation method, the performance of the model on both the Dev and Test datasets decreased, probably due to the noise caused by the synthesized virtual samples. However, as a regularization method in essence, mixup makes our model have better generalization ability. When faced with the more challenging test datasets in phase2, the performance of the ACE model with the mixup method is improved.

3.5 Case Study

We have selected a typical example to analyze some of the problems in our model. When the model needs to choose a correct Reply for the original Quotation, it tends to choose the reply that contains the words that appear in the Quotation context when the Quotation contains very little information. As shown in Table 5, Quotation only contains 10 words, and the information that can be provided is very limited. The context information provides information about the murder, which makes the model choose the wrong option. This shows that context information is not all gain information, and sometimes it can be misleading. Sometimes the relationship between two pairs needs to be identified by logical reasoning, etc.

Table 5. An example of our model's performance on the test dataset.

Quotation	...if they knew they would be processed and charged with murder, i would argue that there would a significant decline in police shootings. *it gives the victim a fair chance at their side...*
Wrong choice	*...surely the officer would draw his/ her weapon at the murderer and scream to get down.* they would proceed to cuff them and bring them to the station for booking/ questioning...
Correct reply	... if they do not die, then it reduces the police force as officers are detained for unnecessary trials. *even if they are dead, treating an officer like a criminal discredits any story they come up with giving a better chance for the true series of events to surface...*

4 Related Work

4.1 Dialogical Argumentation

Dialogical argumentation focuses on the exchange of arguments between users. It is not only necessary to consider how the arguments are generated and evaluated, but also how the agents interact, i.e., what kinds of dialogical moves they can make [3]. In online forums, users' arguments are often expressed in confusing and poorly worded ways. [4] pointed out that when identifying a user's argument, one needs to first come up with a set of prominent arguments and in their work, semantic textual similarity was adopted to identifying prominent arguments. [15] pay attention to micro-level argument mining studies on discussion threads and proposed a novel end-to-end technique to discriminate sentence types, inner-post relations, and inter-post interactions simultaneously. Most of the existing work focuses on studying the text features of reply comments, ignoring the interaction between participants. In order to better model the process of dialogical argumentation, [10] proposed a novel neural network based on the co-attention mechanism to capture the interaction between participants.

4.2 Pre-trained Language Model

The emergence of pre-trained language models has promoted the development of NLP. Most of these pre-training language models use the advanced transformers framework and adopt various pre-training tasks to train a model on a large scale of corpus. A large number of studies [1,2,12,13] have shown that pre-trained models on a large corpus can learn universal language representations, which is beneficial to downstream natural language processing tasks and can avoid training a new model from scratch. We can fine-tune these pre-trained models for specific tasks, and a large number of experiments have shown that the proper use of these pre-trained language models can achieve SOTA effects in different NLP tasks.

5 Conclusion

In this paper, we propose a context-enhanced model based on a pre-trained language model to deal with the task of identifying interactive argument pairs. Through comparison with other methods, as well as official evaluation results, we have achieved the best performance among all participating teams, which proves the effectiveness of our work. In the course of the experiment, we found that identifying two pairs of arguments not only requires good context modeling, but sometimes also requires modeling the logical reasoning information between them. Therefore, in future work, we will pay more attention to this aspect.

Acknowledgements. This work was supported by Beijing Natural Science Foundation (4192057) and Science Foundation of Beijing Language and Culture University (the Fundamental Research Funds for the Central Universities: 21YJ040005).

References

1. Akbik, A., Blythe, D., Vollgraf, R.: Contextual string embeddings for sequence labeling. In: Proceedings of the 27th International Conference on Computational Linguistics, pp. 1638–1649 (2018)
2. Bataa, E., Wu, J.: An investigation of transfer learning-based sentiment analysis in Japanese. arXiv preprint arXiv:1905.09642 (2019)
3. Besnard, P., et al.: Introduction to structured argumentation. Argument Comput. **5**(1), 1–4 (2014)
4. Boltužić, F., Šnajder, J.: Identifying prominent arguments in online debates using semantic textual similarity. In: Proceedings of the 2nd Workshop on Argumentation Mining, pp. 110–115 (2015)
5. Chakrabarty, T., Hidey, C., Muresan, S., McKeown, K., Hwang, A.: Ampersand: Argument mining for persuasive online discussions. arXiv preprint arXiv:2004.14677 (2020)
6. Chen, D., Du, J., Bing, L., Xu, R.: Hybrid neural attention for agreement/disagreement inference in online debates. In: Proceedings of the 2018 Conference on Empirical Methods in Natural Language Processing, pp. 665–670 (2018)
7. Devlin, J., Chang, M.W., Lee, K., Toutanova, K.: Bert: Pre-training of deep bidirectional transformers for language understanding. arXiv preprint arXiv:1810.04805 (2018)
8. Gao, Y., Eger, S., Kuznetsov, I., Gurevych, I., Miyao, Y.: Does my rebuttal matter? insights from a major nlp conference. arXiv preprint arXiv:1903.11367 (2019)
9. Hua, X., Nikolov, M., Badugu, N., Wang, L.: Argument mining for understanding peer reviews. arXiv preprint arXiv:1903.10104 (2019)
10. Ji, L., Wei, Z., Hu, X., Liu, Y., Zhang, Q., Huang, X.J.: Incorporating argument-level interactions for persuasion comments evaluation using co-attention model. In: Proceedings of the 27th International Conference on Computational Linguistics, pp. 3703–3714 (2018)
11. Ji, L., Wei, Z., Li, J., Zhang, Q., Huang, X.: Discrete argument representation learning for interactive argument pair identification. arXiv preprint arXiv:1911.01621 (2019)
12. Ju, Y., Zhao, F., Chen, S., Zheng, B., Yang, X., Liu, Y.: Technical report on conversational question answering. arXiv preprint arXiv:1909.10772 (2019)

13. Lample, G., Conneau, A.: Cross-lingual language model pretraining. arXiv preprint arXiv:1901.07291 (2019)
14. Lawrence, J., Reed, C.: Argument mining: a survey. Comput. Linguist. **45**(4), 765–818 (2020)
15. Morio, G., Fujita, K.: End-to-end argument mining for discussion threads based on parallel constrained pointer architecture. arXiv preprint arXiv:1809.00563 (2018)
16. Persing, I., Ng, V.: End-to-end argumentation mining in student essays. In: Proceedings of the 2016 Conference of the North American Chapter of the Association for Computational Linguistics: Human Language Technologies, pp. 1384–1394 (2016)
17. Sridhar, D., Foulds, J., Huang, B., Getoor, L., Walker, M.: Joint models of disagreement and stance in online debate. In: Proceedings of the 53rd Annual Meeting of the Association for Computational Linguistics and the 7th International Joint Conference on Natural Language Processing (Volume 1: Long Papers), pp. 116–125 (2015)
18. Tan, C., Niculae, V., Danescu-Niculescu-Mizil, C., Lee, L.: Winning arguments: interaction dynamics and persuasion strategies in good-faith online discussions. In: Proceedings of the 25th International Conference on World Wide Web, pp. 613–624 (2016)
19. Vaswani, A., et al.: Attention is all you need. In: Advances in Neural Information Processing Systems, pp. 5998–6008 (2017)
20. Wei, J., Zou, K.: Eda: Easy data augmentation techniques for boosting performance on text classification tasks. arXiv preprint arXiv:1901.11196 (2019)
21. Wei, Z., Liu, Y., Li, Y.: Is this post persuasive? ranking argumentative comments in online forum. In: Proceedings of the 54th Annual Meeting of the Association for Computational Linguistics (Volume 2: Short Papers), pp. 195–200 (2016)
22. Xie, Q., Dai, Z., Hovy, E., Luong, M.T., Le, Q.V.: Unsupervised data augmentation for consistency training. arXiv preprint arXiv:1904.12848 (2019)

Context-Aware and Data-Augmented Transformer for Interactive Argument Pair Identification

Yuanling Geng, Shuqun Li, Fan Zhang, Shaowu Zhang(✉), Liang Yang(✉), and Hongfei Lin

School of Computer Science and Technology, Dalian University of Technology, Dalian 116024, Liaoning Province, China
{gylthornhill,shuqunli,fanz,zhangsw,liang,hflin}@mail.dlut.edu.cn

Abstract. Interactive argument identification is an important research field in dialogical argumentation mining. This task aims to identify the argument pairs with the interactive relationship in the online forum. In this paper, we tackle the task as sentence pair matching. We build our model based on the pre-trained language model (LM) RoBERTa due to its strong ability in modeling semantic information. Based on the peculiarities of the argument texts, we combine the arguments and their corresponding contexts to better identify the interactive relationship of the argument pair. Besides, we adopt data augmentation and vote strategy based on cross-validation to further enhance the performance. Our system rank 5th on track2 of the NLPCC-2021 shared task on Argumentative Text Understanding for AI Debater.

Keywords: Interactive argument identification · Dialogical argumentation · Pre-trained language model · Data augmentation

1 Introduction

With the rapid development of the Internet and online forums, people are more and more inclined to post online comments to express their perspectives, thus mining the arguments and stances in the comments has commercial and academic significance. Argument mining is the automatic identification and extraction of the structure of inference and reasoning expressed as arguments presented in natural language [1]. It is widely used in legal decision-making [2,3], education [4,5], politics [6] and other fields to provide people with convenient automation tools.

Argument mining is usually analyzed into monologue argumentation and dialogical argumentation from the structural level. Existing research on monologue argumentation mainly focuses on the argument structure prediction [7,8],

Y. Geng and S. Li—Contributed equally to this paper.

L. Wang et al. (Eds.): NLPCC 2021, LNAI 13029, pp. 579–589, 2021.
https://doi.org/10.1007/978-3-030-88483-3_47

persuasiveness evaluation [9,10], argument summarization [11,12], essay scoring [13], etc. Recently, with the popularity of online forums such as idebate[1] and changemyview[2], dialogical argumentation has become an active research topic. Interactive argument pair identification is a fundamental step towards understanding dialogical argumentative structure. Hence, interactive argument pair identification is essential in the research of dialogical argumentation mining. Moreover, it can benefit downstream tasks, such as debate summarization [14] and logical chain extraction in debates [15].

Existing methods for addressing the task of interactive argument pair identification are usually to jointly model the dialogical arguments and their context information to identify the argument pair using deep learning models [16]. Furthermore, Yuan et al. [17] leverage external knowledge to construct an argumentation knowledge graph for the discussion process of target topics in online forums. With the ability of deeply model text context information, pre-trained LMs (BERT [18], ERNIE [19], XLnet [20], etc.), obtained from large scale of unlabelled data, are fine-tuned by researchers to satisfy various tasks with different targets. Pre-trained LMs also show promising performance on dialogical argumentation [17,21,22]. Hence, we introduce pre-trained LMs into our system.

In this paper, based on the interactive peculiarity of dialogical argumentation, we tackle interactive argument pair identification as a problem of sentence pairs matching. We adopt the pre-trained language model, RoBERTa, as the feature extractor, and leverage voting strategies to merge the cross-validation results into the final result. In addition, according to the characteristics of constructing the dataset from interactive argument pair identification, we carry out data augmentation to strengthen the generalization and robustness of the model. The implementation of data augmentation is similar to the drop acts in the contrastive learning method *SimCSE* [23], which can construct positive and negative instances from unsupervised data. Therefore, our contributions can be summarized as follows:

- We choose the pre-trained LMs with better performance and use the context of arguments for a better understanding of them.
- We introduce the drop acts of the contrastive learning method *SimCSE* to construct data augmentation, which further improves the performance, generalization, and robustness of our system.
- The final evaluation result shows that our system achieves fifth place, which indicates the effectiveness of our method.

2 Related Work

Dialogical argumentation mining is an important research field that attracts growing attention in recent years. Most research on dialogical argumentation focuses on evaluating the quality of debating or persuasive content. Tan et

[1] https://idebate.org/.
[2] https://www.reddit.com/r/changemyview/.

al. [24] and Wei et al. [25] respectively construct datasets[3] from the *ChangeMyView*[2], an active community on Reddit, for evaluating persuasiveness and summarize the key factors of persuasive arguments. Chakrabarty et al. [21] combine the micro-level (argument as a product) and macro-level (argument as a process) models of argumentation to realize argument mining in online persuasive discussion forums. To better explore the process of dialogical argumentation, Ji et al. [26] propose a co-attention mechanism-based neural network to capture the interactions between participants on the argument level, which achieve significant progress on persuasiveness evaluation.

What's more, some researchers pay increasing attention to the extraction and identification of argument pairs in dialogical arguments. Cheng et al. [22] propose a multi-task learning framework based on a sequence labeling task and a text relation classification task, which effectively extract argument pairs from the peer review and rebuttal interaction dataset. Ji et al. [16] propose an interactive argument pair identification task and release the related dataset. They formulate this task as a problem of sentence pair scoring and compute the textual similarity between the two arguments. Yuan et al. [17] build an argumentation knowledge graph for the discussion process by introducing external knowledge, which achieves state-of-the-art performance on the task of interactive argument pair identification.

The pre-trained language model BERT [18] significantly improves the results of classic tasks in natural language processing. Hence, some research [17,21,22] on dialogical argumentation mining adopt BERT to obtain the representation of input sentences, which Fully contain textual context information. With the development of computing power and the large-scale available corpora, in addition to BERT, more and more variants of pre-trained LMs are proposed, such as XLnet [20], Ernie [19], Albert [27], and RoBERTa [28], constantly improving the performance on many NLP tasks. Among them, RoBERTa takes dynamic masks, removes the task of predicting the next sentence, and is fully trained on 160G data, which has made great progress compared to BERT. Therefore, we leverage RoBERTa as the feature extractor to obtain sentence representations.

Data argumentation is an effective way to improve models' performance. The general methods for data augmentation in NLP tasks include Lexical Substitution [29], Random Noise Injection [30], Text Surface Transformation [31], etc. Recently, Gao et al. [23] present a framework SimCSE, introducing the idea of comparative learning into sentence embedding, which greatly advances the state-of-the-art sentence embeddings. The superiority of the drop acts they adopt in the *SimCSE* to perform data augmentation on unsupervised data attracts our attention. Hence, we employ the drop acts to construct positive instances in the interactive argument pair identification, and generate corresponding proportions of negative examples to achieve data argumentation and improve the model performance.

[3] https://chenhaot.com/pages/changemyview.html.

3 Methodology

The method we proposed is mainly divided into two stages: data augmentation and argument pair identification, as shown in Fig. 1. In the data augmentation phase, we generate corresponding positive and negative examples according to the construction peculiarities of the original dataset. Then, the generated data is added to the original dataset as the training dataset. The sentence pairs built from the dataset are fed into the RoBERTa model in the argument pair identification phase. Finally, the results of the cross-validation are voted and merged to obtain the prediction result.

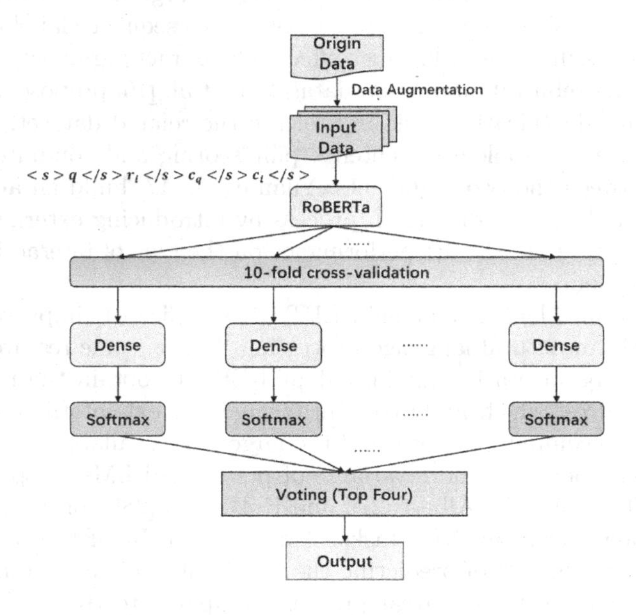

Fig. 1. Overall architecture of our system

3.1 Problem Definition

Given an original argument q and its corresponding argumentative context c_q, We aim to identify the opposite argument from the five candidate arguments $\{r_1, ..., r_5\}$ and their contexts $\{c_1, ..., c_5\}$. Among the five candidate arguments, the argument opposed to the original argument is regarded as the positive argument r^+ with a label of 1, and the remaining arguments are regarded as negative arguments $r_i^-, ..., r_u^-$ with labels of 0. Essentially, this task is a binary classification problem.

3.2 Argument Pair Identification with Context

In this paper, we treat the interactive argument pair identification as the task of sentence pairs matching. We use RoBERTa as our baseline model and take the form of $\{< s > q < /s > r_i < /s >\}$ as input, where $< s >$ is the token for classification and $< /s >$ is the token for separating different sentences.

In practice, we discover that considering the contextual information of the arguments can extend the semantic meaning of arguments and help the model understand what they argue for. Hence, we change the input form to $\{< s > q < /s > r_i < /s > c_q < /s > c_i < /s >\}$, where r_i is one of the five candidate arguments and c_i is its corresponding context. Then, we denote the output hidden states of $< s >$ as h_i.

$$h_i =< s >_i= RoBERTa(q, r_i, c_q, c_i) \tag{1}$$

We leverage a dense layer following the RoBERTa model to project the representation of sentence matching to the classification space. Then, a softmax layer is adopted to perform the prediction. Finally, the training loss of classification is obtained by the cross entropy between the predicted probability $\hat{y}_i$ and the ground truth label y_i.

$$\hat{y}_i = Softmax(Dense(h_i)) \tag{2}$$

$$L = -\sum y_i \log \hat{y}_i + (1 - y_i) log(1 - \hat{y}_i) \tag{3}$$

Furthermore, we adopt the training method of n-fold cross-validation and then obtain the final prediction result through the voting strategy. The training details are described in Sect. 4.2.

3.3 Data Augmentation

At this stage, we follow the input form $\{< s > q < /s > r_i < /s > c_q < /s > c_i < /s >\}$ in Sect. 3.2 for data augmentation.

Firstly, we directly exchange the candidate argument and the original argument symmetrically to form new training data, the input form of which can be represented by $\{< s > r_i < /s > q < /s > c_i < /s > c_q < /s >\}$. Their labels remain the same as before the swap. This step aims to reduce the sensitivity of the model to the sentence order.

Then, we find that Ji et al. [16] regard the quotation-reply pairs extracted as positive samples, and randomly select four replies related to the original post from other posts as negative samples and paired with the quotation to construct the dataset. Therefore, we concatenate the positive argument and the negative arguments as $\{< s > r^+ < /s > r_i^- < /s > c_{r^+} < /s > c_{r_i^-} < /s >\}$, whose label is 0 because it is irrelevant between positive and negative arguments.

In addition, we adopt the method that resembles the dropout acts of *SimCSE* [23] to construct the "positive" example for the quotation argument q. In detail, the quotation argument is fed to the model twice by applying different dropout masks $h^z = f_\theta(q; z)$, where z is a random mask for dropout. The two sentences are only different in embedding, which is still the same argument, so its label is 0. In the detailed implementation of our model, we generate the argument pair by forming the input as $\{< s > q < /s > q < /s > c_q < /s > c_q < /s >\}$. This type of input could avoid the model just learning to distinguish whether the two arguments are relevant but to recognize the opposite views by showing to the model that two relevant and same views are negative samples.

4 Experiments and Details

4.1 Dataset

The dataset for interactive argument pair identification is proposed by Ji et al. [16] They extract all these quotation-reply pairs q,r from posts in ChangeMyView dataset [24], and form each instance with the quotation, one positive reply, four negative replies, and the posts where they exist. The dataset[4] contains 11,565 and 1,481 instances in training set and test set respectively, whose statistic information is shown in Table 1.

Table 1. Overview statistics of the constructed dataset (mean and standard deviation). *arg.*, q, p_r, $n_r$represent *argument*, *quotation*, *positive reply* and negative reply respectively. $q - p_r$ represents the quotation-reply pair between posts.

	Training set	Test set
# of arg. per post	11.8 ± 6.6	11.4 ± 6.2
# of token per post	209.7 ± 117.2	205.9 ± 114.6
# of token per q	20.0 ± 8.6	20.0 ± 8.6
# of token per p_r	16.9 ± 8.1	17.3 ± 8.4
# of token per n_r	19.0 ± 8.0	19.1 ± 8.1
Max # of $q - p_r$ pairs	12	9
Avg. # of $q - p_r$ pairs	1.5 ± 0.9	1.4 ± 0.9

4.2 Implement Details

We divide the experimental stage into two phases according to the different training strategies. In the first stage, we randomly split 10% of the training set as the validation set to fine-tune the models, and find the best model and the best input form through the results of the offline test dataset. In the second stage, after determining the best model and the best input form, we use a 10-fold cross-validation training strategy based on data augmentation to further improve the performance, and finally obtain the prediction result through voting strategy.

[4] http://www.sdspeople.fudan.edu.cn/zywei/data/arg-pairs-fudanU.zip.

In the first stage, inspired by the offical baseline model[5], we try several pretrained language models such as BERT, Albert, and RoBERTa with input form as $\{< s > q < /s > r_i < /s >\}$. By comparing the experimental results, we choose RoBERTa as the base model in our method. Furthermore, we find that introducing the argument context into the model can more accurately identify argument pairs. Hence, the input form is $\{< s > q < /s > r_i < /s > c_q < /s > c_i < /s >\}$.

In the second stage, we adopt the method as described in Sect. 3.3 for data augmentation according to the above input form. Based on this, we carry out cross-validation experiments with different folds, including 5-fold, 7-fold, and 10-fold, to explore the sensitivity of the model to different folds and receive models with the best results on the verification set. Taking the limitation of the upload model size into account, we select the top four models with the best performance in the offline test set, and then take the addition of their results as the final result.

We chose BERT-base, ALBERT, and RoBERTa-base, which are implemented on the lib of Transformers developed by Huggingface[6], to construct our models. The batch size is set to 8. We use Adam optimizer to train our models. The learning rate is 1e–5.

4.3 Results and Analysis

The experimental results of different models on the offline test set in the two stages of training are shown in Table 2.

From the Table 2, we can see the result of using RoBERTa as the feature extractor is better than the results of BERT and Albert in the first stage of training. Compared with BERT and Albert, it has increased by 2.2% and 1.3% respectively. The result of *RoBERTa+context* is 14.7% higher than that of *RoBERTa-base*, which indicates that introducing argument context into the model can significantly improve the result of interactive argument pair identification.

In the second experimental stage, we carry out experiments based on data augmentation and cross-validation. The experiment represented by *RoBERTa+data-aug* is that the RoBERTa model is directly trained on the enhanced dataset without the context of the argument. It improves the result of *RoBERTa-base* by 1.4%, proving the effectiveness of data augmentation. *RoBERTa+context+data-aug* outperforms *RoBERTa+data-aug* and *RoBERTa+context* by 15.5% and 2.2% respectively, indicating data augmentation and context of the argument can further improve the result. For the cross-validation of different folds, the results of 5-fold, 7-fold, and 10-fold are close, which indicates that the results are not sensitive to the number of folds. Plus,

[5] https://github.com/AIDebater/Argumentative-Text-Understanding-for-AI-Debater-NLPCC2021.

[6] https://huggingface.co/.

Table 2. The accuracy of different models on the offline test set

Model	Acurracy
BERT-base	0.656
ALBERT	0.665
RoBERTa-base	0.678
RoBERTa+data-aug	0.692
RoBERTa+context	0.825
RoBERTa+context+data-aug	0.847
Vote-RoBERTa-Top4(5-fold)	0.852
Vote-RoBERTa-Top4(7-fold)	0.851
Vote-RoBERTa-Top4(10-fold)	0.852

the result of *Vote-RoBERTa-Top4(10-fold)* increases 0.5% compared with the single model and obtain the best result of 85.2% on the offline test set.

Hence, we upload the top-4 RoBERTa models of 10-fold cross-validation to the evaluation system to obtain the prediction results and merge them to get the final result by the voting strategy. The official evaluation result shows that we achieve an accuracy of 65.5% on the online test set. With this result, we won a fifth place on the track2 of NLPCC 2021 Shared Task 1 (Argumentative Text Understanding for AI Debater).

However, the decline of accuracy on the online test set catch our attention. By analyzing some instances of the training set that are even difficult for humans to judge, we suspect the reasons are as follows:

1) In some cases, the specific term in the quotation is hard to understand, while the reply does not refer it, resulting in a prediction error.
2) In dialogical arguments, the interaction of opinions usually involves the background of the topic, while the original context of the argument does not include background knowledge, which leads to misjudgment.

5 Conclusion

In this paper, based on the pre-trained LM RoBERTa, we simultaneously model argument pairs and their context to identify interactive argument pairs. Through the experimental results on the offline test set, we discover that in the interactive argument pair Identification, combining the context of the argument can significantly improve the performance. Furthermore, we also implement data augmentation by symmetrically swapping argument pairs and dropout masks to further develop the offline test result. The official evaluation result shows that the performance of our system ranks fifth, proving the effectiveness of the model.

From the analysis of some examples in the training data set, we ascertain that the lack of specific terms processing or background information is a huge challenge to tackle this task. Therefore, in the future, introducing external knowledge to construct a knowledge graph and supplementing Specific knowledge based on thesis topics will be used to explore this task.

Acknowledgements. This work is partially supported by grants from the National Key Research and Development Program of China (No.2018YFC0832101), the Natural Science Foundation of China (No.61632011, 61702080, and 61602079).

References

1. Lawrence, J., Reed, C.: Argument mining: a survey. Comput. Linguist. **45**(4), 765–818 (2020)
2. Mochales, R., Moens, M.F.: Argumentation mining. Artif. Intell. Law **19**(1), 1–22 (2011)
3. Poudyal, P.: A machine learning approach to argument mining in legal documents. In: Pagallo, U., Palmirani, M., Casanovas, P., Sartor, G., Villata, S. (eds.) AICOL 2015-2017. LNCS (LNAI), vol. 10791, pp. 443–450. Springer, Cham (2018). https://doi.org/10.1007/978-3-030-00178-0_30
4. Wachsmuth, H., Al Khatib, K., Stein, B.: Using argument mining to assess the argumentation quality of essays. In: Proceedings of COLING 2016, The 26th International Conference on Computational Linguistics: Technical Papers, pp. 1680–1691 (2016)
5. Zhang, F., Litman, D.: Using context to predict the purpose of argumentative writing revisions. In: Proceedings of the. Conference of the North American Chapter of the Association for Computational Linguistics: Human Language Technologies, vol. 2016, pp. 1424–1430 (2016)
6. Florou, E., Konstantopoulos, S., Koukourikos, A., et al.: Argument extraction for supporting public policy formulation. In: Proceedings of the 7th Workshop on Language Technology for Cultural Heritage, Social Sciences, and Humanities, pp. 49–54 (2013)
7. Morio, G., Ozaki, H., Morishita, T., et al.: Towards better non-tree argument mining: proposition-level biaffine parsing with task-specific parameterization. In: Proceedings of the 58th Annual Meeting of the Association for Computational Linguistics, pp. 3259–3266 (2020)
8. Li, J., Durmus, E., Cardie, C.: Exploring the role of argument structure in online debate persuasion. arXiv preprint arXiv:2010.03538 (2020)
9. Al Khatib, K., Völske, M., Syed, S., et al.: Exploiting personal characteristics of debaters for predicting persuasiveness. In: Proceedings of the 58th Annual Meeting of the Association for Computational Linguistics, pp. 7067–7072 (2020)
10. El Baff, R., Wachsmuth, H., Al Khatib, K., et al.: Analyzing the persuasive effect of style in news editorial argumentation. In: Association for Computational Linguistics (2020)
11. Bar-Haim, R., Kantor, Y., Eden, L., et al.: Quantitative argument summarization and beyond: cross-domain key point analysis. In: Proceedings of the. Conference on Empirical Methods in Natural Language Processing (EMNLP), vol. 2020, pp. 39–49 (2020)
12. Bar-Haim, R., Eden, L., Friedman, R., et al.: From arguments to key points: towards automatic argument summarization. In: Proceedings of the 58th Annual Meeting of the Association for Computational Linguistics, pp. 4029–4039 (2020)

13. Taghipour, K., Ng, H.T.: A neural approach to automated essay scoring. In: Proceedings of the 2016 Conference on Empirical Methods in Natural Language processing, pp. 1882–1891 (2016)
14. Sanchan, N., Aker, A., Bontcheva, K.: Automatic summarization of online debates. In: Proceedings of the 1st Workshop on Natural Language Processing and Information Retrieval associated with RANLP 2017, pp. 19–27. INCOMA Inc. (2017)
15. Botschen, T., Sorokin, D., Gurevych, I.: Frame-and entity-based knowledge for common-sense argumentative reasoning. In: Proceedings of the 5th Workshop on Argument Mining, pp. 90–96 (2018)
16. Ji, L., Wei, Z., Li, J., et al.: Discrete argument representation learning for interactive argument pair identification. arXiv preprint arXiv:1911.01621 (2019)
17. Yuan, J., Wei, Z., Zhao, D., et al.: Leveraging argumentation knowledge graph for interactive argument pair identification (2021)
18. Devlin, J., Chang, M.W., Lee, K., et al.: BERT: pre-training of deep bidirectional transformers for language understanding. In: Proceedings of the 2019 Conference of the North American Chapter of the Association for Computational Linguistics: Human Language Technologies, vol. 1 (Long and Short Papers), pp. 4171–4186 (2019)
19. Sun, Y., Wang, S., Li, Y., et al.: Ernie 2.0: a continual pre-training framework for language understanding. In: Proceedings of the AAAI Conference on Artificial Intelligence, vol. 34, no. 05, pp. 8968–8975 (2020)
20. Yang, Z., Dai, Z., Yang, Y., et al.: Xlnet: generalized autoregressive pretraining for language understanding. In: Advances in Neural Information Processing Systems, vol. 32 (2019)
21. Chakrabarty, T., Hidey, C., Muresan, S., et al.: AMPERSAND: argument mining for PERSuAsive oNline discussions. In: Proceedings of the 2019 Conference on Empirical Methods in Natural Language Processing and the 9th International Joint Conference on Natural Language Processing (EMNLP-IJCNLP), pp. 2933–2943 (2019)
22. Cheng, L., Bing, L., Yu, Q., et al.: Argument pair extraction from peer review and rebuttal via multi-task learning. In: Proceedings of the Conference on Empirical Methods in Natural Language Processing (EMNLP), vol. 2020, pp. 7000–7011 (2020)
23. Gao, T., Yao, X., Chen, D.: SimCSE: simple contrastive learning of sentence embeddings. arXiv preprint arXiv:2104.08821 (2021)
24. Tan, C, Niculae, V., Danescu-Niculescu-Mizil, C., et al.: Winning arguments: interaction dynamics and persuasion strategies in good-faith online discussions. In: Proceedings of the 25th International Conference on World Wide Web, pp. 613–624 (2016)
25. Wei, Z., Liu, Y., Li, Y.: Is this post persuasive? ranking argumentative comments in online forum. In: Proceedings of the 54th Annual Meeting of the Association for Computational Linguistics, vol. 2: Short Papers, pp. 195–200 (2016)
26. Ji, L., Wei, Z., Hu, X., et al.: Incorporating argument-level interactions for persuasion comments evaluation using co-attention model. In: Proceedings of the 27th International Conference on Computational Linguistics, pp. 3703–3714 (2018)
27. Lan, Z., Chen, M., Goodman, S., et al.: Albert: a lite bert for self-supervised learning of language representations. arXiv preprint arXiv:1909.11942 (2019)
28. Liu, Y., Ott, M., Goyal, N., et al.: RoBERTa: a robustly optimized bert pretraining approach. arXiv preprint arXiv:1907.11692 (2019)

29. Wei, J., Zou, K.: EDA: easy data augmentation techniques for boosting performance on text classification tasks. In: Proceedings of the 2019 Conference on Empirical Methods in Natural Language Processing and the 9th International Joint Conference on Natural Language Processing (EMNLP-IJCNLP), pp. 6382–6388 (2019)
30. Xie, Z., Wang, S.I., Li, J., et al.: Data noising as smoothing in neural network language models. arXiv preprint arXiv:1703.02573 (2017)
31. Coulombe, C.: Text data augmentation made simple by leveraging nlp cloud apis. arXiv preprint arXiv:1812.04718 (2018)

ARGUABLY @ AI Debater-NLPCC 2021 Task 3: Argument Pair Extraction from Peer Review and Rebuttals

Guneet Singh Kohli[1(✉)], Prabsimran Kaur[1], Muskaan Singh[2], Tirthankar Ghosal[2], and Prashant Singh Rana[1]

[1] Department of Computer Science and Engineering, Thapar Institute of Engineering and Technology, Patiala, India
{gkohli_be18,pkaur_be18,prashant.singh}@thapar.edu
[2] Institute of Formal and Applied Linguistics, Faculty of Mathematics and Physics, Charles University, Prague, Czech Republic
{singh,ghosal}@ufal.mff.cuni.cz

Abstract. This paper describes our participating system run to the argumentative text understanding shared task for AI Debater at NLPCC 2021 (http://www.fudan-disc.com/sharedtask/AIDebater21/tracks.html). The tasks are motivated towards developing an autonomous debating system. We make an initial attempt with Track-3, namely, *argument pair extraction from peer review and rebuttal* where we extract arguments from peer reviews and their corresponding rebuttals from author responses. Compared to the multi-task baseline by the organizers, we introduce two significant changes: (i) we use ERNIE 2.0 token embedding, which can better capture lexical, syntactic, and semantic aspects of information in the training data, (ii) we perform double attention learning to capture long-term dependencies. Our proposed model achieves the state-of-the-art results with a relative improvement of 8.81% in terms of F1 score over the baseline model. We make our code available publicly at https://github.com/guneetsk99/ArgumentMining_SharedTask. Our team ARGUABLY is one of the third prize-winning teams in Track 3 of the shared task.

Keywords: Argument pair extraction · Peer review · Deep learning

1 Introduction

Arguments or debates are common in societies and are omnipresent in a wide range of day-to-day human activities. Arguments deal with facts, propositions for a specific topic, usually with a motive/direction to converge to a conclusion. Logical argumentation is one of the defining attributes of a thinking mind. However, automatically processing arguments is a challenging task. Argumentation closely approximates *explainability* in Artificial Intelligence (AI), which is a desired characteristic to build trustworthy AI systems. Over the years, argument

L. Wang et al. (Eds.): NLPCC 2021, LNAI 13029, pp. 590–602, 2021.
https://doi.org/10.1007/978-3-030-88483-3_48

mining [16] has gained significant interest of the Natural Language Processing (NLP) community for various real-world problems such as online debates, legal and medical documents, persuasive essays [1,18–20,24]. Quite recently, argumentation mining techniques are applied to peer reviews [6,14]. Peer review texts can be seen as arguments made by reviewers in favor/against the paper under scrutiny. Authors write a response (arguments) to the reviewer's arguments to establish the merit of their work. These argument pairs (Reviews-Rebuttals) help the editors/program-chairs to arrive at a conclusion regarding the paper under scrutiny. The Task 3 of the AI Debater Shared Task at NLPCC 2021 is motivated to extract these argument pairs from peer-review texts. In this work, we attempt to automatically extract arguments from peer reviews [6,14] and their corresponding rebuttals [10,30] from peer-review texts following an open reviewing model.

Peer review texts and rebuttals contain rich argumentative structure. A review and its rebuttal provide interaction between the reviewer and the author with a motive to persuade each other to their corresponding stance. Hence, *Argument Pair Extraction* from peer reviews could be useful to understand the discourse structure of peer reviews and rebuttals and also aid in decision-making. Our team ARGUABLY made the following contributions to this problem as part of the shared task effort.

- We use ERNIE 2.0 token embeddings, which can better capture the lexical, syntactic, and semantic aspects of information in the training data.
- We perform double attention learning to capture long-range relations.
- Our proposed model achieves the state-of-the-art results with a relative improvement of 8.81% in terms of F1 score over the baseline.

2 Related Work

Research in argumentation has been gaining a significant interest recently, ranging from argument mining [23,28,29], quality assessment [8,27], relation detection [9,21] and its generation [10,11,22]. Some works [19,24] propose pipeline approaches to identify argumentative discourse structures in persuasive essays on a single passage in two steps (1) extracting argument components and (2) identifying relations. Swanson et al. [26] and Chakrabarty et al. [2] extract arguments from large corpora of online forums and persuasive discussions on the CMV dataset [4]. Recently for the task of argument mining and sentence pairing, Cheng et al. [4] proposed a multi-task learning framework with hierarchical LSTM networks. We adopted this work as a baseline, improved the representation learning with ERNIE 2.0 and use double attention mechanism to pair the review comment with its corresponding rebuttal.

The work in the field of the scholarly domain, for peer review and rebuttal, is in progressing phase. There are datasets for understanding the structure[15] and the content of peer reviews. [13], proposed a PeerRead dataset, collected from various venues including ACL, NIPS, and ICLR consisting of 14.7K paper drafts with their corresponding accept/reject decisions with 10.7K peer reviews for a subset of papers. Later, a partially annotated dataset, AMPERE [10] collected 14,202 reviews from ICLR, UAI, ACL, and NeurIPS.

This work also presents proposition detection and classification of peer reviews.[7] proposed a way to score after-rebuttal experience from initial review and author responses along with a dataset of 4K reviews and 1.2K author responses collected from ACL. [4] produced a fully annotated dataset with 4,764 pairs of reviews and rebuttals collected from openreview.net. The work studies internal structure and relation between review-rebuttal.

3 Task Description

The Argumentative Text Understanding for AI Debater (NLPCC2021) consists of three tasks. However, we only focus on the third task, i.e., Argument Pair Extraction (APE) from Peer Reviews and Rebuttals. Figure 1 shows a sample instance from the dataset that explains the task. The task is to automatically extract the text spans and classify them into ARGUMENTS and NON-ARGUMENTS. The organizers used the IOB annotation scheme to specify the Beginning, Inside, and Outside of the argumentation text span. ARGUMENTS in reviews are labeled as REVIEW, while ARGUMENTS in rebuttals are labeled as REPLY. In this task, we intend to automatically extract argument pairs, from each passage pair of review-rebuttal. We formally state the problem as, given two passage, a review with x sentences $Rev_1 = [s_1^1, s_1^2, \ldots, s_1^x]$ and corresponding rebuttal with y sentence $Reb_1 = [s_2^1, s_2^2, \ldots, s_2^y]$, the goal is to automatically fetch arguments from these pairs. Particularly, in this work, we first perform argumentation mining, consist of one or more sentences (see Fig. 1 REVIEW-1,

Review:

[The paper trains wide ResNets for 1-bit per weight deployment. **O O Review 430 <sep>** The experiments are conducted on CIFAR-10, CIFAR-100, SVHN and ImageNet32. **O O Review 430 <sep> <sep>** +the paper reads well **O O Review 430**]NON-ARG [+the reported performance is compelling **B-Review B-1 Review 430**]REVIEW-1 **<sep>** [Perhaps the authors should make it clear in the abstract by replacing: **B-Review B-2 Review 430** "Here, we report methodological innovations that result in large reductions in error rates across multiple datasets for deep convolutional neural networks deployed using a single bit for each weight" **I-Review I-2 Review 430** with **I-Review I-2 Review 430** "Here, we report methodological innovations that result in large reductions in error rates across multiple datasets for wide ResNets deployed using a single bit for each weight" **I-Review I-2 430**]REVIEW-2]

Rebuttal:

[**<sep> <sep> <sep>** Thankyou for your comments. **O O Reply 430 <sep> <sep> *** O O Reply 430 <sep>** **Reviewer comment:** "+the reported performance is compelling": **O O Reply 430**]NON-ARG **<sep>**[**Author Response:** To reinforce this aspect, since initial submission we have found the following ways to surpass the performance we initially reported: **B-Reply B-1 Reply 430 <sep>** 1. **I-Reply I-1 Reply 430** We now have conducted experiments on the full Imagenet dataset and have surpassed all previously published results for a single-bit per weight. **I-Reply I-1 Reply 430** Indeed, we provide the first report, to our knowledge, of a top-5 error rate under 10% for this case. **I-Reply I-1 Reply 430 <sep>** 2. **O O Reply 430** For Imagenet32, we realised that the weight decay we used was set to the larger CIFAR value of 0.0005. **B-Reply**

Fig. 1. A Review-Rebuttal pair in the challenge dataset

REVIEW-2 and REPLY-1, REPLY-2) from the passage, notably with no overlapping arguments. Each of these mined arguments are later generated in pairs (for instance REVIEW-1, REPLY-1 and REVIEW-2, REPLY-2).

4 Dataset Description

The dataset provided by the shared task was collected with peer review and rebuttal of ICLR 2013–2020 (except for 2015)[1] shown in Table 1. The collected information mainly can be categorized in two parts: (1) information about the submissions (such as title, authors, year, track, acceptance decision, original submission separated by sections) (2) information about reviews and author responses (such as review passages, rebuttal passages, review scores, reviewer confidence). The dataset was annotated by filtering out peer reviews and rebuttals. As per the annotation in reviews, arguments are sentences that express non-positive comments relating to specific suggestions, questions, or challenges. While in the case of rebuttals, arguments are contents that are answering or explaining that particular review argument. Sentences labeled as non-arguments are comprised of courtesy expressions or general conclusions. The data annotation is based on sentence granularity, i.e., an argument boundary is limited to a sentence. Each sentence is then assigned a label according to the standard IOBES tagging scheme specified in the format:

$$< review\_comments/author\_reply >< BIO\_tag > - < review/reply >< BIO\_tag > - < Pairing\_Index >< review/reply >< Paper\_ID >$$

where each entry of review-reply is separated by a tab/blank space, and the $< SEP >$ tag is used for newline character in the data, usually added at the starting of a new paragraph. In the BIO tagging, B and I are followed by $< review/reply >$ or $< pairing\_number >$, while O tags have no following tags.

- **Example 1**: the sample complexity of the problem is rather difficult. \t B-review \t B-2 \t Review \t 20484.
- **Example 2**: Thank you for your careful review and thoughtful comments \t O \t O \t Reply \t 20484.

As presented in Fig. 1 an instance of the dataset, statements with the label 'O O' represent the non-argument. It proceeds with the type of passage (either review/rebuttal) and the submission number. As far as the arguments are concerned, the statement with the label 'B' represents the beginning of an argument, and the label 'I' represents the continuation of the same argument. The number associated with the 'B' and 'I' tags are the corresponding pair number (between the review and rebuttal), followed by the type of passage and submission number. Single line arguments are labeled as 'B' tags, and multiple lines arguments are labeled as B, I in sequence. We have added an extra notation of NON-ARG, REVIEW, and REPLY in Fig. 1 for a better representation of data for inferring insights from it.

[1] https://openreview.net/.

5 Methodology

This section proposes our architecture to extract argument pairs from the peer reviews in the shared task. As shown in Fig. 2, we propose a multi-task learning framework with hierarchical LSTM networks with A2-nets (double Attention), namely MT-H-LSTM-CRF-A2-nets, to tackle the problem of review-rebuttal pair argument extraction. We extended the work proposed by [4] to tackle argument mining and pairing. The sentence encoder represented in the left with a red dotted box uses pre-trained ERNIE 2.0 [25] token embeddings. Further, these embeddings are fed as input to bi-LSTM to generate sentence embeddings. These encoded sentences are passed to the second layer of LSTM to encode information from subsequent review-response passages further. In addition, we introduce A2-nets [3] for highlighting relevant features of sentences dynamically. It provides an aid in the analysis of the argumentative structure and generating review-response pairs. Finally, the second LSTM network classifies the argument pairs into IOBES and 1/0. We further describe each component of our proposed multi-task pipeline architecture in detail.

Table 1. Statistics of the dataset from [4]

# instances (i.e., review-rebuttal pairs)	4,764
Review	
# sentences	99.8K
# arguments	23.2k
# argument sentences	58.5k
Avg. # sentences per passage	21.0
Avg. # sentences per argument	2.5
Rebuttal	
# sentences	94.9K
# arguments	17.7K
# argument sentences	67.5k
Avg. # sentences per passage	19.9
Avg. # sentences per argument	3.8
Vocab.	
Total vocab size	165k
Review vocab size	103k
Rebuttal vocab size	105k
Avg. % vocab shared in each passage pair	16.6%
Avg. % vocab shared in each argument pair	9.9%

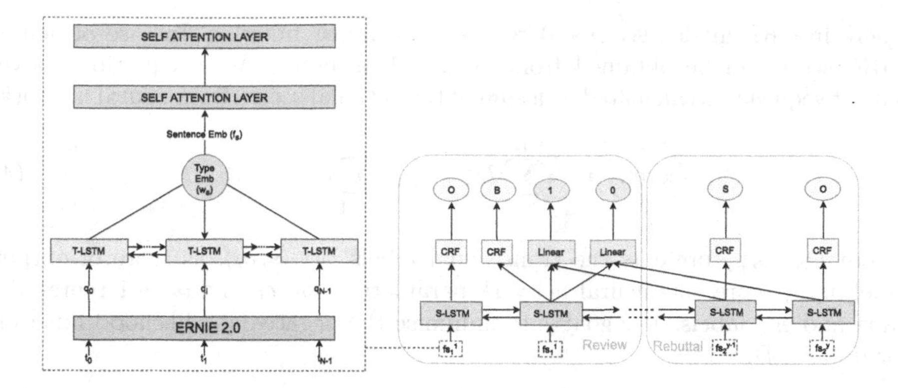

Fig. 2. Overview of our MT-H-LSTM-CRF+ERNIE2.0+ATT2 model architecture. The sentence encoder (in the red dotted box on the left) shows the process of obtaining sentence embeddings from pre-trained ERNIE 2.0 token embeddings with T-LSTM. The multi-task learning based framework (on the right) shows the process of generating labels for both subtasks with the shared S-LSTM along with double attention.

5.1 Sentence Encoder

As in Fig. 2, we input sentence S with N tokens for sentence encoding.

$$S = [t_0, t_1, \cdot t_{N-1}] \tag{1}$$

We employ pre-trained ERNIE 2.0 [25] for generating all token embeddings for a sentence $[q_0, q_1, ...q_{N-1}]$. We pass these token embeddings into the token-level bi-LSTM layer(t-LSTM) for capturing sentence embeddings. We randomly initialize the w_s trainable type embeddings to distinguish between a review or reply sentences. Finally, these randomly initialized embeddings are passed through A2-nets[3] to obtain final sentence embeddings.

5.2 Bi-LSTM-CRF

As we obtain the sentence embedding from the previous layer, we feed them to the Bi-LSTM network to perform sentence-level IOBES tagging for argument mining. The right-hand side of Fig. 2 presents S-LSTM, which encodes the sentence sequence for the CRF layer, which uses the feature of each position to predict tags. In the passage, we are given with a sequence of sentence for review-rebuttal pair,

$$s = s_1^1, s_1^2, \cdot s_1^x, s_2^1, s_2^2, \cdot s_2^y \tag{2}$$

The task is to predict the output sequence label $O \in B, I, E, S, O$. The probability of the prediction, given sentence s,

$$p(\mathbf{O} \mid \mathbf{s}) = \frac{\exp(\text{score}(\mathbf{s}, \mathbf{0}))}{\sum_{\mathbf{0}} \exp(\text{score}(\mathbf{s}, \mathbf{0}))} \tag{3}$$

where in CRF model, score(s,o) represents a linear function. In case of neural CRF, score can be obtained from bi-LSTM encoders. As in equation above, output sequence is calculated as a sum of transitional scores from neural network.

$$\text{score}(\mathrm{s}, \mathbf{o}) = \sum_{i=o}^{n} D_{o_i, o_{i+1}} + \sum_{i=1}^{n} \phi_{\theta_1}(\mathbf{s}, o_i), \tag{4}$$

where D_{oi}, o_{i+1} presents the transitional labels, $\phi_{\theta_1}(\mathbf{s}, o_i)$, as score of output label o_i, θ_1 represents neural network parameters and o_0 and $o_n + 1$ represents start and end labels. The goal is to minimise the negative log-likehood function for dataset D_s,

$$\mathcal{L}_1(D, \theta_1 \mid \mathcal{D}_{f1}) = - \sum_{(\mathbf{s}, \mathbf{o}) \in \mathcal{D}_{f1}} \log p(\mathbf{o} \mid \mathbf{s}) \tag{5}$$

The gradient for the parameters is calculated with forward-backward propagation algorithm of CRF neural network. Finally, for decoding the sequence labels we use viterbi algorithm [12],

$$\mathbf{o}^* = \arg\max_{\mathbf{o} \in \wr} p(\mathbf{o} \mid \mathbf{s}) \tag{6}$$

5.3 Bi-LSTM-Linear

After the argument mining of review-response pairs, the next step involves pairing these arguments. We propose a Bi-LSTM-Linear framework for this binary classification task, for prediction of sentence pairs. It predicts if two sentences belong to the same argument pair. Particularly, it adds up S-LSTM outputs of review-rebuttal pair sentences as in equation,

$$f_{\text{pair}}^{i,j} = h_{f_1^i} + f_{s_2^j}, \forall s_1^i \in R_1, s_2^j \in R_2 \tag{7}$$

where, $f_{i,j}$ pair is passed to linear layer for prediction of binary label $w_{i,j} \in 0, 1$. While the negative log-likehood is minimised for D_{s2} dataset is parameterized with ϕ_2.

$$\mathcal{L}_2(\theta_2 \mid \mathcal{D}_{s2}) = - \sum_{(f_{pair}, w) \in \mathcal{D}_{s2}} (w \log p(w = 1 \mid f_{\text{pair}}) + (1 - w) \log p(w = 0 \mid f_{\text{pair}})). \tag{8}$$

Finally, in decoding each tag is classified in binary,

$$w^* = \arg\max_{w \in \{0,1\}} p(w \mid f_{\text{pair}}) \tag{9}$$

here, the Bi-LSTM-linear and Bi-LSTM-CRF share the sentence representation, to achieve the efficient performance of argument pair extraction.

Table 2. Phase-I trial results

Models	Precision	Recall	F1 Score
BERT [5]	32.99	19.05	24.16
BERT(Tuned)	32.69	20.01	24.82
XLNet(Tuned) [31]	30.61	22.02	25.61
RoBERTa (Tuned) [17]	**34.63**	**22.02**	**26.92**

5.4 Multi-task Training

Firstly, we perform argument mining with sequence labeling task, and further, we feed these labeled arguments into the binary classifier for pairing into review-response pairs. To perform the multi-task learning, we sum up the losses for both as,

$$\mathcal{L} = w_{i_1}\mathcal{L}_1\left(A, \theta_1 \mid \mathcal{D}_{s1}\right) + w_2\mathcal{L}_2\left(\theta_2 \mid \mathcal{D}_{s2}\right) \tag{10}$$

where w_1 and w_2 are weights of corresponding losses for review-rebuttal. During the training, best model was selected from entire development task for argument-pair extraction.

6 Experiments and Results

In this section, we elaborate our experiment and results in two phases: Phase-I and Phase-II, exactly as how we made our submissions and progressed in the challenge.

6.1 Phase-I

Inspired from Cheng et al. [4], our focus was to introduce double-attention mechanism [3] encoding layer so that the architecture could selectively understand the key points of a sentence for inferring the argument pairs later in the model. Table 2 and Table 3 shows our results using the various language models we employ for the task on top of the Cheng et al. [4] architecture for Phase-I and Phase-II, respectively. In Phase-I of the contest, we focus on changing the baseline by experimenting with pre-trained token embeddings, simultaneously experimenting with the number of LSTM layers in the Bi-LSTM encode. We increase the number of successive LSTM's required to carry out the binary sentence (argument) pairing task. We start with the baseline approach to understand and produce gradual improvements on top of it. The use of language models like RoBERTa [17] and XLNet [31], which are pre-trained on large data, improved performance. We achieve improvement over BERT, which could achieve an F1 score of 24.8 compared to RoBERTa [17] and XLNet [31], which reached an

F1 score of 25.48% and 26.3%, respectively. We summarize our attempts for Phase-I trials in Table 2. We could see that RoBERTa [17] became the better performer and selected that model as the ARGUABLY submission for Phase-I. The replacement of Linear CRF for IOBES tagging and BiLSTM encoder for encoding didn't yield positive results. We also tried with CNN-BiLSTM encoder for encoding. However, the F1 score dropped to as low as 23.47. Hence we did not proceed along this line.

6.2 Phase-II

Table 3. Phase-II trial results

Models	Precision	Recall	F1 Score
RoBERTa+Attention	35.93	21.12	27.32
ERNIE2.0 [25]	26.30	24.45	25.34
ERNIE 2.0+(Attention)	34.35	24.39	26.45
ERNIE 2.0+A2-nets	**37.35**	**23.39**	**28.79**

In this phase, we focus on two important aspects to improve our performance from Phase-I.

1) Use ERNIE 2.0 pre-trained embeddings [25]
2) Use double attention mechanism [3] after the BiLSTM Encoder

Table 3 shows the advantage of using the attention mechanism on the existing models increases the F1 score for all the models we use in Phase-I. Using ERNIE 2.0 as the underlying language model paired with double attention was crucial to improving our results. ERNIE, a Continual Pre-training Framework for Language Understanding, outperforms RoBERTa [17] in combination with attention layer pair. We experimented with ERNIE because it had earlier outperformed BERT and XLNET on certain NLP Tasks on the GLUE Benchmark [25]. The use of a single attention layer after encoding results in the improvement of RoBERTa up to an F1 score of 0.273, whereas for ERNIE 2.0, the result was up to 0.2654 from 0.2534 when we did not use any attention layer. The use of two successive layers as a self-attention for sentence encoding is then passed through Linear CRF for tagging. We further test with increasing the

Table 4. Comparison of baseline with the proposed system

Model	Precision	Recall	F1 Score
Baseline [4]	26.69	26.24	26.46
Proposed Work	**37.35**	**23.39**	**28.79**

number of attention layers at each stage. The improvement in the F1 score was significant in the case of ERNIE compared to when it was being applied to the RoBERTa-based model. However, adding the attention layers beyond two results in no improvement. Also, we observe that the use of the attention layer improved the convergence of CRF-loss, which helped infer that the model can understand and tag the respective Reply-Rebuttal arguments. Our final submission was thus ERNIE + Double Attention layer with the baseline model. The comparison analysis on attention/double attention mechanism added in the proposed work is presented in Table 5. The results from Table 4 shows significant improvement over the baseline with our proposed model. We rank fifth in the shared task leaderboard http://121.40.48.157:8866/leaderboard.

Table 5. Comparison based on Attention/double attention mechanism added in the proposed work

Model	Precision	Recall	F1 Score
Baseline Model+ERNIE2.0 [4]	30.24	21.56	25.18
Baseline+Attention	38.62	19.32	25.75
Baseline+Double Attention	36.53	21.12	26.76

Table 6. Hyperparameter details

Hyperparameter	BERT [5]	XLNet[31]	RoBERTa[17]
# Linear layers	5	7	5
Dropout rate	0.4	0.3	0.3
# T-LSTM layers	3	5	5
Batch size	1	1	1
Hidden dimensions	350	300	250
# S-LSTM layers	3	3	3
Learning rate	2e–4	2e–4	2e–4
Pair weight	0.5	0.4	0.4
Python	3.6		
GPU Type	Tesla P100		
GPU RAM	16 GB		
M/c Ram	25 GB with high memory VMs		
CPU	2 × vCPU		

Table 7. Error analysis

Instances	Gold labels	Our model	Baseline
Using scheme B and C student model with low precision could achieve accuracy close to teacher while compressing the model	O	B–1	B
Comments on the experiment section	O	B	B
I am not an expert in this area; I had trouble following the details of the theoretical developments	B–6	O	O
The sample complexity of the problem is rather prohibitive	B–2	O	O
About equation(1):	O	O	B
This paper studies the problem of learning both the distance metric and a linkage rule from clustering examples	O	O	O
The paper is somewhat interesting contribution to the area, but I think can only be seen as a basic step in the general direction	B–1	B–1	I–1
Most of the interest merge functions used for HAC boil down to simple 2-point-merge rules: average-linkage, Ward's method etc.	I–1	I–1	I–1

6.3 Analysis

The instance 1 in Table 7 is a part of the review that was labeled as a beginning of an argument by both our model and the baseline model. However, according to the gold labels, a sentence does not portray any argument. If one considers the preceding statement, The experiments are clear, and the three different schemes provide good analytical insights, it would be clear why instance one is labeled as O in gold labels. However, if we focus on the statement alone, it implies that the reviewer is suggesting the author use a particular technique. This shows that the model does not focus on discourse integration and instead focuses on the current statement. The second instance shows that the model focuses more on the words rather than the sentence meaning as a whole. According to the gold label, the third instance is an argument, but the model, on the other hand, labels it as a non-argument statement. This happens because the model focuses more on words and phrases such as 'expert', 'theoretical', 'developments' rather than understanding the full context of the sentence. The fourth instance in the table shows that the label should have been a separate argument. However, the model considers it to be the continuation of the previous argument because there are no words or context in this statement that show that a new problem is initiated. Thus it assumes that the discussion of the previous problem is continued. The subsequent four instances, i.e. (5th, 6th, 7th, 8th), show that our model mimics the golden label. This indicates that the model was able to correctly understand the context and structure of the sentences and arguments to a reasonable extent and was able to make accurate predictions.

7 Conclusion and Future Work

In this work, we improvise the existing model for extracting argument pairs from peer reviews and rebuttal passages by introducing a double layer of attention with the ERNIE 2.0 language model in the representation layer. Our model ranked fifth in the challenge. In the future, we will explore the effect of attention layers on word-level. We will also explore graph-based argument pair extraction methods leveraging the argument dependencies in reviews and rebuttals.

References

1. Abbott, R., Ecker, B., Anand, P., Walker, M.: Internet argument corpus 2.0: an sql schema for dialogic social media and the corpora to go with it. In: Proceedings of the Tenth International Conference on Language Resources and Evaluation (LREC'16), pp. 4445–4452 (2016)
2. Chakrabarty, T., Hidey, C., Muresan, S., McKeown, K., Hwang, A.: Ampersand: argument mining for persuasive online discussions. arXiv preprint arXiv:2004.14677 (2020)
3. Chen, Y., Kalantidis, Y., Li, J., Yan, S., Feng, J.: $a^\wedge$2-nets: double attention networks. arXiv preprint arXiv:1810.11579 (2018)
4. Cheng, L., Bing, L., Yu, Q., Lu, W., Si, L.: Argument pair extraction from peer review and rebuttal via multi-task learning. In: Proceedings of the 2020 Conference on Empirical Methods in Natural Language Processing (EMNLP), pp. 7000–7011 (2020)
5. Devlin, J., Chang, M.W., Lee, K., Toutanova, K.: Bert: pre-training of deep bidirectional transformers for language understanding. arXiv preprint arXiv:1810.04805 (2018)
6. Falkenberg, L.J., Soranno, P.A.: Reviewing reviews: an evaluation of peer reviews of journal article submissions. Limnol. Oceanogr. Bull. **27**(1), 1–5 (2018)
7. Gao, Y., Eger, S., Kuznetsov, I., Gurevych, I., Miyao, Y.: Does my rebuttal matter? insights from a major nlp conference. arXiv preprint arXiv:1903.11367 (2019)
8. Gretz, S., et al.: A large-scale dataset for argument quality ranking: construction and analysis. In: Proceedings of the AAAI Conference on Artificial Intelligence, vol. 34, pp. 7805–7813 (2020)
9. Hou, Y., Jochim, C.: Argument relation classification using a joint inference model. In: Proceedings of the 4th Workshop on Argument Mining, pp. 60–66 (2017)
10. Hua, X., Nikolov, M., Badugu, N., Wang, L.: Argument mining for understanding peer reviews. arXiv preprint arXiv:1903.10104 (2019)
11. Hua, X., Wang, L.: Neural argument generation augmented with externally retrieved evidence. arXiv preprint arXiv:1805.10254 (2018)
12. Kaji, N., Fujiwara, Y., Yoshinaga, N., Kitsuregawa, M.: Efficient staggered decoding for sequence labeling. In: Proceedings of the 48th Annual Meeting of the Association for Computational Linguistics, pp. 485–494 (2010)
13. Kang, D., et al.: A dataset of peer reviews (peerread): collection, insights and nlp applications. arXiv preprint arXiv:1804.09635 (2018)

14. Kelly, J., Sadeghieh, T., Adeli, K.: Peer review in scientific publications: benefits, critiques, & a survival guide. Ejifcc **25**(3), 227 (2014)
15. Kovanis, M., Porcher, R., Ravaud, P., Trinquart, L.: The global burden of journal peer review in the biomedical literature: strong imbalance in the collective enterprise. PloS one **11**(11), e0166387 (2016)
16. Lawrence, J., Reed, C.: Argument mining: a survey. Comput. Linguist. **45**(4), 765–818 (2020)
17. Liu, Y., et al.: Roberta: a robustly optimized bert pretraining approach. arXiv preprint arXiv:1907.11692 (2019)
18. Mochales, R., Moens, M.F.: Argumentation mining. Artif. Intell. Law **19**(1), 1–22 (2011)
19. Persing, I., Ng, V.: End-to-end argumentation mining in student essays. In: Proceedings of the 2016 Conference of the North American Chapter of the Association for Computational Linguistics: Human Language Technologies, pp. 1384–1394 (2016)
20. Poudyal, P.: A machine learning approach to argument mining in legal documents. In: Pagallo, U., Palmirani, M., Casanovas, P., Sartor, G., Villata, S. (eds.) AICOL 2015-2017. LNCS (LNAI), vol. 10791, pp. 443–450. Springer, Cham (2018). https://doi.org/10.1007/978-3-030-00178-0_30
21. Rocha, G., Stab, C., Cardoso, H.L., Gurevych, I.: Cross-lingual argumentative relation identification: from english to portuguese. In: Proceedings of the 5th Workshop on Argument Mining, pp. 144–154 (2018)
22. Schiller, B., Daxenberger, J., Gurevych, I.: Aspect-controlled neural argument generation. arXiv preprint arXiv:2005.00084 (2020)
23. Shnarch, E., et al.: Will it blend? blending weak and strong labeled data in a neural network for argumentation mining. In: Proceedings of the 56th Annual Meeting of the Association for Computational Linguistics, vol. 2: Short Papers, pp. 599–605 (2018)
24. Stab, C., Gurevych, I.: Identifying argumentative discourse structures in persuasive essays. In: Proceedings of the 2014 Conference on Empirical Methods in Natural Language Processing (EMNLP), pp. 46–56 (2014)
25. Sun, Y., et al.: Ernie 2.0: a continual pre-training framework for language understanding. In: Proceedings of the AAAI Conference on Artificial Intelligence, vol. 34, pp. 8968–8975 (2020)
26. Swanson, R., Ecker, B., Walker, M.: Argument mining: extracting arguments from online dialogue. In: Proceedings of the 16th Annual Meeting of the Special Interest Group on Discourse and Dialogue, pp. 217–226 (2015)
27. Toledo, A., et al.: Automatic argument quality assessment-new datasets and methods. arXiv preprint arXiv:1909.01007 (2019)
28. Trautmann, D., Daxenberger, J., Stab, C., Schütze, H., Gurevych, I.: Fine-grained argument unit recognition and classification. In: Proceedings of the AAAI Conference on Artificial Intelligence, vol. 34, pp. 9048–9056 (2020)
29. Trautmann, D., Fromm, M., Tresp, V., Seidl, T., Schütze, H.: Relational and fine-grained argument mining. Datenbank-Spektrum **20**(2), 99–105 (2020)
30. Xiong, W., Litman, D.: Automatically predicting peer-review helpfulness. In: Proceedings of the 49th Annual Meeting of the Association for Computational Linguistics: Human Language Technologies, pp. 502–507 (2011)
31. Yang, Z., Dai, Z., Yang, Y., Carbonell, J., Salakhutdinov, R.R., Le, Q.V.: Xlnet: generalized autoregressive pretraining for language understanding. In: Advances in Neural Information Processing Systems, vol. 32 (2019)

Sentence Rewriting for Fine-Tuned Model Based on Dictionary: Taking the Track 1 of NLPCC 2021 Argumentative Text Understanding for AI Debater as an Example

Pan He, Yan Wang, and Yanru Zhang(✉)

University of Electronic Science and Technology of China, Chengdu, China
202021080734@std.uestc.edu.cn, {yanbo1990,yanruzhang}@uestc.edu.cn

Abstract. Pre-trained language model for Natural Language Inference (NLI) is hard to understand rare words and terminology in natural language because of the limitation of data size. The size of data in many domain is greatly small but valuable. Considering that the dataset incapacitated the model from learning, we propose a sentence rewriting method based on dictionary to tackle this issue and expand the application boundary of pretrained language model from special domain to general fields. We prove the effectiveness of this method in Track 1 of identifying Topic and Claim in the Argumentative Text Understanding for AI Debater of NLPCC 2021 and have achieved excellent result.

Keywords: NLI · Pre-trained language model · Dictionary

1 Introduction

Natural language processing (NLP) includes several down-stream tasks, such as natural language inference (NLI), text classification, reading comprehension and so on. In contrast to text classification or text similarity calculation, NLI and reading comprehension is in need of models to understand the semantics of text. The leap from text classification to NLI is the model understanding text semantics, while the leap from NLI to Q&A (Question and Answer) is the overall grasp of the short sentence semantics to the short document semantics. Hence, NLI is regarded as the transition of natural language understanding (NLU) according recent research.

Not only has NLI great research value, but also widely applications, such as dialogue purpose recognition and AI debater. Debate and argumentation are important capabilities of human intelligence by using background knowledge combined with text understanding to carry out inference. One of the shared tasks in NLPCC 2021 is to let AI debater understand the debating text, which aims to make AI understand all kinds of texts in the process of debating. The

L. Wang et al. (Eds.): NLPCC 2021, LNAI 13029, pp. 603–613, 2021.
https://doi.org/10.1007/978-3-030-88483-3_49

task of Track 1 is to identify the argumentation materials, which gives a pair of topic and claim. The participants of this task need to judge claims to the topic and classify them into three category which includes supports, neutral, against. This task belongs to the category of NLI, and needs to understand the semantics of topic and claim for inference.

The contributions of our paper are as follows:

* Aiming at the problem of large claim data and less topic data, we augment training set based on the connectivity between claims.
* We propose to solve the problem of rare words by constructing a dictionary for fine-tuned model to query, which includes three steps: (1) build the dictionary of rare words by using simpler synonyms or explanations to express them. (2) fine-tune pretrained language model. (3) sentence rewriting for test set.
* In Task 1 of NLPCC 2021, only the training data of the Phase 1 was used to get 5th.

The second part of our paper briefly describes the recent research of NLI, including some effective methods and models. Section 3 mainly describes the data augmentation, model construction and sentence rewriting. And then, in the Sect. 4, we explore the performance of the proposed method in Track 1 of shared Task 1 of NLPCC 2021. At the end of this paper, we give the conclusion and expectation.

2 Related Work

Since the birth of BERT [1], NLI or NLP methods can be divided into pre-training methods (model based on pre-training) and non pre-training methods (ABCNN [3], SiaGRU [4]). The two types of methods have both advantages and disadvantages. The pre-trained language model can get better results at the cost of higher computational power and training time, while non pre-training model has less training and inference time, but the performance worse than the pre-training model.

2.1 Non Pre-training Mode

In 2017, Qian Chen et al. [5] used the attention mechanism between sentences, combined with local inference and global inference, and only used the sequence inference model of LSTM chain to achieve the effect of state of art on SNLI (The Stanford Natural Language Inference) dataset. Besides, it also achieved excellent results in many sentence matching competitions. Zhiguo Wang et al. [6] solved the problem of insufficient matching of previous interactive networks (ABCNN [3], SiaGRU [4], ESIM [5]) from multiple perspectives (Premise to Hypothesis, Hypothesis to Premise), and integrated the model to achieve better performance in SNLI than ESIM. Inspired by DenseNet [7], Seonhoon Kim [8] combined attention mechanism to learn the logical and emotional relationship between Premise and Hypothesis. As a result, they achieved better results on multiple tasks of NLI.

2.2 Pre-training Model

In 2019, Jacob Devlin and Ming Wei Chang et al. [1] used the encoder of Transformer [2] to design masked language model (MLM) and next sentence predict (NSP) tasks to pre-train model, and then fine-tuned model in different tasks, obtained the results of state-of-art in many down-stream tasks of NLP. Since they put forward the BERT model, there have been more pre-trained language models in different languages (Zh, Da, Fren etc.), and constantly improve the benchmark of each shared task. In 2019, Huawei Noah Ark laboratory [9] proposed the NEZHA Chinese pre-trained language model, which modified the original absolute position of BERT to relative position, and combined with whore word masked(WWM) pre-training, and got better results compared with BERT in NLP tasks of Chinese. Shizhi Diao, Jiaxin Bai et al. [10] used N-gram to enhance the BERT structure, which can capture the semantics of Chinese phrases. They also get good results in many NLP tasks.

3 Methodology

In NLPCC 2021, the training data of Track 1 of Argumentative Text Understanding for AI Debater is less. Additionally, the distribution of topic and claim is unbalanced, and the composition of words in test set and training set are different. In this paper, we mainly explore how to solve these problems by employing with pre-training models including NEZHA and RoBERTa [11,12] for Chinese as the baseline.

3.1 Dataset

In NLPCC 2021, the training data of Track 1 of Argumentative Text Understanding for AI Debater is as follows (Fig. 1):

	Topic	Claim	label
0	同性婚姻应当获得法律认可	一男一女的传统的婚姻制度已是一种压迫性制度，不应扩大这一压迫范围。	Against
1	同性婚姻应当获得法律认可	合法婚姻是一种世俗制度，不应因宗教异议而受到限制。	Support
2	同性婚姻应当获得法律认可	同性伙伴关系上建立的家庭可以为人道主义作贡献。	Support
3	同性婚姻应当获得法律认可	同性伴侣也应享受法律赋予他们的相同权利。	Support
4	同性婚姻应当获得法律认可	同性伴侣可以形成稳定、坚定的关系，与异性伴侣无本质上的区别。	Support
...	...	...	...
6411	应该投资可再生技术	可再生资源的使用增加了企业和个人的支出。	Against
6412	应该投资可再生技术	可再生资源的设备一般安装在农田或者无人区，对人类的影响最小。	Support
6413	应该投资可再生技术	可再生，指可以重新利用的资源	Neutral
6414	应该投资可再生技术	可能生能源比化石燃料或核能发电的价格便宜。	Support
6415	应该投资可再生技术	建设和运营水力发电库可能会对环境造成有害影响。	Against

Fig. 1. Original training data.

The size of training set is 6416×3. For NLP down-stream tasks, this dataset belongs to few-shot learning. According to this given dataset, there are two main challenges to train the model. One is the small scale of this dataset, the other is each topic corresponding to multiple claims. If the training data is simply feed into the model, the text of each topic will be trained repeatedly, which will obviously affect the generalization of the model. To deal with the two problems, the connectivity of data is utilized for enhancing the closure of data. Since each topic corresponds to more than one claim, topic can be regard as a bridge between such claims. Thus, the potential relationship between different claims can be obtained. The relationship between the constructed claims is shown in the following Fig. 2:

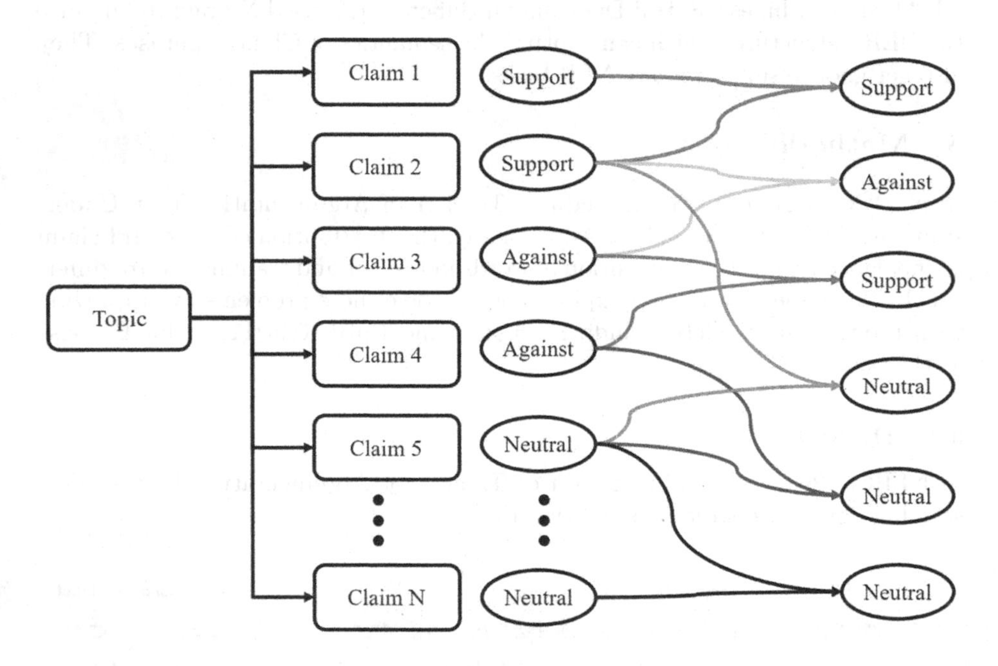

Fig. 2. Data augmentation.

The relationships among claims is inexplicit due to the independence between claims correspond to one topic, which means a few sentence pairs obtaining from data augmentation are unreasonable. This error can be described simply as 1 is equal to 1, 1 is not equal to 2 and 3, but we can't say 2 is equal to 3. To diminish this impact, we make some improvements during the data augmentation which shows in Fig. 3.

The advantage of this data augmentation method is that model can learn the implicit logical relationship between sentence pairs. For the complexity of language, it will bring a little error data. There is no doubt that the benefits and efficiencies brought by this method makes this method acceptable.

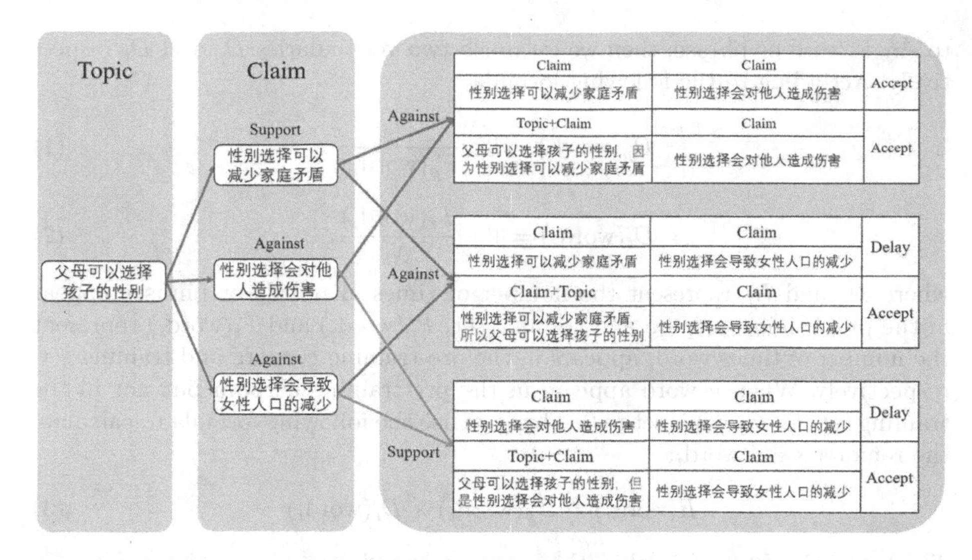

Fig. 3. In the construction of sentence pairs, we replace the form Claim-Claim (CC)with (Topic+Conjunction+Claim)-Claim (TCC)or (Claim+Conjunction+Topic)-Claim (CTC). Causal conjunctions are used to connect those sentence pairs which claims support topics, and adversative conjunctions are used to connect topics and claims with against relationship, which can make the relationship of sentence pairs more reasonable.

3.2 Sentence Rewriting Based Dictionary for Fine-Tuned Model

The first difficulty for the fine-tuned model in the test set is the uncertain features. Furthermore, it is difficult to obtain high quality generalization in the case of large difference between the test set and training set. In the realm of NLP, the difficulty is the diversity of vocabulary. The words or phrases appearing in the test set and rarely appearing in the training set is frequently occurred in few-shot learning. Fine-tuning can alleviate this problem to a certain extent until the emerging of the large corpora pre-training language model. However, these pre-trained language models still have huge potential to promote for the understanding text. Therefore, a sentence rewriting for fine-tuned model based on dictionary construction is proposed in this work.

Dictionary Construction of Rare and Difficult Words. When people do not know words or phrases, looking up the dictionary is an efficient way. Inspired by this idea, we establish a dictionary of rare words and phrases, which can be used to query by fine-tuned model (such as BERT, NEZHA and RoBERTa). Before the model can understand rare words, it has to know which words are difficult to understand. Therefore, the evaluation of rare words based on pre-training corpora and training set is also proposed here. By employing with the word segmentation tool, the pre-training corpora and the training set are divided to single word or phrase, then we establish two vocabularies D_p and D_t respectively. According to the following formula,

$$D_p(\text{word}_i) = \log \frac{N_p}{F_p(\text{word}_i)}. \tag{1}$$

$$D_t(\text{word}_i) = 1 - \frac{F_t(\text{word}_i)}{N_t}. \tag{2}$$

where N_p and N_t represent the number of times all words or phrases appear in the pre-training corpora and training set, $F_p(\text{word}_i)$ and $F_t(\text{word}_i)$ represent the number of times word_i appears in the pre-training corpora and training set, respectively. When a word appears in the pre-training corpora but not in the training set, $F_t(\text{word}_i)$ is set to 0. Then, we use the following formula to calculate the remoteness of word_i:

$$R(\text{word}_i) = D_p(\text{word}_i) \times D_t(\text{word}_i) \tag{3}$$

The larger the $R(\text{word}_i)$ value, the more rare word_i is.

Based on the vocabularies constructed above, synonyms(S) and explanations(E) of words are used as the corresponding contents of dictionary query, and requires,

$$\sum_{\text{word}_j \in S} R(\text{word}_j) \quad or \sum_{\text{word}_j \in E} R(\text{word}_j) < R(\text{word}_i) \tag{4}$$

That is to use simpler words which appear more frequently in the training set and pre-training corpora instead of rare words. word_i, $R(\text{word}_i)$, and S or E are finally constructed as dictionaries of rare words that can be queried by fine-tuned models.

The advantage of calculating remoteness of words and phrases at the same time is that the phrase containing a word is naturally rarer than the single word, which ensures the accuracy of the sentence after rewriting and helps to maintain a high degree of consistency between the meaning before and after rewriting.

We preliminarily obtain synonyms or explanations from Thesaurus [16], Chinese dictionary and Baidupedia [17], and select the explanation with the least remoteness according to the formula. Due to the complexity of Chinese, changing words randomly will bring grammatical errors, which is contrary to the purpose of pre-training model that finds the rationality and logic of words in sentences. We find that most of the words or phrases that are difficult to understand in the model are concentrated in nouns (including rare words, terminology and network words etc.), and only a few adjectives, verbs, prepositions and others have little influence. When establishing the dictionary, nouns are replaced by their explanations because noun replacement generally does not damage the structure and semantics of sentences, and ensures the rationality after rewriting.

Model Building. Track1 of Argumentative Text Understanding for AI Debater in NLPCC 2021 is a classic NLI task, which belongs to the category of sentence matching. Therefore, the fine-tuning model we built is aimed at sentence matching. Inspired by some ideas put forward by Chi Sun, Xipeng Qiu et al. [13], we construct a fine-tuned model based on their ideas as shown in Fig. 4.

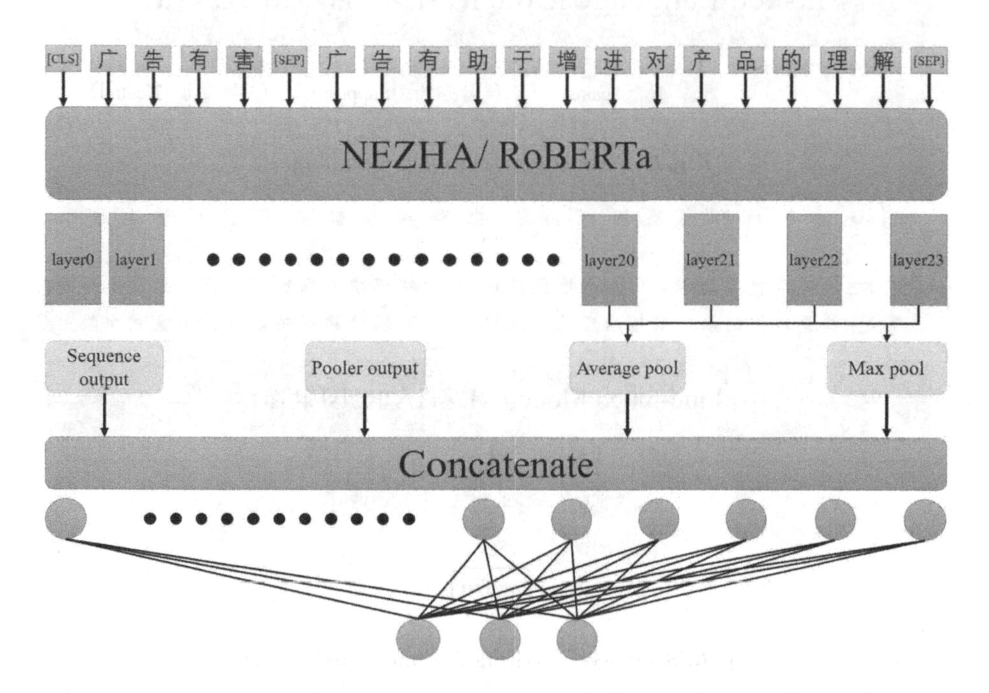

Fig. 4. The relationship between words inferred from the BERT structure contains more information in the last layers, so we select the last four layers of hidden state layers, obtains the maximum pooling and average pooling after splicing, and then concatenate with the sequence output and pooler output to connect to the FC layer for classification.

According to the structure of BERT, it can be found that the embedding layer is only used to denote words, while the internal relationship between words is in the encoder of BERT. It is obviously unreasonable to directly use the vocabulary content of a word for simple operation (SUM, AVG etc.) as embedding, This is also the reason why semantic rewriting is undirectly in the embedding layer. At the same time, we also construct model based on P-tuning [14] strategy to transform sentence matching into MLM (Masked Language Model) task. According to the relationship between Topic and Claim, we add related words between Topci and Claim as pattern and mask them where the model make prediction. The structure of P-tuning model is shown in Fig. 5.

Sentence Rewriting Based on Dictionary. After fine-tuning the pre-trained language model according to the above strategy, we rewrite the text of the test

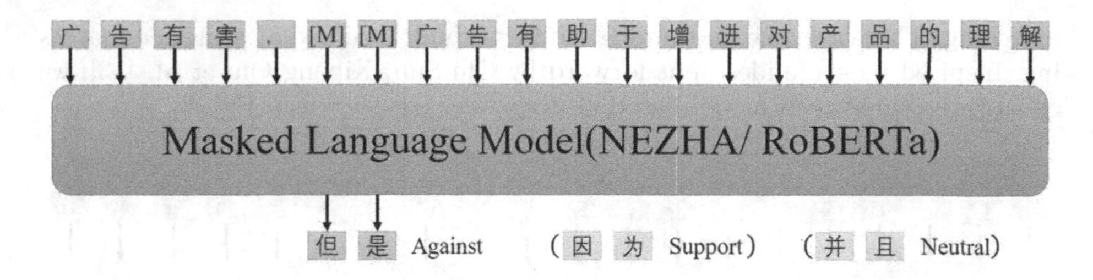

Fig. 5. P-tuning for sentence matching.

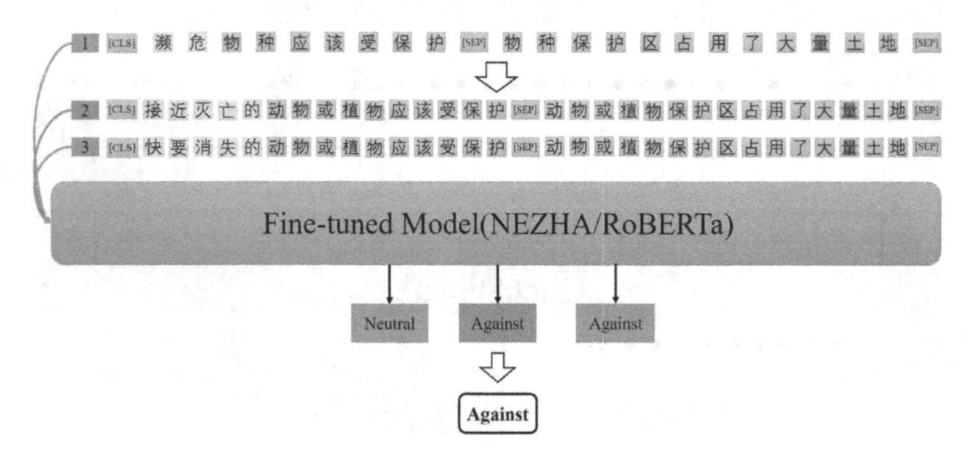

Fig. 6. Sentences rewriting for fine-tuned model.

set through the previously established dictionary, and then feed it into the model for inference (see Fig. 6).

After the rewriting of words and phrases, many rare words can be represented by simple words, and improve the performance in the test set. Moreover, for those words that appear in the test set but not in the training set, rewriting can also show effective performance at most time. In order to avoid the irrationality caused by rewriting, we feed sentences pair into the model through different rewriting and get the final prediction through voting.

4 Experiment

Firstly, the data augmentation method proposed in our paper is used to augment the training set. To reduce the effect of the error sentences pair, We sample the three categories Support, Against, Neutral respectively from augmented data corresponds to the probability distribution of these three categories in the training set. Among the augmented data, the Topic-Claim derived from claims is not the real Topic-Claim as well as training set. In order to keep consistent with the training set, we construct the Topic-Claim by taking the shorter claim as the topic. We adopt the format Topic-Claim pair to pre-train our model. This method enables the model to learn the relationship of Topic-Claim text in a specific down-stream task in the fine-tuning stage. In most cases, we can get better results in the sentence matching. Secondly, we use the model structure proposed in the Sect. 3 to fine-tune the model, and add PGD [15] adversarial training to improve the generosity of the model. Finally, we use the dictionary constructed in the above section to check the performance of the fine-tuned model in test set. The results in validation set and test set are summarized as follows:

Table 1. Performance of different methods.

Method	Dev	Test
Original Train Data	0.90	0.790
Data Augmentation (CC)	0.95	0.835
Data Augmentation (TCC)	**0.95**	**0.843**
Data Augmentation (CC) + (Max, AVG and Concatenate)	0.96	0.855
Data Augmentation (TCC) + (Max, AVG and Concatenate)	**0.96**	**0.865**
Data Augmentation (TCC) + (Max, AVG and Concatenate) + P-tuning	0.98	0.871
Data Augmentation (TCC) + (Max, AVG and Concatenate) + P-tuning + **Dictionary**	**0.98**	**0.891**

Experiments show that the data set after using the connectivity between sentence pairs to augment is much better than the original data in the validation set and the test set, which proves the effectiveness of the data augmentation method. Although sampled the augmentation data, the data set still contains many sentence pairs that are not in line with the reality. On this basis, we use TCC to replace the original CC to construct data. Both fine-tuning base model and fine-tuning the model in Sect. 3 prove that TCC is superior to CC. In addition, P-tuning combined with pattern construction also brings a certain improvement. By observing the error reasoning of the fine-tuned model in the test set, it is found that there are many rare words in the data with wrong prediction results. Experiments also show that our dictionary is conducive to fine-tuned the reasoning of rare words and difficult words, and achieves good results in the test set.

According to the dictionary, the rewritten sentences of rare words in the test set do not necessarily conform to the real natural language rules, and the model based on BERT is depend on the large corpora to learn the relationships among words, so rewriting will lead to sentence errors. In this case, we adopt ensemble learning method to achieve the best results.

5 Conclusion and Expect

In order to solve the problem that the trained model in NLP has poor effect in the face of rare words, we propose to construct a dictionary for the model to query and then convert the difficult text to simpler text, which improves the generalization of the model. And in the Track 1 of Argumentative Text Understanding for AI Debater in NLPCC 2021, combined with the data augmentation based on closure proposed in our paper, excellent results have been achieved in the test set. There are still some errors in dictionary rewriting strategy, such as the rewritten text fail to conform to the rules of language. The future work urges us to continue to look for more effective rewriting methods, and study the influence of dictionary construction in the process of fine-tuning and pre-training.

References

1. Devlin, J., Chang, M.-W., Lee, K., Toutanova, K.: BERT: pre-training of deep bidirectional transformers for language understanding. In Proceedings of the 2019 Conference of the North American Chapter of the Association for Computational Linguistics: Human Language Technologies, Volume 1 (Long and Short Papers), pp. 4171–4186 (2019)
2. Vaswani, A., et al.: Attention is all you need. In: Advances in Neural Information Processing Systems, pp. 6000–6010 (2017)
3. Yin, W., Schutze, H., Xiang, B., Zhou, B.: ABCNN: attention-based convolutional neural network for modeling sentence pairs. arXiv preprint arXiv:1512.05193 (2016)
4. Mueller, J., Thyagarajan, A.: Siamese recurrent architectures for learning sentence similarity. In: Proceedings of the 30th AAAI Conference on Artificial Intelligence, AAAI 2016 (2016)
5. Chen, Q., Zhu, X., Ling, Z., Wei, S., Jiang, H., Inkpen, D.: Enhanced LSTM for natural language inference. arXiv preprint arXiv:1609.06038 (2017)
6. Wang, Z., Hamza, W., Florian, R.: Bilateral multi-perspective matching for natural language sentences. arXiv preprint arXiv:1702.03814 (2017)
7. Huang, G., Liu, Z., van der Maaten, L.: Densely connected convolutional networks. arXiv preprint arXiv:1608.06993 (2018)
8. Kim, S., Kang, I., Kwak, N.: Semantic sentence matching with densely-connected recurrent and co-attentive information. arXiv preprint arXiv:1805.11360 (2018)
9. Wei, J., et al.: Nezha: neural contextualized representation for Chinese language understanding. arXiv preprint arXiv:1909.00204 (2019)
10. Diao, S., Bai, J., Song, Y., Zhang, T., Wang, Y.: ZEN: pre-training Chinese text encoder enhanced by n-gram representations. arXiv preprint arXiv:1911.00720 (2019)
11. Cui, Y., et al.: Pre-training with whole word masking for Chinese BERT. arXiv preprint arXiv:1906.08101 (2019)
12. Liu, Y., et al.: RoBERTa: a robustly optimized BERT pretraining approach. arXiv preprint arXiv:1907.11692 (2019)
13. Sun, C., Qiu, X., Xu, Y., Huang, X.: How to fine-tune BERT for text classification? arXiv preprint arXiv:1905.05583 (2020)
14. Liu, X., et al.: GPT understands, too. arXiv preprint arXiv:2103.10385 (2021)

15. Madry, A., Makelov, A., Schmidt, L., Tsipras, D., Vladu, A.: Towards deep learning models resistant to adversarial attacks. arXiv preprint arXiv:1706.06083 (2017)
16. Synonym. https://github.com/guotong1988
17. Baidupedia. https://baike.baidu.com

Knowledge Enhanced Transformers System for Claim Stance Classification

Xiangyang Li[1], Zheng Li[1], Sujian Li[1(✉)], Zhimin Li[2], and Shimin Yang[2]

[1] Key Laboratory of Computational Linguistics(MOE), Department of Computer Science, Peking University, Beijing, China
{xiangyangli,1800017744,lisujian}@pku.edu.cn
[2] Meituan, Beijing, China
{lizhimin03,yangshimin}@meituan.com

Abstract. In this paper, we present our system for the NLPCC 2021 shared task 1 of "argumentative text understanding for AI debater", where we achieved the 3rd place with the accuracy score of 0.8925 on Track 1, Task 1. Specifically, we proposed a rapid, simple, and efficient ensemble method which uses different pre-trained language models such as BERT, Roberta, Ernie, etc. with various training strategies including warm-up, learning rate schedule and k-fold cross-validation. In addition, we also propose a knowledge enhancement approach, which makes it possible for our model to achieve the first place without introducing external data.

Keywords: Natural language processing · Pre-trained language model · Claim stance classification · Bert

1 Introduction

Sentiment classification is an important research issue in practical application, especially with the development of online posting through various channels such as social media sites, news portals, and forums.

Claim stance classification for debating, as a subtask in the sentiment classification task, plays a pivotal role in many areas, and the task is defined differently in different scene settings, where its most common definition is to automatically classify the stance of a text producer toward a goal into one of the following three categories: support, against, neutrality. A large number of relevant scientific publications reveal the increasing impact and significance of solutions to issues such as sentiment analysis, mockery, controversy, authenticity, rumour, fake news detection, and argument mining. Thus, it is important to build a system that can automatically detect claim stance. [9] proposed to use LSTM model to identify claim stance, [2] improves the robustness of the model by introducing additional data, [22] proposed to use the method of ensemble learning for stance classification.

L. Wang et al. (Eds.): NLPCC 2021, LNAI 13029, pp. 614–623, 2021.
https://doi.org/10.1007/978-3-030-88483-3_50

However, in this evaluation task, the size of the model is limited, so the approach of ensemble dozens of pre-trained language models was not feasible. In addition, because of the lack of Chinese corpus detection of stance classification, it is difficult to conduct knowledge enhancement by introducing additional data.

Therefore, in this paper, we propose two claim stance detection models, one is text-transformers based on efficient ensemble learning method, and the other is through knowledge-enhanced text-transformers model without introducing additional data.

The rest of the paper is organized as follows: Sect. 2 introduces the dataset of this task. Section 3 details the architecture of our system (features, models and ensembles). Section 4 offers result and an analysis of the performance of our models. Section 5 describes the related Work. Finally, Sect. 6 presents our conclusions for this task.

2 Data Analyzation and Process

The dataset is published by NLPCC 2021 evaluation task1 track1. The whole dataset consists of 7406 samples collected from online forums and topic-related articles from Wikipedia (for English) and Baidu Encyclopedia (for Chinese). Human annotators were recruited to annotate claims from the collected articles and identify their stances with regards to the given topics. After that, human translators were recruited to translate the English claims into Chinese. In the first phase, we test the performance of the model on the test set already given. In the end stage of the competition, we need to upload the trained model to the server, making predictions on the unknown dataset in case where the network cannot connect.

Each sample of dataset is as follows: Each line includes three items: topic, claim, and the label (support, against, neutral) as the Fig. 1 demonstrates.

Topic	**Claim**	Label
同性婚姻应当获得法律认可	一男一女的传统的婚姻制度已是一种压迫性制度，不应扩大这一压迫范围。	Against
同性婚姻应当获得法律认可	合法婚姻是一种世俗制度，不应因宗教异议而受到限制。	Support
明星吸毒应该被封杀	"毒品"、"吸毒"国际上习惯只讲麻醉品、精神药品的滥用	Neutral

Fig. 1. Data formats in datasets.

In order to have a better understanding of the dataset, we first perform some exploratory analyses on the dataset, which helps us see the hidden laws in the data at a glance and find a model most suitable for the data.

We first explore the distribution of positive and negative samples in the train sets and test sets, as shown in Fig. 2. From Fig. 2, we can see that in the train and test sets, the distribution of the data in the data set was uneven, which illustrates that our dataset is unbalanced, so we can consider a data balanced sampling method when process data.

In order to analyze the characteristics of the words in the sentence, we calculate the word frequencies of the train and test set respectively, remove the stop words, and make the corresponding word cloud diagram as shown in Fig. 3.

From the Fig. 3, we can see that 'human', 'culture' and 'would' are the words with the highest frequency in the dataset. This suggests that most of the arguments are of humane, social relevance, suggesting that we should add the data relevant to humanities if additional data are introduced.

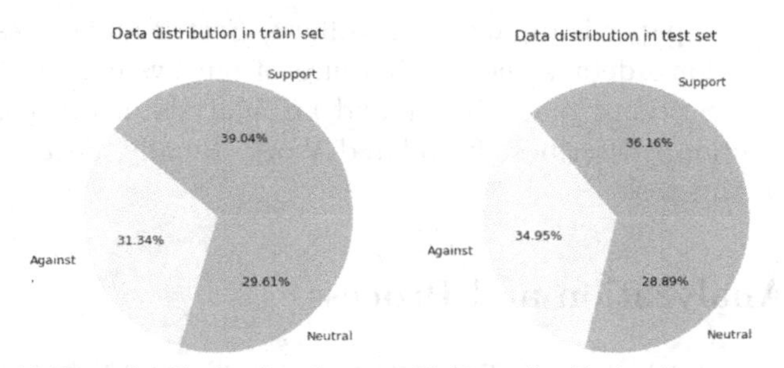

(a) Data distribution in tain set (b) Data distribution in test set

Fig. 2. The distribution of samples in the train and test set.

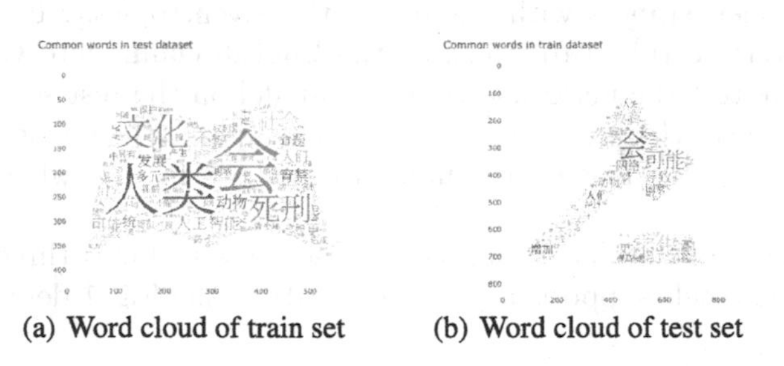

(a) Word cloud of train set (b) Word cloud of test set

Fig. 3. The word cloud diagram of the training set and the validation set. We determine the size of the word in the word cloud according to the frequency of the word.

2.1 Data Process

In the first phase, we split the dataset into train sets and test sets. We use four metrics accuracy, precision, recall and weighted F1-score to validate our model. In the final phase, we merge the train sets and test sets and validate the model with five fold cross validation method.

3 Methodology

We propose two claim stance classification models: one is the Text-Transformers model based on transformers, and the other is knowledge enhanced text-transformers. The description of the two models is as follows.

3.1 Text-Transformers

Contextualized language models such as ELMo and Bert trained on large corpus have demonstrated remarkable performance gains across many NLP tasks recently. In our experiments, we use various architectures of language models as the backbone of our model.

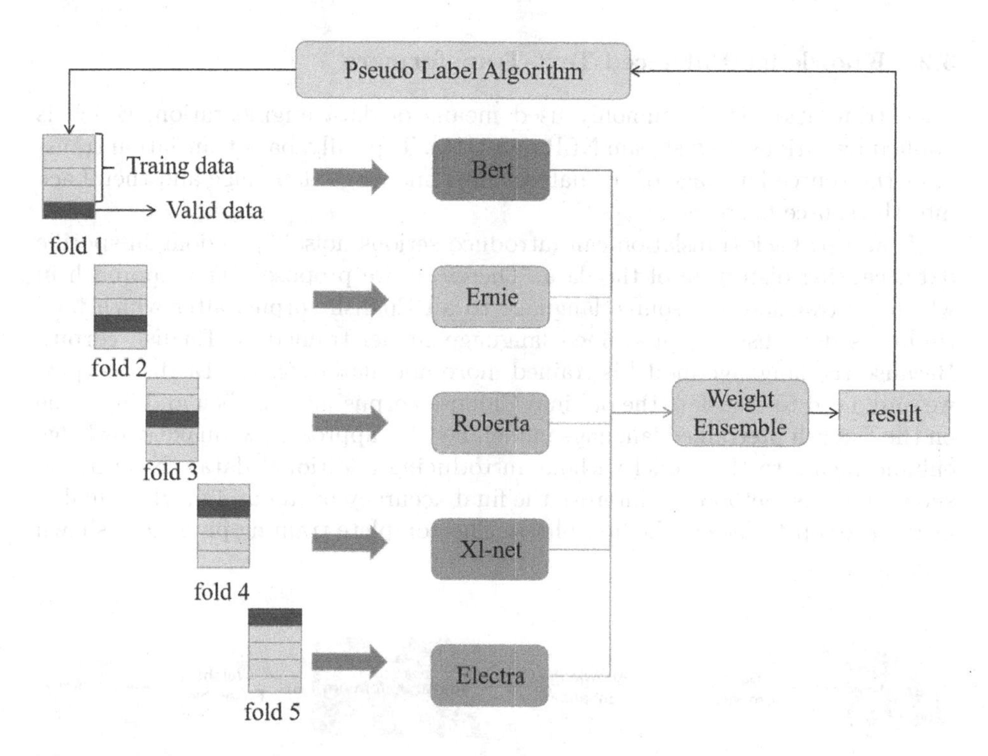

Fig. 4. The architecture of text-transformers. Text transformers ensemble five pretrained language models in total.

Previous approaches [10] have shown that ensemble multiple pre-trained language models is effective for downstream NLP tasks, but it is not possible to ensemble dozens of pre-trained language models when the model size is limited. Therefore, we propose an efficient method to ensemble the pre-training language model. As shown in the Fig. 4, for the architecture of the language model, we use five different language models including Bert, Ernie, Roberta, XL-net, and Electra trained with the five-fold five-model cross-validation. Generally, the model in each fold using the cross validation method should be the same. But in this evaluation task, the size of the model was limited. Therefore, to be able to ensemble as many models as possible, and speed up the training, we use a different model at each fold. By this approach, we constructed a light-weight, rapid, and efficient transformer structure, which we named text-transformers.

With the five-fold five-mode cross validation method, the robustness of the model is improved and better learning results can be obtained. In the final phase, we need to merge only 5 language models. If five-fold single-mode cross validation approach is adopted, in the final phase, we must ensemble 25 pre-trained language models, such a model size is unacceptable for this evaluation task.

3.2 Knowledge Enhanced Text-Transformers

Back translation is a commonly used means of data augmentation, which is applied by various downstream NLP tasks [16]. Typically, back translation translates the source language of the dataset into one target language and then back into the source language.

However, back translation can introduce serious noise [3] to domain specific data, causing distortion of the data. Therefore, we propose a new approach in which we translate the source language to an English corpus, after which fine-tuning is done using a pretrained language model trained on English corpus. Because the language model is trained more adequately on the English corpus, we only need to translate the original Chinese corpus into English and fine-tune on the English pretrained language model. By this approach, we make knowledge enhancements to the model without introducing additional data. Experiments show that this method can improve the final accuracy by about 4%, which makes us get the top 1 place in the first phase. The complete training pipeline is shown in Fig. 5

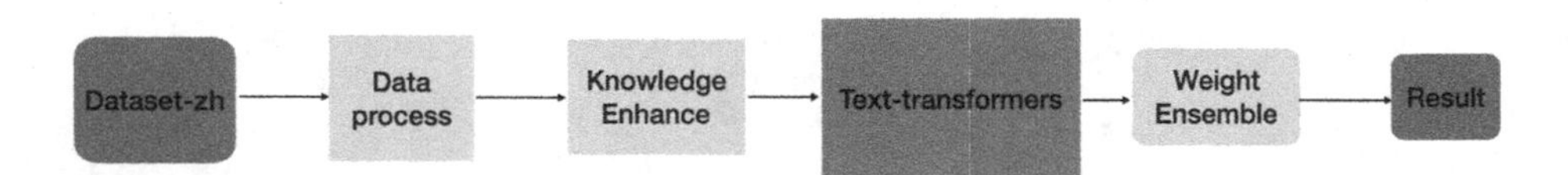

Fig. 5. Final pipeline of knowledge enhanced transformers system for Claim Stance Classification.

4 Experiments

4.1 Experimental Settings

- **Text-Transformers**: We use five pre-training language models BERT [6], Ernie [20], Roberta [11], XL-net [24] and Electra [5], with a hidden size of 768, 12 Transformer blocks and 12 self-attention heads from huggingface[1]. We further fine-tune the models on Google TPU v3-8, with a batch size of 256, max sequence length of 140, learning rate of 1e$-$6 $\sim$ 5e$-$5, number of epochs 12, and warm-up steps of five epoch. We use Adam with $\beta 1 = 0.9$ and

[1] https://github.com/huggingface/transformers.

$\beta 2 = 0.999$. We empirically set the max number of the epoch to 6 and save the best model on the validation set for testing. For the Text-transformers model, due to the complexity of transformer model, we adopt the training strategy as shown in 4.2.

- **Knowledge Enhanced Text-Transformers**: We use Baidu Translation API[2] to translate the dataset from Chinese to English and replace the Chinese pre-trained language model in the text-transformer with the English pre-trained language model. The rest of the experimental settings were the same as text-transformers.
- **Back-Translation Text-Transformers**: We translated the original dataset from Chinese to English and back from English to Chinese using Baidu Translation API, and this dataset was added to the original dataset for use as data augmentation. The rest of the experimental settings were the same as text-transformers.
- **Bert**: To validate the effectiveness of our model, we use Bert as a baseline. The rest of the experimental settings were the same as text-transformers.

4.2 Training Strategy

- **Pseudo Label Algorithm**: Because the amount of data is too small, we propose a pseudo-label algorithm to do data augmentation. If a test data is predicted with a probability greater than 0.95, we think that the data is predicted correctly with a relatively high confidence and add it into the train set.
- **Label Smoothing**: Label smoothing [21] is a regularization technique that introduces noise for the labels. Assuming for a small constant ϵ, the training set label y is correct with a probability or incorrect otherwise. Label Smoothing regularizes a model based on a softmax with output values by replacing the hard 0 and 1 classification targets with targets of $\frac{\epsilon}{k-1}$ and $1 - \epsilon$ respectively. In our strategy, we take ϵ equal to 0.01, k equal to 2.
- **Learning Rate Warm Up**: Using too large learning rate may result in numerical instability especially at the very beginning of the training, where parameters are randomly initialized. The warm up [7] strategy increases the learning rate from 0 to the initial learning rate linearly during the initial N epochs or m batches. In our strategy, we set an initial learning rate of 1e−6, which increased gradually to 5e−5 after 6 epochs.
- **Learning Rate Cosine Decay**: After the learning rate warmup stage described earlier, we typically steadily decrease its value from the initial learning rate. Compared to some widely used strategies including exponential decay and step decay, the cosine decay [12] decreases the learning rate slowly at the beginning, and then becomes almost linear decreasing in the middle, and slows down again at the end. It potentially improves the training progress. In our strategy, after reaching a maximum value of 5e−5, the learning rate decreases to 1e−6 after a cosine decay of 6 epochs

[2] https://api.fanyi.baidu.com/.

- **Domain Pretraining**: Sun et al. [18] demonstrated that pre-trained models such as Bert, which do further domain pretraining on the dataset, can lead to performance gains. Therefore, using the Bert model, we fine-tuned 10 epochs by performing the MLM pre-train task on the original dataset.
- **Weight Ensemble**: We adopt soft voting as an ensemble strategy, which refers to taking the average of the probabilities of all the models predicted to a certain class as the standard and the type of corresponding with the highest probability as the final prediction result. In our method, we take the highest F1-score of each fold model on the validation set as the ensemble weight. Suppose that the test sample is x, the highest scores on their own validation set of five folds respectively are $\{s_i\}_{i=1}^5$, and the five classification models' functions are: $\{f_i\}_{i=1}^5$, then the final result of the classification of this sample given by the model is:

$$f(x) = \frac{\sum_{i=1}^{5} f_i(x)s_i}{\sum_{j=1}^{5} s_j} \tag{1}$$

4.3 Results

Table 1. Results of different models on test set.

Method	Accuracy	Precision	Recall	Weighted F1-score
Text-Transformers	0.812	0.820	0.824	0.821
Knowledge-Enhanced Text-Transformers	**0.867**	**0.862**	**0.854**	**0.860**
Back-Translation Text-Transformers	0.824	0.836	0.830	0.832
Bert	0.784	0.793	0.785	0.790

In Table 1, we presented our results on test sets. For the Text-transformers model, we achieved the weighted F1 scores of 0.821. After adding the knowledge Enhancement, the weighted F1 score of 0.860 was obtained on the test set, achieving the first place in the competition in phase1. In the final phase in order to make full use of the data, we merged the train set and the test set. However, because we can't link to the network in the final phase, so the final model we uploaded was text-transformers, which gave us the third place with accuracy score of 0.8925.

5 Related Work

5.1 Ensemble Learning

Ensemble learning is a machine learning paradigm by which multiple learners are strategically generated, trained, and combined to solve the same problems [15]. Compared to the traditional machine learning method (in which only one model is learned at once), ensemble learning tries to leverage a set of models (also called "weak learners" or "base learners") and combine them into a new stronger learner [25]. [4] used ensemble learning on neural network models to reduce the variance. In the past, ensemble learning is widely used in the computer vision field, Mohammad [1] utilized classifier fusion by ensemble-based methods to enhance action recognition performance. One of the most straightforward but outstanding methods is K-fold cross-validation.

K-fold cross-validation [13] means that the training set is divided into K sub samples, one single sub sample is reserved as the data for validation, and the other K-1 samples are used for training. Cross-validation is repeated K times, and each sub sample is verified once. The average of the results or other combination methods is used to obtain a single estimation. The advantage of this method is that it can repeatedly use the randomly generated sub samples for training and verification, and each time the results are verified, the less biased results can be obtained.

The traditional K-fold cross-validation uses the same model to train each fold and only retains the best results. In our system, we use different models for each fold and keep the models for each fold to fuse the results. Our experiments prove that this method outperforms the common K-fold cross-validation method.

5.2 Pre-training Language Models

Pre-training is originally proposed to solve the lack of parallel corpus. With the development of deep learning, the scale of models grows faster, the model becomes more and more complex, and gradually more and more parameters need to be adjusted. Hence, models need larger parallel datasets. However, for most NLP tasks, it is difficult to find related large-scale labeled data; instead, there are more and more unlabeled data in the NLP area. To solve the difficulty of lacking parallel data and make better use of non-parallel data, pre-training models arise at this moment.

Pre-training and then fine-tuning has become a new paradigm in natural language processing. Through self-supervised learning from a large corpus, the language model can learn general knowledge and then transfer it to downstream tasks by fine-tuning to specific tasks.

Elmo uses Bidirectional LSTM to extract word vectors using context information as demostrated [8,14]. GPT enhances context-sensitive embedding by adjusting the transformer as demostrated [17,23]. Jacob [6] used bidirectional language model BERT to apply close and next sentence prediction to self-supervised learning to strengthen word embeddings. Liu [11] removes the next sentence prediction from self-training, and performs more fully training, getting a better language model named Roberta. Sun [20] strengthened the pre-training language model, completely masking the span in Ernie. Further, Sun [19] proposed continuous multi-task pre-training and several pre-training tasks in Ernie 2.0.

6 Conclusion

In this paper, we presented our approach to claim stance classification. We propose an efficient and rapid, ensemble language model approach when the total size of the model is limited. In addition, we propose a new knowledge enhance method when encountering low resource corpus.

In the future, we plan to further explore the generalization ability of the model we proposed. To be specific, extensive testing and comparison are carried out on other cardinal dataset. In addition, we want to further explore more approaches to the model ensemble. We assume that our model can perform well in other language tasks. More importantly, we believe that our exploration of ensemble approaches to pre-training language models may prove valuable.

References

1. Bagheri, M., et al.: Keep it accurate and diverse: enhancing action recognition performance by ensemble learning. In: Proceedings of the IEEE Conference on Computer Vision and Pattern Recognition Workshops, pp. 22–29 (2015)
2. Bar-Haim, R., Edelstein, L., Jochim, C., Slonim, N.: Improving claim stance classification with lexical knowledge expansion and context utilization. In: Proceedings of the 4th Workshop on Argument Mining, pp. 32–38 (2017)
3. Bogoychev, N., Sennrich, R.: Domain, translationese and noise in synthetic data for neural machine translation. arXiv preprint arXiv:1911.03362 (2019)
4. Brownlee, J.: Ensemble learning methods for deep learning neural networks. In: Machine Learning Mastery (2018)
5. Clark, K., Luong, M.T., Le, Q.V., Manning, C.D.: Electra: pre-training text encoders as discriminators rather than generators. arXiv preprint arXiv:2003.10555 (2020)
6. Devlin, J., Chang, M.W., Lee, K., Toutanova, K.: Bert: pre-training of deep bidirectional transformers for language understanding. arXiv preprint arXiv:1810.04805 (2018)
7. He, K., Zhang, X., Ren, S., Sun, J.: Deep residual learning for image recognition. In: Proceedings of the IEEE Conference on Computer Vision and Pattern Recognition, pp. 770–778 (2016)
8. Hochreiter, S., Schmidhuber, J.: Long short-term memory. Neural Comput. **9**(8), 1735–1780 (1997)
9. Kochkina, E., Liakata, M., Augenstein, I.: Turing at semeval-2017 task 8: sequential approach to rumour stance classification with branch-lstm. arXiv preprint arXiv:1704.07221 (2017)
10. Li, X., Xia, Y., Long, X., Li, Z., Li, S.: Exploring text-transformers in aaai 2021 shared task: Covid-19 fake news detection in english. arXiv preprint arXiv:2101.02359 (2021)
11. Liu, Y., et al.: Roberta: a robustly optimized bert pretraining approach. arXiv preprint arXiv:1907.11692 (2019)
12. Loshchilov, I., Hutter, F.: Sgdr: stochastic gradient descent with warm restarts. arXiv preprint arXiv:1608.03983 (2016)
13. Mosteller, F., Tukey, J.W.: Data analysis, including statistics. Handb. Social Psychol. **2**, 80–203 (1968)

14. Peters, M.E., Neumann, M., Iyyer, M., Gardner, M., Clark, C., Lee, K., Zettlemoyer, L.: Deep contextualized word representations. arXiv preprint arXiv:1802.05365 (2018)
15. Polikar, R.: Ensemble learning. In: Zhang, C., Ma, Y. (eds.) Ensemble Machine Learning, pp. 1–34. Springer, Heidelberg (2012). https://doi.org/10.1007/978-1-4419-9326-7_1
16. Prabhumoye, S., Tsvetkov, Y., Salakhutdinov, R., Black, A.W.: Style transfer through back-translation. arXiv preprint arXiv:1804.09000 (2018)
17. Radford, A., Narasimhan, K., Salimans, T., Sutskever, I.: Improving language understanding by generative pre-training (2018)
18. Sun, C., Qiu, X., Xu, Y., Huang, X.: How to fine-tune BERT for text classification? In: Sun, M., Huang, X., Ji, H., Liu, Z., Liu, Y. (eds.) CCL 2019. LNCS (LNAI), vol. 11856, pp. 194–206. Springer, Cham (2019). https://doi.org/10.1007/978-3-030-32381-3_16
19. Sun, Y., et al.: Ernie 2.0: a continual pre-training framework for language understanding. In: AAAI, pp. 8968–8975 (2020)
20. Sun, Y., et al.: Ernie: enhanced representation through knowledge integration. arXiv preprint arXiv:1904.09223 (2019)
21. Szegedy, C., Ioffe, S., Vanhoucke, V., Alemi, A.: Inception-v4, inception-resnet and the impact of residual connections on learning. arXiv preprint arXiv:1602.07261 (2016)
22. Tutek, M., et al.: Takelab at semeval-2016 task 6: stance classification in tweets using a genetic algorithm based ensemble. In: Proceedings of the 10th International Workshop on Semantic Evaluation (SemEval-2016), pp. 464–468 (2016)
23. Vaswani, A., et al.: Attention is all you need. In: Advances in Neural Information Processing Systems, pp. 5998–6008 (2017)
24. Yang, Z., Dai, Z., Yang, Y., Carbonell, J., Salakhutdinov, R.R., Le, Q.V.: Xlnet: generalized autoregressive pretraining for language understanding. In: Advances in Neural Information Processing Systems, pp. 5753–5763 (2019)
25. Zhou, Z.H.: Ensemble learning. Encycl. Biometrics **1**, 270–273 (2009)

Author Index

Printed in the United States
by Baker & Taylor Publisher Services